Universitext

Springer
New York
Berlin
Heidelberg
Barcelona
Budapest
Hong Kong
London
Milan
Paris
Santa Clara
Singapore
Tokyo

Universitext

Editors (North America): S. Axler, F.W. Gehring, and K.A. Ribet

Aksoy/Khamsi: Nonstandard Methods in Fixed Point Theory
Andersson: Topics in Complex Analysis
Aupetit: A Primer on Spectral Theory
Booss/Bleecker: Topology and Analysis
Borkar: Probability Theory: An Advanced Course
Carleson/Gamelin: Complex Dynamics
Cecil: Lie Sphere Geometry: With Applications to Submanifolds
Chae: Lebesgue Integration (2nd ed.)
Charlap: Bieberbach Groups and Flat Manifolds
Chern: Complex Manifolds Without Potential Theory
Cohn: A Classical Invitation to Algebraic Numbers and Class Fields
Curtis: Abstract Linear Algebra
Curtis: Matrix Groups
DiBenedetto: Degenerate Parabolic Equations
Dimca: Singularities and Topology of Hypersurfaces
Edwards: A Formal Background to Mathematics I a/b
Edwards: A Formal Background to Mathematics II a/b
Foulds: Graph Theory Applications
Friedman: Algebraic Surfaces and Holomorphic Vector Bundles
Fuhrmann: A Polynomial Approach to Linear Algebra
Gardiner: A First Course in Group Theory
Gårding/Tambour: Algebra for Computer Science
Goldblatt: Orthogonality and Spacetime Geometry
Gustafson/Rao: Numerical Range: The Field of Values of Linear Operators
and Matrices
Hahn: Quadratic Algebras, Clifford Algebras, and Arithmetic Witt Groups
Holmgren: A First Course in Discrete Dynamical Systems
Howe/Tan: Non-Abelian Harmonic Analysis: Applications of $SL(2, R)$
Howes: Modern Analysis and Topology
Humi/Miller: Second Course in Ordinary Differential Equations
Hurwitz/Kritikos: Lectures on Number Theory
Jennings: Modern Geometry with Applications
Jones/Morris/Pearson: Abstract Algebra and Famous Impossibilities
Kannan/Krueger: Advanced Analysis
Kelly/Matthews: The Non-Euclidean Hyperbolic Plane
Kostrikin: Introduction to Algebra
Luecking/Rubel: Complex Analysis: A Functional Analysis Approach
MacLane/Moerdijk: Sheaves in Geometry and Logic
Marcus: Number Fields
McCarthy: Introduction to Arithmetical Functions
Meyer: Essential Mathematics for Applied Fields
Mines/Richman/Ruitenburg: A Course in Constructive Algebra
Moise: Introductory Problems Course in Analysis and Topology
Morris: Introduction to Game Theory
Polster: A Geometrical Picture Book
Porter/Woods: Extensions and Absolutes of Hausdorff Spaces
Ramsay/Richtmyer: Introduction to Hyperbolic Geometry
Reisel: Elementary Theory of Metric Spaces
Rickart: Natural Function Algebras

(continued after index)

Robert Friedman

Algebraic Surfaces and Holomorphic Vector Bundles

Springer

Robert Friedman
Department of Mathematics
Columbia University
New York, NY 10027
USA

Mathematics Subject Classification (1991): 14-01, 14A10, 14J60

Library of Congress Cataloging-in-Publication Data
Friedman, Robert, 1955–
   Algebraic surfaces and holomorphic vector bundles / Robert D.
Friedman.
      p.   cm. — (Universitext)
   Includes bibliographical references and index.
   ISBN 0-387-98361-9 (softcover : alk. paper)
   1. Surfaces. Algebraic.   2. Vector bundles.   I. Title.
QA571.F75   1998
515′.35—dc21   97-38108

Printed on acid-free paper.

Production managed by Victoria Evarretta; manufacturing supervised by Thomas King.
Camera-ready copy prepared from the author's $\mathcal{AMS}$-TeX files.
Printed and bound by R.R. Donnelley and Sons, Harrisonburg, VA.
Printed in the United States of America.

9 8 7 6 5 4 3 2 1

ISBN 0-387-98361-9 Springer-Verlag New York Berlin Heidelberg   SPIN 10647715

Preface

This book is based on courses given at Columbia University on vector bundles (1988) and on the theory of algebraic surfaces (1992), as well as lectures in the Park City/IAS Mathematics Institute on 4-manifolds and Donaldson invariants. The goal of these lectures was to acquaint researchers in 4-manifold topology with the classification of algebraic surfaces and with methods for describing moduli spaces of holomorphic bundles on algebraic surfaces with a view toward computing Donaldson invariants. Since that time, the focus of 4-manifold topology has shifted dramatically, at first because topological methods have largely superseded algebro-geometric methods in computing Donaldson invariants, and more importantly because of the new invariants defined by Seiberg and Witten, which have greatly simplified the theory and led to proofs of the basic conjectures concerning the 4-manifold topology of algebraic surfaces. However, the study of algebraic surfaces and the moduli spaces of bundles on them remains a fundamental problem in algebraic geometry, and I hope that this book will make this subject more accessible. Moreover, the recent applications of Seiberg-Witten theory to symplectic 4-manifolds suggest that there is room for yet another treatment of the classification of algebraic surfaces. In particular, despite the number of excellent books concerning algebraic surfaces, I hope that the half of this book devoted to them will serve as an introduction to the subject. There are few references to the general subject of vector bundles on algebraic varieties beyond the book by Okonek, Schneider and Spindler on vector bundles on projective spaces, the Astérisque volume of Seshadri on bundles over curves, and a recent book by Huybrechts and Lehn. I hope that combining the study of surfaces with that of vector bundles on them (and on curves) will be mutually beneficial to both fields. For example, detailed knowledge of a surface X is necessary in order to give a detailed picture of the moduli space of bundles over X, and results about ruled surfaces are an ingredient in the proof of the Bogomolov inequality presented here. On the other hand, the Bogomolov inequality gives important information about linear systems on surfaces, by a theorem of Reider, and in particular gives a short proof of Bombieri's theorem on the behavior of $|nK_X|$ when X is a minimal surface of general type. The original motivation of computing Donaldson invariants has however disappeared except for a brief discussion in Chapter 8 for elliptic surfaces.

It is a pleasure to thank the audience at the lectures which served as the raw material for this book, as well as David Gomprecht, my course assistant for the Park City institute, for an excellent job in proofreading the rough draft of the first part of this book. I would also like to thank Tomás Gómez and Titus Teodorescu for comments on various manuscript versions, and Dave Bayer for doing an excellent job with the figures.

New York, New York   Robert Friedman

Contents

Introduction

The study of algebraic surfaces is by now over one hundred years old. Many of the fundamental results were established by the Italian school of algebraic geometry, for example Castelnuovo's criterion for a surface to be rational (1895), the theorem of Enriques that a surface is rational or ruled if and only if P_4 or P_6 is zero (1905), and in general the role of the canonical divisor in the classification of surfaces. This theory was reworked from the modern perspective of sheaves, cohomology, and characteristic classes in a series of papers by Kodaira (1960–1968) and by the Shafarevich seminar (1961–1963). In particular, new ideas were developed to attack those questions in the classification theory which had proved resistant to the synthetic techniques of the Italian school, for example the classification of elliptic surfaces or the structure of the moduli space of $K3$ surfaces and its relationship with the period map. Another deep result which seems to be inaccessible to the classical methods is the Bogomolov-Miyaoka-Yau inequality $c_1^2 \leq 3c_2$. Moreover, the new methods could be extended to the study of compact complex surfaces (Kodaira) or algebraic surfaces in positive characteristic (Mumford and Bombieri-Mumford). Despite the great progress in understanding algebraic surfaces, many open questions remain. For example, the fundamental problem of whether there exists a classification scheme of some sort for surfaces of general type seems to require a completely new insight.

By contrast, the study of holomorphic vector bundles on algebraic surfaces is much more recent, and effectively dates back to two papers by Schwarzenberger (1961). For the case of algebraic curves, Grothendieck (1956) showed that every holomorphic vector bundle over $\mathbb{P}^1$ is a direct sum of line bundles (a result known in a different language to Hilbert, Plemelj and Birkhoff, and prior to them to Dedekind and Weber). Atiyah (1957) classified all vector bundles over an elliptic curve and made some preliminary remarks concerning vector bundles over curves of higher genus. In 1960, the picture changed radically when Mumford introduced the notion of a stable or semistable vector bundle on an algebraic curve and used geometric invariant theory to construct moduli spaces for all semistable vector bundles over a given curve. Soon thereafter Narasimhan and Seshadri (1965) related the notion of stability to the existence of a unitary flat structure (in the case of trivial determinant) or equivalently a flat connection compatible with an appropriate Hermitian metric. For curves, much recent work has centered on the enumerative geometry of the moduli space of curves. Explicit geometric constructions for the moduli space were given for genus 2 curves by Narasimhan and Ramanan (1969) and for hyperelliptic curves in general by Desale and Ramanan (1976).

In this context, Schwarzenberger made the following contributions to the theory of vector bundles over a surface. In general, for a variety X of dimension greater than 1, a vector bundle on X is not a direct sum of line bundles or an extension of line bundles. Schwarzenberger's first paper studied rank 2 bundles V which are not simple ("almost decomposable" in his terminology), in other words for which the automorphism group is larger than $\mathbb{C}^*$. He showed, using the existence of a rank 1 endomorphism on V, that V is an extension by a line bundle of a coherent sheaf of the form $L \otimes I_Z$, where L is a line bundle and Z is a 2-dimensional local complete intersection subscheme, and in the case of surfaces X he gave a mechanism for describing the set of all such extensions with a fixed Z. To do so, he passed to a blowup $\tilde{X}$ of X in order to be able to replace I_Z by a line bundle of the form $\mathcal{O}_{\tilde{X}}(-\sum_i a_i E_i)$, where the E_i are the components of the exceptional divisor and the a_i are nonnegative integers. As part of the study, he analyzed when a vector bundle on $\tilde{X}$ is the pullback of a bundle on X.

In Schwarzenberger's second paper, he showed that every rank 2 vector bundle on a smooth surface X is of the form $\pi_* L$, where $\pi \colon Y \to X$ is a smooth double cover of X and L is a line bundle on Y. He then applied this construction to construct bundles on $\mathbb{P}^2$ which were not almost decomposable; these turn out to be exactly the stable bundles on $\mathbb{P}^2$. He showed further that, if V is a stable rank 2 vector bundle on $\mathbb{P}^2$, then the Chern classes for V satisfy the basic inequality $c_1(V)^2 < 4c_2(V)$.

In the years after Schwarzenberger's papers, the study of bundles over surfaces diverged into two streams. In the first, there were various attempts to generalize Mumford's definition of stability to surfaces and higher-dimensional varieties and to use this definition to construct moduli spaces of vector bundles. Takemoto (1972, 1973) gave the straightforward generalization to higher-dimensional (polarized) smooth projective varieties that we have simply called stability here (this definition is also called Mumford-Takemoto stability, μ-stability, or slope stability). Aside from proving boundedness results for surfaces, he was unable to prove the existence of a moduli space with this definition (and in fact it is still an open question whether the set of all semistable bundles forms a moduli space in a natural way). Shortly thereafter, Gieseker (1977) introduced the notion of stability now called Gieseker stability or Gieseker-Maruyama stability. Gieseker showed that the set of all Gieseker semistable torsion free sheaves on a fixed algebraic surface X (modulo a suitable equivalence relation) formed a projective variety, containing the set of all Mumford stable vector bundles as a Zariski open set. This result was generalized by Maruyama (1978) to the case where X has arbitrary dimension. The differential geometric meaning of Mumford stability is the Kobayashi-Hitchin conjecture, that every stable vector bundle has a Hermitian-Einstein connection, unique in an appropriate sense. This result, the higher-dimensional analogue of the theorem of Narasimhan and Seshadri, was proved by Donaldson (1985) for surfaces, by Uhlenbeck and Yau (1986) for general Kähler manifolds, and also by Donaldson (1987) in the case of a smooth projective variety. (The easier converse, that an irreducible Hermitian-Einstein connection defines a holomorphic structure for which the bundle is stable, was established previously by Kobayashi and Lübke.) The geometric meaning of Gieseker stability is more mysterious, although Leung (1993) has obtained results in this direction. A related general result is Bogomolov's inequality for stable vector bundles, which follows from the Donaldson-Uhlenbeck-Yau theorem as well as from various purely algebraic arguments (Bogomolov, 1977).

The other stream in studying vector bundles consists in analyzing moduli spaces for specific classes of surfaces (and perhaps specific choices of the Chern classes). The case of $\mathbb{P}^2$ and more generally $\mathbb{P}^n$ has received a great deal of attention, and moduli spaces of vector bundles on $\mathbb{P}^2$ have been described quite explicitly by the method of monads (Barth, Hulek and Maruyama). Because this subject has been well described elsewhere (see for example [117]), we do not discuss monads in this book. The case of ruled surfaces has been analyzed by Hoppe and Spindler and also by Brosius. Takemoto briefly treated the case of abelian surfaces, but the study of vector bundles (not necessarily of rank 2) over $K3$ and abelian surfaces really got off the ground with a series of papers by Mukai. This was the state of the art until about 1985, when Donaldson theory gave a powerful impetus to the study of rank 2 vector bundles over surfaces. We shall describe some of the developments arising after 1985 at the end of Chapter 10.

There is perhaps a third stream which should be mentioned, that of the enumerative geometry of the moduli space. By now these questions have been well studied for bundles over curves (Verlinde formula, cohomology ring of the moduli space), and in some sense Donaldson theory is simply a question about the enumerative geometry of the moduli space of bundles over a surface. Deep structure theorems and conjectures in gauge theory, due to Kronheimer and Mrowka and Witten, suggest that there is a very simple enumerative structure to this moduli space, but as yet there is no way to see why this should be true purely within the context of algebraic geometry.

The goal of this book is to provide a unified introduction to the study of algebraic surfaces and of holomorphic vector bundles on them. I have tried to keep the prerequisites to a good working knowledge of Hartshorne's book on algebraic geometry [61] as well as standard commutative algebra (see for example Matsumura's book [87]). Aside from what is contained in [61], we freely use the exponential sheaf sequence on a complex manifold and the Leray spectral sequence (typically when it degenerates) as well as basic properties of Chern classes which are summarized in Chapter 2, and for which Fulton's book [45] is a standard reference. For the most part, we use the Riemann-Roch theorem only for vector bundles on a curve or surface, for which proofs are given in the exercises to Chapter 2. However, we use the Grothendieck-Riemann-Roch theorem once in Chapter 8 and the Riemann-Roch theorem for a divisor on a threefold in Chapter 10, without recalling the general statements. There is also a brief appeal to relative duality in Chapter 7 and to the existence of a relative Picard scheme for smooth fibrations of relative dimension 1 in Chapter 9. The appendix to Chapter 9 uses a little Galois theory, and some results which are not used in the rest of the book use standard facts about group cohomology. The last section of Chapter 4 assumes some basic familiarity with differential geometry on a complex manifold, for example as described in the book by Griffiths and Harris [55], and can be skipped. In Chapter 8, there is a brief discussion of Donaldson invariants which motivates some of the enumerative calculations in the rest of the chapter, but which can otherwise be omitted. Of necessity, I have largely limited myself to the part of the study of vector bundles which does not involve the heavy machinery of deformation theory or geometric invariant theory; a few descriptive sections outline the main results.

For the first eight chapters, the plan has been to alternate between the study of surfaces and the study of bundles on them. This has the pedagogical advantage that, for example, vector bundles over curves are studied in Chapter 4, then used to describe ruled surfaces in Chapter 5. In Chapter 6, we use the knowledge of ruled surfaces to describe vector bundles over them, and in Chapter 9 they reappear as part of the proof of Bogomolov's inequality. Similarly, ruled surfaces are described in Chapter 5 and elliptic surfaces in Chapter 7, and the structure of the moduli space of vector bundles over such surfaces is then described in Chapters 6 and 8. I have tried to emphasize how the internal geometry of the surface is reflected in the birational geometry of the moduli space. In the last two chapters, we drop the strict division of material: Chapter 9 gives a proof of Bogomolov's inequality, which belongs to the theory of vector bundles, as well as applications to the study of linear systems (in particular pluricanonical systems) on an algebraic surface. In Chapter 10, we prove the main theorems on the classification of algebraic surfaces and outline the current state of knowledge concerning moduli spaces of rank 2 vector bundles over algebraic surfaces. The proofs of the classification results for surfaces are old-fashioned, in the sense that they do not appeal to Mori theory. On the other hand, the old-fashioned proofs may be better adapted to handling the classification of symplectic 4-manifolds. The point of view of Mori theory and the classification results for threefolds are briefly described toward the end of the chapter. Because of the way we alternate between surfaces and vector bundles, it may be a little disorienting to try to read the book chronologically, and certainly the chapters on surfaces can be, for the most part, read independently of the chapters on vector bundles. On the other hand, the later chapters on vector bundles over ruled or elliptic surfaces draw heavily on the description of the corresponding surfaces in the chapters that precede them.

Constraints of length and time dictated that many topics had to be left out. For surfaces, I would have liked to devote more time to rational and minimally elliptic singularities and to the classification of surfaces of small degree. For vector bundles, without the main tool of deformation theory, we are only able to scratch the surface of this rapidly evolving field. Because this theory does not seem to be close to a definitive state, it seems worthwhile to focus on many concrete examples.

Finally, there are many exercises at the end of each chapter, and they are an integral part of the book. In particular, many results are left to the exercises and they are frequently used in later chapters. I hope that the emphasis on examples, both in the text and the exercises, will help to serve as an introduction to this rich and beautiful field of mathematics.

1

Curves on a Surface

Introduction

In this book, unless otherwise specified, by *surface* we shall always mean a connected compact complex manifold of complex dimension 2 which is a holomorphic submanifold of $\mathbb{P}^N$ for some N. Thus, "surface" is short for smooth (connected) complex algebraic surface. By Chow's theorem, a surface is also described as the zero set in $\mathbb{P}^N$ of a finite number of homogeneous polynomials in $N+1$ variables. The study of surfaces is concerned both with the intrinsic geometry of the surface and with the geometry of the possible embeddings of the surface in $\mathbb{P}^N$. Just as with curves, we could organize this study in order of increasing complexity. In terms of the extrinsic (synthetic) geometry of a surface in $\mathbb{P}^N$, we could for instance try to study and eventually classify surfaces in $\mathbb{P}^N$ of relatively small degree. Or we could attempt to order surfaces by complexity via some intrinsic invariants, by analogy with the genus of a curve. This is the aim of the Kodaira classification, which orders surfaces by their Kodaira dimension. For this scheme, we have a fairly complete understanding of surfaces except in the case of Kodaira dimension 2, general type surfaces. We will cover the broad outlines of the general theory of surfaces. In this chapter, we will discuss the basic invariants, intersection theory and Riemann-Roch, and the structure of the set of ample divisors. In Chapter 3, we will discuss birational geometry. Chapters 5 and 7 will concern some of the main examples of surfaces: rational and ruled surfaces, $K3$ surfaces, as well as an introduction to elliptic surfaces. Finally, in Chapter 10, we shall give a general overview of the classification of algebraic surfaces.

We begin with the description of the basic numerical and topological invariants of a surface.

Invariants of a surface

A surface X is in particular a complex manifold, and always carries a canonical orientation from its complex structure. Viewing X as an oriented 4-manifold, its main topological invariants are its fundamental group $\pi_1(X, *)$, the Betti numbers $b_i(X) = b_{4-i}(X)$, and the intersection pairing on $H_2(X; \mathbb{Z})$. Here by Poincaré duality $H_2(X; \mathbb{Z}) \cong H^2(X; \mathbb{Z})$ and intersection pairing corresponds under this isomorphism to cup product from $H^2(X; \mathbb{Z}) \otimes H^2(X; \mathbb{Z})$ to $H^4(X; \mathbb{Z}) \cong \mathbb{Z}$ by taking the canonical orientation. Over $\mathbb{R}$, the intersection pairing is specified by $b_2(X)$ and by $b_2^+(X)$, the number of positive entries along the diagonal when the form is diagonalized over $\mathbb{R}$. We also let $b_2^-(X) = b_2(X) - b_2^+(X)$. If $X = \mathbb{P}^2$ or if X is one of an unknown but finite number of surfaces of general type whose universal cover is the unit ball in $\mathbb{C}^2$, then $H^2(X; \mathbb{R}) \cong \mathbb{R}$. If X does not belong to this finite list of examples, then $H_2(X; \mathbb{R})$ is always indefinite (cf. for example [40, p. 29, Lemma 2.4]). It then follows from the classification of quadratic forms over $\mathbb{Z}$ [138], [92] that the intersection pairing on $H_2(X; \mathbb{Z})$ mod torsion is specified by its rank, signature, and type, i.e., whether or not there exists an element $\alpha \in H_2(X; \mathbb{Z})$ with $\alpha^2 \equiv 1 \mod 2$ or not. (If there exists such an α the form is *odd* or of Type I; otherwise it is *even* or of Type II.) To decide if a surface is of Type I or Type II, we use the Wu formula, which says that $\alpha^2 \equiv \alpha \cdot [K_X] \mod 2$. Here $[K_X]$ denotes the homology class associated to the canonical line bundle K_X via $c_1(K_X)$ and Poincaré duality. Thus, again by Poincaré duality, there exists an α with $\alpha^2 \equiv 1 \mod 2$ if and only if the image of $[K_X]$ in $\bar{H}_2(X; \mathbb{Z}) = H_2(X; \mathbb{Z})$ modulo torsion is not divisible by two.

There are also the holomorphic invariants of X. The most basic ones are the *irregularity* $q(X)$ of X and the *geometric genus* $p_g(X)$ of X, defined by

$$q(X) = \dim_{\mathbb{C}} H^0(X; \Omega_X^1) = \dim_{\mathbb{C}} H^1(X; \mathcal{O}_X),$$
$$p_g(X) = \dim_{\mathbb{C}} H^0(X; \Omega_X^2) = \dim_{\mathbb{C}} H^2(X; \mathcal{O}_X).$$

Thus, $q(X)$ is the number of independent holomorphic 1-forms on X and $p_g(X)$ is the number of holomorphic 2-forms on X. We note that the fact that the two different expressions above for $q(X)$ are equal follows from Hodge theory, since X is an algebraic surface over $\mathbb{C}$, and do not hold for an arbitrary compact complex surface or for a surface defined over a field of positive characteristic; in either case the "correct" definition of $q(X)$ is $\dim H^1(X; \mathcal{O}_X)$. (That the two expressions for $p_g(X)$ are equal follows from Serre duality which holds in general.) Additional invariants are given by $h^{1,1}(X) = \dim H^1(X; \Omega_X^1)$ and $c_1(X)^2 = [K_X]^2$. The relation of these invariants to the topological ones is as follows:

$$b_1(X) = 2q(X),$$

$$b_2(X) = 2p_g(X) + h^{1,1}(X),$$
$$b_2^+(X) = 2p_g(X) + 1.$$

Here the first two equalities follow by Hodge theory and the last is one form of the Hodge index theorem for a surface. We also have the Euler characteristic

$$\chi(X) = 1 - b_1(X) + b_2(X) - b_3(X) + 1$$
$$= 2 - 2b_1(X) + b_2(X) = 2 - 4q + 2p_g(X) + h^{1,1}(X)$$

and the holomorphic Euler characteristic

$$\chi(\mathcal{O}_X) = h^0(\mathcal{O}_X) - h^1(\mathcal{O}_X) + h^2(\mathcal{O}_X) = 1 - q(X) + p_g(X).$$

There is also Noether's formula (in some sense a special case of the Riemann-Roch theorem for surfaces) which says that

$$c_1^2(X) + c_2(X) = 12\chi(\mathcal{O}_X),$$

or in other words that $[K_X]^2 + \chi(X) = 12(1 - q(X) + p_g(X))$. An easy manipulation of the formulas (Exercise 1) shows that Noether's formula is equivalent to the Hirzebruch signature theorem

$$b_2^+(X) - b_2^-(X) = \tfrac{1}{3}(c_1^2(X) - 2c_2(X)).$$

Beyond this there are the "higher" holomorphic invariants of X, the *plurigenera* $P_n(X) = \dim H^0(X; K_X^{\otimes n})$, defined for $n \geq 1$. Thus, $P_1(X) = p_g(X)$. It is by now well known [39] that the plurigenera are not in general homotopy or homeomorphism invariants of X. It has recently been shown via new invariants introduced by Seiberg and Witten that the plurigenera are diffeomorphism invariants of X (see for example [16] and [41]). We shall discuss some of these developments further in Chapter 10.

Divisors on a surface

We recall that a (reduced irreducible) *curve* C on X is an irreducible holomorphic subvariety of complex dimension 1. Thus, locally C is described as $\{f(z_1, z_2) = 0\}$, where f is a holomorphic function of z_1, z_2. Of course, C need not be a (holomorphic) submanifold of X; if it is we say that C is a *smooth* curve. A *divisor* D on X is a finite formal sum $\sum_i n_i C_i$ of distinct irreducible curves C_i, where the $n_i \in \mathbb{Z}$. The set of all divisors $\operatorname{Div} X$ is thus the free abelian group generated by the irreducible curves on X. The divisor D is *effective* if the $n_i \geq 0$ for all i. An effective divisor $D \neq 0$ will also be called a curve. We write $D \geq 0$ if D is effective and $D_1 \geq D_2$ if $D_1 - D_2 \geq 0$. If f_i is a local equation for the curve C_i, then D is locally described by the meromorphic function $\prod_i f_i^{n_i}$, which is in fact holomorphic if and only if D is effective. Conversely, a meromorphic function f on X has an associated divisor (f), which is the curve of zeros of f minus the curve of poles of f. All of these constructions also make sense locally on open subsets U of X. Given a divisor D on X, there is an associated line bundle $\mathcal{O}_X(D)$ whose associated sections on an open subset U of X are meromorphic functions g on U such that $(g) + D|U \geq 0$, where $D|U$ is the restriction of the divisor D to U in the obvious sense. In particular, if D is effective, then the constant function 1 defines a global section of D. The map $D \mapsto \mathcal{O}_X(D)$ is then a homomorphism from the free abelian group $\operatorname{Div} X$ to the group $\operatorname{Pic} X$ of line bundles on X under tensor product.

We recall that two divisors D_1 and D_2 are *linearly equivalent* (which we shall write as $D_1 \equiv D_2$) if and only if $\mathcal{O}_X(D_1) \cong \mathcal{O}_X(D_2)$ if and only if $D_1 - D_2 = (f)$ for a globally defined meromorphic function f on X. A linear equivalence class of divisors will be called a *divisor class*. Every holomorphic line bundle L on X is of the form $\mathcal{O}_X(D)$ for some divisor D. In fact, defining a *meromorphic section* of a line bundle via local trivializations, if s is a meromorphic section of L, then s has associated to it a well-defined divisor $(s) = D$ and it is straightforward to show that $L \cong \mathcal{O}_X(D)$. Thus, if L has a holomorphic section s, then $L = \mathcal{O}_X(D)$ for an effective divisor D. For example, given an effective D, the global section 1 of $\mathcal{O}_X(D)$ described above vanishes exactly along D (viewed as a section of $\mathcal{O}_X(D)$, of course). The group of divisor classes may be naturally identified with $\operatorname{Pic} X$. We shall call the divisor (or corresponding divisor class) D ample or very ample if the line bundle $\mathcal{O}_X(D)$ is ample or very ample. Divisors are functorial in the following sense: if $\pi \colon X \to Y$ is a surjective map, then pullback π^* induces a homomorphism $\operatorname{Div} Y \to \operatorname{Div} X$. Given a divisor D, it defines a homology class. One way to see this, for an irreducible effective divisor D, is to choose a triangulation of X for which the support of D is a subcomplex. Another way is to use the homomorphism $\operatorname{Div} X \to H^2(X; \mathbb{Z})$ given by $D \mapsto c_1(\mathcal{O}_X(D))$, followed by Poincaré duality. Here, for a holomorphic line bundle L, we can define the first Chern class $c_1(L)$ directly via the exponential sheaf sequence

$$0 \to \mathbb{Z} \to \mathcal{O}_X \xrightarrow{\exp(2\pi i \cdot)} \mathcal{O}_X^* \to 1,$$

by taking c_1 to be the coboundary map $H^1(\mathcal{O}_X^*) = \operatorname{Pic} X \to H^2(X; \mathbb{Z})$. In any case, we shall denote by $[D]$ the homology class associated to D.

Our goal now will be to give an algebraically defined intersection pairing on $\operatorname{Div} X$ which agrees in a natural sense with the topological intersection form under the induced map $\operatorname{Div} X \to H_2(X; \mathbb{Z})$. We begin by defining a local intersection number for two curves C_1, C_2 which have no component in common.

Definition 1. Let C_1, C_2 be two curves with no component in common and let $x \in X$. Define $C_1 \cdot_x C_2 = \dim_{\mathbb{C}} \mathcal{O}_{X,x}/(f_1, f_2)$, where f_i is a local equation for C_i at x.

Note that to say that C_1, C_2 have no component in common is exactly to say that f_1 and f_2 are relatively prime in the ring $\mathcal{O}_{X,x}$ for every x and so define an ideal such that the quotient ring is a finite-dimensional $\mathbb{C}$-algebra. It is clear that this quotient is independent of the choice of local equation and is the part of the scheme-theoretic intersection $C_1 \cap C_2$ supported at x. Of course this is empty if $x \notin C_1 \cap C_2$, and in this case it is easy to see that $C_1 \cdot_x C_2 = 0$. Note also:

Lemma 2. *The curves C_1 and C_2 meet transversally at the point x if and only if $C_1 \cdot_x C_2 = 1$.*

Proof. By definition C_1 and C_2 meet transversally at x if and only if $(f_1, f_2) = \mathfrak{m}_x$, the maximal ideal of $\mathcal{O}_{X,x}$ at x, and since $\mathcal{O}_{X,x}/(f_1, f_2)$ is a nonzero $\mathbb{C}$-algebra and thus always has dimension at least 1, this condition is equivalent to the condition that $C_1 \cdot_x C_2 = 1$. $\square$

To define an intersection pairing for divisors, we proceed as follows: suppose as before that C_1, C_2 have no component in common and define

$$C_1 \cdot C_2 = \sum_{x \in X} C_1 \cdot_x C_2.$$

By hypothesis this is a finite sum.

Lemma 3. *If C_1 is a smooth irreducible curve and C_1 is not a component of C_2, then $C_1 \cdot C_2 = \deg \mathcal{O}_X(C_2)|C_1$ (here deg is the degree of a complex line bundle on the curve C_1). In particular, in this case $C_1 \cdot C_2$ only depends on the linear equivalence class of C_2.*

Proof. Take the exact sequence

$$0 \to \mathcal{O}_X(-C_2) \to \mathcal{O}_X \to \mathcal{O}_{C_2} \to 0$$

and tensor it with $\mathcal{O}_{C_1}$. Let f_1 and f_2 be local equations for C_1 and C_2, respectively, at x. An easy calculation using the fact that f_1 and f_2 are relatively prime shows that the resulting sequence

$$0 \to \mathcal{O}_X(-C_2) \otimes \mathcal{O}_{C_1} \to \mathcal{O}_{C_1} \to \bigoplus_{x \in C_1 \cap C_2} \mathcal{O}_{X,x}/(f_1, f_2) \to 0$$

is still exact (in other words, $\mathrm{Tor}_1^{\mathcal{O}_{X,x}}(\mathcal{O}_{X,x}/(f_1), \mathcal{O}_{X,x}/(f_2)) = 0$). If we tensor the second exact sequence with $\mathcal{O}_X(C_2)|C_1$, we obtain instead

$$0 \to \mathcal{O}_{C_1} \to \mathcal{O}_{C_1}(C_2) \to \bigoplus_{x \in C_1 \cap C_2} \mathcal{O}_{X,x}/(f_1, f_2) \to 0.$$

Thus, $\mathcal{O}_{C_1}(C_2)$ has a section vanishing at exactly $C_1 \cdot C_2$ points, counted with multiplicity, and so $\deg \mathcal{O}_{C_1}(C_2) = C_1 \cdot C_2$. $\square$

Theorem 4. *There is a unique symmetric bilinear pairing from* $\mathrm{Div}\,X$ *to* $\mathbb{Z}$, *denoted by* $\langle D_1, D_2 \rangle$ *which factors through linear equivalence and has the property that* $\langle C_1, C_2 \rangle = C_1 \cdot C_2$ *for* C_1 *and* C_2 *distinct smooth curves meeting transversally.*

Proof. Note the following standard lemma:

Lemma 5. *Let* L *be a line bundle on* X. *Then there exist two very ample divisors* H' *and* H'' *on* X *such that* $L = \mathcal{O}_X(H' - H'')$. *In particular every divisor* $D \in \mathrm{Div}\,X$ *is linearly equivalent to a difference of two very ample divisors.* $\square$

To prove Theorem 4, we begin with the uniqueness. Given D_1 and D_2 in $\mathrm{Div}\,X$ write $D_i \equiv H_i' - H_i''$. We may assume that all of the H_i', H_i'' are distinct and smooth and meet transversally. Thus, necessarily we must have

$$(1.1) \qquad \langle D_1, D_2 \rangle = H_1' \cdot H_2' - H_1' \cdot H_2'' - H_1'' \cdot H_2' + H_1'' \cdot H_2''.$$

To see the existence, note that the above formula can be written as

$$(1.2) \qquad \langle D_1, D_2 \rangle = \deg \mathcal{O}_X(D_1)|H_2' - \deg \mathcal{O}_X(D_1)|H_2''.$$

Fix D_2 and choose smooth curves H_2', H_2'' meeting transversally with D_2 linearly equivalent to $H_2' - H_2''$. For an arbitrary divisor D_2, we can define $\langle D_1, D_2 \rangle$ by formula (1.1). Using (1.2), we see that $\langle D_1, D_2 \rangle$ only depends on the linear equivalence class of D_1, and in particular does not depend on the choice of H_1' and H_1''. By symmetry the same is true for D_2. Thus, $\langle D_1, D_2 \rangle$ is well defined by (1.2), and is clearly symmetric. It follows from (1.2) that $\langle D_1, D_2 \rangle$ is bilinear, and we are done. $\square$

We shall usually denote $\langle D_1, D_2 \rangle$ by $D_1 \cdot D_2$. As a corollary of the above proof, note that, if D_2 is smooth, then $D_1 \cdot D_2 = \deg \mathcal{O}_X(D_1)|D_2$. In fact, a similar formula is true for an irreducible curve D_2, noting that we can define the degree of a line bundle L on an irreducible curve C in several equivalent ways:

(i) As the degree of the pullback of L to the normalization $\tilde{C}$ of C;
(ii) By writing $L = \mathcal{O}_C(\sum_i n_i p_i)$, where the p_i are points in the smooth part of C, and taking $\deg L = \sum_i n_i$;
(iii) Via the exponential sheaf sequence

$$0 \to \mathbb{Z} \to \mathcal{O}_C \to \mathcal{O}_C^* \to 0$$

and the fact that for an irreducible curve we have $H^2(C; \mathbb{Z}) \cong \mathbb{Z}$.

The uniqueness part of the proof of Theorem 4 shows that $D_1 \cdot D_2 = [D_1] \cdot [D_2]$, where $[D_i]$ is the homology class associated to D_i and we use intersection product in homology. Finally, two remarks that we shall often use are the following: if H is ample and D is effective and nonzero, then $H \cdot D > 0$. Moreover, if C_1 and C_2 are distinct irreducible curves, then $C_1 \cdot C_2 \geq 0$, and $C_1 \cdot C_2 = 0$ if and only if C_1 and C_2 are disjoint.

In the rest of this chapter we shall use intersection theory to analyze curves and linear systems on X.

Adjunction and arithmetic genus

Suppose that C is a smooth curve on X. Then $C \cdot C = \deg \mathcal{O}_X(C)|C$. By general results on smooth divisors (see for example Hartshorne [61, p. 182]), $\mathcal{O}_X(C)|C = N_{C/X}$ is the normal bundle to C in X. For an effective divisor C, not necessarily smooth or reduced, we shall sometimes define the normal bundle of C in X to be simply $\mathcal{O}_X(C)|C$. We shall also usually abbreviate $C \cdot C$ by C^2.

For a smooth curve C, we also have the *adjunction formula*

$$(1.3) \qquad K_C = K_X \otimes \mathcal{O}_X(C)|C.$$

This follows from the normal bundle sequence

$$0 \to T_C \to T_X|C \to N_{C/X} \to 0,$$

which gives $\det(T_X|C) = (K_X|C)^{-1} = K_C^{-1} \otimes \mathcal{O}_X(C)|C$. Thus, if $g(C) = g$ is the genus of C, then

$$(1.4) \qquad 2g - 2 = \deg K_C = (K_X + C) \cdot C.$$

Here we use the same symbol K_X to denote the canonical line bundle and the canonical divisor (class).

For C any nonzero effective divisor on X, we can still define the *dualizing sheaf* ω_C by the same formula as (1.3)

$$(1.5) \qquad \omega_C = K_X \otimes \mathcal{O}_X(C)|C.$$

The significance of ω_C is that it is the unique line bundle on the (possibly singular) scheme C for which Serre duality holds: there is a trace map $H^1(C; \omega_C) \to \mathbb{C}$ such that, for every line bundle L on C, the induced map

$$H^0(C; L) \otimes H^1(C; \omega_C \otimes L^{-1}) \to \mathbb{C}$$

is a perfect pairing (Serre [136] for the case where C is reduced and Barth-Peters-Van de Ven [7] in general). In particular $H^1(C; \omega_C)$ is dual to $H^0(C; \mathcal{O}_C)$. **Warning**: If C is reduced and connected, $H^0(C; \mathcal{O}_C) \cong \mathbb{C}$ and the trace map is an isomorphism. In general, however, $H^0(C; \mathcal{O}_C)$ may be larger than $\mathbb{C}$ (Exercise 3) and thus $H^1(C; \omega_C)$ may have dimension larger than 1 also.

For a general nonzero effective divisor C we define the *arithmetic genus* of C by the same formula as before

$$(1.6) \qquad 2p_a(C) - 2 = (K_X + C) \cdot C.$$

In case C is reduced and irreducible, $2p_a(C)-2$ is therefore equal to $\deg \omega_C$. In general an application of the Riemann-Roch theorem on X (see Exercise 8) shows that

$$(1.7) \qquad\qquad p_a(C) = 1 - \chi(C; \mathcal{O}_C).$$

Thus, if $h^0(C; \mathcal{O}_C) = 1$, for example, if C is reduced and connected, then $p_a(C) = h^1(\mathcal{O}_C) = h^0(C; \omega_C)$. As a result we have:

Corollary 6. *If C is an irreducible curve on X, then $p_a(C) \geq 0$. Thus, $(K_X + C) \cdot C \geq -2$. Moreover, $(K_X + C) \cdot C = -2$ if and only if $p_a(C) = 0$, and otherwise $(K_X + C) \cdot C \geq 0$.* □

In fact, we can say more than the statement that $(K_X + C) \cdot C \geq 0$ if $p_a(C) \geq 1$:

Proposition 7. *If C is an irreducible curve on X with $p_a(C) \geq 1$, then ω_C has no base locus.* □

For a proof of Proposition 7, see, for example, Catanese [17].

In general it will be useful to have various ways to calculate $p_a(C)$. We begin with the case where C is reduced and irreducible. In this case the normalization $\tilde{C}$ is a smooth connected curve, and has a well-defined genus $g(\tilde{C})$. Let $\nu : \tilde{C} \to C$ be the normalization map. Now consider the exact sequence

$$(1.8) \qquad\qquad 0 \to \mathcal{O}_C \to \nu_*\mathcal{O}_{\tilde{C}} \to \nu_*\mathcal{O}_{\tilde{C}}/\mathcal{O}_C \to 0.$$

For $x \in C$, we define the *local genus drop* at x to be the nonnegative integer

$$\delta_x = \dim_\mathbb{C} [\nu_*\mathcal{O}_{\tilde{C}}/\mathcal{O}_C]_x .$$

Thus, $\delta_x = 0$ if and only if x is a smooth point of C. For example, if x is an ordinary double point of C, then $\delta_x = 1$. Likewise, if x is a cusp point, so that, locally analytically near x, C is described by the equation $y^2 = x^3$, $\tilde{C}$ is a smooth curve with coordinate t, and $\nu(t) = (t^2, t^3)$ in local coordinates, we again have $\delta_x = 1$. We leave it as an exercise to show that conversely, if $\delta_x = 1$, then x is either an ordinary double point of C or a cusp. The *genus drop* of the curve C is the nonnegative integer $\delta = \sum_{x \in C} \delta_x$. Note that we can still define the local invariant δ_x, and hence δ, if C is only assumed reduced, but not necessarily irreducible.

Lemma 8. *If C is reduced and irreducible, then $p_a(C) = g(\tilde{C}) + \delta$. Thus, $p_a(C) = 0$ if and only if C is a smooth rational curve. More generally, let C be a reduced but not necessarily irreducible curve on X, and let $C_1, \ldots, C_n$ be the connected components of $\tilde{C}$, with $g(C_i) = g_i$. Then $p_a(C) = \sum_i g_i + \delta + 1 - n$.*

Proof. We shall just check the first statement, as the proof of the second is similar. Since $\tilde{C}$ is connected, $H^0(\mathcal{O}_C) \cong H^0(\mathcal{O}_{\tilde{C}})$ and the long exact cohomology sequence of the sheaf sequence (1.8) gives

$$0 \to H^0(\nu_*\mathcal{O}_{\tilde{C}}/\mathcal{O}_C) \to H^1(\mathcal{O}_C) \to H^1(\mathcal{O}_{\tilde{C}}) \to 0.$$

A dimension count gives $h^1(\mathcal{O}_C) = h^1(\mathcal{O}_{\tilde{C}}) + \delta = g(\tilde{C}) + \delta$.   □

We will return to the study of δ in Chapter 3. We now briefly discuss the nonreduced case. The main tool for studying C in this case is the following exact sequence: suppose that C and D are two nonzero effective divisors, not necessarily reduced or irreducible. Here we will allow C and D to have components in common. Then, by Exercise 9, there is an exact sequence

(1.9) $$0 \to \mathcal{O}_D(-C) \to \mathcal{O}_{C+D} \to \mathcal{O}_C \to 0.$$

Thus, (1.9) allows us to work out $\chi(\mathcal{O}_{C+D})$ from the knowledge of $\chi(\mathcal{O}_D(-C))$ and $\chi(\mathcal{O}_C)$, and would allow us to work out $h^0(\mathcal{O}_{C+D})$ and thus $h^1(\mathcal{O}_{C+D})$ if we could work out the coboundary maps in the associated long exact sequence (which is usually impossible in general). An important application of (1.9) is to work out $\mathcal{O}_{nC_0}$, where C_0 is a reduced and irreducible curve. In this case, for $n \geq 2$, the exact sequence of (1.9) applied to $C = (n-1)C_0$ and $D = C_0$ becomes

(1.10) $$0 \to \mathcal{O}_{C_0}(-(n-1)C_0) \to \mathcal{O}_{nC_0} \to \mathcal{O}_{(n-1)C_0} \to 0.$$

The Riemann-Roch formula

Let D be a divisor on X. The Riemann-Roch formula is then:

Theorem 9. *The Euler characteristic $\chi(X; \mathcal{O}_X(D))$ is given by the following formula*

$$\chi(X; \mathcal{O}_X(D)) = \tfrac{1}{2}D \cdot (D - K_X) + \chi(\mathcal{O}_X).$$

Proof. The formula is trivially valid for $D = 0$. Next, for $D = C$ a smooth curve, use the exact sequence

$$0 \to \mathcal{O}_X \to \mathcal{O}_X(C) \to \mathcal{O}_X(C)|C \to 0.$$

By additivity of the Euler characteristic,

$$\chi(\mathcal{O}_X(C)) = \chi(\mathcal{O}_X) + \chi(\mathcal{O}_X(C)|C).$$

On the other hand, $g(C) = (C^2 + K_X \cdot C)/2 + 1$ and $\deg \mathcal{O}_X(C)|C = C^2$. Thus, an application of the Riemann-Roch theorem for curves gives

$$\chi(\mathcal{O}_X(C)|C) = C^2 - \left(\frac{C^2 + K_X \cdot C}{2} + 1\right) + 1 = \frac{C^2 - K_X \cdot C}{2},$$

verifying the formula in this case. In the general case, write $D = C_1 - C_2$ where C_1 and C_2 are smooth, which is possible by Lemma 5, and use the exact sequence

$$0 \to \mathcal{O}_X(C_1 - C_2) \to \mathcal{O}_X(C_1) \to \mathcal{O}_X(C_1)|C_2 \to 0.$$

Thus, $\chi(\mathcal{O}_X(C_1-C_2)) = \chi(\mathcal{O}_X(C_1))-\chi(\mathcal{O}_X(C_1)|C_2)$. By the previous case $\chi(\mathcal{O}_X(C_1)) = \frac{1}{2}(C_1^2 - C_1 \cdot K_X)+\chi(\mathcal{O}_X)$, and adjunction and Riemann-Roch on C_2 give

$$\chi(\mathcal{O}_X(C_1)|C_2) = (C_1 \cdot C_2) - \tfrac{1}{2}(C_2^2 + C_2 \cdot K_X) - 1 + 1.$$

Combining these gives the following formula for $\chi(\mathcal{O}_X(C_1 - C_2))$:

$$\tfrac{1}{2}(C_1^2 - C_1 \cdot K_X) + \chi(\mathcal{O}_X) - (C_1 \cdot C_2) + \tfrac{1}{2}(C_2^2 + C_2 \cdot K_X)$$
$$=\tfrac{1}{2}(C_1^2 + C_2^2 - 2(C_1 \cdot C_2) - C_1 \cdot K_X + C_2 \cdot K_X) + \chi(\mathcal{O}_X)$$
$$=\tfrac{1}{2}((C_1 - C_2)^2 - (C_1 - C_2) \cdot K_X) + \chi(\mathcal{O}_X),$$

as desired. $\square$

Note that as a consequence of the Riemann-Roch formula, we recover the Wu formula for divisors: $K_X \cdot D \equiv D^2 \mod 2$.

Closely tied in with the Riemann-Roch formula is Serre duality :

Theorem 10. *If D is a divisor on X, then the vector space $H^i(X; \mathcal{O}_X(D))$ is naturally dual to $H^{2-i}(X; \mathcal{O}_X(K_X - D))$.* $\square$

Notice that the Riemann-Roch formula is indeed invariant under the substitution $D \mapsto K_X - D$.

Finally, the Riemann-Roch formula is most effective when we have some criteria for the vanishing of $H^i(X; \mathcal{O}_X(D))$. The most famous of these is the Kodaira vanishing theorem: if D is ample, then $H^i(X; \mathcal{O}_X(-D)) = 0$ for $i = 0, 1$. Dually $H^i(X; \mathcal{O}_X(K_X+D)) = 0$ for $i = 1, 2$. We shall discuss a generalization of the Kodaira vanishing theorem at the end of this chapter.

Algebraic proof of the Hodge index theorem

We shall give Grothendieck's elegant proof of the algebraic version of the Hodge index theorem for divisors [57].

Theorem 11. *Let H be an ample divisor on X, and let D be a divisor such that $D \cdot H = 0$. Then $D^2 \leq 0$, and if $D^2 = 0$, then $D \cdot E = 0$ for all divisors E.*

Proof. We begin with the following lemma:

Lemma 12. *Let H be an ample divisor on X, and let D be a divisor such that $D^2 > 0$ and $D \cdot H > 0$. Then for all $n \gg 0$, the divisor nD is nonzero and effective.*

Proof. Applying the Riemann-Roch formula and Serre duality to nD gives

$$h^0(\mathcal{O}_X(nD)) + h^0(\mathcal{O}_X(K_X - nD)) \geq \chi(\mathcal{O}_X(nD)) = \tfrac{1}{2}D^2 n^2 + O(n).$$

Thus, for all $n \gg 0$ either nD or $K_X - nD$ is effective. However, $H \cdot (K_X - nD) < 0$ if $n > (H \cdot K_X)/(H \cdot D)$, so that $K_X - nD$ cannot be effective as soon as n is sufficiently large. Thus, nD is effective for all $n \gg 0$, and it is nonzero since $(nD)^2 = n^2 D^2 > 0$. $\square$

Returning to the proof of the Hodge index theorem, let D be a divisor such that $D \cdot H = 0$. Suppose first that $D^2 > 0$. By Serre's criterion (see, for example, [61, II §7]), $mH + D$ is ample for all $m \gg 0$. Now $D^2 > 0$ and $D \cdot (mH + D) = D^2 > 0$, so applying Lemma 12 to the divisor D with H replaced by $mH + D$, it follows that nD is effective and nonzero for all $n \gg 0$. But then $(nD) \cdot H > 0$, and so $D \cdot H \neq 0$, contradicting our hypothesis on D.

Thus, we must have $D^2 \leq 0$. Suppose now that $D^2 = 0$. If there exists a divisor E with $D \cdot E \neq 0$, after replacing E by $-E$ we may assume that $D \cdot E > 0$. Next replace E by the divisor $E' = (H^2)E - (H \cdot E)H$. Then $E' \cdot H = 0$ and $D \cdot E' = (H^2)(D \cdot E) > 0$. If we now set $D' = mD + E'$, then $D' \cdot H = 0$ and $(D')^2 = 2m(D \cdot E') + (E')^2$. For $m \gg 0$, D' is thus a divisor satisfying $(D')^2 > 0$, $D' \cdot H = 0$, contradicting the first part of the proof. It follows that either $D^2 < 0$ or $D^2 = 0$ and $D \cdot E = 0$ for all divisors E, as claimed. $\square$

Definition 13. A divisor D is *numerically equivalent to* 0 if $D \cdot E = 0$ for all divisors E. Two divisors D_1 and D_2 are *numerically equivalent* if $D_1 - D_2$ is numerically equivalent to 0. Note that linear equivalence implies numerical equivalence. We let $\operatorname{Num} X$ be the quotient of $\operatorname{Div} X$ by the equivalence relation of numerical equivalence.

Let $\bar{H}^2(X; \mathbb{Z}) = H^2(X; \mathbb{Z})$ modulo torsion. Then $\bar{H}^2(X; \mathbb{Z})$ is a subgroup of $H^2(X; \mathbb{C})$. According to Hodge theory, $H^2(X; \mathbb{C}) \cong H^{2,0}(X) \oplus H^{1,1}(X) \oplus H^{0,2}(X)$, where $H^{p,q}(X) \cong H^q(X; \Omega_X^p)$ and $H^{p,q}(X)$ is the complex conjugate of $H^{q,p}(X)$. Thus, the subspaces $H^{2,0}(X) \oplus H^{0,2}(X)$ and $H^{1,1}(X)$ are invariant under complex conjugation and are therefore the complexifications of subspaces of $H^2(X; \mathbb{R})$, which we denote by $\left(H^{2,0}(X) \oplus H^{0,2}(X)\right)_{\mathbb{R}}$ and $H^{1,1}(X)_{\mathbb{R}}$, respectively. If D is a divisor, then the image of its associated cohomology class in $H^2(X; \mathbb{C})$ lies in $H^{1,1}(X)$ as well as in the image of $H^2(X; \mathbb{Z})$, which we can write as $\bar{H}^2(X; \mathbb{Z}) \cap H^{1,1}(X)$. The usual Hodge index theorem then says that the intersection pairing is positive definite on $\left(H^{2,0}(X) \oplus H^{0,2}(X)\right)_{\mathbb{R}}$ and negative definite on the complement of an ample divisor in $H^{1,1}(X)_\mathbb{R}$. Combining the algebraic Hodge index theorem above with the usual Hodge index theorem, we find:

Lemma 14. *The natural map from* $\operatorname{Num} X$ *to* $\bar{H}^2(X;\mathbb{Z}) \cap H^{1,1}(X) \subseteq H^2(X;\mathbb{C})$ *is an isomorphism.*

Proof. There is the natural map from $\operatorname{Div} X$ to $\bar{H}^2(X;\mathbb{Z}) \cap H^{1,1}(X)$, which is surjective by the Lefschetz theorem on $(1,1)$-classes. Clearly, if D is not numerically equivalent to zero, then the image of D in $\bar{H}^2(X;\mathbb{Z}) \cap H^{1,1}(X)$ is nonzero. Conversely, suppose that the image of D in $\bar{H}^2(X;\mathbb{Z}) \cap H^{1,1}(X)$ is nonzero. Let H be an ample divisor on X. If $[D] \cdot H \neq 0$, then D is not numerically trivial. If $[D] \cdot H = 0$, then by the usual Hodge index theorem $[D]^2 < 0$, and so D is again not numerically trivial. Thus, the kernel of the map from $\operatorname{Div} X$ to $\bar{H}^2(X;\mathbb{Z}) \cap H^{1,1}(X)$ is exactly the subgroup of numerically trivial divisors. $\square$

As a corollary, we have:

Corollary 15. *The group* $\operatorname{Num} X$ *is a finitely generated torsion free abelian group, of rank at most* $h^{1,1}(X)$. $\square$

We denote the rank of $\operatorname{Num} X$ by $\rho(X) = \rho$, the *Picard number* of X. Over $\mathbb{R}$, the signature of the nondegenerate pairing on $\operatorname{Num} X \otimes \mathbb{R}$ is $2 - \rho$. The following is then an immediate consequence of the fact that the signature is well defined:

Corollary 16. *Let* H *be any divisor on* X *with* $H^2 > 0$, *not necessarily ample. Then the intersection pairing on*

$$H^\perp = \{D \in \operatorname{Num} X : D \cdot H = 0\}$$

is negative definite.

Ample and nef divisors

We begin with the statement of the Nakai-Moishezon criterion for ampleness ([61, p. 365]):

Theorem 17. *A divisor* H *is ample on* X *if and only if* $H^2 > 0$ *and* $H \cdot C > 0$ *for all irreducible curves* C *on* X. $\square$

There is a generalization of Theorem 17 to arbitrary, possibly singular compact (proper) schemes [98], [69].

Corollary 18. *If the divisor* H *is numerically equivalent to an ample divisor, then* H *is ample.* $\square$

It is easy to see that the corollary fails if we replace ample by very ample.

The next question is to describe in general terms the set of all ample divisors in $\operatorname{Num} X$. It is easy to see (Exercise 11) that this set is closed under positive integer linear combinations. It is convenient to work in $\operatorname{Num} X \otimes \mathbb{R}$. Note that $\operatorname{Num} X \otimes \mathbb{R} = \mathbb{R}^{\rho}$ has an intersection form q, the natural extension of intersection pairing on $\operatorname{Num} X$. There is a real basis $e_1, \ldots, e_{\rho}$ of $\operatorname{Num} X \otimes \mathbb{R}$ such that

$$q\left(\sum_i x_i e_i\right) = x_1^2 - \sum_{i>1} x_i^2.$$

Let $P \subset \operatorname{Num} X \otimes \mathbb{R}$ be the subset of vectors $\xi = \sum_i x_i e_i$ such that $q(\xi) > 0$. Then P consists of two pieces P_+, P_- where $P_+ = \{\sum_i x_i e_i : x_1 > 0\}$ and $P_- = \{\sum_i x_i e_i : x_1 < 0\}$.

Lemma 19. *With notation as above:*

(i) $P_{\pm}$ *are convex.*

(ii) *Given $\xi \in P_+$, suppose that η lies in the closure $\bar{P}$ of P. Then $\xi \cdot \eta = 0$ if and only if $\eta = 0$. Otherwise, η lies in the closure of P_+ if and only if $\xi \cdot \eta > 0$.*

Proof. Part (i) is an easy consequence of the Cauchy-Schwarz inequality and is left to the reader. For (ii), first suppose that $\xi \cdot \eta = 0$. Since $\eta \in \bar{P}$, $\eta^2 \geq 0$. Thus, ξ and η span a positive semidefinite subspace of $\operatorname{Num} X \otimes \mathbb{R}$, which can have rank at most 1, and so $\eta = 0$. Now consider the function $P \to \{\pm 1\}$ defined by $\eta \mapsto \operatorname{sign}(\xi \cdot \eta)$, which is well defined by the first part of the proof since $\xi \cdot \eta \neq 0$. This function is constant on the connected components of P and thus on P_+, P_-. Since $\xi \cdot \xi > 0$ and $\xi \cdot (-\xi) < 0$, we see that the sign must always be positive on P_+ and always negative on P_-. $\square$

Fix an ample divisor H. We can choose our basis above so that $H \in P_+$. Since $H \cdot H' > 0$ for every ample divisor H', $H' \in P_+$ for every ample H'. We can take the convex hull of the ample divisors in $\operatorname{Num} X \otimes \mathbb{R}$. It forms a cone in $P_+ \subset \operatorname{Num} X \otimes \mathbb{R}$, the *ample cone* $\mathcal{A}(X)$. If $H \in \mathcal{A}(X) \cap \operatorname{Num} X$, then H is ample by the Nakai-Moishezon criterion, since $H^2 > 0$ and $H \cdot C > 0$ for every irreducible curve C.

Lemma 20. $\mathcal{A}(X)$ *is open.*

Proof. Let H be an ample line bundle. For an integral basis $d_1, \ldots, d_{\rho}$ of $\operatorname{Num} X$, let D_i be a divisor corresponding to d_i. By Serre's criterion for ampleness, there is a positive integer N_i such that $N_i H \pm D_i$ is ample for every i. Thus, $\mathcal{A}(X)$ contains the convex hull of the points $H \pm (1/N_i) d_i$ for every i, and hence contains an open set around H. Now every element of $\mathcal{A}(X)$ is a finite sum $\sum_k \lambda_k H_k$, where $\lambda_k \in \mathbb{R}^+$ and H_k is ample, and for every k there is an open neighborhood U_k of $\lambda_k H_k$ contained in $\mathcal{A}(X)$. It follows that $\mathcal{A}(X)$ contains $\sum_k U_k$, which contains an open set around $\sum_k \lambda_k H_k$. $\square$

Lemma 21. *Let*

$$\mathcal{A}'(X) = \{x \in P_+ : x \cdot C > 0 \text{ for all irreducible curves } C\};$$
$$\overline{\mathcal{A}(X)} = \{x \in \overline{P_+} : x \cdot C \geq 0 \text{ for all irreducible curves } C\},$$

where $\overline{P_+}$ is the closure of P_+. Then $\mathcal{A}(X) \subseteq \mathcal{A}'(X) \subseteq \overline{\mathcal{A}(X)}$ and $\overline{\mathcal{A}(X)}$ is the closure of $\mathcal{A}(X)$. Moreover, $\mathcal{A}(X)$ is the interior of $\mathcal{A}'(X)$ and of $\overline{\mathcal{A}(X)}$.

Proof. We have already seen that $\mathcal{A}(X)$ is an open convex cone. Clearly, $\mathcal{A}(X) \subseteq \mathcal{A}'(X) \subseteq \overline{\mathcal{A}(X)}$ and all three sets are convex. Moreover, $\overline{\mathcal{A}(X)}$ is closed and thus contains the closure of $\mathcal{A}(X)$. Conversely, let $\lambda \in \overline{\mathcal{A}(X)}$. Let $\mu_1, \ldots, \mu_\rho$ be a basis for $\mathrm{Num}\, X \otimes \mathbb{R}$ consisting of ample divisors, which exists because $\mathcal{A}(X)$ is open. Then, for every $n > 0$, the set $\{\lambda + \sum_i t_i \mu_i : 0 < t_i < 1/n\}$ is open in $\mathrm{Num}\, X \otimes \mathbb{R}$, and thus contains a rational point h_n. Clearly, $h_n^2 > 0$ and $h_n \cdot C > 0$ for every irreducible curve C, so that some integral multiple of h_n is ample by the Nakai-Moishezon criterion. Thus, $h_n \in \mathcal{A}(X)$, and clearly $\lim_{n\to\infty} h_n = \lambda$. It follows that λ is in the closure of $\mathcal{A}(X)$. The closure of $\mathcal{A}(X)$ therefore contains $\overline{\mathcal{A}(X)}$ and so is equal to $\overline{\mathcal{A}(X)}$. Finally, it is a general fact that an open convex subset of $\mathbb{R}^n$ is the interior of its closure (this follows easily from, for example, [131, p. 81 ex. 1]), and so $\mathcal{A}(X)$ is the interior of $\overline{\mathcal{A}(X)}$ and hence of $\mathcal{A}'(X)$. $\square$

Note that, in the definition of $\mathcal{A}'(X)$ or $\overline{\mathcal{A}(X)}$, it is enough to consider only those irreducible curves C with $C^2 < 0$. Indeed, if $C^2 \geq 0$, then as $C \cdot H > 0$ for every ample divisor H, $C \in P_+$ and thus $\lambda \cdot C > 0$ for every $\lambda \in P_+$. Despite the fact that $\mathcal{A}(X)$ is described by a countable number of inequalities defined by integral elements of $\mathrm{Num}\, X \otimes \mathbb{R}$, its boundary can be very complicated, in the sense that the boundary can be far from being a finite polyhedron, even locally. For example, the boundary can be "round." We also note that there are surfaces X where $\mathcal{A}'(X)$ is neither open nor closed.

Motivated by Lemma 21, we make the following definition:

Definition 22. A divisor D is *nef* if $D \cdot C \geq 0$ for all irreducible curves C. A divisor D is *big* if $D^2 > 0$.

In earlier terminology, a nef divisor is also called *numerically effective* or *pseudo-ample*. According to some authors "nef" stands for "numerically eventually free," and we will discuss the reason for this shortly.

Lemma 21 then implies that the divisors in $\overline{\mathcal{A}(X)}$ are exactly the nef divisors with $D^2 \geq 0$. In fact this last condition is redundant:

Lemma 23. *Suppose that D is a nef divisor. Then $D^2 \geq 0$.*

Proof. Fix an ample divisor H. If $D^2 < 0$, there exists a $t_0 > 0$ such that $(D + t_0 H)^2 = 0$ and $(D + tH)^2 > 0$ for $t > t_0$. By the Nakai-Moishezon criterion, $D + tH$ is then ample for $t > t_0, t \in \mathbb{Q}$. For such t, some multiple of $D + tH$ is then effective, so that $D \cdot (D + tH) = D^2 + t(D \cdot H) \geq 0$. Likewise, $D \cdot H \geq 0$. By continuity, $D^2 + t_0(D \cdot H) \geq 0$, and so

$$0 = (D + t_0 H)^2 = D^2 + 2t_0(D \cdot H) + t_0^2 H^2$$
$$= D^2 + t_0(D \cdot H) + t_0(D \cdot H) + t_0^2 H^2$$
$$\geq t_0(D \cdot H) + t_0^2 H^2 \geq t_0^2 H^2 > 0,$$

a contradiction. Thus, $D^2 \geq 0$. □

We note, however, that an example due to Mumford [60] shows that there exists a surface X and a divisor D such that $D \cdot C > 0$ for every irreducible curve C on X but $D^2 = 0$.

A standard example of a nef and big divisor on X is obtained by taking a divisor D such that the morphism defined by the complete linear system $|D|$ has no base points and is generically finite onto its image. More generally, by analogy with ampleness we make the following:

Definition 24. A divisor D is *eventually base point free* if for all $n \gg 0$, the linear systen $|nD|$ has no base points.

Suppose that D is eventually base point free. Let φ_{nD} be the morphism defined by $|nD|$ for n such that $|nD|$ is base point free, and let X_0 be the image of X under φ_{nD}. We assume that φ_{nD} is generically finite, or equivalently that D is big. From the embedding $X_0 \subset \mathbb{P}^N$, there is an ample line bundle L_0 on X_0, the restriction of $\mathcal{O}_{\mathbb{P}^N}(1)$, which pulls back to $\mathcal{O}_X(nD)$. The morphism φ_{nD} has a Stein factorization $X \to \bar{X} \to X_0$, where $\bar{X}$ is normal and $\bar{X} \to X_0$ is finite. Let $\bar{L}$ be the pullback of L_0 to $\bar{X}$. Since $\bar{X} \to X_0$ is finite, $\bar{L}$ is ample, and it pulls back to $\mathcal{O}_X(nD)$. If $\pi \colon X \to \bar{X}$ is the natural map, then since $\bar{X}$ is normal (or by the construction of the Stein factorization) $\pi_* \mathcal{O}_X = \mathcal{O}_{\bar{X}}$. Thus, $\pi_* \mathcal{O}_X(nD) = \pi_* \pi^* \bar{L} = \bar{L} \otimes \pi_* \mathcal{O}_X = \bar{L}$. It follows that $H^0(X; \mathcal{O}_X(nD)) = H^0(\bar{X}; \bar{L})$ and similarly for $H^0(X; \mathcal{O}_X(mnD))$. Hence, for all $k \gg 0$ and divisible by n, the image of X under the morphism φ_{kD} defined by $|kD|$ is the normal projective surface $\bar{X}$. Moreover, if C is an irreducible curve on X, then $\varphi_{kD}(C)$ is a single point if and only if $C \cdot D = 0$. It follows that $\bar{X}$ is the uniquely specified normal surface obtained by contracting all such curves C. By the Hodge index theorem, the curves C for which $C \cdot D = 0$ span a negative definite sublattice of $\operatorname{Num} X$. Moreover, they must be independent in $\operatorname{Num} X \otimes \mathbb{Q}$:

Lemma 25. *Let $C_1, \ldots, C_r$ be distinct curves spanning a negative definite sublattice of* $\operatorname{Num} X$. *Then the classes of the* C_i *are independent in* $\operatorname{Num} X \otimes \mathbb{Q}$.

Proof. If not, there exists a relation $\sum_i n_i C_i = 0$ with the $n_i \in \mathbb{Z}$, not all 0, and after deleting some of the C_i we can assume that none of the n_i is 0. Since the C_i are effective, not all of the n_i are positive (otherwise intersect with an ample divisor). Collecting the negative terms, and possibly relabeling, there is a relation of the form

$$\sum_{i=1}^{s} n_i C_i = \sum_{j=s+1}^{r} m_j C_j,$$

where $m_j = -n_j > 0$. Now $(\sum_{i=1}^{s} n_i C_i)^2 \le 0$ by assumption. On the other hand,

$$\left(\sum_{i=1}^{s} n_i C_i \right)^2 = \left(\sum_{i=1}^{s} n_i C_i \right) \cdot \left(\sum_{j=s+1}^{r} m_j C_j \right) \ge 0,$$

which is only possible if $(\sum_{i=1}^{s} n_i C_i)^2 = 0$. Since the lattice spanned by the C_i is negative definite, it follows that $\sum_{i=1}^{s} n_i C_i$ is numerically equivalent to zero, which is impossible since the n_i are all positive. $\square$

Despite the apparent similarities between being eventually base point free and being ample, they are very different properties. The major difference is that there is no numerical criterion for when a nef and big divisor is eventually base point free (Exercise 7 in Chapter 3). On the other hand, Mumford has proved the following generalization of the Kodaira vanishing theorem:

Theorem 26. *Let D be a nef and big divisor on the smooth surface X. Then $H^i(X; \mathcal{O}_X(-D)) = 0$ for $i = 0, 1$. Dually $H^i(X; \mathcal{O}_X(K_X + D)) = 0$ for $i = 1, 2$.*

We will give a proof of Theorem 26 based on Bogomolov's inequality in Chapter 9.

Exercises

1. Show that Noether's formula is equivalent to the Hirzebruch signature theorem:

$$b_2^+(X) - b_2^-(X) = \tfrac{1}{3}(c_1^2(X) - 2c_2(X)).$$

2. Let X be a smooth surface in $\mathbb{P}^3$ of degree d. Show using standard facts about the cohomology of $\mathbb{P}^3$ that $q(X) = 0$. (A stronger statement follows from the Lefschetz theorem on hyperplane sections: X is simply connected.) Using the fact that $K_X = \mathcal{O}_X(d - 4)$ by adjunction and

the fact that $q(X) = 0$, determine $c_1^2(X)$ and $p_g(X)$. Apply Noether's formula to find $b_2(X)$ and thus $h^{1,1}(X)$. (Of course, you could also, with somewhat more effort, find $h^{1,1}(X)$ directly and then find $b_2(X)$.) When is the intersection form on X of Type I?

3. Let C be a smooth rational curve on X. For $n > 0$, find $\dim H^0(nC; \mathcal{O}_{nC})$ and $p_a(nC)$ in terms of C^2. What can you say if instead $g(C) > 0$?

4. Let C be a reduced irreducible curve and x a point of C. Show that $\delta_x = 1$ if and only if x is an ordinary double point or a cusp. (One direction has already be done.)

5. Calculate δ_x for a singularity of the form $y^2 = x^{2k+1}$ and also for a singularity of the form $y^2 = x^{2k}$. What is the local picture of these singularities?

6. Let $\lambda_1, \ldots, \lambda_n$ be distinct complex numbers and consider the local singularity C given by

$$\prod_{i=1}^{n}(y - \lambda_i x) = 0.$$

Thus, C is a union of n distinct lines meeting at the origin. Compute δ_0 directly from the definition in this case. (Note: C is **not** obtained from its normalization by identifying the common point on the n branches if $n > 2$.) We will see an easier way to compute δ_0 for this example in Chapter 3.

7. Let C be an irreducible curve whose singular locus is a single ordinary double point x, and let $p, q \in \tilde{C}$ be the preimage under the normalization map ν of x. Show that $\nu^* \omega_C = K_{\tilde{C}} \otimes \mathcal{O}_{\tilde{C}}(p + q)$. Which local sections of this line bundle are the pullbacks of sections of ω_C? Analyze a cusp singularity similarly.

8. Use the Riemann-Roch theorem applied to $\chi(\mathcal{O}_X(-C))$ to show that, for every effective nonzero divisor C, $p_a(C) = 1 - \chi(C; \mathcal{O}_C)$. Also show that, for two such effective divisors C and D,

$$p_a(C + D) = p_a(C) + p_a(D) + C \cdot D - 1.$$

9. Verify that, in the notation of (1.9), the natural map $\mathcal{O}_{C+D} \to \mathcal{O}_C$ is surjective (this is obvious) and that its kernel is $\mathcal{O}_D(-C)$.

10. Let D_1 and D_2 be two divisors with $D_1^2 \geq 0$. Show that

$$(D_1)^2(D_2)^2 \leq (D_1 \cdot D_2)^2$$

and analyze the case where equality holds. (This is an easy consequence of the Hodge index theorem which applies to all nondegenerate symmetric bilinear forms whose intersection matrix has just one positive eigenvalue.)

11. Show directly that, if H_1 is ample and H_2 is eventually base point free, then $H_1 + H_2$ is ample. Using the Nakai-Moishezon criterion, show that if H_1 is ample and H_2 is nef, then $H_1 + H_2$ is ample.

12. Let $\pi\colon X \to Y$ be a surjective morphism of surfaces. Given a divisor D on X, there is a divisor $\pi_* D$ on Y defined as follows: For C irreducible, $\pi_* C = 0$ if $\pi(C)$ is a point, and otherwise $\pi_* C = d\pi(C)$, where d is the degree of the map from C to $\pi(C)$. For general D we extend π_* by linearity. Prove the projection formula $\pi^* D \cdot E = D \cdot \pi_* E$. Conclude that π^* induces a map $\operatorname{Num} Y \to \operatorname{Num} X$, also denoted π^*, and that π_* induces a map $\operatorname{Num} X \to \operatorname{Num} Y$ which we continue to denote by π_*.

13. Let H be a nef and big divisor on the surface X. Suppose that $H = A + B$, where A and B are effective. Show that $A \cdot B \geq 0$, with equality holding if and only if one of A or B is 0. We say that H is *numerically connected*. (Let $E = aH - A$ with $a = (A \cdot H)/H^2$. Since $0 \leq A \cdot H \leq (A + B) \cdot H = H^2$, $0 \leq a \leq 1$. By the Hodge index theorem, $E^2 \leq 0$. Now estimate $A \cdot B = (aH - E) \cdot ((1 - a)H + E)$, and analyze the case of equality.)

2

Coherent Sheaves

What is a coherent sheaf?

Let X be a scheme (or analytic space) with $x \in X$ and let $R = \mathcal{O}_{X,x}$ be the local ring at x. For our purposes X will usually be regular, and we could as well work in the analytic category, so that the reader can for the moment take $R = \mathbb{C}\{z_1, \ldots, z_n\}$ to be the ring of convergent power series at the origin if so desired. There are two paradigms for what a coherent sheaf $\mathcal{F}$ on X should look like:

(i) A locally free sheaf, locally modeled on the free module R^N;
(ii) An ideal sheaf, locally modeled on an ideal $I \subseteq R$.

The general torsion free coherent sheaf, roughly speaking, is a blend of these two models, and torsion, as we shall see, essentially corresponds to some torsion free sheaf supported on a proper subvariety of X. Another way to think of coherent sheaves is the following: begin with a vector bundle V on X. Its sheaf of regular (or holomorphic) sections is locally isomorphic to $\mathcal{O}_X^N$, in any open set where the bundle is trivialized. Moreover, there is a natural functor from the category of algebraic or holomorphic vector bundles over X to the category of locally free sheaves of $\mathcal{O}_X$-modules. Every locally free sheaf arises from a vector bundle, and two vector bundles are isomorphic if and only if the corresponding locally free sheaves are isomorphic. However, not every morphism of locally free $\mathcal{O}_X$-modules comes from a vector bundle morphism. In a local trivialization, a morphism $\mathcal{O}_X^N \to \mathcal{O}_X^M$ is just given by a matrix of functions, which need not have constant rank. However, vector bundle morphisms are required to have constant rank, so that the cokernel is again a vector bundle. From this point of view, coherent sheaves are the smallest category of $\mathcal{O}_X$-modules we can obtain by locally enlarging the category of locally free $\mathcal{O}_X$-modules so that every morphism has a cokernel, and then gluing these local models together. This then is the definition that a coherent sheaf of $\mathcal{O}_X$-modules $\mathcal{F}$ locally has a presentation

$$\mathcal{O}_X^N \to \mathcal{O}_X^M \to \mathcal{F} \to 0.$$

The local model of such a sheaf is then a finite R-module M. The Noetherian properties of R are reflected in the statement (not trivial in the analytic case) that the kernel of a map of $\mathcal{O}_X$-modules $\mathcal{O}_X^N \to \mathcal{O}_X^M$ is also coherent. The definition of a coherent sheaf allows us to bring in all the complexities and beauty of commutative algebra, and it is to be hoped that the chapters on vector bundles will amply illustrate both of these features. Let us give some very simple illustrations here of this picture.

Example 1: The meaning of torsion. We suppose for simplicity that X is smooth. A torsion section s of $\mathcal{F}$ corresponds to an element m of the R-module M which is annihilated by some $f \in R$. Thus, s is (at least near $x \in X$) supported on the subvariety $\{f = 0\}$. Conversely, if s is a section with support on a proper subvariety $V \subset X$, let $f \neq 0 \in R$ be an element corresponding to the ideal of V, and let $V' = \{f = 0\} \supseteq V$. The statement that s has support on V means exactly that the restriction of s to $X - V$ is zero, and thus that s restricts to 0 on $X - V'$. Algebraically, this means that m is in the kernel of the natural map $M \to M_f = M \otimes R_f$, where R_f is the localization of R at f, and then it follows from the definition that there exists an $n \geq 1$ such that $f^n m = 0$. Thus, m is a torsion section, annihilated by some power of f.

In particular (using finite generation properties associated to coherence) a torsion coherent sheaf $\mathcal{F}$ is precisely one which is supported on a proper subvariety.

Example 2: Generic rank. Let X again be a smooth (and connected) scheme, which for simplicity we shall also assume to be affine: $X = \operatorname{Spec} R$. Let $\mathcal{F}$ be a coherent sheaf on X, corresponding to the R-module N. Then as R is an integral domain it has a field of fractions K, and the *rank* of N is the dimension of the K-vector space $N \otimes_R K$. Let r be the rank, so that $N \otimes_R K = K^r$. Choose a basis $v_1, \ldots, v_r$ of $N \otimes_R K$. After clearing denominators we may assume that v_i is the image of $n_i \in N$. Thus, there is a well-defined map from the free module R^r to N which makes the following diagram commute:

$$
\begin{array}{ccc}
N & \longrightarrow & N \otimes_R K = K^r \\
\uparrow & & \| \\
R^r & \longrightarrow & R^r \otimes_R K = K^r,
\end{array}
$$

and so $R^r \to N$ is injective. If T is the cokernel, using the fact that

$$R^r \otimes_R K \to N \otimes_R K \to T \otimes_R K \to 0$$

is still exact and that the first map is an isomorphism we see that $T \otimes_R K = 0$, and thus T is a torsion sheaf. It follows by Example 1 that there is an open subset $U = \operatorname{Spec} R_f$ of X such that $\mathcal{F}|U$ is free. Its rank is then the *generic rank* of $\mathcal{F}$. Pursuing this idea further, it is not hard to show that we may stratify X by locally closed subvarieties S_α such that the restriction of $\mathcal{F}$ to each S_α is free.

Coherent but non-locally free sheaves play two different roles in the study of the moduli of vector bundles:

1. As a way to dismantle a vector bundle V into canonically defined pieces;
2. As the necessary ingredient for compactifying moduli spaces of vector bundles.

In general, one problem with studying holomorphic vector bundles from the point of view of algebraic geometry is that a vector bundle is not a very geometric object in the sense of algebraic geometry, for example in terms of special configurations of points or divisors on a variety. On the other hand, the existence of many interesting vector bundles implies in some way that there is a lot of interesting hidden projective geometry on, say, an algebraic surface: for example, points in special position or systems of curves in special position. One goal of trying to break up a vector bundle into more canonical pieces will be to make explicit the link between the classification of bundles and the algebraic geometry of the variety.

In this chapter, we will discuss some of the basics of the theory of vector bundles and methods for constructing them. After reviewing Chern classes, we discuss the construction techniques of extensions, elementary modifications, and double covers. The chapter ends with some commutative algebra. In Chapter 4, we will introduce the notion of stability and derive some of its basic properties. Chapters 6 and 8 describe vector bundles over ruled and elliptic surfaces. Finally, in Chapter 9, we return to the general theory and prove Bogomolov's inequality for a stable rank 2 vector bundle.

A rapid review of Chern classes for projective varieties

Let V be a vector bundle on a quasiprojective variety X. We can define the Chern classes of V as follows:

$$(2.1) \qquad c_1(V) = c_1(\det V) \in H^2(X; \mathbb{Z})$$

if X is defined over $\mathbb{C}$. In general we could use an algebraic substitute for $H^2(X; \mathbb{Z})$ such as $\operatorname{Pic} X$ in the algebraic case (in which case we take $c_1(V)$ to be actually equal to $\det V$), or $\operatorname{Num} X$. To define $c_i(V)$ in general, we first define it for a direct sum of line bundles $V = L_1 \oplus \cdots \oplus L_n$. In this case, formally

$$(2.2) \qquad 1 + c_1(V) + \cdots + c_n(V) = (1 + c_1(L_1)) \cdots (1 + c_1(L_1));$$

the actual formula is obtained by equating the terms that lie in $H^{2i}(X; \mathbb{Z})$. A similar formula holds if V, instead of being a direct sum of line bundles, has a filtration by subbundles V_i such that $V_i/V_{i-1} = L_i$ is a line bundle.

To define $c_i(V)$ for a general V, one shows that there exists a variety Y and a morphism $\pi \colon Y \to X$ for which $\pi^* \colon H^i(X; \mathbb{Z}) \to H^i(Y; \mathbb{Z})$ is injective, for all i, and such that $\pi^* V$ has a filtration as described above. We have then defined $c_i(\pi^* V)$, and one shows that these classes are in the image of π^* for all i. Then we can set $c_i(V)$ to be the unique class such that $\pi^* c_i(V) = c_i(\pi^* V)$. For our purposes, the Chern classes $c_i(V)$ will lie in $H^{2i}(X; \mathbb{Z})$. However, for a smooth projective variety X, there exist algebraic analogues of $H^{2i}(X; \mathbb{Z})$, namely the higher Chow groups $A^i(X)$ consisting of algebraic cycles of codimension i modulo rational equivalence. For more about the Chow groups, see Fulton's book [45] or Appendix A in Hartshorne's book [61]. Essentially by definition, $A^1(X) = \operatorname{Pic} X$. But for $i > 1$, the groups $A^i(X)$ are poorly understood and can be quite large. In any case, one can define Chern classes $c_i(V) \in A^i(X)$, which are a much finer invariant of V. For X defined over $\mathbb{C}$, there is is a natural homomorphism $A^i(X) \to H^{2i}(X; \mathbb{Z})$ which sends a codimension i cycle to its fundamental class, and the image of $c_i(V)$ under this homomorphism is the usual Chern class. Unless otherwise specified, we will always take $c_i(V) \in H^{2i}(X; \mathbb{Z})$.

The Chern classes so constructed are functorial: $c_i(f^* V) = f^* c_i(V)$, and satisfy the Whitney product formula: for an exact sequence

$$0 \to V' \to V \to V'' \to 0$$

of vector bundles, we have

(2.3) $$c(V) = c(V')c(V''),$$

where $c(V)$ is the *total Chern class*

(2.4) $$c(V) = 1 + c_1(V) + \cdots + c_n(V).$$

The idea behind the construction of the Chern classes is called the *splitting principle* and has the following useful extension: every "universal" formula for Chern classes which holds for direct sums of line bundles holds in general. For example, for the dual bundle $V^\vee$ of a vector bundle V we have the formula

(2.5) $$c(V^\vee) = c(V)^\vee,$$

where given a class $\alpha = \sum_i \alpha_i \in \bigoplus H^{2i}(X; \mathbb{Z})$, $\alpha^\vee$ is by definition the class $\sum_i (-1)^i \alpha_i$. Similar formulas can be given for $\operatorname{Sym}^k V$, $\bigwedge^k V$, $\operatorname{Hom}(V, W)$, $V \otimes W$, ... For example, if L is a line bundle and V is a bundle of rank r, then

(2.6)
$$c_1(V \otimes L) = c_1(V) + rc_1(L),$$
$$c_2(V \otimes L) = c_2(V) + (r-1)c_1(V) \cdot c_1(L) + \binom{r}{2} c_1(L)^2.$$

On a smooth quasiprojective variety, we can define the Chern classes of any coherent sheaf. This is because, by a theorem of Serre, every coherent sheaf $\mathcal{F}$ on a smooth quasiprojective variety admits a finite resolution by locally free sheaves: there exists an exact sequence of sheaves

$$0 \to \mathcal{E}^n \to \cdots \to \mathcal{E}^0 \to \mathcal{F} \to 0,$$

where the $\mathcal{E}^i$ are locally free. We can then define the total Chern class of $\mathcal{F}$ by the formula

(2.7)
$$c(\mathcal{F}) = \prod_i c(\mathcal{E}^i)^{(-1)^i}.$$

It is not difficult to show that this definition is independent of the choice of the resolution and satisfies the Whitney product formula. However, other properties (for example, the correct definition of pullback or tensor product) require using the derived functors of tensor product, i.e., the Tor sheaves, and we refer to Fulton [45] or Borel-Serre [12] for more details. One important example which we shall use often is the following: let Z be a reduced irreducible subvariety of X of codimension r, and let $j\colon Z \to X$ be the inclusion map. Then, as a consequence of the Grothendieck-Riemann-Roch theorem, $c_i(j_*\mathcal{O}_Z) = 0$ for $i < r$ and

(2.8)
$$c_r(j_*\mathcal{O}_Z) = (-1)^{r-1}(r-1)!\,[Z],$$

where $[Z] \in H^{2r}(X;\mathbb{Z})$ is the cycle defined by Z. A similar result holds for a vector bundle V of rank n on Z: $c_i(j_*V) = 0$ for $i < r$ and

$$c_r(j_*V) = (-1)^{r-1}(r-1)!\,n[Z].$$

From this it is easy to deduce that $c_i(j_*\mathcal{O}_Z) = 0$ for $i < r$ and $c_r(j_*\mathcal{O}_Z) = (-1)^{r-1}(r-1)!\,[Z]$ for an arbitrary closed subscheme Z of X of pure codimension r, where $[Z]$ is again the associated cycle. Here if the irreducible components of Z are $Z_1,\dots,Z_a$ and the length of $\mathcal{O}_Z$ along Z_i is m_i, then $[Z] = \sum_i m_i[Z_i]$. For example, suppose that X is a smooth projective surface and that Z is a 0-dimensional subscheme of X supported at $p_1,\dots,p_k$. Let $j\colon Z \to X$ be the inclusion. Then $\mathcal{O}_{X,p_i}/I_{Z,p_i}$ is a finite-dimensional $\mathbb{C}$-vector space. Define $\ell(Z_{p_i})$ to be its dimension and let $\ell(Z) = \sum_i \ell(Z_{p_i})$. Then

$$c_2(j_*\mathcal{O}_Z) = -\ell(Z) \in H^4(X;\mathbb{Z}) \cong \mathbb{Z}.$$

Applying the Whitney product formula to the exact sequence

$$0 \to I_Z \to \mathcal{O}_X \to j_*\mathcal{O}_Z \to 0,$$

we see that $c_2(I_Z) = \ell(Z)$. More generally, if X is a smooth surface and V is a rank 2 vector bundle on X for which there is an exact sequence

$$0 \to L \to V \to L' \otimes I_Z \to 0,$$

where Z has dimension 0, then we see that

(2.9)
$$c_1(V) = c_1(L) + c_1(L'),$$
$$c_2(V) = c_1(L) \cdot c_1(L') + \ell(Z).$$

For an effective divisor D, the formula for $c_1(j_*\mathcal{O}_D)$ is easy to verify: using the exact sequence

$$0 \to \mathcal{O}_X(-D) \to \mathcal{O}_X \to j_*\mathcal{O}_D \to 0,$$

and the Whitney product formula, we see that

(2.10)
$$c(j_*\mathcal{O}_D) = c(\mathcal{O}_X(-D))^{-1} = \sum_{i=1}^{\infty}[D]^i.$$

We will need an extension of this formula to the case of a line bundle L on Z:

Lemma 1. *Let L be a line bundle on the effective divisor $D \subset X$, and let $j: D \to X$ be the inclusion. Then*

$$c_1(j_*L) = [D],$$
$$c_2(j_*L) = [D]^2 - j_*c_1(L).$$

Proof. We may suppose that $L = \mathcal{O}_D(V_2-V_1)$ is the line bundle associated to a difference of two effective Cartier divisors V_1, V_2 on D. Begin with the exact sequence

$$0 \to j_*\mathcal{O}_D(-V_1) \to j_*\mathcal{O}_D \to k_{1*}\mathcal{O}_{V_1} \to 0,$$

where k_{i*} is the inclusion of V_i in X. For simplicity of notation we shall often drop the j_* or k_{1*} and understand that all sheaves are sheaves on X, and Chern classes are to be taken in this sense. The Whitney product formula and the calculation (2.10) of $c(j_*\mathcal{O}_D)$ gives

$$(1 + c_1(\mathcal{O}_D(-V_1)) + c_2(\mathcal{O}_D(-V_1)) + \cdots) =$$
$$= (1 + [D] + [D]^2 + \cdots)(1 - [V_1] + \cdots)^{-1}.$$

Equating terms gives

$$c_1(\mathcal{O}_D(-V_1)) = [D],$$
$$c_2(\mathcal{O}_D(-V_1)) = [D]^2 + [V_1].$$

A similar calculation with the exact sequence

$$0 \to j_*\mathcal{O}_D(-V_1) \to j_*\mathcal{O}_D(V_2 - V_1) \to k_{2*}\mathcal{O}_{V_2}(V_2 - V_1) \to 0$$

finishes the argument. $\square$

We shall also need the Riemann-Roch theorem for vector bundles on curves and surfaces (see Exercise 17):

Theorem 2.

(i) *Let V be a vector bundle of rank r on the smooth curve C of genus g, and let $\deg \det V = d$. Then*

$$\chi(C; V) = d + r(1 - g).$$

(ii) *Let V be a vector bundle of rank r on the smooth surface X. Then*

$$\chi(X; V) = \frac{c_1(V) \cdot (c_1(V) - K_X)}{2} - c_2(V) + r\chi(\mathcal{O}_X). \qquad \square$$

Rank 2 bundles and sub-line bundles

Our goal for the rest of this chapter will be to describe ways to construct rank 2 vector bundles V over a smooth projective variety X. The simplest method is to take a direct sum of two line bundles: $V = L_1 \oplus L_2$. Of course, we don't expect to obtain especially interesting bundles in this way. One way to modify this idea is to consider extensions of line bundles, in other words rank 2 vector bundles V such that there is an exact sequence

$$0 \to L_1 \to V \to L_2 \to 0.$$

If X is a curve, then all rank 2 bundles can in fact be obtained in this way. However, for a surface X, most interesting bundles do not have such a description.

To classify such bundles V, we recall that all such extensions are classified by the group $\mathrm{Ext}^1(L_2, L_1) = H^1(X; (L_2)^{-1} \otimes L_1)$. (See Exercise 1 for a more concrete picture of this.) Here an isomorphism between two extensions V and V' (in the strong sense) is an isomorphism of bundles $\alpha \colon V \to V'$ such that the following diagram commutes:

$$
\begin{array}{ccccccccc}
0 & \longrightarrow & L_1 & \longrightarrow & V & \longrightarrow & L_2 & \longrightarrow & 0 \\
  &   & \| &   & \downarrow{\scriptstyle \alpha} & & \| & & \\
0 & \longrightarrow & L_1 & \longrightarrow & V' & \longrightarrow & L_2 & \longrightarrow & 0.
\end{array}
$$

A *weak isomorphism* of extensions is similarly defined, except that we do not require that the maps $L_i \to L_i$ be the identity. As we are primarily interested in V and not in the strong isomorphism class of the extension, we shall just care about weak isomorphism classes of extensions. Since L_i is a line bundle, its only automorphisms are $\mathbb{C}^*$, and so $\mathbb{C}^* \times \mathbb{C}^*$ acts on the set of all isomorphism classes of extensions. Since the scalars are endomorphisms of $V = V'$, the diagonal subgroup of $\mathbb{C}^* \times \mathbb{C}^*$ acts trivially, and the weak equivalence classes of extensions are the quotient of $\mathrm{Ext}^1(L_2, L_1)$ by an action of $\mathbb{C}^*$. It is easy to check that this action is just scalar multiplication, so that the weak equivalence classes are either the trivial extension $L_1 \oplus L_2$ or are parametrized by a projective space $\mathbb{P}(\mathrm{Ext}^1(L_2, L_1))$.

For example, if $X = \mathbb{P}^2$, L_i is necessarily $\mathcal{O}_{\mathbb{P}^2}(a_i)$ and

$$\text{Ext}^1(L_2, L_1) = H^1(\mathbb{P}^2; \mathcal{O}_{\mathbb{P}^2}(a_1 - a_2)) = 0.$$

Thus, we cannot obtain any interesting bundles on the very simple surface $\mathbb{P}^2$ in this way.

To make an interesting construction along these lines, consider, instead of a rank 1 subbundle L_1 of V, the following object:

Definition 3. A *sub-line bundle* of a rank 2 vector bundle on X is a rank 1 subsheaf which is a line bundle.

Note that every rank 2 vector bundle has a sub-line bundle: by Serre's theorem, $V \otimes H$ has a global section (indeed is generated by global sections) if H is a sufficiently ample line bundle. Thus, there is an inclusion $H^{-1} \to V$. However, unless V has a canonically defined sub-line bundle, the fact that it has some such is not very helpful.

Let us give a description of the local picture of a sub-line bundle. Let $R = \mathcal{O}_{X,x}$ be the local ring of X at x; it is a UFD since X is regular. Let $\varphi \colon L \to V$ be a sub-line bundle. After choosing local trivializations φ corresponds to an inclusion $R \to R \oplus R$. Thus, φ is (locally) determined by $\varphi(1) = (f, g) \in R \oplus R$. Now either f and g are relatively prime elements of R or they are not relatively prime. If they are relatively prime, consider the map $\psi \colon R \oplus R \to R$ given by $\psi(a, b) = ag - bf$. Clearly, $\text{Im } \psi = (f, g)R = I$ is the ideal generated by f and g. The following is an easy special case of the exactness of the Koszul complex:

Claim 4. *If f and g are relatively prime, the sequence*

(2.11) $$0 \to R \xrightarrow{\varphi} R \oplus R \xrightarrow{\psi} I \to 0$$

is exact.

Proof. If $(a, b) \in \text{Ker } \psi$, then $ag = bf$. Since f and g are relatively prime, $f|a$ and $g|b$. Thus, $a = fh$ and $b = gh'$. On the other hand, $fgh = fgh'$, so that $h = h'$. Thus, $(a, b) = \varphi(h) \in \text{Im } \varphi$. $\square$

It is also possible that f and g are not relatively prime. Let $t = \gcd(f, g)$, where t is not a unit in R, and write $f = tf'$, $g = tg'$. Note that $(R \oplus R)/\text{Im } \varphi$ has torsion, since $(f', g') \notin \text{Im } \varphi$ but $t(f', g') \in \text{Im } \varphi$. There is an induced map $\varphi' \colon R \to R \oplus R$ defined by $\varphi'(1) = (f', g')$, and φ is the composition of this map with multiplication by t. Alternatively, φ extends to a map from $(1/t)R$ to $R \oplus R$. Globally, t defines an effective divisor D on X and we can summarize these calculations in a coordinate free way as follows:

Proposition 5.

(i) Let $\varphi\colon L \to V$ be a sub-line bundle. Then there exists a unique effective divisor D on X, possibly 0, such that the map φ factors through the inclusion $L \to L \otimes \mathcal{O}_X(D)$ and such that $V/(L \otimes \mathcal{O}_X(D))$ is torsion free.

(ii) In the above situation, if V/L is torsion free, i.e., if $D = 0$, then there exists a local complete intersection codimension two subscheme Z of X and an exact sequence

$$0 \to L \to V \to L' \otimes I_Z \to 0.$$ □

In particular, if $X = C$ is a curve, then $Z = \emptyset$ and V is an extension of line bundles in Case (ii). Thus, since every rank 2 vector bundle has a sub-line bundle, every rank 2 bundle over a curve can be written as an extension of line bundles. Let us give a simple application of this, by classifying all the rank 2 vector bundles over a curve of genus 0 or 1. (The case of genus 0 is a special case of a theorem of Grothendieck [56], and the case of genus 1 is due to Atiyah [4].)

Theorem 6.

(i) Let $C = \mathbb{P}^1$ and let V be a rank 2 vector bundle over $\mathbb{P}^1$. Then $V \cong \mathcal{O}_{\mathbb{P}^1}(a) \oplus \mathcal{O}_{\mathbb{P}^1}(b)$ for integers a, b, unique up to order.

(ii) Let C be a smooth curve of genus 1 and let V be a rank 2 bundle on C. Then exactly one of the following holds:
   (a) V is a direct sum of line bundles;
   (b) V is of the form $\mathcal{E} \otimes L$, where L is a line bundle on C and $\mathcal{E}$ is the (unique) extension of $\mathcal{O}_C$ by $\mathcal{O}_C$ which does not split into the direct sum $\mathcal{O}_C \oplus \mathcal{O}_C$;
   (c) V is of the form $\mathcal{F}_p \otimes L$, where L is a line bundle on C, $p \in C$, and $\mathcal{F}_p$ is the unique nonsplit extension of the form

$$0 \to \mathcal{O}_C \to \mathcal{F}_p \to \mathcal{O}_C(p) \to 0.$$

Proof. (i) Let V be a rank 2 vector bundle on $\mathbb{P}^1$. Since

$$\deg \det(V \otimes \mathcal{O}_{\mathbb{P}^1}(a)) = \deg \det(V) + 2a,$$

we may assume that $\deg \det(V) = 0$ or -1. Let us first assume that $\deg \det(V) = 0$, i.e., $\det V = \mathcal{O}_{\mathbb{P}^1}$. Then by the Riemann-Roch theorem applied to vector bundles on $\mathbb{P}^1$, we have $\chi(V) = 2$. Hence $h^0(V) \geq 2$. Choose a 2-dimensional subspace of $H^0(V)$ and consider the associated map $\mathcal{O}_{\mathbb{P}^1}^2 \to V$. If this map is injective, then its determinant $\det \mathcal{O}_{\mathbb{P}^1}^2 = \mathcal{O}_{\mathbb{P}^1} \to \det V = \mathcal{O}_{\mathbb{P}^1}$ is nonzero. Hence the determinant map is an isomorphism. It follows that $V \cong \mathcal{O}_{\mathbb{P}^1}^2$, which is certainly a direct sum of line bundles. Otherwise, the image of the map $\mathcal{O}_{\mathbb{P}^1}^2 \to V$ is a line bundle L

which has the property that the image of $H^0(L)$ contains the 2-dimensional subspace of $H^0(V)$ that we chose. Thus, $H^0(L) \geq 2$, and so $L \cong \mathcal{O}_{\mathbb{P}^1}(k)$ for $k \geq 1$. The map $L \to V$ factors through a map $L \otimes \mathcal{O}_{\mathbb{P}^1}(\ell) \to V$, where $\ell \geq 0$ and where the quotient is torsion free and thus a line bundle. So in this case we can write

$$0 \to \mathcal{O}_{\mathbb{P}^1}(a) \to V \to \mathcal{O}_{\mathbb{P}^1}(-a) \to 0,$$

with $a = k + \ell \geq 1$. But the extensions of $\mathcal{O}_{\mathbb{P}^1}(-a)$ by $\mathcal{O}_{\mathbb{P}^1}(a)$ are classified by $H^1(\mathcal{O}_{\mathbb{P}^1}(2a))$, which is zero since $a \geq 0$. Thus, the extension splits: $V \cong \mathcal{O}_{\mathbb{P}^1}(a) \oplus \mathcal{O}_{\mathbb{P}^1}(-a)$.

In case $\deg c_1(V) = -1$, then again by the Riemann-Roch theorem for vector bundles, $\chi(V) \geq 1$. So there is a nonzero map $\mathcal{O}_{\mathbb{P}^1} \to V$, which as above factors through a map $\mathcal{O}_{\mathbb{P}^1}(a) \to V$ with $a \geq 0$ and the quotient is a line bundle. Thus, there is an exact sequence

$$0 \to \mathcal{O}_{\mathbb{P}^1}(a) \to V \to \mathcal{O}_{\mathbb{P}^1}(-a - 1) \to 0.$$

As $H^1(\mathcal{O}_{\mathbb{P}^1}(2a - 1)) = 0$ for all $a \geq 0$, this extension must again split and $V \cong \mathcal{O}_{\mathbb{P}^1}(a) \oplus \mathcal{O}_{\mathbb{P}^1}(-a - 1)$. We leave the uniqueness as an exercise.

Next we prove (ii). Let C be a curve of genus 1 and V a rank 2 vector bundle on C. As before, after twisting V by a line bundle we may assume that $\deg \det V = 0$ or 1. First assume that $\deg \det V = 0$. Note the following:

Claim. *Suppose that $\deg \det V = 0$ and that there is a nonzero map $L_0 \to V$ with $\deg L_0 \geq 0$. Then either V splits or V is of the form $\mathcal{E} \otimes L$, where $\deg L = 0$. In particular, V satisfies (a) or (b) of the statement of the proposition.*

Proof of the Claim. As usual, we can factor the map $L_0 \to V$, so that there is an exact sequence

$$0 \to L_1 \to V \to L_2 \to 0,$$

with $\deg L_1 \geq 0$ and $\deg L_2 = -\deg L_1$. As we have seen, such extensions are classified by $H^1((L_2)^{-1} \otimes L_1)$, which is Serre dual to $H^0(L_2 \otimes (L_1)^{-1})$. If $\deg L_1 > 0$, then $\deg L_2 \otimes (L_1)^{-1} = -2 \deg L_1 < 0$. Thus, $H^0(L_2 \otimes (L_1)^{-1}) = 0$ and the extension splits, i.e., $V \cong L_1 \oplus L_2$. If $\deg L_1 = 0$, then either $L_2 \otimes (L_1)^{-1}$ is nontrivial, in which case $H^0(L_2 \otimes (L_1)^{-1}) = 0$ again, or $L_2 \otimes (L_1)^{-1} = \mathcal{O}_C$, i.e., $L_1 = L_2$. In this last case, $\dim H^0(\mathcal{O}_C) = 1$, and there is correspondingly a nonsplit extension $\mathcal{E}$ of $\mathcal{O}_C$ by $\mathcal{O}_C$, unique up to weak isomorphism. Clearly, in this case, $V = \mathcal{E} \otimes L_1$.   □

Returning to the proof of (ii), with $\deg \det V = 0$, suppose that $h^0(V) \neq 0$. Then there is a nonzero map $\mathcal{O}_C \to V$, so that the hypotheses of the claim are verified. Thus, we see that V satisfies (a) or (b) of the proposition.

So we may assume that $h^0(V) = 0$. Choose a point $p \in C$, and consider the vector bundle $V \otimes \mathcal{O}_C(p)$. By the Riemann-Roch theorem, $\chi(V \otimes \mathcal{O}_C(p)) = 2$ and thus $h^0(V \otimes \mathcal{O}_C(p)) \geq 2$. Choose a 2-dimensional subspace of $H^0(V \otimes \mathcal{O}_C(p))$ and consider the map $\mathcal{O}_C^2 \to V \otimes \mathcal{O}_C(p)$. If the image of this map is a line bundle, then as in the proof of (1) there is a line subbundle L_0 of $V \otimes \mathcal{O}_C(p)$ with $\deg L_0 \geq 2$. Thus, V has a subbundle $L_1 = L_0 \otimes \mathcal{O}_C(-p)$ with $\deg L_1 \geq 1$. So we can apply the claim (and in fact V splits). In the remaining case the induced map $\varphi \colon \mathcal{O}_C^2 \to V \otimes \mathcal{O}_C(p)$ is an inclusion. By comparing determinants, since we have $\det \mathcal{O}_C^2 = \mathcal{O}_C$ and $\deg \det[V \otimes \mathcal{O}_C(p)] = 2$, the determinant $\det \varphi$ must vanish at some point $x \in C$. Thus, there is a nonzero v in the fiber $\mathbb{C} \oplus \mathbb{C}$ of $\mathcal{O}_C^2$ at x with $\varphi_x(v) = 0$. But since the induced map from $H^0(\mathcal{O}_C^2)$ to the fiber of $\mathcal{O}_C^2$ over x is an isomorphism, there exists a section $s \in H^0(\mathcal{O}_C^2)$ whose restriction to the fiber over x is v. Hence the induced map $\varphi|\mathcal{O}_C \cdot s$ vanishes at x. Thus, the induced map $\mathcal{O}_C \to V \otimes \mathcal{O}_C(p)$ vanishes at x to order at least 1, and perhaps elsewhere. So there is a subbundle of $V \otimes \mathcal{O}_C(p)$ of the form $\mathcal{O}_C(\mathbf{d})$, where $\mathbf{d}$ is an effective divisor on C containing x in its support. Thus, V contains the subbundle $\mathcal{O}_C(\mathbf{d}) \otimes \mathcal{O}_C(-p)$, which has degree ≥ 0. Again by the claim, V satisfies (a) or (b).

Finally, we must consider the case where $\deg \det V = 1$. By the Riemann-Roch theorem, $\chi(V) = 1$. Thus, $h^0(V) \geq 1$ and there is a nonzero map $\mathcal{O}_C \to V$. If this map vanishes at some point, then we have an exact sequence

$$0 \to L_1 \to V \to L_2 \to 0,$$

with $\deg L_1 = d \geq 1$ and $\deg L_2 = 1 - d$. So $h^1((L_2)^{-1} \otimes L_1) = h^0((L_1)^{-1} \otimes L_2) = 0$ since

$$\deg(L_1)^{-1} \otimes L_2) = 1 - 2d < 0.$$

Thus, $V = L_1 \oplus L_2$. The remaining case is where $\mathcal{O}_C$ is a subbundle of V. In this case we have an exact sequence

$$0 \to \mathcal{O}_C \to V \to \mathcal{O}_C(q) \to 0,$$

where q is a point of C. This extension either splits, in which case V satisfies (a), or it does not split, in which case V satisfies (c). $\square$

Remark. Suppose that V satisfies (ii)(c) above. Then it is a straightforward exercise (Exercise 2) to show that $V \cong V \otimes F$ for every line bundle F on C with $F^{\otimes 2} = \mathcal{O}_C$. More generally, for every p and q in C, there is a line bundle L, unique up to multiplying by a 2-torsion line bundle, such that $\mathcal{F}_p \otimes L = \mathcal{F}_q$. Otherwise, the descriptions of V in the three cases of (ii) above are unique, up to permuting the factors of a direct sum of line bundles.

We return to the general problem of understanding rank 2 vector bundles. The above discussion suggests that we should reverse the analysis of Proposition 5 and try to construct vector bundles as extensions

$$0 \to L \to V \to L' \otimes I_Z \to 0,$$

where L and L' are line bundles on X and Z is a local complete intersection codimension 2 subscheme. Thus, we must analyze $\text{Ext}^1(L' \otimes I_Z, L)$. (It follows from (ii) of Lemma 7 below that the nontrivial weak isomorphism classes of extensions will be parametrized by $\mathbb{P}\,\text{Ext}^1(L' \otimes I_Z, L)$.) Now in general there is a local to global spectral sequence for Ext groups, which gives in our case a spectral sequence with E_2 term

$$E_2^{p,q} = H^p(X; \mathcal{E}xt^q(L' \otimes I_Z, L)) \implies \text{Ext}^{p+q}(L' \otimes I_Z, L).$$

This spectral sequence is really just a long exact sequence, because of the following:

Lemma 7. *Let R be a regular local ring and let $I = (f, g)R$ be an ideal of R generated by two relatively prime elements. Then:*

(i) $\text{Hom}_R(I, R) \cong R$ *and the isomorphism is induced by the natural restriction map* $\text{Hom}_R(R, R) \to \text{Hom}_R(I, R)$;

(ii) $\text{Hom}_R(I, I) \cong R$ *and again the isomorphism is induced by the natural restriction map* $\text{Hom}_R(R, R) \to \text{Hom}_R(I, R)$;

(iii) $\text{Ext}^1_R(I, R) \cong \text{Ext}^2_R(R/I, R) \cong R/I$;

(iv) $\text{Ext}^k_R(I, R) = 0$ *for* $k \geq 2$.

Proof. Apply the functor $\text{Hom}_R(\cdot, R)$ to the resolution (2.11) of I. We obtain the short complex

$$0 \leftarrow \text{Ext}^1_R(I, R) \leftarrow R \xleftarrow{t_\varphi} R \oplus R \leftarrow \text{Hom}_R(I, R).$$

Here the transpose of φ is the map $(a, b) \mapsto af + bg$. By the arguments used to prove the exactness of (2.11), it follows that $\text{Hom}_R(I, R) = R$ and that $\text{Ext}^1_R(I, R) = R/I$. A check left to the reader shows that the restriction map $\text{Hom}_R(R, R) \to \text{Hom}_R(I, R)$ in fact is an isomorphism. From the exact sequence

$$0 \to I \to R \to I/R \to 0,$$

we see that there is an inclusion

$$\text{Hom}_R(I, I) \subseteq \text{Hom}_R(I, R) \cong \text{Hom}_R(R, R) = R.$$

On the other hand, $R \subseteq \text{Hom}_R(I, I)$ by multiplication, and thus $\text{Hom}_R(I, I) = R$ acting by multiplication. The higher Ext groups are zero since I has a short free resolution. Finally, the isomorphism $\text{Ext}^1_R(I, R) \cong$

$\mathrm{Ext}_R^2(R/I, R)$ is an immediate consequence of the above by applying the long exact Ext sequence to

$$0 \to I \to R \to R/I \to 0. \qquad \square$$

Note that, just as in Claim 4, the lemma is again a special case of the calculation of $\mathrm{Ext}_R^\bullet(I, R)$ for an ideal I generated by a regular sequence.

In the global case, $Hom(L' \otimes I_Z, L) = (L')^{-1} \otimes L$ and $Ext^1(L' \otimes I_Z, L)$ is a line bundle supported on Z. Tracing through the above construction, it is not to hard to identify this line bundle with $\det(I_Z/I_Z^2)^\vee \otimes (L')^{-1} \otimes L$. Here, since Z is a local complete intersection, I_Z/I_Z^2 is a locally free rank 2 sheaf on Z and its dual is by definition the *normal bundle* of Z in X. This definition agrees with the usual definition in case Z is smooth, and is the generalization of the definition of the normal bundle of a divisor D on X.

We can now replace the Ext spectral sequence by a long exact sequence

$$0 \to H^1((L')^{-1} \otimes L) \to \mathrm{Ext}^1(L' \otimes I_Z, L) \to$$
$$\to H^0(Ext^1(L' \otimes I_Z, L)) \to H^2((L')^{-1} \otimes L).$$

In case X is a surface, $Ext^1(L' \otimes I_Z, L) = \mathcal{O}_Z$.

Next we need to decide when an extension V, which *a priori* is just a coherent sheaf, is in fact locally free. If V corresponds to an extension class $\xi \in \mathrm{Ext}^1(L' \otimes I_Z, L)$, we have the image of ξ in $H^0(Ext^1(L' \otimes I_Z, L))$, and there is the following theorem of Serre:

Theorem 8. *The extension corresponding to ξ is locally free if and only if the section ξ generates the sheaf $Ext^1(L' \otimes I_Z, L)$, i.e., the natural map*

$$\mathcal{O}_X \to Ext^1(L' \otimes I_Z, L)$$

defined by ξ is onto.

Proof. This is a local question: Let M be the R-module V_x, where as usual $R = \mathcal{O}_{X,x}$. Thus, there is an exact sequence

$$0 \to R \to M \to I \to 0.$$

Applying $\mathrm{Hom}_R(\cdot, R)$ to this sequence, there is a long exact sequence

$$\mathrm{Hom}_R(R, R) \xrightarrow{\partial} \mathrm{Ext}_R^1(I, R) \to \mathrm{Ext}_R^1(M, R) \to \mathrm{Ext}_R^1(R, R) = 0,$$

and the image of $\mathrm{Id} \in \mathrm{Hom}_R(R, R)$ in $\mathrm{Ext}_R^1(I, R)$ corresponds to the value of the extension class in the stalk over x. Moreover, for $k \geq 2$ this sequence shows that $\mathrm{Ext}_R^k(M, R) \cong \mathrm{Ext}_R^k(I, R) = 0$. Thus, $\mathrm{Ext}_R^k(M, R) = 0$ for all $k \geq 1$ if and only if $\mathrm{Ext}_R^1(M, R) = 0$ if and only if ξ generates $\mathrm{Ext}_R^1(I, R)$. To conclude, we use the following theorem of Serre whose proof is deferred to Theorem 17 below:

Theorem 9. *Let R be a regular Noetherian local ring and let M be a finite R-module. Then M is free if and only if $\operatorname{Ext}^i_R(M, R) = 0$ for all $i \geq 1$.*

Corollary 10. *Suppose in the above situation that X is a surface and that $H^2((L')^{-1} \otimes L) = 0$. Then there exist locally free extensions V of $L' \otimes I_Z$ by L.* $\square$

Example. On $\mathbb{P}^2$, take $L = L' = \mathcal{O}_{\mathbb{P}^2}$. Then $H^2(\mathcal{O}_{\mathbb{P}^2}) = 0$, and so there exist locally free extensions V on $\mathbb{P}^2$ of the form

$$0 \to \mathcal{O}_{\mathbb{P}^2} \to V \to I_Z \to 0.$$

We leave it as an exercise to show that, if $Z \neq \emptyset$, then V is not a direct sum of line bundles.

In many situations, the corollary is not sufficient, and we will need to analyze the Ext exact sequence further. We shall only consider the case where X is a smooth surface.

Claim 11. *There is a commutative diagram*

$$
\begin{array}{ccccc}
H^1((L')^{-1} \otimes L) & \longrightarrow & \operatorname{Ext}^1(L' \otimes I_Z, L) & \longrightarrow & H^0(\mathcal{E}xt^1(L' \otimes I_Z, L)) \\
\downarrow{\scriptstyle\cong} & & \| & & \downarrow{\scriptstyle\cong} \\
\operatorname{Ext}^1(L', L) & \longrightarrow & \operatorname{Ext}^1(L' \otimes I_Z, L) & \longrightarrow & \operatorname{Ext}^2(\mathcal{O}_Z, L),
\end{array}
$$

where the bottom row is the exact sequence obtained by applying the long exact Ext sequence to the exact sequence

$$0 \to L' \otimes I_Z \to L' \to \mathcal{O}_Z \to 0.$$

Proof. The spectral sequence for Ext gives an isomorphism

$$\operatorname{Ext}^2(\mathcal{O}_Z, L) \cong H^0(\mathcal{E}xt^2(\mathcal{O}_Z, L)) \cong H^0(\mathcal{E}xt^1(L' \otimes I_Z, L)).$$

From this, the commutativity of the above diagram is a straightforward consequence of the compatibility of the Ext spectral sequences with the long exact sequences associated to the $\mathcal{E}xt$ sheaves. $\square$

Thus, it will suffice to analyze the image of $\operatorname{Ext}^1(L' \otimes I_Z, L)$ in $\operatorname{Ext}^2(\mathcal{O}_Z, L)$, noting that the isomorphism

$$H^0(\mathcal{E}xt^2(\mathcal{O}_Z, L)) \cong H^0(\mathcal{E}xt^1(L' \otimes I_Z, L))$$

has the property that a section of $\mathcal{E}xt^1(L' \otimes I_Z, L)$ is a generating section if and only if the corresponding section of $\mathcal{E}xt^2(\mathcal{O}_Z, L)$ is generating. Now we can apply Serre duality on the surface X to the Ext groups above, in the form which says that, for a coherent sheaf $\mathcal{F}$ on X, $\text{Ext}^i(\mathcal{F}, K_X)$ is dual to $H^{2-i}(X; \mathcal{F})$. Thus, the sequence

$$\text{Ext}^1(L' \otimes I_Z, L) \to \text{Ext}^2(\mathcal{O}_Z, L) \to \text{Ext}^2(L', L) \to \text{Ext}^2(L' \otimes I_Z, L)$$

is dual to the sequence

$$H^1(L' \otimes L^{-1} \otimes K_X \otimes I_Z) \leftarrow H^0(\mathcal{O}_Z) \leftarrow$$
$$\leftarrow H^0(L' \otimes L^{-1} \otimes K_X) \leftarrow H^0(L' \otimes L^{-1} \otimes K_X \otimes I_Z).$$

We now suppose that $Z = \{p_1, \ldots, p_n\}$ consists of distinct (reduced) points. Thus, $\mathcal{O}_Z = \bigoplus_i \mathcal{O}_{p_i}$ and the duality pairing between $\text{Ext}^2(\mathcal{O}_Z, L)$ and $H^0(\mathcal{O}_Z)$ is local, in the sense that it is induced by a direct sum of local nondegenerate pairings $\text{Ext}^2(\mathcal{O}_{p_i}, L) \otimes H^0(\mathcal{O}_{p_i}) \to \mathbb{C}$. Choosing a trivialization of L at each p_i, we may identify $\text{Ext}^2(\mathcal{O}_Z, L) \cong \mathbb{C}^n$ and similarly for $H^0(\mathcal{O}_Z)$. Thus, the pairing between $\text{Ext}^2(\mathcal{O}_Z, L)$ and $H^0(\mathcal{O}_Z)$ is of the form $\langle x, y \rangle = \sum_i \lambda_i x_i y_i$, where λ_i is a nonzero complex number. We then have the following result:

Theorem 12. *A locally free extension of $L' \otimes I_Z$ by L exists if and only if every section of $L' \otimes L^{-1} \otimes K_X$ which vanishes at all but one of the p_i vanishes at the remaining point as well.*

Proof. Let s be a section of $\text{Ext}^2(\mathcal{O}_Z, L)$. Then s is the image of a section of $\text{Ext}^1(L' \otimes I_Z, L)$ if and only if $\partial(s) = 0$, where ∂ is the connecting homomorphism in the Ext long exact sequence. Now $\partial(s) = 0$ if and only if $\langle \partial(s), e \rangle = 0$ for all $e \in H^0(L' \otimes L^{-1} \otimes K_X)$ if and only if $\langle s, \partial(e) \rangle = 0$ for all $e \in H^0(L' \otimes L^{-1} \otimes K_X)$, where $\partial(e)$ is the image of e in

$$H^0(\mathcal{O}_Z) = H^0((L' \otimes L^{-1} \otimes K_X) \otimes \mathcal{O}_Z).$$

(Here we denote by $\langle , \rangle$ any of the Serre duality pairings involved.) Choosing local coordinates for X and local trivializations of the bundles L and L' at the p_i, we may identify $\partial(e)$ with a vector $(e_1, \ldots, e_n) \in \mathbb{C}^n$, and we may similarly identify s with a vector $(s_1, \ldots, s_n)$. Finally, as we have seen above, the pairing $\langle , \rangle$ may be identified with the diagonal form $\langle x, y \rangle = \sum_{i=1}^n \lambda_i x_i y_i$, where the $\lambda_i \neq 0$, since it is a sum of local nondegenerate pairings. Now a locally free extension exists if and only if there exists an s as above with $s_i \neq 0$ for all i which is liftable to $\text{Ext}^1(L' \otimes I_Z, L)$, or equivalently if and only if $\langle s, \partial(e) \rangle = 0$ for all $e \in H^0(L' \otimes L^{-1} \otimes K_X)$. The proposition is now an immediate consequence of the following linear algebra lemma:

Lemma 13. *Let $\langle , \rangle$ be the bilinear form on $\mathbb{C}^n$ given by*

$$\langle x, y \rangle = \sum_i \lambda_i x_i y_i,$$

where $\lambda_i \neq 0$. Let W be a vector subspace of $\mathbb{C}^n$. Then there exists an $s = (s_1, \ldots, s_n) \in W^\perp$ with $s_i \neq 0$ for all i if and only if W does not contain $\delta_i = (0, \ldots, 1, \ldots, 0)$ for any i.

Proof. If W contains δ_i for some i, then $\langle s, \delta_i \rangle = \lambda_i s_i$. Thus, if $s \in W^\perp$, then $s_i = 0$. Conversely, if every $s \in W^\perp$ satisfies $s_i = 0$ for some i, then $W^\perp \subseteq \bigcup_{i=1}^n H_i$, where H_i is the hyperplane $\{s \in \mathbb{C}^n : s_i = 0\}$. As $\mathbb{C}$ is infinite, there exists an i such that $W^\perp \subseteq H_i$. But then $W \supseteq H_i^\perp = \mathbb{C}\delta_i$, and so $\delta_i \in W$ for some i. □

Definition 14. We say that $Z = \{p_1, \ldots, p_n\}$ has the *Cayley-Bacharach property* relative to the linear system $L' \otimes L^{-1} \otimes K_X$ if every section of $L' \otimes L^{-1} \otimes K_X$ which vanishes at all but one of the p_i vanishes at the remaining point as well.

For example, it is well known that every cubic passing through eight of the nine points of intersection of two cubics meeting transversally passes through the ninth intersection point as well [55] (this is also a consequence of Lemma 17 in Chapter 5). Thus, nine such points have the Cayley-Bacharach property relative to $\mathcal{O}_{\mathbb{P}^2}(3)$. Taking $L = \mathcal{O}_{\mathbb{P}^2}(-3)$ and $L' = \mathcal{O}_{\mathbb{P}^2}(3)$, so that $L' \otimes L^{-1} \otimes K_X = \mathcal{O}_{\mathbb{P}^2}(3)$, the above arguments show that, for Z the set of nine points of intersection of two transverse cubics, there is a unique locally free extension V up to isomorphism which sits in an exact sequence

$$0 \to \mathcal{O}_{\mathbb{P}^2}(-3) \to V \to \mathcal{O}_{\mathbb{P}^2}(3) \otimes I_Z \to 0.$$

We leave it as exercise to show that in fact V is the trivial bundle.

For a less trivial example, if we take sufficiently many points $\{p_1, \ldots, p_n\}$ in general position on X, then there will be no section of $L' \otimes L^{-1} \otimes K_X$ which vanishes at all but one of the p_i. Thus, vacuously Z has the Cayley-Bacharach property relative to $L' \otimes L^{-1} \otimes K_X$. In this way we can construct rank 2 vector bundles V on X with $\det V = L \otimes L'$ such that $c_2(V)$ is arbitrarily large.

More generally, a set of n points $\{p_1, \ldots, p_n\}$ in general position on X will impose independent conditions on sections of $L' \otimes L^{-1} \otimes K_X$. For such points, we can count how many vector bundles we can construct by Theorem 12. However, this number is usually much smaller than the dimension of the moduli space of vector bundles. In this way, the existence of many vector bundles implies that, for infinitely many linear series $|D|$ there must be many configurations of points in special position with respect to $|D|$.

Elementary modifications

Definition 15. Suppose that X is regular and that D is an effective divisor on X. Denote by $j\colon D \to X$ the inclusion. Let V be a rank 2 bundle on X and L a line bundle on D, and suppose that we are given a surjection $V \to j_*L$. Define W as the kernel of the given surjection $V \to j_*L$; thus there is an exact sequence

$$0 \to W \to V \to j_*L \to 0.$$

We say that W is a *elementary modification* of V.

Lemma 16. *An elementary modification is locally free. Its Chern classes are given by*

$$c_1(W) = c_1(V) - [D],$$
$$c_2(W) = c_2(V) - c_1(V) \cdot [D] + j_*c_1(L).$$

Proof. Locally on X, with $R = \mathcal{O}_{X,x}$, D is defined by an equation t and the surjection $V \to j_*L$ is given by a surjection $\bar{\varphi}\colon R \oplus R \to R/tR$. Since $R \oplus R$ is free, $\bar{\varphi}$ lifts to a map $\varphi\colon R \oplus R \to R$ which reduces mod t to $\bar{\varphi}$. Since $\varphi\colon R \oplus R \to R$ is surjective mod t, it is surjective by Nakayama's lemma. Since R is free, the surjection φ is split. The kernel of φ is a rank 1 summand of $R \oplus R$, and since R is local it is a free rank 1 R-module and so isomorphic to R. This gives a possibly different direct sum decomposition of $R \oplus R$ for which φ is of the form $(x, y) \mapsto y \pmod{t}$. Thus, locally W is isomorphic to $R \oplus tR \cong R \oplus R$ and so W is locally free. The statement about the Chern classes of W follows from the Whitney product formula and Lemma 1:

$$
\begin{aligned}
1 &+ c_1(W) + c_2(W) + \cdots \\
&= (1 + c_1(V) + c_2(V) + \cdots)(1 + [D] + ([D]^2 - j_*c_1(L)) + \cdots)^{-1} \\
&= (1 + c_1(V) + c_2(V) + \cdots)(1 - [D] - [D]^2 + j_*c_1(L) + [D]^2 + \cdots) \\
&= 1 + (c_1(V) - [D]) + (c_2(V) - c_1(V) \cdot [D] + j_*c_1(L)) + \cdots.
\end{aligned}
$$

Equating terms gives the statement of Lemma 16. □

Given an elementary modification $0 \to W \to V \to j_*L \to 0$, the surjection $V \to j_*L$ gives an exact sequence

$$0 \to L' \to V|D \to L \to 0,$$

where L' is a line bundle on D, defined as the kernel of the map $V|D \to L$. Thus, there is an induced surjection $W \to j_*L'$. Since $\det W = \det V \otimes \mathcal{O}_X(-D)$, it follows that

$$\det(W|D) = \det(V|D) \otimes \mathcal{O}_X(-D) = L \otimes L' \otimes \mathcal{O}_X(-D).$$

Thus, $\mathrm{Ker}(W|D \to L')$ is the line bundle $L \otimes \mathcal{O}_X(-D)$. We leave it as an exercise to show that if U is given by the elementary modification

$$0 \to U \to W \to j_*L' \to 0,$$

then $U \cong V \otimes \mathcal{O}_X(-D)$. In other words, repeating an elementary modification in the obvious way essentially brings us back to where we started.

Elementary modifications are the first case in understanding the structure of an inclusion of a rank 2 bundle W in another rank 2 bundle V, by analogy with the case of sub-line bundles. Needless to say, the general case is much more subtle (see Exercise 10).

We can also reverse the construction of elementary modifications as follows: given a rank 2 bundle W on X and a line bundle L on the divisor $D \subset X$, we can try to construct extensions V of j_*L by W. Such extensions are classified by the Ext group $\mathrm{Ext}^1(j_*L, W)$, and an easy exercise gives

$$\mathrm{Ext}^1(j_*L, W) \cong H^0(D; (W \otimes \mathcal{O}_X(D))|D \otimes L^{-1}).$$

We could also see this by noting that if W is an elementary modification of V, then V, or equivalently $V^{\vee}$, is also obtained as an elementary modification

$$0 \to V^{\vee} \to W^{\vee} \to j_*(L^{-1}) \otimes \mathcal{O}_X(D) \to 0,$$

and these are classified by

$$\mathrm{Hom}(W^{\vee}, j_*(L^{-1}) \otimes \mathcal{O}_X(D)) = H^0(D; (W \otimes \mathcal{O}_X(D))|D \otimes L^{-1}).$$

Singularities of coherent sheaves

In this section, we will try to determine when a coherent sheaf on a regular scheme X is locally free, and, in general, try to attach some invariants to a coherent sheaf in order to determine how bad its singularities are. As these questions are local, it suffices to consider the corresponding problem for a regular local Noetherian ring R and a finite R-module M. We shall denote the maximal ideal of R by $\mathfrak{m}$. We begin with the following theorem due to Serre (already stated as Theorem 9):

Theorem 17. *Let R be a regular local Noetherian ring and let M be a finite R-module. Then M is free if and only if $\mathrm{Ext}^i_R(M, R) = 0$ for all $i \geq 1$ if and only if $\mathrm{Ext}^i_R(M, N) = 0$ for all $i \geq 1$ and all R-modules N.*

Proof. Clearly, if M is free, then $\mathrm{Ext}^i_R(M, N) = 0$ for all $i \geq 1$ and all R-modules N and, in particular, $\mathrm{Ext}^i_R(M, R) = 0$ for all $i \geq 1$. Conversely, suppose that $\mathrm{Ext}^i_R(M, R) = 0$ for all $i \geq 1$. First we claim that $\mathrm{Ext}^1_R(M, N) = 0$ for all finite R-modules N. Indeed choose a surjection $R^{n_1} \to N \to 0$ with kernel N_1. Thus, there is an exact sequence

$$0 \to N_1 \to R^{n_1} \to N \to 0.$$

The long exact Ext sequence and the hypothesis that $\mathrm{Ext}^i_R(M, R) = 0$ for all $i \geq 1$ give an exact sequence

$$0 = \mathrm{Ext}^1_R(M, R^{n_1}) \to \mathrm{Ext}^1_R(M, N) \to \mathrm{Ext}^2_R(M, N_1) \to \mathrm{Ext}^2_R(M, R^{n_1}) = 0.$$

Thus, $\mathrm{Ext}^1_R(M, N) \to \mathrm{Ext}^2_R(M, N_1)$ is an isomorphism. Continuing in this way, we choose a surjection $R^{n_2} \to N_1$ with kernel N_2 and see that $\mathrm{Ext}^2_R(M, N_1) \cong \mathrm{Ext}^3_R(M, N_2)$. Eventually we obtain $\mathrm{Ext}^d_R(M, N_{d-1}) \cong \mathrm{Ext}^{d+1}_R(M, N_d)$, where $d = \dim R$. By the Hilbert syzygy theorem [87], $\mathrm{Ext}^{d+1}_R(M, N_d) = 0$. Thus, $\mathrm{Ext}^1_R(M, N) = 0$.

To finish the proof of Theorem 17, choose a surjection $R^n \to M$ and thus an exact sequence

$$0 \to N \to R^n \to M \to 0.$$

We claim that this exact sequence splits, i.e., that there is a map $M \to R^n$ lifting the identity map $M \to M$. If so, then M is a summand of a free R-module, and thus it is projective. Since R is local, this will imply that M is free.

To see that the above exact sequence splits consider the associated exact Ext sequence

$$\mathrm{Hom}_R(M, R^n) \to \mathrm{Hom}_R(M, M) \to \mathrm{Ext}^1_R(M, N).$$

Since $\mathrm{Ext}^1_R(M, N) = 0$, we can indeed lift $\mathrm{Id} \in \mathrm{Hom}_R(M, M)$ to an element of $\mathrm{Hom}_R(M, R^n)$, as desired. $\square$

Given an R-module M, recall that the *support* $\mathrm{Supp}\, M$ is defined to be $\{\mathfrak{p} \in \mathrm{Spec}\, R : M_\mathfrak{p} \neq 0\} = V(\mathrm{Ann}_R(M))$, where, for $\mathfrak{a}$ an ideal of R, $V(\mathfrak{a})$ is the closed subset of $\mathrm{Spec}\, R$ of prime ideals containing $\mathfrak{a}$, corresponding to the support of the subscheme $R/\mathfrak{a}$. In particular, $\mathrm{Supp}\, M$ is a closed subscheme of $\mathrm{Spec}\, R$, and is the smallest closed subscheme S such that the sheaf $\tilde{M}$ on $\mathrm{Spec}\, R$ corresponding to M is zero on $\mathrm{Spec}\, R - S$.

Definition 18. With M as above, the *singular support* of M is the set

$$S(M) = \{\mathfrak{p} \in \mathrm{Spec}\, R : M_\mathfrak{p} \text{ is not locally free over } R_\mathfrak{p}\}$$

$$= \bigcup_{i=1}^{d} \mathrm{Supp}\, \mathrm{Ext}^i_R(M, R).$$

Thus, $S(M)$ is a closed subset of codimension at least 1 in $\mathrm{Spec}\, R$, the coherent sheaf $\tilde{M}$ corresponding to M is locally free on $\mathrm{Spec}\, R - S(M)$, and $S(M)$ is the smallest closed subset with this property.

Torsion free and reflexive sheaves

Our goal in this section will be to describe two classes of sheaves with rather mild singularities. As in the previous section, we shall only consider the local case where $R = \mathcal{O}_{X,x}$, or more generally R is a regular local Noetherian ring, and M is a finite R-module.

Definition 19. Let M_{tors} be the torsion submodule of M and define the rank of M to be the dimension of the K-vector space $M \otimes_R K$, where K is the fraction field of R. M is *torsion free* if $M_{\text{tors}} = 0$. A coherent sheaf $\mathcal{F}$ on a regular scheme X is torsion free if and only if the R-modules $\mathcal{F}_x$ are torsion free for every $x \in X$.

Proposition 20. *The following are equivalent:*

(i) $M_{\text{tors}} = 0$.
(ii) *There exists an inclusion $M \subseteq R^n$ for some n.*
(iii) *There exists an inclusion $M \subseteq R^n$ where $n = \operatorname{rank} M$.*

Proof. Clearly, (iii) implies (ii) implies (i). To see that (i) implies (iii), note that $M_{\text{tors}} = 0$ is equivalent to the natural map $M \to M \otimes_R K = K^n$ being injective, where $n = \operatorname{rank} M$. Let $e_1, \ldots, e_n$ be a K-basis for $M \otimes_R K$ and let $m_1, \ldots, m_k$ generate M over R. Write $m_i = \sum_{j=1}^{n} r_{ij} e_j$, where the $r_{ij} \in K$. If $r \in R$ is a common denominator for the r_{ij}, i.e., $r r_{ij} \in R$ for all i, j, then

$$M \subseteq \bigoplus_{i=1}^{n} R \cdot r^{-1} e_i \subseteq M \otimes_R K,$$

proving (iii). $\square$

Thus, all torsion free modules are given as submodules of R^n for some n. The standard example is the case of an ideal I, i.e., a submodule of R. In general, ideals I will not be free. Indeed, since $\operatorname{rank} I = 1$, I is free if and only if it is projective (recall that R is local) if and only if it is principal.

Corollary 21. $M_{\text{tors}} = \operatorname{Ker}(M \to M^{\vee\vee})$.

Proof. Clearly, $M_{\text{tors}} \subseteq \operatorname{Ker}(M \mapsto M^{\vee\vee})$. To see the opposite containment, we may as well replace M by M/M_{tors} and show that in this case (M torsion free) M injects into $M^{\vee\vee}$. Using Lemma 20, choose an inclusion $M \subseteq R^n$ for some n, and consider the following diagram:

$$
\begin{array}{ccc}
M & \longrightarrow & R^n \\
\downarrow & & \| \\
M^{\vee\vee} & \longrightarrow & (R^n)^{\vee\vee}.
\end{array}
$$

From this, the injectivity of $M \to M^{\vee\vee}$ is clear. $\square$

The proof of the following, which is somewhat technical, is deferred to the last section of this chapter (see also [117]).

Proposition 22. *If M is torsion free, then* codim $S(M) \geq 2$. *In particular, if R has dimension 1, then every torsion free R-module is free.*

For example, if $M = (f, g)R$ is an ideal generated by two relatively prime elements of R, neither one of which is a unit, then it is easy to see that $S(M)$ is the scheme Spec $R/(f,g)R$ which has codimension exactly 2.

There is an obvious restatement of Proposition 22 for coherent sheaves: let X be a regular scheme and let $\mathcal{F}$ be a torsion free sheaf on X. Then there exists a closed subscheme Y of X of codimension at least 2 such that $\mathcal{F}|X - Y$ is locally free.

Definition 23. A finite R-module M is *reflexive* if the natural map $M \to M^{\vee\vee}$ is an isomorphism. Hence a reflexive module is torsion free. Reflexive sheaves on a regular scheme are similarly defined.

We have the following results concerning reflexive modules, whose proofs are given at the end of the chapter (see also [117]).

Proposition 24. *The following are equivalent:*

(i) *M is reflexive.*
(ii) *There exists a finite R-module N such that $M \cong N^{\vee}$.*
(iii) *M is torsion free, and, for all ideals $\mathfrak{a}$ of R such that* codim $V(\mathfrak{a}) \geq 2$, *the natural map $H^0(\operatorname{Spec} R, \tilde{M}) \to H^0(\operatorname{Spec} R - V(\mathfrak{a}), \tilde{M})$ is an isomorphism, where $\tilde{M}$ is the coherent sheaf on Spec R naturally associated to M.*

Proposition 25. *If M is reflexive, then* codim $S(M) \geq 3$. *Hence if* dim $R \leq 2$, *every reflexive R-module is free.*

We leave to the reader the formulation of the above results for a reflexive sheaf on a regular scheme.

Using Corollary 21 for a regular local ring of dimension 2, we see that every torsion free R-module M canonically sits inside a free rank 2 R-module, namely $M^{\vee\vee}$. Conversely, we can obtain every torsion free R-module as follows: starting with R^n, choose a submodule M such that R/M has finite length. Then M is a torsion free R-module with $M^{\vee\vee} = R^n$. The simplest example of this construction is $\bigoplus_{i=1}^n I_{Z_i} \subseteq R^n$, where I_{Z_i} is the ideal of a 0-dimensional subscheme of Spec R. Of course, not every M is of this form.

Our final result about reflexive modules is the following.

Lemma 26. *A rank 1 reflexive R-module M is free.*

Proof. Let M be a rank 1 reflexive R-module and set $X = \operatorname{Spec} R$ and $Y = S(M) \subseteq X$. If $\tilde{M}$ is the sheaf on X induced by M, then $\tilde{M}|X - Y$ is a locally free $\mathcal{O}_{X-Y}$ module of rank 1 and is thus a line bundle. Since $\operatorname{codim} Y \geq 2$ by Proposition 25, $\operatorname{Pic}(X - Y) = \operatorname{Pic} R = 0$, for example, by [61, II.6.5 and 6.16]. Thus, $\tilde{M}|X - Y \cong \mathcal{O}_{X-Y}$. Since M is reflexive, it follows from (iii) of Proposition 24 that $M = H^0(X - Y, \tilde{M}|X - Y) \cong H^0(X - Y, \mathcal{O}_{X-Y}) = R$. $\square$

Thus, to find examples of reflexive modules which are not free, we must consider R-modules of rank at least 2, where $\dim R \geq 3$. One way to produce such examples is the following: start with R a regular local ring of dimension at least 3, and choose a regular sequence (t, f, g). Consider the R-module M defined by the short exact sequence

$$0 \to R \to R^3 \to M \to 0,$$

where the map $R \to R^3$ sends 1 to (t, f, g). We leave it as an exercise to show that M is reflexive. If $S = R/tR$ and $\bar{f}, \bar{g}$ are the induced elements of S, we may think of M as a typical 1-parameter deformation of the torsion free S-module $S \oplus J$, where $J = (\bar{f}, \bar{g})S$.

Double covers

In this section, we give a method for constructing rank 2 vector bundles via double covers. All of the results described here are essentially due to Schwarzenberger [134].

Let X and Y be two smooth varieties and let $f: X \to Y$ be a finite morphism of degree 2. If D is a divisor on X and $\mathcal{O}_X(D)$ is the associated line bundle, then $f_*\mathcal{O}_X(D)$ is a rank 2 vector bundle on Y. This procedure is, apart from taking direct sums of line bundles, perhaps the simplest method for constructing rank 2 bundles. To analyze the direct image, we begin by recalling some standard facts about double covers.

Let Y be a smooth variety and let B be a reduced effective divisor on Y. Suppose that $\mathcal{O}_Y(B)$ is divisible by 2 in $\operatorname{Pic} Y$ and let L be a choice of a square root, i.e., $L^{\otimes 2} = \mathcal{O}_Y(B)$. Then we construct the double cover X of Y branched along B associated to L as a subvariety of the total space of the bundle L:

$$X = \{x : x^{\otimes 2} = s\},$$

where s is any section of $L^{\otimes 2}$ corresponding to B. It is easy to see that X is smooth if and only if B is smooth. In this case $f^*L = \mathcal{O}_X(B)$, where we have identified B with its inverse image on X. Conversely, given $f : X \to Y$, we can recover B and L as follows. Let $\iota : X \to X$ be the sheet interchange involution. Then B is the image under f of the fixed set of ι and $f_*\mathcal{O}_X = \mathcal{O}_Y \oplus L^{-1}$, where the direct sum decomposition corresponds to taking the $+1$ and -1 eigenspaces of ι.

Let $\operatorname{Div} X$ be the group of divisors on X and similarly for $\operatorname{Div} Y$. There is the natural map $f_*\colon \operatorname{Div} X \to \operatorname{Div} Y$ which is defined on a reduced irreducible effective divisor D as follows: $f_*(D) = rf(D)$, where r is the degree of $f|D\colon D \to f(D)$. Hence r is either 1 or 2. In general we extend f_* to all divisors by linearity. (Of course, we can define f_* for any finite morphism by the same procedure; see also Exercise 12 in Chapter 1.) It is easy to see that this f_* is compatible with the map $f_*\colon H^2(X) \to H^2(Y)$ induced by Poincaré duality. In addition, f_*f^* is multiplication by 2 in $\operatorname{Div} Y$ and $f^*f_*(D) = D + \iota(D)$ for all $D \in \operatorname{Div} X$. We can similarly define f_* on cycles of any degree. It is easy to check directly that the projection formula $f_*(D \cdot f^*E) = f_*D \cdot E$ holds if, for example, D and f^*E are effective divisors meeting transversally; the general case is done in [45].

Note that $f_*(D)$ is a *divisor* whereas $f_*\mathcal{O}_X(D)$ is a rank 2 bundle. The relationship between the two is given by the following (cf. also [61, IV, p. 306, Ex. 2.6]):

Proposition 27. *We have the following equality in* $\operatorname{Pic} Y$:

$$c_1(f_*\mathcal{O}_X(D)) = [\det f_*\mathcal{O}_X(D)] = [f_*D] - [L].$$

In particular the linear equivalence class of f_*D *depends only on the linear equivalence class of* D, *and so* f_* *induces a homomorphism from* $\operatorname{Pic} X$ *to* $\operatorname{Pic} Y$.

Proof. For $D = 0$, we have $c_1(f_*\mathcal{O}_X) = [\det(\mathcal{O}_Y \oplus L^{-1})] = [L^{-1}]$, which verifies the formula in this case since $f_*0 = 0$. Next consider a general divisor $D = \sum_i n_i D_i$, where the D_i are distinct reduced irreducible divisors on X. The proof will proceed by induction on $n = \sum_i |n_i|$. If $n = 0$, $D = 0$ and we have verified the proposition in this case. Otherwise, choose some $n_j \neq 0$. If $n_j < 0$, then we set $D' = D + D_j$. Thus, $D' = \sum_i n_i' D_i$, where $n_i' = n_i$ if $i \neq j$, and $n_j' = n_j + 1$, so that $|n_j'| = |n_j| - 1$. Consider the exact sequence

$$0 \to \mathcal{O}_X(D) \to \mathcal{O}_X(D') \to \mathcal{O}_{D_j}(D') \to 0.$$

Now f_* is an exact functor since f is finite and therefore affine. Thus, there is an induced exact sequence

$$0 \to f_*\mathcal{O}_X(D) \to f_*\mathcal{O}_X(D') \to f_*\mathcal{O}_{D_j}(D') \to 0.$$

The restriction of f to D_j is a morphism of degree $r = 1$ or 2, and $f_*\mathcal{O}_{D_j}$ is a vector bundle of rank r on $f(D_j)$. By the comments after (2.8), $c_1(f_*\mathcal{O}_{D_j}) = r[f(D_j)] = [f_*D_j]$. By the Whitney product formula we thus have $c_1(f_*\mathcal{O}_X(D)) = c_1(f_*\mathcal{O}_X(D')) - [f_*D_j]$. By induction,

$c_1(f_*\mathcal{O}_X(D')) = [f_*D'] - [L]$. Thus,

$$c_1(f_*\mathcal{O}_X(D)) = [f_*D'] - [L] - [f_*D_j] = [f_*D] - [L].$$

This concludes the argument in case some $n_j < 0$, and a similar argument handles the case where all n_j are > 0. $\square$

A straightforward argument shows that the homomorphism from $\mathrm{Pic}\, X$ to $\mathrm{Pic}\, Y$ in Proposition 27 is just the norm homomorphism

$$H^1(X; \mathcal{O}_X^*) \to H^1(Y; \mathcal{O}_Y^*).$$

To illustrate Proposition 27, let $f \colon \mathbb{P}^1 \to \mathbb{P}^1$ be the degree 2 map given by taking the double cover of $\mathbb{P}^1$ branched at two points. Then in the above notation B consists of two points and $L = \mathcal{O}_{\mathbb{P}^1}(1)$. Thus, $c_1(f_*\mathcal{O}_{\mathbb{P}^1}(d)) = d - 1$. We leave it as an exercise to show that, for $d = 2k$ even, $f_*\mathcal{O}_{\mathbb{P}^1}(d) = \mathcal{O}_{\mathbb{P}^1}(k) \oplus \mathcal{O}_{\mathbb{P}^1}(k-1)$, whereas for $d = 2k+1$ odd, $f_*\mathcal{O}_{\mathbb{P}^1}(d) = \mathcal{O}_{\mathbb{P}^1}(k) \oplus \mathcal{O}_{\mathbb{P}^1}(k)$.

Our next result is a calculation of $c_2(f_*\mathcal{O}_X(D))$, originally proved by Schwarzenberger via the Grothendieck-Riemann-Roch theorem.

Proposition 28. *With $f \colon X \to Y$ as above and D a divisor on X, the following equality holds in $H^4(Y; \mathbb{Z}[\frac{1}{2}])$:*

$$c_2(f_*\mathcal{O}_X(D)) = \tfrac{1}{2}((f_*D)^2 - f_*(D^2) - f_*D \cdot L).$$

Proof. To begin with, we note the following:

Lemma 29. *There is a natural exact sequence*

$$0 \to \mathcal{O}_X(\iota(D)) \otimes f^*L^{-1} \to f^*f_*\mathcal{O}_X(D) \to \mathcal{O}_X(D) \to 0.$$

Proof. The natural map $f^*f_*\mathcal{O}_X(D) \to \mathcal{O}_X(D)$ is surjective because f is an affine map. The kernel is thus the line bundle $\det f^*f_*\mathcal{O}_X(D) \otimes \mathcal{O}_X(-D)$. Using Proposition 27, $\det f^*f_*\mathcal{O}_X(D) = f^*(\mathcal{O}_Y(f_*D) \otimes L^{-1}) = \mathcal{O}_X(f^*f_*D) \otimes f^*L^{-1} = \mathcal{O}_X(D + \iota(D)) \otimes f^*L^{-1}$. Combining gives Lemma 29. $\square$

Returning to the proof of Proposition 28, the Whitney product formula applied to the exact sequence in Lemma 29 gives $f^*c_2(f_*\mathcal{O}_X(D)) = c_2(f^*f_*\mathcal{O}_X(D)) = D \cdot \iota(D) - D \cdot f^*L$. On the other hand, we have

$$(D + \iota(D))^2 = (f^*f_*D)^2 = f^*(f_*D)^2$$
$$= D^2 + \iota(D)^2 + 2D \cdot \iota(D)$$

and thus, in $H^4(X; \otimes \mathbb{Z}[\frac{1}{2}])$, $D \cdot \iota(D) = \frac{1}{2} f^*(f_* D)^2 - \frac{1}{2} D^2 - \frac{1}{2} \iota(D)^2$. Thus, (with all coefficients in $\mathbb{Z}[\frac{1}{2}]$)

$$\begin{aligned} 2c_2(f_* \mathcal{O}_X(D)) &= f_* f^* c_2(f_* \mathcal{O}_X(D)) = f_* c_2(f^* f_* \mathcal{O}_X(D)) \\ &= f_*[D \cdot \iota(D) - D \cdot f^* L] \\ &= \tfrac{1}{2} f_* f^*(f_* D)^2 - \tfrac{1}{2} f_*(D^2) - \tfrac{1}{2} f_* \iota(D)^2 - f_*(D \cdot f^* L) \\ &= (f_* D)^2 - f_*(D^2) - (f_* D) \cdot L, \end{aligned}$$

where at the last stage we have used the projection formula and the fact that $f_* \iota(D)^2 = f_*(D^2)$. Dividing by 2 then gives the formula in Proposition 28. $\square$

Example. Let $Q \cong \mathbb{P}^1 \times \mathbb{P}^1$ be a smooth quadric in $\mathbb{P}^3$ and let $\pi \colon Q \to \mathbb{P}^2$ be the projection of Q onto a plane. Then Q is a double cover of $\mathbb{P}^2$ via π; the branch locus is a smooth conic. Hence, in the notation of this section, $L = \mathcal{O}_{\mathbb{P}^2}(1)$. A basis for $\operatorname{Pic} Q$ is given by $\{f_1, f_2\}$, where the f_i are lines on Q belonging to the two distinct rulings. Thus, $\pi_*(f_i)$ is a line in $\mathbb{P}^2$. Let $V_{a,b}$ be the rank 2 bundle $\pi_* \mathcal{O}_Q(a f_1 + b f_2)$ on $\mathbb{P}^2$. Using Propositions 27 and 28, $\det V_{a,b} = \mathcal{O}_{\mathbb{P}^2}(a + b - 1)$ and $c_2(V_{a,b}) = \frac{1}{2}((a^2 - a) + (b^2 - b))$. We leave it as an exercise to determine when $V_{a,b}$ is a direct sum of line bundles.

The following result of Schwarzenberger says that every rank 2 bundle over a curve or surface arises as the push forward of a line bundle on a double cover. However, the nonuniqueness of the construction makes this result mainly of theoretical interest.

Proposition 30. *Let Y be a smooth curve or surface and let V be a rank 2 vector bundle on Y. Then there exists a smooth double cover $f \colon X \to Y$ and a line bundle M on X such that $V = f_* M$.*

Proof. Let $P = \mathbb{P}(V^\vee) \xrightarrow{\pi} Y$ be the projectivized bundle associated to V (here our sign conventions are the opposite ones to those in, for example, [61] or EGA). Then for the tautological bundle $\mathcal{O}_P(1)$ on P, we have [61, III, p. 253, Ex. 8] $\pi_* \mathcal{O}_P(1) = V$. Let L be a sufficiently ample line bundle on Y and let $L' = \mathcal{O}_P(2) \otimes \pi^* L$. Then L' is very ample on P [61, II, 7.10]. If $\mathcal{X}$ is the linear system corresponding to L', then for $X \in \mathcal{X}$ and ℓ a fiber of π, either $\ell \subset X$ or X meets ℓ in exactly two points, counted with multiplicity.

Claim 31. *There exists a smooth $X \in \mathcal{X}$ such that the induced map $f \colon X \to Y$ is finite and so is a double cover.*

Proof of the Claim. Consider the incidence correspondence $I \subset \mathcal{X} \times Y$ given by

$$I = \{(X, p) : \pi^{-1}(p) \subset X\}$$

and let π_1, π_2 be the projections of I to $\mathcal{X}$, Y, respectively. Then $\pi_1(I)$ is the set of X which contain some fiber of π, and we will be done by Bertini's theorem if we show that $\pi_1(I)$ is a proper subset of $\mathcal{X}$. Let $\dim \mathcal{X} = N$. Then we claim that $\dim I \leq N - 1$ (it is here that we use the assumption $\dim Y \leq 2$). To see this, let $p \in Y$ and let $\ell = \pi^{-1}(p)$. Then $\dim \pi_2^{-1}(p) = \dim\{X \in \mathcal{X} : \ell \subset X\} = h^0(P; L' \otimes I_\ell) - 1$, where I_ℓ is the ideal sheaf of ℓ. Consider the exact sequence of sheaves

$$0 \to L' \otimes I_\ell \to L' \to \mathcal{O}_\ell(2) \to 0.$$

For L sufficiently ample on Y, for example, $L = L_0^{\otimes 2}$ where $\mathcal{O}_P(1) \otimes \pi^* L_0$ is very ample on P, we claim that the map $H^0(P; L') \to H^0(\ell; \mathcal{O}_\ell(2))$ is surjective. Indeed, since $\mathcal{O}_P(1) \otimes \pi^* L_0$ is very ample on P and restricts to $\mathcal{O}_\ell(1)$ on ℓ, the map $H^0(P; \mathcal{O}_P(1) \otimes \pi^* L_0) \to H^0(\ell; \mathcal{O}_\ell(1))$ is surjective. But then $L' = (\mathcal{O}_P(1) \otimes \pi^* L_0)^{\otimes 2}$, so that $H^0(L')$ contains the image of $\mathrm{Sym}^2 H^0(\mathcal{O}_P(1) \otimes \pi^* L_0)$, and thus the image of $H^0(L')$ in $H^0(\ell; \mathcal{O}_\ell(2))$ contains the image of $\mathrm{Sym}^2 H^0(\ell; \mathcal{O}_\ell(1)) = H^0(\ell; \mathcal{O}_\ell(2))$.

As $\dim H^0(\ell; \mathcal{O}_\ell(2)) = 3$, it follows that $\dim \pi_2^{-1}(p) = h^0(P; L' \otimes I_\ell) = (h^0(L') - 1) - 3 = N - 3$. Hence $\dim I = \dim \pi_2^{-1}(p) + \dim Y \leq N - 1$. Thus, $\pi_1(I)$ must be a proper subset of $\mathcal{X}$, proving the claim. $\square$

Returning to the proof of Proposition 30, choose X as in the claim. It suffices to prove that $f_*(\mathcal{O}_P(1)|X) = V$. Apply π_* to the exact sequence

$$0 \to \mathcal{O}_P(1) \otimes \mathcal{O}_P(-X) \to \mathcal{O}_P(1) \to \mathcal{O}_P(1)|X \to 0.$$

We thus obtain a map $V = \pi_* \mathcal{O}_P(1) \to \pi_*(\mathcal{O}_P(1)|X) = f_*(\mathcal{O}_P(1)|X)$. To see that this map is an isomorphism, note that for all $y \in Y$, $\mathcal{O}_P(1) \otimes \mathcal{O}_P(-X)|\pi^{-1}(y) = \mathcal{O}_{\mathbb{P}^1}(-1)$ and that $H^0(\mathcal{O}_{\mathbb{P}^1}(-1)) = H^1(\mathcal{O}_{\mathbb{P}^1}(-1)) = 0$. Thus,

$$R^0 \pi_*(\mathcal{O}_P(1) \otimes \mathcal{O}_P(-X)) = R^1 \pi_*(\mathcal{O}_P(1) \otimes \mathcal{O}_P(-X)) = 0,$$

proving that the map $V \to f_* \mathcal{O}_P(1)|X$ is an isomorphism. $\square$

Consider the example $Y = \mathbb{P}^2$. Every smooth double cover $\pi : X \to \mathbb{P}^2$ is branched along a smooth plane curve of degree $2d$, and if $d > 2$, then the generic such double cover X has $\mathrm{Pic}\, X \cong \mathbb{Z}$, with a generator pulled back from $\mathcal{O}_{\mathbb{P}^2}(1)$. But $\pi_* \pi^* \mathcal{O}_{\mathbb{P}^2}(n) \cong \mathcal{O}_{\mathbb{P}^2}(n) \oplus \mathcal{O}_{\mathbb{P}^2}(n - d)$, and in particular it is a direct sum of line bundles. Conversely, this construction realizes the direct sum of two line bundles as the direct image of a line bundle on an appropriate double cover. Thus, the existence of nontrivial rank 2 vector bundles on $\mathbb{P}^2$ implies that there are many configurations of plane curves

C_1 and C_2 in special position, in the sense that C_1 is smooth and C_2 is everywhere tangent to C_1.

Appendix: some commutative algebra

Our goal is to prove some of the more technical results used above. We begin by recalling the following definitions (cf. [87] or [139]:

Definition 32. If R is a regular local ring and M is a finite R-module, define the *projective dimension* proj. dim M, by any of the following equivalent ways:

1. the minimal length of a free resolution of M;
2. $\sup\{k$: there exists a finitely generated R-module N such that $\operatorname{Ext}_R^k(M, N) \neq 0\}$;
3. $\sup\{k : \operatorname{Ext}_R^k(M, R) \neq 0\}$.

Here the equivalence of (2) and (3) is an argument analogous to the proof of Theorem 17.

For $M \neq 0$, let depth M be the maximal length of an M-sequence $x_1, \ldots, x_k \in R$, i.e., $x_i \in \mathfrak{m}$ and for all i the map $M/(x_1, \ldots, x_{i-1})M \to M/(x_1, \ldots, x_{i-1})M$ induced by multiplication by x_i is injective, where depth 0 means that there is no such x_1. More generally, for $\mathfrak{a}$ an ideal of R, we set $\operatorname{depth}_\mathfrak{a} M$ to be the maximal length of an M-sequence $x_1, \ldots, x_k$ such that $x_i \in \mathfrak{a}$ for all i. Let us collect some salient facts about depth:

Lemma 33. *Let R be a regular local ring of dimension d.*

(i) $\operatorname{depth}_\mathfrak{a} M = \inf\{\operatorname{depth} M_\mathfrak{p} : \mathfrak{p} \in V(\mathfrak{a})\}$.
(ii) proj. dim $M + \operatorname{depth} M = d$.
(iii) *The local cohomology group $H_\mathfrak{a}^i(M)$ is 0 for all $i \leq k-1$ if and only if $\operatorname{depth}_\mathfrak{a} M \geq k$.*
(iv) *If $\mathfrak{a}$ is an ideal of R with codim $\mathfrak{a} \geq k$, then $H_\mathfrak{a}^i(R) = 0$ for all $i \leq k-1$.*

Proof. (i) This is [87, p.105].

(ii) This is the dimension formula of Auslander-Buchsbaum [87].

(iii) This is [61, p. 217, Ex. 3.4].

(iv) It suffices by (i) and (iii) to show that depth $R_\mathfrak{p} \geq k$ for all primes $\mathfrak{p} \in V(\mathfrak{a})$. Since R is regular, $R_\mathfrak{p}$ is regular, and thus

$$\operatorname{depth} R_\mathfrak{p} = \dim R_\mathfrak{p} = d - \dim V(\mathfrak{p}) = \operatorname{codim} V(\mathfrak{p}).$$

Since $\mathfrak{a}$ has codimension at least k, $\operatorname{codim} V(\mathfrak{p}) \geq k$ and thus depth $R_\mathfrak{p} \geq k$. □

Lemma 34. *For every R-module M,*

$$\dim \operatorname{Supp} \operatorname{Ext}^i_R(M, R) \leq d - i,$$

i.e., codim $\operatorname{Supp} \operatorname{Ext}^i_R(M, R) \geq i$.

Proof. If $\mathfrak{p} \in \operatorname{Supp} \operatorname{Ext}^i_R(M, R)$, then the localization $\operatorname{Ext}^i_R(M, R)_\mathfrak{p}$ is not 0. As there is a natural isomorphism

$$\operatorname{Ext}^i_R(M, R)_\mathfrak{p} \cong \operatorname{Ext}^i_{R_\mathfrak{p}}(M_\mathfrak{p}, R_\mathfrak{p}),$$

$\operatorname{Ext}^i_{R_\mathfrak{p}}(M_\mathfrak{p}, R_\mathfrak{p}) \neq 0$, and hence proj. dim $M_\mathfrak{p} \geq i$, where the proj. dim is as an $R_\mathfrak{p}$-module. As $R_\mathfrak{p}$ is again a regular local ring of dimension $d - \dim V(\mathfrak{p})$, from

$$\text{proj. dim } M_\mathfrak{p} + \text{depth } M_\mathfrak{p} = d - \dim V(\mathfrak{p}),$$

we obtain $\dim V(\mathfrak{p}) \leq d - i$ for all $\mathfrak{p} \in \operatorname{Supp} \operatorname{Ext}^i_R(M, R)$. Thus,

$$\dim \operatorname{Supp} \operatorname{Ext}^i_R(M, R) \leq d - i. \qquad \square$$

Definition 35. *M is a kth syzygy module if there exists an exact sequence*

$$0 \to M \to R^{n_1} \to R^{n_2} \to \cdots \to R^{n_{k-1}} \to R^{n_k}.$$

Lemma 36. *Let M be a kth syzygy module. Then:*

(i) *proj. dim $M \leq d - k$.*
(ii) *codim $S(M) \geq k + 1$.*
(iii) *For all ideals $\mathfrak{a}$ of R, $\text{depth}_\mathfrak{a} M \geq k$.*

Proof. We claim that, for all $i \geq 1$, $\dim \operatorname{Supp} \operatorname{Ext}^i_R(M, R) \leq d - k - i$ and that, in particular, $\operatorname{Supp} \operatorname{Ext}^i_R(M, R) = \emptyset$ if $i > d - k$. Thus, by definition $\dim S(M) \leq d - k - 1$. Moreover, we must have $\operatorname{Ext}^i_R(M, R) = 0$ for $i > d - k$, so that proj. dim $M \leq d - k$ as well. We will prove the claim by induction on k. For $k = 0$, i.e., no condition on M, this is just the syzygy theorem. If now M is a $(k + 1)^{\text{st}}$ syzygy module, then there exists an exact sequence

$$0 \to M \to R^{n_1} \to M' \to 0,$$

where M' is a kth syzygy module. An argument with the Ext exact sequence shows that $\operatorname{Ext}^i_R(M, R) \cong \operatorname{Ext}^{i+1}_R(M', R)$ for all $i \geq 1$. Thus, by the inductive hypothesis $\dim \operatorname{Supp} \operatorname{Ext}^i_R(M, R) = \dim \operatorname{Supp} \operatorname{Ext}^{i+1}_R(M', R) \leq d - k - (i + 1) = d - (k + 1) - i$. This completes the inductive step and thus the proof of (i) and (ii).

To see the last statement, it suffices by (i) of Lemma 33 to prove it for prime ideals $\mathfrak{p}$. Now since $R_\mathfrak{p}$ is a flat R-module, $M_\mathfrak{p}$ is a kth syzygy module as well (for $R_\mathfrak{p}$). Thus,

$$\text{proj. dim } M_\mathfrak{p} \geq \dim R_\mathfrak{p} - k$$

and so, since

$$\text{proj. dim } M_{\mathfrak{p}} + \text{depth } M_{\mathfrak{p}} = \dim R_{\mathfrak{p}},$$

we have depth $M_{\mathfrak{p}} \geq k$. $\square$

Proposition 37. *The following are equivalent:*

(i) M *is torsion free.*
(ii) M *is a first syzygy module.*
(iii) *For all proper ideals* $\mathfrak{a}$, $\text{depth}_{\mathfrak{a}} M \geq 1$.

Proof. The equivalence of (i) and (ii) follows from Proposition 20. Next note that, if $f \cdot m = 0$ for $f \in R, f \neq 0$ and $m \in M, m \neq 0$, then for $\mathfrak{a} = (f)$ we have $\text{depth}_{\mathfrak{a}} M = 0$. Conversely, if M is torsion free, then it is a first syzygy module and so $\text{depth}_{\mathfrak{a}} M \geq 1$ by (iii) of Lemma 36. $\square$

Corollary 38. *If M is torsion free, then* $\text{codim } S(M) \geq 2$. *In particular, if R has dimension 1, then every torsion free R-module is free.*

Proof. This is immediate from Lemma 36. $\square$

Proposition 39. *The following are equivalent:*

(i) M *is reflexive.*
(ii) *There exists a finite R-module N such that* $M \cong N^{\vee}$.
(iii) M *is a second syzygy module.*
(iv) *For all ideals $\mathfrak{a}$ of R such that* $\text{codim } V(\mathfrak{a}) \geq 1$, $\text{depth}_{\mathfrak{a}} M \geq 1$, *and for all ideals $\mathfrak{a}$ of R such that* $\text{codim } V(\mathfrak{a}) \geq 2$, $\text{depth}_{\mathfrak{a}} M \geq 2$.
(v) M *is torsion free, and, for all ideals $\mathfrak{a}$ of R such that* $\text{codim } V(\mathfrak{a}) \geq 2$, *the natural map* $H^{0}(\text{Spec } R, \tilde{M}) \to H^{0}(\text{Spec } R - V(\mathfrak{a}), \tilde{M})$ *is an isomorphism, where $\tilde{M}$ is the coherent sheaf on $\text{Spec } R$ naturally associated to M.*

Proof. (i) $\Longrightarrow$ (ii) is trivial (take $N = M$). (ii) $\Longrightarrow$ (iii): Choose a presentation

$$R^{n_2} \to R^{n_1} \to N \to 0.$$

Dualizing gives an exact sequence

$$0 \to N^{\vee} = M \to R^{n_1} \to R^{n_2},$$

i.e., M is a second syzygy module. (iii) $\Longrightarrow$ (iv): Since a second syzygy module is torsion free, $H^0_{\mathfrak{a}}(M) = 0$ for all proper ideals $\mathfrak{a}$. Thus, if M is a second syzygy module we have an exact sequence

$$0 \to M \to R^{n_1} \to M' \to 0$$

with $M' \subseteq R^{n_2}$ and hence torsion free. Applying the long exact sequence of local cohomology, we obtain

$$0 = H_{\mathfrak{a}}^0(M') \to H_{\mathfrak{a}}^1(M) \to H_{\mathfrak{a}}^1(R^{n_1}).$$

From (iv) of Lemma 33, if codim $V(\mathfrak{a}) \geq 2$, then $H_{\mathfrak{a}}^1(R^{n_1}) = 0$ and hence $H_{\mathfrak{a}}^1(M) = 0$. The statement (iv) then follows from (iii) of Lemma 33.

To see that (iv) $\Longrightarrow$ (v), note that depth$_{\mathfrak{a}} M \geq 1$ for all $\mathfrak{a}$ is equivalent to M being torsion free. Let $\mathfrak{a}$ be an ideal with codim $V(\mathfrak{a}) \geq 2$. Then by hypothesis depth$_{\mathfrak{a}} M \geq 2$, and so $H_{\mathfrak{a}}^1(M) = 0$ by (iii) of Lemma 33 again. Setting $\tilde{M} =$ the sheaf on Spec R naturally associated to M, we have a long exact sequence

$$0 \to H_{\mathfrak{a}}^0(M) \to H^0(\mathrm{Spec}\, R, \tilde{M}) \to H^0(\mathrm{Spec}\, R - V(\mathfrak{a}), \tilde{M}) \to H_{\mathfrak{a}}^1(M).$$

Thus, $H^0(\mathrm{Spec}\, R, \tilde{M}) \cong H^0(\mathrm{Spec}\, R - V(\mathfrak{a}), \tilde{M})$.

Finally, to see that (v) $\Longrightarrow$ (i), since M is torsion free we have $M \hookrightarrow M^{\vee\vee}$ and codim $S(M) \geq 2$. Let $\tilde{M}$ be the sheaf on Spec R naturally induced by M and $\widetilde{M^{\vee\vee}}$ that induced by $M^{\vee\vee}$. We have a commutative diagram

$$
\begin{array}{ccc}
H^0(\mathrm{Spec}\, R - S(M), \tilde{M}) & = & H^0(\mathrm{Spec}\, R - S(M), \widetilde{M^{\vee\vee}}) \\
\uparrow{\scriptstyle f} & & \uparrow{\scriptstyle g} \\
H^0(\mathrm{Spec}\, R, \tilde{M}) & \xrightarrow{h} & H^0(\mathrm{Spec}\, R, \widetilde{M^{\vee\vee}}).
\end{array}
$$

By hypothesis f is an isomorphism. Since $M^{\vee\vee}$ is the dual of the R-module $M^{\vee}$, by using the implication (ii) $\Longrightarrow$ (iii) g is an isomorphism. Thus, h is an isomorphism, and so $M = M^{\vee\vee}$, i.e., M is reflexive.

Corollary 40. *If M is reflexive, then codim $S(M) \geq 3$. Hence if $\dim R \leq 2$, every reflexive R-module is free.*

Proof. Since a reflexive R-module is a second syzygy module, the corollary follows immediately from Lemma 36.

Remark. (1) The proof that (vi) implies (i) also shows that for a general torsion free R-module M, $M^{\vee\vee} = H^0(\mathrm{Spec}\, R - S(M), \tilde{M})$. For example, if I is an ideal in R such that codim Spec $R/I \geq 2$, then $I^{\vee\vee} = R$.

(2) One can more generally show the following. An R-module M is a kth syzygy module if and only if, for all $j \leq k$ and for all ideals $\mathfrak{a}$ of R with codim $V(\mathfrak{a}) \geq j$, depth$_{\mathfrak{a}} M \geq j$. There is also an equivalent statement in terms of local cohomology.

Exercises

1. Let L_1 and L_2 be line bundles on a scheme X. Give an alternate description of $\mathrm{Ext}^1(L_2, L_1) = H^1((L_2)^{-1} \otimes L_1)$ as follows. If V sits in

the exact sequence

$$0 \to L_1 \to V \to L_2 \to 0,$$

then the transition functions for V can be taken to be upper triangular, with the diagonal entries transition functions for L_1 and L_2. How does the nonzero off-diagonal entry transform? Define in this way a section of $H^1((L_2)^{-1} \otimes L_1)$, independent of choices, and then show that this procedure can be reversed. Generalize this to vector bundles V_1, V_2 of arbitrary rank.

2. Prove the uniqueness assertions of the remark after the proof of Theorem 6: if $\mathcal{F}_p$ is the nonsplit extension of $\mathcal{O}_C(p)$ by $\mathcal{O}_C$, where C is an elliptic curve, then $\mathcal{F}_p \otimes L \cong \mathcal{F}_p$ if and only if L is a line bundle of order 2. Conclude that $Hom(\mathcal{F}_p, \mathcal{F}_p) \cong \mathcal{O}_C \oplus L_1 \oplus L_2 \oplus L_3$, where the L_i are the nontrivial 2-torsion line bundles on C. Moreover, for all $q \in C$, there exists an L such that $\mathcal{F}_p \otimes L \cong \mathcal{F}_q$ (here L is unique up to a 2-torsion line bundle).

3. Let V be a vector bundle on $\mathbb{P}^2$ which sits in an exact sequence

$$0 \to \mathcal{O}_{\mathbb{P}^2} \to V \to I_Z \to 0.$$

If $Z \neq \emptyset$, show that V is not a direct sum of line bundles.

4. Let Z be a set of nine points in $\mathbb{P}^2$ which is the transverse intersection of two smooth cubics. Show that there is a unique locally free extension V up to isomorphism which sits in an exact sequence

$$0 \to \mathcal{O}_{\mathbb{P}^2}(-3) \to V \to \mathcal{O}_{\mathbb{P}^2}(3) \otimes I_Z \to 0.$$

Show that in fact V is the trivial bundle.

5. Show, for a rank 2 vector bundle V, that $V^\vee \cong V \otimes \det V^{-1}$. Now suppose that

$$0 \to W \to V \to j_*L \to 0$$

is an elementary modification of V, where $j \colon D \to X$ is the inclusion of a smooth divisor. Show that $V^\vee$ is obtained as an elementary modification of $W^\vee$, and thus that V is an elementary modification of $W^\vee \otimes \det V$.

6. The line bundle $\mathcal{O}_{\mathbb{P}^1}(1)$ is generated by two global sections, and thus there is a surjection $\mathcal{O}_{\mathbb{P}^2} \oplus \mathcal{O}_{\mathbb{P}^2} \to j_*\mathcal{O}_{\mathbb{P}^1}(1)$, where $j \colon \mathbb{P}^1 \to \mathbb{P}^2$ is the inclusion of a line in $\mathbb{P}^2$. Show that the elementary modification

$$0 \to V \to \mathcal{O}_{\mathbb{P}^2} \oplus \mathcal{O}_{\mathbb{P}^2} \to j_*\mathcal{O}_{\mathbb{P}^1}(1) \to 0$$

defined by this map satisfies $c_1(V) = -[\ell]$, where ℓ is a line in $\mathbb{P}^2$, $c_2(V) = 1$, and $H^0(V) = 0$. Is V a direct sum of line bundles?

7. Let $0 \to W \to V \to j_*L \to 0$ be an elementary modification defined by the line bundle L on the divisor D, and let L' be the kernel of the surjection $V|D \to L$. Thus, there is an induced surjection $W \to L'$.

Show that if U is the induced elementary modification

$$0 \to U \to W \to L' \to 0,$$

then U is just $V \otimes \mathcal{O}_X(-D)$.

8. Let X be regular, let D be a divisor in X, and let V be a rank 2 vector bundle on X. Suppose that there is an exact sequence

$$0 \to H \to V \to H' \to 0,$$

where H and H' are line bundles, and let $L = H'|D$. For the elementary modification W corresponding to the surjection $V \to j_*L$, show that there is an exact sequence

$$0 \to H'(-D) \to W \to H \to 0.$$

9. Using the above exercise, show that for a smooth curve C, if V is an arbitrary rank 2 bundle, then there is a sequence of elementary modifications beginning with V and ending with a bundle of the form $(\mathcal{O}_C \oplus \mathcal{O}_C) \otimes L$, i.e., up to a twist by a line bundle, V is obtained via elementary modifications from the trivial bundle. (Starting with an arbitrary exact sequence $0 \to H \to V \to H' \to 0$ and applying the previous exercise, we can assume that $\deg H'$ is sufficiently negative and thus that the exact sequence splits. Now work with both factors and use the fact that, on a smooth curve of genus g, every divisor of degree at least g is effective.)

10. Let R be a regular local ring and $F: R^2 \to R^2$ a homomorphism of R-modules, corresponding to a 2×2 matrix. Suppose that the matrix is of the form $\begin{pmatrix} f & h_1 \\ g & h_2 \end{pmatrix}$ with $\gcd(f,g) = 1$. Show that $R^2/F(R^2) \cong I_Z/I_E$, where $I_Z = (f,g)R$ and $I_E = tR$, where $t = fh_2 - gh_1$ is the determinant of the matrix.

(Working a little harder, one can show that for every 2×2 matrix whose entries are relatively prime, there exists a choice of bases for the domain and range R^2's for which the first column has relatively prime entries. Note that if E is singular, for example nonreduced, then the modules I_Z/I_E can be very complicated.)

11. Let R be a regular local ring of dimension at least 3, and choose a regular sequence (t, f, g). Consider the R-module M defined by the short exact sequence

$$0 \to R \to R^3 \to M \to 0,$$

where the map $R \to R^3$ sends 1 to (t, f, g). Show that M is reflexive. (Show that there is an exact sequence

$$0 \to M^\vee \to R^3 \to I \to 0,$$

where $I = (t, f, g)$, and that $\operatorname{Ext}^1_R(I, R) = 0$.) What is $S(M)$ in this case?

12. Let M be a complex manifold of dimension at least 2 and let $Y \subset M$ be a closed analytic subspace of codimension at least 2. Suppose that L is a line bundle on $M - Y$. Is it true that L extends uniquely to a line bundle on M? (Compare [55, Example 4, p. 49] for the case $M = \mathbb{C}^2$, using the exponential sheaf sequence.) If L extends to a coherent sheaf on M, then L extends to a line bundle on M. Conclude that a reflexive rank one coherent sheaf on a complex manifold is locally free. (Locally around a point of Y, show that L has a holomorphic section in $M - Y$ and is thus associated to a divisor, which extends across Y.)

13. Let $f \colon \mathbb{P}^1 \to \mathbb{P}^1$ be the degree 2 map given by taking the double cover of $\mathbb{P}^1$ branched at two points. Show that, for $d = 2k$ even, $f_*\mathcal{O}_{\mathbb{P}^1}(d) = \mathcal{O}_{\mathbb{P}^1}(k) \oplus \mathcal{O}_{\mathbb{P}^1}(k-1)$, whereas for $d = 2k+1$ odd, $f_*\mathcal{O}_{\mathbb{P}^1}(d) = \mathcal{O}_{\mathbb{P}^1}(k) \oplus \mathcal{O}_{\mathbb{P}^1}(k)$. (It suffices by the projection formula to do the cases $d = 0, 1$. The case $d = 0$ is done. For $d = 1$, write $f_*\mathcal{O}_{\mathbb{P}^1}(1) = \mathcal{O}_{\mathbb{P}^1}(a) \oplus \mathcal{O}_{\mathbb{P}^1}(b)$ and calculate $H^0(\mathcal{O}_{\mathbb{P}^1}(1)) = H^0(f_*\mathcal{O}_{\mathbb{P}^1}(1))$, and $H^0(f_*\mathcal{O}_{\mathbb{P}^1}(1) \otimes \mathcal{O}_{\mathbb{P}^1}(-1)) = H^0(\mathcal{O}_{\mathbb{P}^1}(1) \otimes f_*\mathcal{O}_{\mathbb{P}^1}(-1))$.)

14. Let $Q \cong \mathbb{P}^1 \times \mathbb{P}^1$ be a smooth quadric in $\mathbb{P}^3$ and let $\pi \colon Q \to \mathbb{P}^2$ be the double cover obtained by projecting Q onto a plane. Let $V_{a,b}$ be the rank 2 bundle $\pi_*\mathcal{O}_Q(af_1 + bf_2)$ on $\mathbb{P}^2$, where f_1 and f_2 are the two rulings on Q, as in the example at the end of the last section. Show that, if $a = b$, then $V_{a,a} = \pi_*(\mathcal{O}_Q \otimes \pi^*\mathcal{O}_{\mathbb{P}^2}(a)) = \mathcal{O}_{\mathbb{P}^2}(a) \oplus \mathcal{O}_{\mathbb{P}^2}(a-1)$. If $|a - b| = 1$, say $b = a + 1$, then show that $V_{a,a+1} = \mathcal{O}_{\mathbb{P}^2}(a) \oplus \mathcal{O}_{\mathbb{P}^2}(a)$. In all other cases, show that $V_{a,b}$ is not a direct sum of line bundles.

15. Let X be a regular scheme. Given a rank 2 bundle W on X and a line bundle L on the divisor $D \subset X$, show that

$$\mathrm{Ext}^1(j_*L, W) \cong H^0(D; (W \otimes \mathcal{O}_X(D))|D \otimes L^{-1}).$$

16. Let X be a regular scheme and Z a codimension 2 local complete intersection subscheme of X. Verify that there is a canonical isomorphism $\mathcal{E}xt^1(I_Z, \mathcal{O}_X) \cong \det N_{Z/X}$, where $N_{Z/X}$ is the dual of the locally free rank 2 sheaf I_Z/I_Z^2.

17. Prove the Riemann-Roch theorem for vector bundles V of rank r on a smooth curve or surface (Theorem 2). (For a curve C, show first that there is some exact sequence

$$0 \to L \to V' \to 0,$$

where L is a line bundle and V' is a vector bundle of rank $r - 1$, and apply induction. For surfaces S, show that there is an exact sequence

$$0 \to L \to V' \to 0,$$

where L is a line bundle and V' is a torsion free sheaf of rank $r - 1$. Thus, there is another exact sequence

$$0 \to V' \to W \to Q \to 0,$$

where W is a vector bundle and Q has finite support. Apply induction and use (2.8).)

18. Let $f\colon X \to Y$ be the double cover constructed by choosing a square root L of the line bundle $\mathcal{O}_Y(B)$, where Y and B and therefore X are smooth. Show that $h^i(X; \mathcal{O}_X) = h^i(Y; \mathcal{O}_Y) \oplus h^i(Y; L^{-1})$ for all i. Also, by using the natural inclusion $f^*K_Y \to K_X$ and local coordinates, show that $K_X = f^*K_Y \otimes \mathcal{O}_X(B) = f^*(K_Y \otimes L)$, viewing B as a smooth divisor on X. Conclude that $f_*K_X = K_Y \oplus (K_Y \otimes L)$ and thus that $p_g(X) = p_g(Y) + h^0(Y; K_Y \otimes L)$. Finally, if X and Y are surfaces, then $K_X^2 = 2K_Y^2 + 4(K_Y \cdot L) + 2L^2$. For example, if $Y = \mathbb{P}^2$ and B is a smooth curve of degree $2d$, show that $p_g(X) = 0$ if and only if $d = 1, 2$, that otherwise $p_g(Y) = \binom{d-1}{2}$, and that $K_Y^2 = 2(d-3)^2$.

3

Birational Geometry

In this chapter, we describe the basic properties of surface birational geometry. After reviewing the operation of blowing up a point on a surface X, and the relationship between the invariants of the blown up surface $\tilde{X}$ and the invariants of X, we prove the Castelnuovo criterion for blowing down a curve. Using the Castelnuovo criterion, we show that every birational morphism between two smooth surfaces is a composition of blowups, and discuss various notions of minimal models. At the end of the chapter, we discuss more general contractions to normal surfaces, with particular attention to rational singularities and rational double points.

Blowing up

The operation of blowing up goes by many names in the literature: σ-process, monoidal transformation, standard quadratic transformation, ... We begin with a review of standard facts about blowing up. Let X be a surface and let p be a point of X. Let $\rho: \tilde{X} \to X$ be the blowup of X at the point p. Let $E = \rho^{-1}(p)$ be the exceptional divisor. Thus, $E \cong \mathbb{P}^1$ and $\deg N_{E/\tilde{X}} = -1$, so that $E^2 = -1$ and $E \cdot K_X = -1$ by the adjunction formula. We may describe $\tilde{X}$ locally as follows: let U be a coordinate neighborhood of p, with coordinates x, y centered at p. Let $\tilde{U} = \rho^{-1}(U)$. Then $\tilde{U}$ is covered by two coordinate charts $\tilde{U}_1$ and $\tilde{U}_2$. Here $\tilde{U}_1$ has coordinates x', y' and the map ρ is given by $x = x'$, $y = x'y'$. The chart $\tilde{U}_2$ has coordinates x'', y'' and in these coordinates ρ is given by $x = x''y''$, $y = y''$. We glue $\tilde{U}_1 - \{y' = 0\}$ to $\tilde{U}_2 - \{x'' = 0\}$ via

$$x'' = \frac{1}{y'},$$

$$y'' = x'y'.$$

The exceptional divisor E is defined in $\tilde{U}_1$ by x' and in $\tilde{U}_2$ by y''. From this it is easy to check that $E \cong \mathbb{P}^1$ and that $N_{E/\tilde{X}}$ is identified with $\mathcal{O}_{\mathbb{P}^1}(-1)$.

Thus, $\deg N_{E/\tilde{X}} = -1$. Also recall the universal property of $\tilde{X}$ [61, p. 164]: if $\varphi: Y \to X$ is a morphism such that $\varphi^{-1}(\mathfrak{m}_x)\mathcal{O}_Y$ is the ideal of a Cartier divisor on Y, then φ factors through $\tilde{X}$: there exists a unique morphism $\tilde{\varphi}: Y \to \tilde{X}$ such that $\varphi = \rho \circ \tilde{\varphi}$.

We will frequently use the following lemma:

Lemma 1. *Let $\mathfrak{m}_p$ be the maximal ideal of p in X. Then*

$$\rho_*\mathcal{O}_{\tilde{X}}(aE) = \begin{cases} \mathfrak{m}_p^n, & \text{if } a = -n < 0, \\ \mathcal{O}_X, & \text{if } a \geq 0. \end{cases}$$

Proof. For a neighborhood U of p, let $f \in \Gamma(\rho^{-1}(U), \mathcal{O}_{\tilde{X}}(aE))$. Thus, f defines a section of $\mathcal{O}_{\tilde{X}}$ over $\rho^{-1}(U) - E = U - \{p\}$ or in other words a holomorphic function on $U - \{p\}$. By Hartogs' theorem f extends to a holomorphic function on U. It follows that there is a natural inclusion $j: \rho_*\mathcal{O}_{\tilde{X}}(aE) \hookrightarrow \mathcal{O}_X$ for all a. If $a \geq 0$, we can reverse this inclusion too: if $g \in \Gamma(U, \mathcal{O}_X)$, then ρ^*g is a section of $\mathcal{O}_{\tilde{X}}$ over $\rho^{-1}(U)$ and thus a section of $\mathcal{O}_{\tilde{X}}(aE)$ over $\rho^{-1}(U)$, i.e., a section of $\rho_*\mathcal{O}_{\tilde{X}}(aE)$ over U. Clearly, $j(\rho^*g) = g$ and so $\rho_*\mathcal{O}_{\tilde{X}}(aE) = \mathcal{O}_X$.

For $a = -n < 0$, we must determine the image of j. Equivalently, given a function g holomorphic in U, it suffices to determine when $\rho^*g \in \mathcal{O}_{\tilde{X}}(aE)$. Now we can write $g(x, y) = \sum_{\nu=m}^{\infty} g_\nu(x, y)$, where g_ν is homogeneous of degree ν and $g_m \neq 0$. Thus, m is the multiplicity $\text{mult}_p\, g$ and $g \in \mathfrak{m}_p^m$; indeed $g \in \mathfrak{m}_p^m - \mathfrak{m}_p^{m+1}$. Working, for example, in $\tilde{U}_1$ and using the coordinates described above, we have

$$\rho^*g(x', y') = g(x', x'y') = (x')^m(g_m(1, y') + x'g'),$$

where $x'g'$ vanishes along E and $g_m(1, y')$ does not vanish identically on E. Thus, we see that

$$\rho^*g \in \Gamma(\rho^{-1}(U), \mathcal{O}_{\tilde{X}}(-mE)) - \Gamma(\rho^{-1}(U), \mathcal{O}_{\tilde{X}}(-(m+1)E)).$$

It follows that $\rho^*g \in \Gamma(\rho^{-1}(U), \mathcal{O}_{\tilde{X}}(-mE))$ if and only if $g \in \mathfrak{m}_p^m$, which was what we needed to complete the proof of Lemma 1. $\square$

Definition 2. Let C be a nonzero effective divisor on X. We define the *multiplicity* $\text{mult}_p\, C$ to be $\text{mult}_p\, g$, where g is any local defining equation for C at p. It is easy to check that this definition is independent of the choice of g. In this case, the proof of Lemma 1 shows that we can write (as effective divisors) $\rho^*C = mE + C'$, where C' is an effective nonzero divisor on $\tilde{X}$ and C' does not contain E as a component ($C' - E$ is not effective). We call C' the *proper transform* of C.

We leave as an exercise the statement that $C' \cdot E = m$ and that the scheme-theoretic intersection $C' \cap E$ can be identified with the projective tangent cone to C at p.

We turn now to the structure of $\operatorname{Pic}\tilde{X}$ and $\operatorname{Num}\tilde{X}$.

Proposition 3.

(i) $\rho^*\colon \operatorname{Pic}X \to \operatorname{Pic}\tilde{X}$ and $\rho^*\colon \operatorname{Num}X \to \operatorname{Num}\tilde{X}$ are injective.

(ii) $\operatorname{Pic}\tilde{X} = \rho^*\operatorname{Pic}X \oplus \mathbb{Z}[E]$.

(iii) $\operatorname{Num}\tilde{X} = \rho^*\operatorname{Num}X \oplus \mathbb{Z}[E]$, and this direct sum is orthogonal with respect to intersection pairing on $\tilde{X}$ and X.

Proof. The proof of (i) in the case of $\rho^*\colon \operatorname{Pic}X \to \operatorname{Pic}\tilde{X}$ follows from the projection formula $\rho_*\rho^*L = L \otimes \rho_*\mathcal{O}_{\tilde{X}} = L$. The proof for $\rho^*\colon \operatorname{Num}X \to \operatorname{Num}\tilde{X}$ also follows from the projection formula and the definition of ρ_*.

To see (ii), let $\operatorname{Pic}'\tilde{X}$ be the set of $L \in \operatorname{Pic}\tilde{X}$ such that $\deg(L|E) = 0$. Thus, $\operatorname{Pic}'\tilde{X}$ is the kernel of the natural restriction map $\operatorname{Pic}\tilde{X} \to \operatorname{Pic}E \cong \mathbb{Z}$. Note that $\deg \mathcal{O}_{\tilde{X}}(aE)|E = -a$, so that, identifying $a[E] \in \mathbb{Z}[E]$ with $\mathcal{O}_{\tilde{X}}(aE)$, we have a splitting $\operatorname{Pic}\tilde{X} = \operatorname{Pic}'\tilde{X} \oplus \mathbb{Z}[E]$. To prove (ii) it therefore suffices to identify $\operatorname{Pic}'\tilde{X}$ with $\rho^*\operatorname{Pic}X$. Clearly, $\rho^*\operatorname{Pic}X$ is contained in $\operatorname{Pic}'\tilde{X}$. Conversely, suppose that $L \in \operatorname{Pic}'\tilde{X}$. Consider $L|\tilde{X} - E = X - \{p\}$. Since $\operatorname{Pic}(X - \{p\}) = \operatorname{Pic}X$. the line bundle L on $X - \{p\}$ extends uniquely to a line bundle on X (compare the proof of Lemma 26 in Chapter 2). Let M be the extension of $L|\tilde{X} - E$ to a line bundle on X; in fact $M = (\rho_*L)^{\vee\vee}$. Then $\rho^*M|\tilde{X} - E \cong L|\tilde{X} - E$. Thus, $\rho^*M \otimes L^{-1}$ has a regular nowhere vanishing section over $\tilde{X} - E = X - \{p\}$. Such a section extends to give a meromorphic section of $\rho^*M \otimes L^{-1}$ which can only have zeros or poles along E. Thus, $\rho^*M = L \otimes \mathcal{O}_{\tilde{X}}(aE)$ for some integer a. As both ρ^*M and L have trivial restriction to E, $a = 0$. This concludes the proof of (ii).

To see the direct sum part of (iii), it again suffices to show that every $D \in \operatorname{Num}\tilde{X}$ with $D \cdot E = 0$ is of the form ρ^*D'. Representing D as the numerical equivalence class of a line bundle L with $\deg L|E = 0$, this follows from (ii). The fact that the direct sum is orthogonal is an easy consequence of the projection formula, which implies that $\rho^*D \cdot \rho^*D' = D \cdot \rho_*\rho^*D' = D \cdot D'$ and that $\rho^*D \cdot E = D \cdot \rho_*E = 0$. $\square$

The proof also shows that the projection of $\operatorname{Num}\tilde{X}$ onto the factor $\rho^*\operatorname{Num}X \cong \operatorname{Num}X$ is given by ρ_*. Likewise, the projection $\operatorname{Pic}\tilde{X} \to \rho^*\operatorname{Pic}X \cong \operatorname{Pic}X$ is given by $L \mapsto (\rho_*L)^{\vee\vee}$. In general we see that $\rho_*(\rho^*M \otimes \mathcal{O}_{\tilde{X}}(aE))$ is M if $a \geq 0$ and is $M \otimes \mathfrak{m}_p^n$ if $a = -n < 0$.

Here are some easy consequences of Proposition 3 (which could also be checked directly, cf. Exercise 1).

Proposition 4.

(i) If C is a curve on X and $\operatorname{mult}_p C = m$, then the proper transform C' of C satisfies $C' \cdot E = m$.

(ii) With notation as in (i), $(C')^2 = C^2 - m^2$.

(iii) $K_{\tilde{X}} = \rho^* K_X + E$.

Proof. Write $\rho^* C = C' + mE$, where C' is the proper transform of C. Since E and $\rho^* C$ are orthogonal,

$$C' \cdot E = (\rho^* C - mE) \cdot E = -mE^2 = m.$$

Next we note that $(C')^2 = C^2 - m^2$. Finally, we can write $K_{\tilde{X}} = \rho^* K_X + aE$ for some integer a. By adjunction $K_{\tilde{X}} \cdot E + E^2 = -2$, and thus $a = 1$. □

Corollary 5. $c_1^2(\tilde{X}) = c_1^2(X) - 1$ and $p_g(\tilde{X}) = p_g(X)$. More generally $P_n(\tilde{X}) = P_n(X)$ for all $n \geq 1$.

Proof. The equality for $c_1^2(\tilde{X})$ follows immediately from (iii) above. To see the other statement, note that

$$H^0(\tilde{X}; K_{\tilde{X}}^{\otimes n}) = H^0(\tilde{X}; \rho^* K_X^{\otimes n} \otimes \mathcal{O}_{\tilde{X}}(nE))$$

$$= H^0(X; \rho_*[\rho^* K_X^{\otimes n} \otimes \mathcal{O}_{\tilde{X}}(nE)]).$$

Now by the projection formula, if $n > 0$,

$$\rho_*[\rho^* K_X^{\otimes n} \otimes \mathcal{O}_{\tilde{X}}(nE)] = K_X^{\otimes n} \otimes \rho_* \mathcal{O}_{\tilde{X}}(nE) = K_X^{\otimes n} \otimes \mathcal{O}_X = K_X^{\otimes n},$$

by Lemma 1, since $n \geq 1$. Thus, $H^0(\tilde{X}; K_{\tilde{X}}^{\otimes n}) \cong H^0(X; K_X^{\otimes n})$, and so $P_n(\tilde{X}) = P_n(X)$. □

Thus, the invariants P_n are unchanged after blowing up. Such a statement would fail if we had considered the equally "natural" bundle $K_X^{-1} = \det T_X$, for example.

Finally, we note that $q(X)$ is also unchanged under blowing up:

Proposition 6. $q(\tilde{X}) = q(X)$.

Proof. One proof uses the topological fact that $H^1(\tilde{X}; \mathbb{Z}) \cong H^1(X; \mathbb{Z})$; indeed an easy Mayer-Vietoris argument shows that $\pi_1(\tilde{X}; *) \cong \pi_1(X; *)$ via $\rho_\#$. A second proof uses the fact that $\mathrm{Pic}^0 \tilde{X} = \mathrm{Pic}^0 X$, where $\mathrm{Pic}^0 X$ is the component containing the identity of the complex Lie group $\mathrm{Pic}\, X$, by Proposition 3, together with the fact that $q(X) = \dim \mathrm{Pic}^0 X$. A third proof uses the Leray spectral sequence for the map ρ_* (or the easy special case that is in [61]) as follows. We shall show that $R^1 \rho_* \mathcal{O}_{\tilde{X}} = 0$. Then by the Leray spectral sequence

$$H^1(\tilde{X}; \mathcal{O}_{\tilde{X}}) = H^1(X; \rho_* \mathcal{O}_{\tilde{X}}) = H^1(X; \mathcal{O}_X) = q(X).$$

To see that $R^1 \rho_* \mathcal{O}_{\tilde{X}} = 0$, we apply the formal functions theorem [61, p. 277] to ρ and $\mathcal{O}_{\tilde{X}}$:

$$R^1 \rho_* \mathcal{O}_{\tilde{X}} = \varprojlim_n H^1(nE; \mathcal{O}_{nE}).$$

To evaluate $H^1(nE; \mathcal{O}_{nE})$, use the exact sequence

$$0 \to \mathcal{O}_E(-(n-1)E) \to \mathcal{O}_{nE} \to \mathcal{O}_{(n-1)E} \to 0$$

of (1.10) of Chapter 1. Since $\mathcal{O}_E(-(n-1)E) \cong \mathcal{O}_{\mathbb{P}^1}(n-1)$, it follows that $H^1(\mathcal{O}_E(-(n-1)E)) = 0$ for all $n \geq 0$. Thus, by induction, starting with the case $n = 1$ where we know that $H^1(E; \mathcal{O}_E) = g(E) = 0$, we see that $H^1(\mathcal{O}_{nE}) = 0$ for every $n \geq 1$. This concludes the proof that $R^1 \rho_* \mathcal{O}_{\tilde{X}} = 0$ and thus of Proposition 6. $\square$

To conclude this section, we will say more about the arithmetic genus of an irreducible curve. Let C be a (reduced) irreducible curve on X and suppose that $p \in C$. Let m be the multiplicity of C at p, let $\tilde{X}$ be the blowup of X at p, and let C' be the proper transform of C. We shall compare $p_a(C)$ with $p_a(C')$.

Proposition 7. $p_a(C') = p_a(C) - \dfrac{m(m-1)}{2}$.

Proof. By our general formulas

$$2p_a(C') - 2 = K_{\tilde{X}} \cdot C' + (C')^2$$
$$= (\rho^* K_X + E) \cdot (\rho^* C - mE) + C^2 - m^2$$
$$= K_X \cdot C + C^2 + m - m^2,$$

and this is a restatement of Proposition 7. $\square$

Note that $p_a(C') < p_a(C)$ unless $m = 1$, i.e., p is a smooth point of C. Since $p_a(C) \geq 0$, we cannot continue blowing up singular points indefinitely: if we successively blow up the singular points of C, then all of the singular points on the proper transform, and so on, we eventually arrive at a surface Y and a proper transform of C which is smooth. This is *embedded resolution* for curves on a surface. Keeping track of the total change in $p_a(C)$, we arrive at the classical formula for δ_p:

$$\delta_p = \sum_{q \to p} \frac{m_q(m_q - 1)}{2},$$

where the q are the "infinitely near" points to p and are defined as follows: they include p, all the points on C' mapping to p via ρ, all the points on blowups of $\tilde{X}$ lying on C' and mapping to p via the composite map, and so on. If C consists (locally analytically) of the union of m distinct lines meeting at a point p (the case of Exercise 6 of Chapter 1), then $\delta_p = m(m-1)/2$.

The Castelnuovo criterion and factorization of birational morphisms

In this section we review the main facts about birational morphisms. We begin by recalling the following theorem, due to Van der Waerden, which is the "easy" special case of Zariski's Main Theorem.

Theorem 8. *Let $\pi\colon Y \to X$ be a birational morphism between two smooth varieties. Let $y \in Y$ and let $x = \pi(y)$. Then either there is a Zariski open subset U of X containing x and a morphism $U \to Y$ which is an inverse to π, or there exists a hypersurface V on Y containing y such that the Zariski closure of $\pi(V)$ has codimension at least 2 on X.*

Proof. There is an inclusion of local rings $\mathcal{O}_{X,x} \subseteq \mathcal{O}_{Y,y}$ and the two rings have the same field of fractions. An easy argument shows that $\mathcal{O}_{X,x} = \mathcal{O}_{Y,y}$ if and only if π is invertible on some Zariski open subset of X containing x. Writing $\mathcal{O}_{Y,y}$ as the localization of $\mathbb{C}[t_1,\ldots,t_n]/I$ for some ideal I, we can write $t_i = f_i/g_i$, where $f_i, g_i \in \mathcal{O}_{X,x}$. By standard commutative algebra [87], $\mathcal{O}_{X,x}$ is a UFD. Thus, we may assume that f_i and g_i are relatively prime in $\mathcal{O}_{X,x}$. If g_i is a unit for all i (i.e., $g_i(y) \neq 0$), then $\mathcal{O}_{X,x} = \mathcal{O}_{Y,y}$ and the first conclusion of the theorem holds. Otherwise, g_i is not a unit for some i, so that $\{\pi^* g_i = 0\}$ defines a hypersurface V on Y containing y. Moreover, in $\mathcal{O}_{Y,y}$ $\pi^* f_i = \pi^* g_i \cdot t_i$, and so both $\pi^* f_i$ and $\pi^* g_i$ vanish on V. Thus, f_i and g_i vanish on $\pi(V)$, and so the closure of $\pi(V)$ is contained in $\{f_i = g_i = 0\}$. Since f_i and g_i were assumed relatively prime, $\{f_i = g_i = 0\}$ has codimension 2 in X, and so the second alternative of Theorem 8 holds. $\square$

We can now prove the following fundamental result, known as the Castelnuovo criterion:

Theorem 9. *Let Y be a smooth surface, and let E be a curve on Y such that $E \cong \mathbb{P}^1$ and $E^2 = -1$. Then there exists a smooth surface X, a point $p \in X$, and an isomorphism from the blowup $\tilde{X}$ of X to Y such that E is the image of the exceptional curve on $\tilde{X}$.*

Proof. The are two steps to the proof. In the first and easier step, we construct the surface X. The second step shows that X is smooth and identifies Y as the blowup of X.

Step I. To find X, choose a very ample divisor H on Y. We may also assume after replacing H by a multiple that $H^1(Y; \mathcal{O}_Y(H)) = 0$. Let $a = H \cdot E > 0$. Consider the linear system corresponding to the divisor $H + aE$. Note that $(H + aE) \cdot E = 0$. We claim that the linear system $|H + aE|$ has no base points and that the image of the corresponding morphism $\varphi\colon Y \to \mathbb{P}^N$ is a surface X such that $\varphi(E)$ is a single point $p \in X$ and $\varphi|Y - E$ is an isomorphism from $Y - E$ to $X - \{p\}$. To see this, note that since $a > 0$, $|H + aE|$ contains the subseries $|H|$ which, since H is very ample, separates points and tangent directions on $Y - E$. Next we claim that $|H + aE|$ has no base points along E. Since $\mathcal{O}_Y(H + aE)|E = \mathcal{O}_E$, it will suffice to show that the map $H^0(\mathcal{O}_Y(H + aE)) \to H^0(\mathcal{O}_Y(H + aE)|E) = H^0(\mathcal{O}_E)$ is surjective. The cokernel of this map is $H^1(\mathcal{O}_Y(H + (a-1)E))$.

Claim. For $0 \le k \le a + 1$, $H^1(\mathcal{O}_Y(H + kE)) = 0$.

Proof of the Claim. Consider the exact sequence

$$0 \to \mathcal{O}_Y(H + (k-1)E) \to \mathcal{O}_Y(H + kE) \to \mathcal{O}_E(H + kE) \to 0.$$

By assumption $H^1(\mathcal{O}_Y(H)) = 0$. Moreover, $H^1(E; \mathcal{O}_E(H + kE)) = H^1(\mathcal{O}_{\mathbb{P}^1}(a - k))$ and this group vanishes for $k \le a + 1$. Thus, by induction on k we see that $H^1(\mathcal{O}_Y(H + kE)) = 0$ for $0 \le k \le a + 1$. $\square$

Thus, $|H + aE|$ has no base locus and so defines a morphism $\varphi \colon Y \to \mathbb{P}^N$ for some N. Since $(H + aE) \cdot E = 0$, $\varphi(E)$ is a point p. Moreover, given a point $q \in Y - E$, since $|H|$ has no base locus, there exist curves in $|H + aE|$ vanishing along E but not along q, by using the subseries $|H|$. Thus, φ separates $\varphi(E) = p$ from the points of $Y - E$ as claimed. The morphism defined by φ maps Y onto a projective surface X_0. Taking X to be the normalization of X_0 gives the candidate for the blowdown of Y. This concludes the proof of Step I.

Step II. Let $p = \varphi(E)$ as above. We must identify Y with the blowup $\rho \colon \tilde{X} \to X$ of X at p. First we claim that X is smooth at p. It suffices to show that the completion $\hat{\mathcal{O}}_{X,p}$ of the local ring of X at p is a formal power series ring. Since X is normal and φ is birational, $\mathcal{O}_X = \varphi_*\mathcal{O}_Y$. By the formal functions theorem,

$$\hat{\mathcal{O}}_{X,p} = \varprojlim H^0(Y; \mathcal{O}_Y/\mathfrak{m}_p^n\mathcal{O}_Y).$$

But the sheaf $\mathcal{O}_Y/\mathfrak{m}_p^n\mathcal{O}_Y$ is supported on E and is thus annihilated by some power of $I_E = \mathcal{O}_Y(-E)$. In particular there is a surjective map $\mathcal{O}_Y/\mathcal{O}_Y(-NE) \to \mathcal{O}_Y/\mathfrak{m}_p^n\mathcal{O}_Y$ and it will in fact suffice to show that $\varprojlim H^0(Y; \mathcal{O}_Y/\mathcal{O}_Y(-nE))$ is a formal power series ring. Consider the exact sequence

$$0 \to \mathcal{O}_Y(-nE)/\mathcal{O}_Y(-(n+1)E) \to \mathcal{O}_Y/\mathcal{O}_Y(-(n+1)E) \to$$
$$\to \mathcal{O}_Y/\mathcal{O}_Y(-nE) \to 0.$$

Note that $\mathcal{O}_Y(-nE)/\mathcal{O}_Y(-(n+1)E)$ is identified with $\mathcal{O}_{\mathbb{P}^1}(n)$ and thus

$$H^1(\mathcal{O}_Y(-nE)/\mathcal{O}_Y(-(n+1)E)) = 0.$$

It follows that the map

$$H^0(\mathcal{O}_Y/\mathcal{O}_Y(-(n+1)E)) \to H^0(\mathcal{O}_Y/\mathcal{O}_Y(-nE))$$

is surjective for every $n \geq 0$. For $n = 1$, we have the inclusion

$$H^0(\mathcal{O}_Y(-E)/\mathcal{O}_Y(-2E)) \subset H^0(\mathcal{O}_Y/\mathcal{O}_Y(-2E)).$$

As $\mathcal{O}_Y(-E)/\mathcal{O}_Y(-2E) \cong \mathcal{O}_{\mathbb{P}^1}(1)$, $\dim H^0(\mathcal{O}_Y(-E)/\mathcal{O}_Y(-2E)) = 2$, and we can choose $z_1^{(1)}, z_2^{(1)}$ a basis for $H^0(\mathcal{O}_Y(-E)/\mathcal{O}_Y(-2E))$. For all $n \geq 1$, we can choose $z_1^{(n)}, z_2^{(n)} \in H^0(\mathcal{O}_Y(-E)/\mathcal{O}_Y(-(n+1)E))$ mapping onto $z_1^{(n-1)}, z_2^{(n-1)}$. The natural map

$$\mathrm{Sym}^n H^0(\mathcal{O}_Y(-E)/\mathcal{O}_Y(-2E)) \to H^0(\mathcal{O}_Y(-nE)/\mathcal{O}_Y(-(n+1)E))$$

can be identified with the isomorphism $\mathrm{Sym}^n H^0(\mathcal{O}_{\mathbb{P}^1}(1)) \to H^0(\mathcal{O}_{\mathbb{P}^1}(n))$. From this, an easy induction shows that the $\mathbb{C}$-algebra map

$$\mathbb{C}[z_1^{(n)}, z_2^{(n)}] \to H^0(\mathcal{O}_Y/\mathcal{O}_Y(-(n+1)E))$$

is surjective for every n. Taking $z_i = \varprojlim z_i^{(n)}$, it also follows that the induced map

$$\mathbb{C}[[z_1, z_2]] \to \varprojlim H^0(Y; \mathcal{O}_Y/\mathcal{O}_Y(-nE))$$

is surjective and hence an isomorphism (since $\mathcal{O}_X$ must have Krull dimension 2). Thus, X is smooth at p.

Let $\mathfrak{m}_p$ be the maximal ideal of p. Clearly, $(\varphi^{-1}\mathfrak{m}_p)\mathcal{O}_Y \subseteq \mathcal{O}_Y(-E) = I_E$. Moreover, the images of z_1 and z_2 generate I_E mod I_E^2, since $\mathcal{O}_{\mathbb{P}^1}(1)$ is generated by its global sections, and thus z_1 and z_2 generate I_E. It follows that $(\varphi^{-1}\mathfrak{m}_p)\mathcal{O}_Y = I_E$. Thus, by the universal property of blowing up the morphism φ factors through the blowup $\rho: \tilde{X} \to X$: there is a morphism $\tilde{\varphi}: Y \to \tilde{X}$ such that $\varphi = \rho \circ \tilde{\varphi}$. Clearly, $\tilde{\varphi}$ is birational. Let $\tilde{E}$ be the exceptional curve on $\tilde{X}$. Since Y is projective, $\tilde{\varphi}$ is surjective, and so we must have $\tilde{\varphi}(E) = \tilde{E}$. In particular, $\tilde{\varphi}(E)$ is a curve on $\tilde{X}$. But by Theorem 8 above, if $\tilde{\varphi}$ is not an isomorphism, then there must exist a curve C on Y such that $\tilde{\varphi}(C)$ is a point. As $\tilde{\varphi}$ is an isomorphism on $Y - E$, the only possibility for such a curve C is $C = E$, but we have seen that $\tilde{\varphi}(E)$ is again a curve. Thus, $\tilde{\varphi}$ is an isomorphism, identifying Y with the blowup of X at p. $\square$

Definition 10. A curve E on a smooth surface Y such that $E \cong \mathbb{P}^1$ and $E^2 = -1$ is called an *exceptional curve*. The smooth surface X obtained from Y via Castelnuovo's criterion is called the *contraction of Y along E*, or the *contraction of E*. We also say that X is obtained from Y by *blowing down E*.

Here are two other characterizations of exceptional curves:

Lemma 11. *E is an exceptional curve if and only if $E^2 = E \cdot K_Y = -1$ if and only if $E^2 < 0$ and $E \cdot K_Y < 0$.*

The proof is left as an exercise.

Next we turn to the factorization of birational morphisms.

Theorem 12. *Let $\pi \colon Y \to X$ be a birational morphism. Then π is a composition of blowups.*

Proof. Given $x \in X$, suppose that π^{-1} is not defined at x. Then by Theorem 8, $\pi^{-1}(x) = C = \bigcup_i C_i$ is a curve on Y, and by Zariski's connectedness theorem [61, p. 279], C is connected. We will find a component C_i of C such that $C_i \cdot K_Y < 0$ and $C_i^2 < 0$; by Lemma 11, C_i is therefore exceptional and we can contract it by Castelnuovo's criterion. The result is a new surface $\bar{Y}$. There is an induced continuous map $\bar{\pi} \colon \bar{Y} \to X$. It is easy to see that $\bar{\pi}$ is a morphism: supposing that U is an affine neighborhood of x, contained in $\mathbb{A}^n$ for some n, the coordinate functions on $\mathbb{A}^n$ define functions on $\pi^{-1}(U)$ and thus functions on $\pi^{-1}(U) - C_i \cong \bar{\pi}^{-1}(U) - \{p\}$, where $p = \bar{\pi}(C_i)$. These then extend to regular functions on $\bar{\pi}^{-1}(U)$ by Hartogs' theorem, and so $\bar{\pi}$ is a morphism. We may reapply the argument to the morphism $\bar{Y} \to X$, noting that since C has only finitely many components this procedure will stop.

To find C_i, we shall first prove the following two claims:

Claim 1. *There exist positive integers r_j with $K_Y = \pi^* K_X + \sum_j r_j C_j + D$, where D is an effective curve disjoint from C.*

Claim 2. *There exists an i such that $\sum_j r_j (C_i \cdot C_j) < 0$ and $C_i^2 < 0$.*

Proof that the claims imply Theorem 12. We have $C_i^2 < 0$ and $C_i \cdot K_Y = \sum_j r_j (C_i \cdot C_j) < 0$. Thus, C_i is exceptional.

Proof of Claim 1. Choose local coordinates z_1, z_2 at $x \in X$. Given j, choose a smooth point $y \in C_j$ and let w_1, w_2 be local coordinates for Y centered at y. Then $z_i = z_i(w_1, w_2)$ and given the local generating section $\omega = dz_1 \wedge dz_2$ of K_X, we have

$$\pi^* \omega = \frac{\partial(z_1, z_2)}{\partial(w_1, w_2)} dw_1 \wedge dw_2.$$

Since C_j is mapped to a point, the coefficient of $dw_1 \wedge dw_2$ vanishes along C_j. Thus, the natural map $\pi^* K_X \to K_Y$ defined by pullback vanishes to positive order along the C_j (and perhaps elsewhere, but not otherwise in a neighborhood of C), and this is the statement of Claim 1.

Proof of Claim 2. Choose a very ample divisor H on X. Thus, $H^2 > 0$ and $\pi^* H \cdot C_j = 0$ for all j. There exists a curve in $|H|$ passing through x, so that $\pi^*H = H' + \sum_j s_j C_j$ with $s_j > 0$ for all j, where H' is a curve in Y which does not contain any of the C_j in its support. Moreover, by Zariski's connectedness theorem π^*H is connected, and so $H' \cdot C_i \neq 0$ for some i. For every j,

$$0 = \pi^*H \cdot C_j = H' \cdot C_j + \sum_{i \neq j} s_i(C_i \cdot C_j) + s_j(C_j)^2.$$

Moreover, $H' \cdot C_j \geq 0$ and $(C_i \cdot C_j) \geq 0$ if $i \neq j$, and by the connectedness theorem at least one of these is > 0. Thus, $s_j C_j^2 < 0$, and therefore we must have $C_j^2 < 0$ for all j.

Suppose now that $\sum_j r_j(C_i \cdot C_j) \geq 0$ for all i. By the Hodge index theorem, since $\sum_j r_j C_j$ is orthogonal to π^*H, which has positive square, $(\sum_j r_j C_j)^2 \leq 0$, and $(\sum_j r_j C_j)^2 = 0$ if and only if $\sum_j r_j C_j$ is numerically trivial. But

$$0 \geq \left(\sum_j r_j C_j\right)^2 = \sum_i r_i \sum_j r_j(C_i \cdot C_j) \geq 0$$

so that $(\sum_j r_j C_j)^2 = 0$. It follows that $\sum_j r_j C_j$ is numerically trivial. But $H' \cdot C_i \geq 0$ for all i and $H' \cdot C_i > 0$ some i. Since the r_i are positive, this implies that $H' \cdot (\sum_j r_j C_j) > 0$, and so $\sum_j r_j C_j$ is not numerically trivial, a contradiction. Thus, there exists an i such that $\sum_j r_j(C_i \cdot C_j) < 0$, as claimed. □

We shall analyze the above arguments for more general contractions at the end of the chapter.

Next, we have the following result on the elimination of indeterminacy of rational maps:

Theorem 13. *Let $f: X \dashrightarrow Y$ be a rational map from the smooth surface X to a projective variety Y. Then there exists a sequence of blowups $X_n \to \cdots \to X_0 = X$ and a morphism $\tilde{f}: X_n \to Y$ such that $\tilde{f}$ and f agree on a Zariski open subset of X_n.*

Proof. Clearly, we may assume that $Y = \mathbb{P}^N$ and that $f(X)$ is nondegenerate, so that f corresponds to a linear system $\mathcal{L}$ on X without fixed curves. We may further assume that $N \geq 1$ and that $\mathcal{L}$ actually has base points. If p is a base point for $\mathcal{L}$, let $\rho: X_1 \to X$ be the blowup of X at p and let E be the exceptional curve. If $D \in \mathcal{L}$ and D' is the proper transform of D, then $\rho^*D = D' + kE$. Let k_0 be the minimum possible value for k as D ranges over the elements of $\mathcal{L}$. Thus, $\rho^*D - k_0 E$ is effective for all $D \in \mathcal{L}$, and there exists a $D_0 \in \mathcal{L}$ such that $\rho^*D_0 - k_0 E = D_1$ is the proper transform of D_0. Since p is a base point, $k_0 \geq 1$. Thus, $(\rho^*D_0 - k_0 E)^2 = D_0^2 - k_0^2 < D_0^2$. Set $\mathcal{L}' = \{\rho^*D - k_0 E : D \in \mathcal{L}\}$. Thus, $\mathcal{L}'$ consists of effective divisors and the base points of $\mathcal{L}'$ are the base points of $\mathcal{L}$ other than p together with some possible base points along E. The only possible fixed curve of $\mathcal{L}'$ would be

E, but, by the choice of k_0, E is not a fixed curve of $\mathcal{L}'$. Hence $\mathcal{L}'$ has no fixed curves. It follows that $(D_1)^2 \geq 0$ for $D_1 \in \mathcal{L}'$. Clearly, $\mathcal{L} \cong \mathcal{L}'$, and the rational map from X_1 to $\mathbb{P}^N$ defined by $\mathcal{L}'$ agrees with f away from E and the points of X_1 corresponding to points of indeterminacy of f other than p. If $\mathcal{L}'$ has no base points, we are done. Otherwise, continue this procedure. If D_k denotes a typical element of the linear system at stage k, then $0 \leq D_k^2 < D_{k-1}^2$. Thus, this procedure cannot continue indefinitely, and eventually we reach a base point free linear series. $\square$

Corollary 14. *Let $f\colon X \dashrightarrow Y$ be a birational map between two smooth surfaces. Then there exists a smooth surface Z and morphisms $\pi_1\colon Z \to X$, $\pi_2\colon Z \to Y$, such that π_1 and π_2 are sequences of blowups and $f \circ \pi_1 = \pi_2$ in the sense of rational maps (i.e., where defined).*

Proof. Blow up X until f becomes a morphism, and let the resulting surface and morphism to X be denoted $\pi_1\colon Z \to X$. Then the induced morphism $\pi_2\colon Z \to Y$ is a birational morphism, and thus by Theorem 12 it is a sequence of blowups. $\square$

Example. Consider the rational map $f\colon \mathbb{P}^2 \dashrightarrow \mathbb{P}^2$ defined by

$$(z_0, z_1, z_2) \mapsto (1/z_0, 1/z_1, 1/z_2)$$

in homogeneous coordinates. This map (called a *Cremona transformation*) is defined as long as $(z_0, z_1, z_2) \notin \{(1,0,0), (0,1,0), (0,0,1)\}$. We leave it as an exercise to show that f becomes a morphism on the blowup of $\mathbb{P}^2$ at these three points and that the birational morphism from the blowup to the target $\mathbb{P}^2$ consists exactly in contracting the proper transforms of the three lines joining pairs of points in $\{(1,0,0), (0,1,0), (0,0,1)\}$.

A base point p of a linear series $\mathcal{L}$ is called a *simple base point* if the following holds: if $\rho\colon \tilde{X} \to X$ is the blowup of X at p, the linear series

$$\mathcal{L}' = \{\rho^* D - E : D \in \mathcal{L}\}$$

has no base points along E. It is easy to see that p is a simple base point if and only if the general element of $\mathcal{L}$ is smooth at p and two general elements have different tangent directions at p. If p is not a simple base point, then we say that $\mathcal{L}$ has an *infinitely near* base point at p. In general, given a complete linear system $|D|$ and a point $p \in X$, we can consider the linear system $\mathcal{L} = |D - p|$ of all curves in D passing through p. We say that p is an *assigned base point* of $\mathcal{L}$. Any other base points of $\mathcal{L}$ (including infinitely near base points at p) are *unassigned base points*. We could likewise consider the linear system of all curves in $|D|$ containing p which either have a given tangent direction at p or are singular at p (this corresponds to looking at $|\rho^* D - E - q|$, where $q \in E$). Likewise, the linear system of all curves in $|D|$ containing p which are singular at p, corresponding to $|\rho^* D - 2E|$, is denoted by $|D - 2p|$. These are examples of assigned infinitely near base points. Note that $|D - p|$ defines a morphism on $\tilde{X}$ if and only if $|D - p|$ has no unassigned base points.

We note the following consequence of Corollary 14:

Corollary 15. *Let* $f \colon X \dashrightarrow Y$ *be a birational map between two smooth surfaces. Then* $P_n(X) = P_n(Y)$ *and* $q(X) = q(Y)$.

Proof. By Corollary 14, there exists a surface Z and morphisms $Z \to X$, $Z \to Y$ which are repeated blowups. Using Corollary 5 repeatedly, $P_n(X) = P_n(Z) = P_n(Y)$. A similar argument handles q. $\square$

Minimal models

Definition 16. An algebraic surface X is *minimal* if it contains no exceptional curves. An algebraic surface X is a *minimal model* of a surface Y if there exists a birational morphism $Y \to X$ (necessarily a blowup of X, by Theorem 12), such that X is minimal.

Lemma 17. *For every surface* Y, *there exists a minimal model* X.

Proof. If Y is minimal, we are done. Otherwise, there exists an exceptional curve E on Y. Let Y_1 be the smooth surface obtained by contracting E. If Y_1 is minimal, we are done. Otherwise, continue. At each stage we obtain a surface Y_n such that $\operatorname{rank} \operatorname{Num} Y_n = \operatorname{rank} \operatorname{Num} Y_{n-1} - 1$, by Proposition 3. Thus, this procedure must terminate. $\square$

On the other hand, the minimal model of a surface Y need not be unique. The simplest example where this fails is the case where X is a surface containing a curve $C \cong \mathbb{P}^1$ with $C^2 = 0$. If Y is the blowup of a point $p \in C$, then the proper transform C' of C is $\cong \mathbb{P}^1$ and satisfies $(C')^2 = -1$. Thus, it can be contracted, and the resulting surface X' is a candidate for a new minimal model for Y. More generally we could construct examples of this type if X contains a smooth rational curve C with $C^2 \geq 0$, by blowing up several points on C until the proper transform has self-intersection -1, or even more generally if C is any curve on X such that the embedded resolution has genus 0 and nonnegative self-intersection. As this last example suggests, the set of all minimal models can be very complicated. For instance, an example of Kodaira which we shall discuss in Chapter 5 shows that there exist blowups of $\mathbb{P}^2$ which have infinitely many exceptional curves; of course, they cannot be pairwise disjoint. So it is important to find cases where the minimal model of Y is unique, in a strong sense:

Definition 18. A surface X is a *strong minimal model* of Y if X is minimal, there exists a birational morphism $f: Y \to X$, and if $\tilde{Y}$ is a blowup of Y and $g: \tilde{Y} \to X'$ is a birational morphism to a smooth surface X', then there exists a morphism $h: X' \to X$ such that $\tilde{f} = h \circ g$, where $\tilde{f}$ is the composition $\tilde{Y} \to Y \to X$.

It is easy to see that the morphism h above is unique, and that a strong minimal model of Y is unique up to isomorphism. More generally, suppose that $Y_1 \dashrightarrow Y_2$ is a birational map. Then if $f_1: Y_1 \to X$ is a strong minimal model of Y_1, there exists a birational morphism $f_2: Y_2 \to X$ commuting (in the sense of rational maps) with the given birational map $Y_1 \dashrightarrow Y_2$. We see this by blowing up Y_1 until the rational map to Y_2 becomes a morphism and applying the definition with $X' = Y_2$.

We then have:

Theorem 19. *Suppose that $P_n(Y) \neq 0$ for some $n \geq 1$. Then every minimal model of Y is a strong minimal model.*

Proof. Let X, $\tilde{Y}$, and X' be as in the statement of Definition 18. Since the hypothesis $P_n(Y) \neq 0$ is unchanged under blowups, we may (in the notation of Definition 18) assume that $\tilde{Y} = Y$. Factor the morphism $f: Y \to X$ into a sequence $Y = X_n \to X_{n-1} \to \cdots \to X_0 = X$, where E_i is an exceptional curve on X_i and X_{i-1} is the contraction of E_i. The morphism $Y \to X'$ may also be factored. Suppose that we have factored it by first blowing down the curve $F \subset Y$, and let $g': Y \to Y'$ be the result of contracting F. If we can show that there is a morphism $h': Y' \to X$ such that $f = h' \circ g'$, then by repeating the argument with Y' replacing Y (and noting that $P_n(Y') = P_n(Y)$) we can continue until we reach a morphism from X' to X.

Suppose that E_n is disjoint from F. Then we can contract E_n on $Y = X_n$ and the image of E_n on X'. Continue in this way until one of the E_i is not disjoint from (the image of) F. Note that this must happen at some stage since X is minimal and so F cannot map isomorphically onto a curve in X of square -1. If $E_i = F$ we are done. Otherwise, $E_i \cdot F \geq 1$. Thus, $E_i \cdot (E_i + F) \geq 0$ and $F \cdot (E_i + F) \geq 0$.

Now suppose that $H \in |nK_{X_i}|$ is effective; we know that such an H exists since $P_n(Y) \neq 0$. Since E_i and F are exceptional $H \cdot E_i = H \cdot F = -n$. Thus, E_i and F are components of H. We may write $H = rE_i + sF + H'$, where $r, s \geq 1$ and H' is effective and does not contain either E_i or F in its support. Suppose for example that $r \geq s$ and write $H = (r-s)E_i + s(E_i + F) + H'$. Then

$$-n = H \cdot F = (r-s)(E_i \cdot F) + s(F \cdot (E_i + F)) + F \cdot H' \geq 0,$$

which is a contradiction. $\square$

As a result of the classification theorems in Chapter 10, it turns out in fact that a surface Y does not have a strong minimal model if and only if there exists a smooth rational curve C on Y with $C^2 \geq 0$, if and only if $P_n(Y) = 0$ for all $n \geq 1$. Thus, the examples we discussed prior to Definition 18 are typical.

More general contractions

Suppose that D is a nef and big divisor on the surface X. What prevents D from being eventually base point free? If we consider the curves C such that $C \cdot D = 0$, then by Lemma 25 of Chapter 1 the C span a negative definite sublattice of Num $X \otimes \mathbb{R}$ and are linearly independent. In particular there are only finitely many such curves C. Now we have the following contraction theorem of Grauert:

Theorem 20. *Suppose that X is a smooth complex surface, not necessarily compact, and that $C_1, \ldots, C_r$ are compact curves on X such that the symmetric matrix $((C_i \cdot C_j))$ is negative definite. (Here, for compact curves C_i on a possibly noncompact surface, we can simply define $(C_i \cdot C_j) = \deg \mathcal{O}_X(C_i)|C_j$.) Then there exists a normal analytic surface X' and a proper map $p \colon X \to X'$ such that $p(C_i)$ is a single point for all i and $p^{-1}(p(x)) = \{x\}$ if $x \notin \bigcup_i C_i$.* $\square$

It is not hard to see that X' is uniquely specified by the above requirements. We say that X' is obtained by *contracting* the C_i on X. We note that the condition that the matrix $((C_i \cdot C_j))$ is negative definite is a necessary condition, as follows from a result of Mumford:

Theorem 21. *Suppose that $p \colon X \to X'$ is a proper birational morphism as in Theorem 20 and that $C_1, \ldots, C_r$ are the 1-dimensional fibers of p. Then the matrix $((C_i \cdot C_j))$ is negative definite.*

Proof. Since X' is normal, the fibers of p are connected, by Zariski's connectedness theorem. It suffices to prove Theorem 21 for a single connected fiber. Following the proof of Claim 2 of Theorem 12, but using, instead of a hypersurface passing through the singular point, a holomorphic function vanishing at the singular point, we see that $C_i^2 < 0$ for all i, that $C_i \cdot C_j \geq 0$ if $i \neq j$, and that there exist positive integers s_i such that, for all j, $\sum_i s_i(C_i \cdot C_j) \leq 0$, with $\sum_i s_i(C_i \cdot C_j) < 0$ for at least one j (consider the intersection $0 = (H' + \sum_i s_i C_i) \cdot C_j$, noting that by the connectedness assumption $H' \cdot C_j > 0$ for at least one j).

The rest of the proof is now a formal argument about symmetric matrices. Replacing for notational simplicity $s_i C_i$ by v_i, we have a lattice spanned by v_i, with the following properties:

1. For every i, $v_i^2 < 0$;
2. For every i and for all $j \neq i$, $v_i \cdot v_j \geq 0$ with $v_i \cdot v_j > 0$ for some j;
3. For every j, $\sum_i (v_i \cdot v_j) \leq 0$, with $\sum_i (v_i \cdot v_j) < 0$ for at least one j.

Given $\lambda_i \in \mathbb{R}$, suppose first that $\lambda_i \geq 0$ for every i. Then

$$\left(\sum_i \lambda_i v_i \right)^2 = \sum_i \lambda_i \left(\sum_j \lambda_j (v_i \cdot v_j) \right)$$

$$= \sum_i \lambda_i \left(\sum_j \lambda_i (v_i \cdot v_j) + \sum_{j \neq i} (\lambda_j - \lambda_i)(v_i \cdot v_j) \right)$$

$$\leq \sum_i \lambda_i \left(\sum_{j \neq i} (\lambda_j - \lambda_i)(v_i \cdot v_j) \right).$$

In this last sum the term $v_i \cdot v_j$ occurs twice, once with coefficient $\lambda_i (\lambda_j - \lambda_i)$ and once with coefficient $\lambda_j (\lambda_i - \lambda_j)$. Collecting these gives a term $-(\lambda_j - \lambda_i)^2 v_i \cdot v_j \leq 0$. Thus, $(\sum_i \lambda_i v_i)^2 \leq 0$ in case $\lambda_i \geq 0$ for all i and thus if $\lambda_i \leq 0$ for all i as well. In the general case, after relabeling the v_i we can write

$$v = \sum_1^{n_1} \lambda_i v_i + \sum_{n_1+1}^{n_2} \lambda_i v_i = w_1 + w_2,$$

say, where $\lambda_i > 0$ for $i \leq n_1$ and $\lambda_i < 0$ for $n_1 < i \leq n_2$. Then

$$v^2 = w_1^2 + w_2^2 + 2w_1 \cdot w_2.$$

We have seen that $w_1^2 \leq 0$, $w_2^2 \leq 0$ and it is easy to see that $w_1 \cdot w_2 \leq 0$. This proves that the form is at least negative semidefinite. To see that it is actually definite, we analyze the above proof more carefully. First, analyzing the arguments above, it is enough to show that, for every nonempty subset A of $\{1, \ldots, r\}$ and every choice of $\lambda_i > 0$, we have $(\sum_{i \in A} \lambda_i v_i)^2 < 0$. This follows from the inequalities above showing that $(\sum_i \lambda_i v_i)^2 \leq 0$, provided that we show that there exists a $j \in A$ such that $\sum_{i \in A} (v_i \cdot v_j) < 0$. We have shown this above if $A = \{1, \ldots, r\}$, and if $A \neq \{1, \ldots, r\}$, then consider

$$\left(\sum_{i \in A} v_i + \sum_{i \notin A} v_i \right) \cdot v_j \leq 0.$$

As $\bigcup_i C_i$ is connected, there must exist a $j \in A$ and an $i \notin A$ such that $C_i \cdot C_j > 0$, and thus $v_i \cdot v_j > 0$. It follows that $\sum_{i \in A} (v_i \cdot v_j) < 0$. $\square$

Returning to our original problem concerning nef and big divisors, starting with a nef and big D on the smooth projective surface X, let $p: X \to X'$ be the normal surface obtained by contracting the curves C with $D \cdot C = 0$. When does $\mathcal{O}_X(D)$ descend to a line bundle on X'? A necessary and sufficient condition is the following: for each singular point x of X', there should exist an analytic neighborhood U of x such that the restriction of the line bundle $\mathcal{O}_X(D)$ to $p^{-1}(U)$ is trivial. Here the necessity is obvious, and the sufficiency follows since, if $\mathcal{O}_X(D)|p^{-1}(U) = \mathcal{O}_X|p^{-1}(U)$, then $p_*(\mathcal{O}_X(D)|p^{-1}(U)) = p_*(\mathcal{O}_X|p^{-1}(U)) = \mathcal{O}_U$, since, as X' is normal, $p_*\mathcal{O}_X = \mathcal{O}_{X'}$. Let $C_1, \ldots, C_r$ be the irreducible components of $p^{-1}(x)$. By the exponential sheaf sequence, we have an exact sequence

$$0 \to H^1(p^{-1}(U); \mathbb{Z}) \to H^1(p^{-1}(U); \mathcal{O}) \to H^1(p^{-1}(U); \mathcal{O}^*)$$
$$\to H^2(p^{-1}(U); \mathbb{Z}) \to H^2(p^{-1}(U); \mathcal{O}),$$

where $H^1(p^{-1}(U); \mathcal{O}^*) = \mathrm{Pic}\left(p^{-1}(U)\right)$ is the group of holomorphic line bundles on $p^{-1}(U)$. If U is a Stein contractible neighborhood of x, then an application of the Leray spectral sequence shows that $H^i(p^{-1}(U); \mathcal{O}) = H^0(U; R^i p_* \mathcal{O})$. (Here the Stein condition means that the higher cohomology of a coherent analytic sheaf on U is zero.) Moreover, $H^2(p^{-1}(U); \mathbb{Z}) \cong \mathbb{Z}^r$, and the map $H^1(p^{-1}(U); \mathcal{O}^*) \to H^2(p^{-1}(U); \mathbb{Z})$ is the obvious map which sends a holomorphic line bundle L to the vector whose i^{th} component is $\deg(L|C_i)$. Finally, as p has relative dimension 1, $R^2 p_* \mathcal{O} = 0$. Thus, for U as above, there is an exact sequence

$$H^0(R^1 p_* \mathcal{O}_X) \to \mathrm{Pic}\left(p^{-1}(U)\right) \to \mathbb{Z}^r.$$

We see that we have proved the following statement:

Lemma 22. *Suppose in the above situation that $R^1 p_* \mathcal{O}_X = 0$. Then there exist arbitrarily small analytic neighborhoods U' of $\bigcup_i C_i$ such that every line bundle L in U' is specified by the integers $L \cdot C_i$. In particular, if $L \cdot C_i = 0$ for all i, then L is trivial in a neighborhood of $\bigcup_i C_i$.* $\square$

Here the neighborhoods U' in Lemma 22 are the sets of the form $p^{-1}(U)$, where U is a Stein contractible neighborhood of x.

Definition 23. An isolated surface singularity $x \in X'$ is *rational* if $R^1 \pi_* \mathcal{O}_X = 0$ for every resolution $\pi: X \to X'$. It is easy to see in fact by the factorization of birational maps that $R^1 \pi_* \mathcal{O}_X = 0$ for some resolution $\pi: X \to X'$ if and only if $R^1 \pi_* \mathcal{O}_X = 0$ for every resolution $\pi: X \to X'$.

To check if a singularity is rational, suppose that $\pi: X \to X'$ is a resolution of the singularity x with $\pi^{-1}(p) = \bigcup_i C_i$. Then we have the following consequence of the formal functions theorem:

Lemma 24. *The singularity x is rational if and only if, for every effective nonzero cycle $Z = \sum_i n_i C_i$, $H^1(Z; \mathcal{O}_Z) = 0$.*

Proof. By the formal functions theorem $R^1 \pi_* \mathcal{O}_X = 0$ if and only if

$$\varprojlim_Z H^1(Z; \mathcal{O}_Z) = 0,$$

where the limit is over the inverse system of effective cycles Z supported in $\bigcup_i C_i$. Thus, if $H^1(Z; \mathcal{O}_Z) = 0$ for all such Z, then $R^1\pi_*\mathcal{O}_X = 0$. Conversely, we claim that the map $\varprojlim_Z H^1(Z; \mathcal{O}_Z) \to H^1(Z; \mathcal{O}_Z)$ is surjective for every Z. It suffices to show that, if $Z' \geq Z$, then the map $H^1(\mathcal{O}_{Z'}) \to H^1(\mathcal{O}_Z)$ is surjective. But writing $Z' - Z = Z'' \geq 0$, there is an exact sequence

$$0 \to \mathcal{O}_{Z''}(-Z) \to \mathcal{O}_{Z'} \to \mathcal{O}_Z \to 0.$$

The corresponding long exact sequence is surjective on H^1's because, as Z'' is a curve, $H^2(Z''; \mathcal{O}_{Z''}(-Z)) = 0$. It follows that $H^1(\mathcal{O}_{Z'}) \to H^1(\mathcal{O}_Z)$ is surjective for all $Z' \geq Z$, and so $\varprojlim_Z H^1(Z; \mathcal{O}_Z) \to H^1(Z; \mathcal{O}_Z)$ is surjective. Since $\varprojlim_Z H^1(Z; \mathcal{O}_Z) = 0$, $H^1(Z; \mathcal{O}_Z) = 0$ for every Z supported on $\bigcup_i C_i$. $\square$

Example. If C is a smooth rational curve with $C^2 < 0$, then by Exercise 3 in Chapter 1 the result of contracting C gives a rational singularity.

If $C = \sum_i C_i$ is the exceptional set of a rational singularity, then since $h^1(\mathcal{O}_C) = 0$, every component C_i of C is smooth rational, two components can meet at at most one point, and there transversally, and the dual complex associated to $\bigcup_i C_i$ has no cycles, i.e., is a tree. Here given a collection of curves C_i on a surface X, we define a complex by letting the vertices be the C_i and connecting two different vertices corresponding to C_i, C_j by exactly $C_i \cdot C_j$ edges. This complex is called the *dual complex* or *dual graph* of $\bigcup_i C_i$. We can also weight the complex by assigning the positive integer $-C_i^2$ to the vertex corresponding to C_i. However, the above necessary conditions for a rational singularity are by no means sufficient (see Exercise 13).

Artin [3] has shown the following criterion for a singularity to be rational:

Theorem 25. *The singularity x is rational if and only if, for every effective nonzero cycle $Z = \sum_i n_i C_i$, $p_a(Z) \leq 0$.*

Proof. If the singularity is rational, then $h^1(\mathcal{O}_Z) = 0$ for all Z supported on $\bigcup_i C_i$. Thus,

$$p_a(Z) = 1 - h^0(\mathcal{O}_Z) + h^1(\mathcal{O}_Z) \leq 1 - h^0(\mathcal{O}_Z) \leq 0.$$

Conversely, suppose that $p_a(Z) \leq 0$ for all Z supported on $\bigcup_i C_i$. We must show that in this case $h^1(\mathcal{O}_Z) = 0$ for all such Z. If $Z = C_i$, then $p_a(C_i) = h^1(\mathcal{O}_{C_i}) = 0$ and, by Corollary 5 of Chapter 1, $p_a(C_i) = 0$ implies that C_i is smooth rational. Choose an arbitrary effective nonzero $Z = \sum_i n_i C_i$. We shall show by induction on $n = \sum_i n_i$ that $h^1(\mathcal{O}_Z) = 0$, the case $n = 1$ having been dealt with above. We claim that there is an i

such that $-C_i^2 + Z \cdot C_i \leq 1$. For, if $-C_i^2 + Z \cdot C_i \geq 2$ for all i, then

$$K_X \cdot C_i + Z \cdot C_i = -2 - C_i^2 + Z \cdot C_i \geq 0,$$

and thus $K_X \cdot Z + Z^2 \geq 0$. It would then follow that

$$p_a(Z) = 1 + \tfrac{1}{2}(K_X \cdot Z + Z^2) \geq 1,$$

contradicting the assumption that $p_a(Z) \leq 0$.

For a choice of i such that $-C_i^2 + Z \cdot C_i \leq 1$, consider $Z' = Z - C_i$ and take the exact sequence (using (1.9))

$$0 \to \mathcal{O}_{C_i}(-Z') \to \mathcal{O}_Z \to \mathcal{O}_{Z'} \to 0.$$

By the inductive hypothesis $H^1(\mathcal{O}_{Z'}) = 0$. Moreover, $\mathcal{O}_{C_i}(-Z')$ is a line bundle on the curve $C_i \cong \mathbb{P}^1$ of degree $-C_i \cdot Z' = C_i^2 - C_i \cdot Z \geq -1$, and so $H^1(\mathcal{O}_{C_i}(-Z')) = 0$. It follows that $H^1(\mathcal{O}_Z) = 0$ as well. $\square$

There is the following analogue of the Castelnuovo criterion for a rational singularity [2]:

Theorem 26. *Let X be a smooth projective surface and let $\{C_1, \ldots, C_n\}$ be a collection of curves spanning a negative definite sublattice of $\operatorname{Num} X$ such that $H^1(Z; \mathcal{O}_Z) = 0$ for every effective nonzero cycle $Z = \sum_i n_i C_i$. Then there exists a normal projective surface $\bar{X}$ and a birational morphism $\pi \colon X \to \bar{X}$ such that the irreducible components of the 1-dimensional fibers of π are exactly the C_i. Moreover, the singularities of $\bar{X}$ are rational.*

Proof. We shall imitate as far as possible the easy part of the proof of Castelnuovo's theorem. Choose a very ample divisor H on X with $H^1(\mathcal{O}_X(H)) = 0$. Let $\Lambda = \operatorname{span}_{\mathbb{Z}}\{C_1, \ldots, C_n\}$ with its induced intersection form. The linear form defined by $C \in \Lambda \mapsto H \cdot C$ is strictly positive on the generators C_i. Since the intersection form on Λ is negative definite and in particular nondegenerate, the homomorphism $\Lambda \to \Lambda^\vee$ induced by the intersection pairing has finite cokernel. Thus, after replacing H by nH we may assume that there exists $Z = \sum_i n_i C_i$ such that $H \cdot C = -Z \cdot C$ for all $C \in \Lambda$. Next we claim that all of the coefficients n_i are strictly positive. Indeed we can write $Z = Z_+ + Z_-$, where Z_+ is the sum of the terms with positive coefficients, Z_- the sum of the terms with negative coefficients. If $C_i \in \operatorname{Supp} Z_-$, then $Z_+ \cdot C_i \geq 0$ and $Z \cdot C_i = -H \cdot C_i < 0$. Thus, $Z_- \cdot C_i < 0$. It follows that $(-Z_-)^2 > 0$, contradicting the negative definiteness. Likewise, there cannot exist a C_i with $n_i = 0$. Thus, all of the coefficients of Z are strictly positive, and Z is effective.

Now by construction $(H + Z) \cdot C_i = 0$ for all i. Consider the exact sequence

$$0 \to \mathcal{O}_X(H) \to \mathcal{O}_X(H + Z) \to \mathcal{O}_Z(H + Z) \to 0.$$

Applying the exponential sheaf sequence to the (possibly nonreduced) divisor Z, we see that the hypothesis $H^1(Z; \mathcal{O}_Z) = 0$ implies that $\operatorname{Pic} Z \cong \mathbb{Z}^n$

via the map $L \mapsto L \cdot C_i$. Thus, $\mathcal{O}_Z(H + Z) = \mathcal{O}_Z$. Consider the long exact cohomology sequence associated to

$$0 \to \mathcal{O}_X(H) \to \mathcal{O}_X(H + Z) \to \mathcal{O}_Z(H + Z) \to 0.$$

It follows from the assumption that $H^1(\mathcal{O}_X(H)) = 0$ that the section $1 \in H^0(\mathcal{O}_Z) = H^0(\mathcal{O}_Z(H + Z))$ lifts to a section of $H^0(\mathcal{O}_X(H + Z))$. Hence $|H + Z|$ has no base locus along Z. Since $|H + Z|$ contains the subseries $|H|$, it has no base locus elsewhere and so defines a morphism $X \to X'$ from X' to a projective variety X', necessarily a surface since

$$(H + Z)^2 = (H + Z) \cdot H + (H + Z) \cdot Z = H^2 + H \cdot Z > H^2 > 0.$$

Taking the Stein factorization of the morphism $X \to X'$ as in Chapter 1, discussion after Definition 25, gives the normal surface $\bar{X}$. Clearly, all of the singularities of $\bar{X}$ are rational. $\square$

The most important examples of rational singularities are the *rational double points*, whose dual graphs are depicted in Figure 1 below (the numbers above the vertices are the coefficients of the corresponding curves in the fundamental cycle, which is defined in Exercise 8).

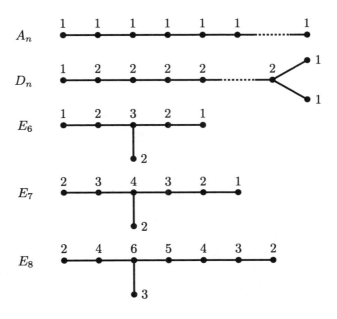

Figure 1

To define the rational double points, suppose that $\{C_1, \ldots, C_n\}$ is a collection of n distinct curves such that $C_i \cong \mathbb{P}^1$ and $C_i^2 = -2$ for all i, and such that the span of the C_i is negative definite. In this case, the dual complex of the set $\{C_1, \ldots, C_n\}$ has connected components which form a graph of type $A_n, n \geq 1$, $D_n, n \geq 4$, or $E_n, n = 6, 7, 8$; this is just a statement about when lattices spanned by vectors of square -2 are negative definite (Exercise 16). Assuming for simplicity that the dual complex is connected, we can then contract the curves C_i and the resulting singularities are by definition the rational double points. In fact the analytic type of the singularity is determined by the dual graph, and all of these singularities are locally hypersurface singularities in $\mathbb{C}^3$. Note that $K_X \cdot C_i = 0$ for all i. Indeed this is one of the real reasons for the importance of the rational double points: the components of a resolution of a rational double point are among the curves which are orthogonal to K_X. We will discuss the properties of the rational double points in greater detail in the exercises. At the moment, we note:

Theorem 27. *The rational double points are rational singularities.*

Proof. Since $K_X \cdot Z = 0$ for every effective nonzero Z whose support is contained in $\bigcup_i C_i$, $2p_a(Z) - 2 = Z^2$ for every such Z. As the intersection pairing is negative definite for a rational double point, $Z^2 < 0$ if $Z \neq 0$. Since $Z^2 < 0$ and is even, $2p_a(Z) = 2 + Z^2 \leq 0$ as well. Thus, by Artin's criterion (Theorem 25) rational double points are rational singularities. $\square$

As a consequence we can prove the following theorem of Mumford:

Theorem 28. *Suppose that K_X is nef and big. Then it is eventually base point free, and the normal surface corresponding to the image of the morphism defined by $|nK_X|$, $n \gg 0$, is the surface with rational double points obtained by contracting all the (finitely many) smooth rational curves C on X with $C^2 = -2$.*

Proof. Let C be an irreducible curve such that $C \cdot K_X = 0$. Then since $K_X^2 > 0$, $C^2 < 0$ by the Hodge index theorem and thus $2p_a(C) - 2 < 0$. It follows that $p_a(C) = 0$ and that C is a smooth rational curve. If we consider the set of all such C, then they are independent in $\operatorname{Num} X$, by Lemma 25 of Chapter 1, and so there are only finitely many. Again by the Hodge index theorem, the lattice they span is negative definite, and thus there exists, by Theorems 26 and 27, a normal surface $\bar{X}$ with just rational double point singularities which is exactly the contraction of the C's. By Lemma 22 and the discussion preceding it, K_X descends to a line bundle on $\bar{X}$, which we shall denote by $\omega_{\bar{X}}$ (in fact it is the dualizing sheaf of $\bar{X}$). Also $\omega_{\bar{X}}^2 = K_X^2 > 0$, and if D is any irreducible 1-dimensional subvariety on $\bar{X}$, and D' is the unique irreducible curve on X mapping onto D, then

$\omega_{\bar{X}} \cdot D = K_X \cdot D' > 0$, since K_X is nef and D' is by construction not one of the curves C with $K_X \cdot C = 0$. By the general version of the Nakai-Moishezon criterion for a singular scheme, $\omega_{\bar{X}}$ is ample. In particular, $\omega_{\bar{X}}$ is eventually base point free and the same must be true for K_X. Moreover, the morphism defined by $|nK_X|$ factors through $\bar{X}$ and has the same image as the morphism defined by $\omega_{\bar{X}}^{\otimes n}$. Since $\omega_{\bar{X}}$ is ample, this image is exactly $\bar{X}$. $\square$

A theorem of Bombieri shows that it suffices to take $n = 5$ in the above result. We will give a proof of this theorem in Chapter 9.

Exercises

1. Let C be a curve on X and let $\tilde{X} \to X$ be the blowup of X at p. Let C' be the proper transform of C. Show that $C' \cdot E = m$ and that the scheme-theoretic intersection $C' \cap E$ is the projective tangent cone to C at p.

2. Apply the method of proof of Theorem 13 to the linear system $|H - p|$, where H is a line in $\mathbb{P}^2$ and p is a point. Show that if H' is the proper transform of H on the blowup of $\mathbb{P}^2$ at p, then $|H'|$ has no base points and defines a morphism from the blowup to $\mathbb{P}^1$. What are the fibers of this morphism?

3. For the examples $y^2 = x^{2k+1}$ and $y^2 = x^{2k}$ of Exercise 5 of Chapter 1, calculate the infinitely near points and their multiplicities and obtain another calculation of δ_0.

4. Prove Lemma 11, that E is an exceptional curve if and only if $E^2 = E \cdot K_Y = -1$ if and only if $E^2 < 0$ and $E \cdot K_Y < 0$.

5. Verify the statements about the Cremona transformation, and show that the rational map f corresponds to the linear system of quadrics passing simply through the points $(1,0,0)$, $(0,1,0)$, and $(0,0,1)$.

6. Show that the germ of an isolated normal surface singularity has a strong minimal model in the sense of Definition 18. (Follow the proof of Theorem 19 until you reach a contradiction to the negative definiteness theorem.)

7. Let C be a smooth cubic in $\mathbb{P}^2$ and let D be a smooth plane quartic meeting C transversally in 12 points $\{p_1, \ldots, p_{12}\}$. Let X be the blowup of $\mathbb{P}^2$ at $\{p_1, \ldots, p_{12}\}$ and let D' be the proper transform of D. Show that $|D'|$ is base point free and that $\mathcal{O}_X(D') = \pi^*\mathcal{O}_{\mathbb{P}^2}(4) \otimes \mathcal{O}_X(-\sum_i E_i)$, where the E_i are the exceptional curves on X. On the other hand, show that, for a general choice of points $\{p_1, \ldots, p_{12}\}$, the line bundle $L = \pi^*\mathcal{O}_{\mathbb{P}^2}(4) \otimes \mathcal{O}_X(-\sum_i E_i)$ restricts to a line bundle of degree 0 but infinite order on the proper transform C' of C, and so the linear system corresponding to $L^{\otimes n}$ has C' as a fixed component for every $n > 0$. In fact, for a general choice of the p_i, there does not exist any line bundle on X for which the restriction to C' is the trivial line bundle.

8. (The fundamental cycle.) Suppose that $C_1, \ldots, C_n$ are irreducible curves spanning a negative definite lattice and that $\bigcup_i C_i$ is connected. Show that there is a unique effective nonzero cycle Z_0 such that $Z_0 \cdot C_i \leq 0$ for all i, $Z_0 \cdot C_i < 0$ for at least one i, and Z_0 is minimal with respect to the above property. (If $Z_1 = \sum_i n_i C_i$ and $Z_2 = \sum_i m_i C_i$ both have the above property, so does

$$Z = \sum_i \min(n_i, m_i) C_i.$$

Thus, there is a minimal such cycle if any exist. To show that it exists, we can use the argument of the first part of the proof of Theorem 21.) This Z_0 is called the *fundamental cycle*.

9. Show that we can find the fundamental cycle as follows: choose $Z_1 = C_i$ for any i. Now suppose that we have inductively found $Z_1, \ldots, Z_i$. If $Z_i \cdot C_\alpha \leq 0$ for all α, then stop. Otherwise, there is a C_α with $Z_i \cdot C_\alpha > 0$. Set $Z_{i+1} = Z_i + C_\alpha$. Show by induction that $Z_0 - Z_i$ is effective for every i, and thus that this procedure must terminate with Z_0. Such a sequence of curves $Z_1, \ldots, Z_n = Z_0$ is called a *computation sequence*. Conversely, if $C_1, \ldots, C_n$ are a collection of curves such that $C_i^2 < 0$ for all i and such that this sequence terminates, then the C_i span a negative definite lattice. (Use the second part of the proof of Theorem 21.)

10. If Z_0 is the fundamental cycle of $\bigcup_i C_i$ as above, where $\bigcup_i C_i$ is connected, show that $H^0(Z_0; \mathcal{O}_{Z_0}) \cong \mathbb{C}$. (Induct on a computation sequence, using (1.9).) Thus, $p_a(Z_0) \geq 0$.

11. (Artin.) With notation as in Exercise 8, the configuration $\{C_1, \ldots, C_n\}$ contracts to a rational singularity if and only if $H^1(Z_0; \mathcal{O}_{Z_0}) = 0$, where Z_0 is the fundamental cycle, if and only if $p_a(Z_0) = 0$. (The equivalence of the two versions follows from the last exercise. To see that $H^1(Z_0; \mathcal{O}_{Z_0}) = 0$ implies that $H^1(Z; \mathcal{O}_Z) = 0$ for every Z supported in $\bigcup_i C_i$, it suffices to show that $H^1(nZ_0; \mathcal{O}_{nZ_0}) = 0$ for all $n > 0$. Use the condition $H^1(Z_0; \mathcal{O}_{Z_0}) = 0$ to show that all the C_i are smooth rational, and that for a computation sequence we must have $Z_i \cdot C_\alpha \leq 1$. Now consider the sequence

$$0 \to \mathcal{O}_{C_\alpha}(-nZ_0 - Z_i) \to \mathcal{O}_{nZ_0 + Z_{i+1}} \to \mathcal{O}_{nZ_0 + Z_i} \to 0,$$

using induction on i and n.)

12. (A dual form of Artin's criterion [43].) With notation as in Exercise 8, suppose that $\bigcup_i C_i$ contracts to a nonrational singularity. Then there exist nonnegative integers n_i with $n_i > 0$ for at least one i such that $(K_X + \sum_i n_i C_i) \cdot C_j \geq 0$ for all j. (Consider a minimal element under the partial ordering $\geq$ in the set of all curves $C = \sum_i m_i C_i$ such that $h^1(\mathcal{O}_C) \neq 0$. If $C = C_i$ for some i, then C_i is not a smooth rational curve, and we can apply adjunction to $K_X + C$. Otherwise, for $C_i \leq C$,

let $C' = C - C_i$. Then $h^1(\mathcal{O}_{C'}) = 0$. From the exact sequence

$$0 \to \mathcal{O}_{C_i}(-C') \to \mathcal{O}_C \to \mathcal{O}_{C'} \to 0,$$

we see that $h^1(\mathcal{O}_{C_i}(-C')) \neq 0$. By duality

$$h^1(\mathcal{O}_{C_i}(-C')) = h^0(\omega_{C_i} \otimes \mathcal{O}_{C_i}(C'))$$

and so $\deg \omega_i + C' \cdot C_i \geq 0$. Now apply adjunction.)

13. Let $C_1, \ldots, C_n$ be a chain of curves, so that $C_i \cdot C_j \neq 0$ if and only if $i = j \pm 1$. Suppose that $C_i^2 \leq -2$ for all i. Show that the C_i span a negative definite lattice, by finding the fundamental cycle of C and using Exercise 9. Show also that, in this case, if the C_i are all smooth rational curves, then the singularity is rational. Next consider the case of curves C_1, C_2, C_3, D, with $C_i \cdot C_j = 0$ for $i \neq j$ and $C_i \cdot D = 1$ for all i. Suppose that $C_i^2, D^2 \leq -2$, and show that the C_i and D span a negative definite lattice. Moreover, if the C_i and D are all smooth rational curves, then the singularity is a rational singularity. Finally, consider the case of curves C_1, C_2, C_3, C_4, D, with $C_i \cdot C_j = 0$ for $i \neq j$ and $C_i \cdot D = 1$ for all i. Suppose that $C_i^2, D^2 \leq -2$, and show that the C_i and D span a negative definite lattice if and only if $C_i^2 \leq -3$ for some i or $D^2 \leq -3$. Moreover, if the C_i and D are all smooth rational curves, then the singularity is a rational singularity if and only if $D^2 \leq -3$.

14. Let Z_0 be the fundamental cycle of a rational double point. . Show that $Z_0^2 = -2$ (the terminology from the theory of root systems is that Z_0 is the *highest root*.) Conversely, if Z_0 is the fundamental cycle of a rational singularity, then the singularity is a rational double point if and only if $Z_0^2 = -2$.

15. Compute the fundamental cycles for the rational double points as follows. The rational double point of type A_n corresponds to a chain of curves $C_1, \ldots, C_n$ with $C_i \cdot C_{i+1} = 1$, $C_i \cdot C_j = 0$ if $j \neq i \pm 1$ and $C_i^2 = -2$ for all i, and the fundamental cycle for such configurations has been computed in Exercise 13. Otherwise, consider curves $A_1, \ldots, A_{p-1}, B_1, \ldots, B_{q-1}, C_1, \ldots, C_{r-1}, D$ such that each of the configurations $A_1, \ldots, A_{p-1}, B_1, \ldots, B_{q-1}, C_1, \ldots, C_{r-1}$ is a chain (i.e., $A_i \cdot A_{i+1} = 1$, $A_i \cdot A_j = 0$ if $j \neq i \pm 1$, and similarly for the B and C curves), $A_i \cdot B_j = 0$ for all i and j, and similarly for $A_i \cdot C_k$ and $B_j \cdot C_k$, and $A_{p-1} \cdot D = B_{q-1} \cdot D = C_{r-1} \cdot D = 1$, whereas D meets no other curves. Lastly we assume that each curve in the list is smooth rational and of square -2. Here D_n corresponds to the case $(p, q, r) = (2, 2, n-2)$, and E_n to the case $(p, q, r) = (2, 3, n-3)$. The lattice spanned by the curves is called a $T_{p,q,r}$ lattice and the dual graph is called a $T_{p,q,r}$ graph. (See Figure 2 on the next page.) Show that the fundamental cycle Z_0 is given by

$$Z_0 = \begin{cases} C_1 + 2C_2 + \cdots + 2C_{n-3} + 2D + A_1 + B_1, & \text{in case } D_n, \\ B_1 + 2B_2 + 3D + 2C_2 + C_1 + 2A_1, & \text{in case } E_6, \\ C_1 + 2C_2 + 3C_3 + 4D + 3B_2 + 2B_1 + 2A_1, & \text{in case } E_7, \\ 2C_1 + 3C_2 + 4C_3 + 5C_4 + 6D + 4B_2 + 2B_1 + 3A_1, & \text{in case } E_8. \end{cases}$$

In particular, using the last part of Exercise 9, conclude that the lattices in question really are negative definite.

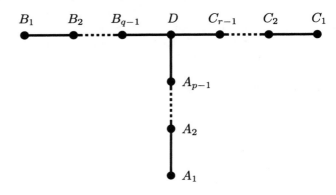

Figure 2

16. Every connected configuration of curves of square -2 spanning a negative definite lattice spans a lattice of type A_n, D_n, E_6, E_7, E_8. (Every sublattice of such a lattice is negative definite. Thus, if we are given a chain $A_1, \ldots A_k$ which is a subset of the given set of curves, then $(A_1 + \cdots + A_k)^2 = -2$ and the lattice spanned by $A = A_1 + \cdots + A_k$ and the remaining curves is still negative definite. More generally, if X and C are elements of square -2, then $X \cdot C = 0$ or ± 1. Show that the dual graph of the set of curves is a tree and (by considering the configuration $C_1 + C_2 + C_3 + C_4 + D$, where $C_i \cdot C_j = 0$ if $i \neq j$ and $C_i \cdot D = 1$ for $1 = 1, \ldots, 4$ and showing that it is not negative definite) that in fact the dual graph is either a chain (case A_n) or that of a $T_{p,q,r}$ graph. Show that such a graph must contain $T_{2,2,n-2}$ or $T_{2,3,n-3}$, and in fact must contain $T_{2,3,n-3}$ except in case D_n. Now argue that if a $T_{p,q,r}$ graph properly contains a $T_{2,3,n-3}$ graph and the corresponding $T_{p,q,r}$ lattice is negative definite, then there has to exist a curve C, meeting either $A_1, B_1,$ or $C_1,$ with $C \cdot Z_0 = 1$. Using the previous problem, show that we must be in case E_6, E_7, E_8. In particular, the $T_{p,q,r}$ lattice is negative definite only for $(p, q, r) = (2, 2, n - 2), (2, 3, 3), (2, 3, 4), (2, 3, 5)$.)

17. Let $\{C_1, \ldots, C_n\}$ be the components of the resolution of a rational singularity with fundamental cycle Z_0 such that $Z_0^2 = -1$ (and such

that $\bigcup_i C_i$ is connected). Show that one of the C_i is an exceptional curve, and that the blown down configuration also has fundamental cycle of square -1. Conclude that the C_i can be successively contracted in some order to a smooth point.

More generally, Artin has shown [3] that, with C_i and Z_0 as above, if the singularity is rational, then its multiplicity is given by $-Z_0^2$ and its embedding dimension by $-Z_0^2 + 1$. In particular, by using Exercise 14, we see that a rational singularity is a rational double point if and only if it is a hypersurface singularity, if and only if its multiplicity is 2 (hence the name rational double point). From this, one can work out the analytic type of the equations for the rational double points: they are

$$
\begin{array}{lll}
A_n : & x^2 + y^2 + z^{n+1}, & \\
D_n : & x^2 + y^2 z + z^{n-1} & (n \geq 4), \\
E_6 : & x^2 + y^3 + z^4, & \\
E_7 : & x^2 + y^3 + yz^3, & \\
E_8 : & x^2 + y^3 + z^5. &
\end{array}
$$

Finally, there is yet another approach to rational double points: they are locally analytically isomorphic to the quotient of the germ of $\mathbb{C}^2$ at the origin by a finite subgroup of $SL(2, \mathbb{C})$, and as such were studied by Klein. For a very readable discussion of these and other properties, we refer the reader to [28] as well as [20].

4

Stability

In this chapter, we shall define stability and investigate some of its elementary properties. After giving simple examples of stable bundles over curves and $\mathbb{P}^2$, we describe unstable and strictly semistable bundles carefully and look at what happens when we change polarizations. The final section, which is not necessary for the rest of this book, describes the differential geometry of stable bundles.

Definition of Mumford-Takemoto stability

Let X be a smooth projective variety of dimension d and let H be an ample line bundle on X.

Definition 1. For V a torsion free coherent sheaf on X, the *normalized degree* $\mu_H(V)$ of V with respect to H is the rational number

$$\mu_H(V) = \frac{1}{\operatorname{rank} V}(c_1(V) \cdot H^{d-1}).$$

We shall omit the subscript when the line bundle H is clear from the context. Note that if we replace H by aH, $a \in \mathbb{Z}^+$, then $\mu_{aH}(V) = a^{d-1}\mu_H(V)$. We can also define the normalized degree with respect to a nef and big line bundle H, as well as with respect to a Kähler metric, by replacing H by the Kähler form ω.

The idea is that the normalized degree is a measure of the "degree" of V with respect to H. For example, if $V = \mathcal{O}_X(D)$ is the line bundle associated to an effective divisor D, then $\mu(\mathcal{O}_X(D))$ is indeed just $\deg D$. The normalized degree has the following convexity property with respect to exact sequences.

Lemma 2. *Suppose that*

$$0 \to V' \to V \to V'' \to 0$$

is an exact sequence of nonzero torsion free sheaves on X and H is an ample line bundle on X. Let $\mu = \mu_H$. Then

$$\min(\mu(V'), \mu(V'')) \leq \mu(V) \leq \max(\mu(V'), \mu(V'')),$$

and equality holds at either end if and only if $\mu(V') = \mu(V'') = \mu(V)$.

Proof. Let $n' = \operatorname{rank} V'$ and $n'' = \operatorname{rank} V''$. By the Whitney product formula, $c_1(V) = c_1(V') + c_1(V'')$. Thus,

$$\mu(V) = \frac{n'}{n' + n''}\mu(V') + \frac{n''}{n' + n''}\mu(V'').$$

So $\mu(V) = \lambda\mu(V') + (1 - \lambda)\mu(V'')$, where $0 < \lambda < 1$. Thus, if say $\mu(V') \leq \mu(V'')$, then $\mu(V') \leq \mu(V) \leq \mu(V'')$, with equality if and only if $\mu(V') = \mu(V'') = \mu(V)$. $\square$

We also have the following property of normalized degree for subsheaves of equal rank.

Lemma 3. *Suppose that W is a subsheaf of the torsion free sheaf V, with* rank $W = $ rank V. *Then $\mu(W) \leq \mu(V)$. Moreover, if V and W are vector bundles, then either $\mu(W) < \mu(V)$ or $W = V$.*

Proof. We shall just consider the case where W and V are vector bundles of rank r, and leave the general case to the exercises. In this case there is the associated map $\det W \to \det V$, which is a nonzero map of line bundles. Thus, $\det V = \det W \otimes \mathcal{O}_X(D)$, where D is effective, and $D = \emptyset$ if and only if $\det V \cong \det W$, in other words if and only if the inclusion map $W \to V$ is an isomorphism. In this case $W = V$. Otherwise, D is an effective nonzero divisor, and so

$$\mu(V) = \frac{1}{r}H^{d-1} \cdot c_1(V) = \frac{1}{r}[H^{d-1} \cdot (c_1(W) + D)]$$

$$> \frac{1}{r}H^{d-1} \cdot c_1(W) = \mu(W).$$

Thus, $\mu(W) < \mu(V)$. $\square$

We now give the definition of Mumford-Takemoto stability which for the purposes of this book will simply be denoted stability.

Definition 4. For V as above, V is *H-stable* (resp., *H-semistable*) if, for all coherent subsheaves W of V with $0 < \operatorname{rank} W < \operatorname{rank} V$, we have $\mu(W) < \mu(V)$ (resp., $\mu(W) \leq \mu(V)$). We call V *unstable* if it is not semistable and *strictly semistable* if it is semistable but not stable. (The convention that unstable is the opposite of semistable is unfortunately by now well established.) Finally, a subsheaf W of a torsion free sheaf V with $0 < \operatorname{rank} W < \operatorname{rank} V$ is *destabilizing* if $\mu(W) \geq \mu(V)$.

Thus, for example, line bundles are (vacuously) stable. Moreover, V is H-stable if and only if it is aH-stable, for $a \in \mathbb{Z}^+$, and similarly for semistability. Furthermore stability and semistability only depend on the numerical equivalence class of H. If X is a curve, then $\mu_H(V) = \deg c_1(V)$ is independent of the choice of H. However, for $\dim X \geq 2$, if $\operatorname{rank} \operatorname{Num} X \geq 2$, then the definition of stability depends on the choice of the numerical equivalence class of H. We shall return to this point later.

Next let us work out the meaning of stability in the case where V is a rank 2 vector bundle on a surface X. In this case, we need only check stability with respect to rank 1 torsion free sheaves W. But since W is torsion free, $W^{\vee\vee}$ is a reflexive rank 1 sheaf on X and is thus a line bundle L (either since $\dim X = 2$ or since $\operatorname{rank} W = 1$). Hence W itself is of the form $L \otimes I_Z$ for some ideal sheaf I_Z, where $\dim Z = 0$, and $c_1(W) = c_1(W^{\vee\vee})$. Moreover, since $V = V^{\vee\vee}$, there is an inclusion of $W^{\vee\vee} = L$ in V. Thus, we see that it suffices to check the degree μ_H for sub-line bundles of V. Now suppose that the quotient V/L is not torsion free. Then as we have seen in Proposition 5 of Chapter 2, the map $L \to V$ factors through the inclusion $L \to L \otimes \mathcal{O}_X(E)$, where E is an effective, nonzero divisor, and $V/(L \otimes \mathcal{O}_X(E))$ is torsion free. As $H \cdot (L \otimes \mathcal{O}_X(E)) = H \cdot L + H \cdot E > H \cdot L$, the normalized degree of L can only increase. Thus, it suffices to check the degree of all sub-line bundles L such that V/L is torsion free.

Lemma 5. *Let V be a torsion free sheaf on a smooth projective variety X.*

(i) *V is stable if and only if, for all coherent subsheaves W of V with $0 < \operatorname{rank} W < \operatorname{rank} V$ such that V/W is torsion free, $\mu(W) < \mu(V)$.*

(ii) *V is stable if and only if there exists a line bundle F such that $V \otimes F$ is stable if and only if for all line bundles F, $V \otimes F$ is stable.*

(iii) *V is stable if and only if $V^{\vee\vee}$ is stable.*

(iv) *V is stable if and only if for all torsion free quotients Q of V with $0 < \operatorname{rank} Q < \operatorname{rank} V$, we have $\mu(Q) > \mu(V)$.*

(v) *V is stable if and only if $V^{\vee}$ is stable.*

Moreover, all of the above statements hold if we replace stable by semistable.

Proof. We shall only check these statements in the case where V is a rank 2 bundle on a surface X. In this case, we have already seen (i), and (ii) is a consequence of the fact that $\mu(L \otimes F) = \mu(L) + H \cdot F$ and $\mu(V \otimes F) = \mu(V) + H \cdot F$ by (2.6) of Chapter 2. Part (iii) is obvious in our case since $V = V^{\vee\vee}$, and (iv) is a consequence of Lemma 2. Finally, (v) holds in our case as a result of (ii), since $V^{\vee} = V \otimes \det V^{-1}$. $\square$

Lemma 6. *Let*

$$0 \to V' \to V \to V'' \to 0$$

be an exact sequence of nonzero torsion free sheaves, with $\mu(V') = \mu(V) = \mu(V'')$. Then V is semistable if and only if V' and V'' are semistable, and V is never stable. In particular, if V' and V'' both have rank 1, then V is semistable.

Proof. Suppose that V is semistable. Let W is a subsheaf of V'. Then W is a subsheaf of V and so $\mu(W) \le \mu(V) = \mu(V')$. Thus, V' is semistable. Likewise, if $\bar{W}$ is a subsheaf of V'', then the inverse image W of $\bar{W}$ in V satisfies $\mu(W) \le \mu(V) = \mu(V')$. Using the exact sequence

$$0 \to V' \to W \to \bar{W} \to 0,$$

it follows from Lemma 2 that $\mu(\bar{W}) \le \mu(V') = \mu(V'')$ as well. Thus, V'' is also semistable.

Conversely, suppose that V' and V'' are semistable. Let W be a nonzero subsheaf of V. Let $p \colon W \to V''$ be the projection and let $W' = \operatorname{Ker} p$, $W'' = \operatorname{Im} p$. Thus, there is an exact sequence $0 \to W' \to W \to W'' \to 0$. We may assume that both W' and W'' are nonzero, since otherwise $W \cong W'$, say, and thus $\mu(W) = \mu(W') \le \mu(V') = \mu(V)$. But if $W'' \ne 0$, then either $\operatorname{rank} W'' = \operatorname{rank} V''$, in which case $\mu(W'') \le \mu(V'')$ by Lemma 3, or $\operatorname{rank} W'' < \operatorname{rank} V''$. By hypothesis V'' is semistable, and so $\mu(W'') \le \mu(V'') = \mu(V)$. A similar argument handles $\mu(V')$. Hence, by Lemma 2, $\mu(W) \le \max(\mu(W'), \mu(W'')) \le \mu(V)$ as well. $\square$

Proposition 7. *If $\varphi \colon V_1 \to V_2$ is a nonzero homomorphism between two stable torsion free sheaves V_1 and V_2 with $\mu(V_1) = \mu(V_2)$, then φ is injective and is an isomorphism if V_1 and V_2 are vector bundles or if $V_1 = V_2$.*

Proof. Suppose that φ is not injective. Then $\operatorname{Im} \varphi = W$ is a proper torsion free quotient of V_1 and thus $\mu(W) > \mu(V_1) = \mu(V_2)$. Thus, W must have the same rank r as V_2, since V_2 is stable. By Lemma 3, however, $\mu(W) \le \mu(V_2)$, a contradiction. Hence φ is injective. Thus, W is a subsheaf of V_2 with $\mu(W) = \mu(V_2)$, and so by stability $\operatorname{rank} W = \operatorname{rank} V_2$. Again using Lemma 3, if $V_1 \cong W$ and V_2 are vector bundles and $\mu(V_1) = \mu(V_2)$, then $W = V_2$ and φ is an isomorphism. The case where $V_1 = V_2$ is not necessarily locally free follows from the general fact that an injective map from a coherent sheaf (on a projective scheme) to itself is necessarily an isomorphism, which we leave as Exercise 1. $\square$

Corollary 8. *If V is a stable torsion free sheaf, then V is simple, i.e., $\operatorname{End} V = \{\lambda \cdot \operatorname{Id} : \lambda \in \mathbb{C}\}$.*

Proof. If φ is a nonzero endomorphism of V, then it is an isomorphism, by Proposition 7. The proof of Corollary 8 is then a standard Schur's lemma argument: every nonzero element of $\operatorname{End} V$ is invertible, so that $\operatorname{End} V$ is a finite-dimensional division algebra over $\mathbb{C}$. Thus, $\operatorname{End} V \cong \mathbb{C}$. Alternatively choose $p \in X$ such that V is locally free at p and consider the induced map on the stalk $\varphi_p \colon V_p \to V_p$. If λ is an eigenvalue of φ_p, then the map $\varphi - \lambda \cdot \mathrm{Id}$ cannot be an isomorphism and so must be 0. Hence $\varphi = \lambda \cdot \mathrm{Id}$. $\square$

Remark. An argument similar to the proof of Proposition 7 shows that, if $\varphi \colon V_1 \to V_2$ is a nonzero homomorphism between two semistable torsion free sheaves V_1 and V_2 with $\mu(V_1) = \mu(V_2)$ and such that at least one of V_1, V_2 is stable, then either φ is injective or the rank of its image is equal to the rank of V_2.

Examples for curves

Let us give some examples of stable and semistable bundles for the case of curves.

Theorem 9.

(i) *Over* $\mathbb{P}^1$, *there are no stable rank 2 bundles. The only semistable rank 2 bundles are the bundles* $\mathcal{O}_{\mathbb{P}^1}(a) \oplus \mathcal{O}_{\mathbb{P}^1}(a)$.

(ii) *Let* C *be a curve of genus 1. The only stable rank 2 bundles over* C *are the bundles of the form* $\mathcal{F}_p \otimes L$, *where* L *is a line bundle on* C, $p \in C$, *and* $\mathcal{F}_p$ *is the unique nonsplit extension of* $\mathcal{O}_C(p)$ *by* $\mathcal{O}_C$, *as in Theorem 6 of Chapter 2. The only strictly semistable rank 2 bundles over* C *are either of the form* $L_1 \oplus L_2$, *where* $\deg L_1 = \deg L_2$, *or of the form* $\mathcal{E} \otimes L$, *where* L *is a line bundle on* C *and* $\mathcal{E}$ *is the unique nonsplit extension of* $\mathcal{O}_C$ *by* $\mathcal{O}_C$.

Proof. These statements all follow easily from Theorem 6 of Chapter 2 and from Lemma 6, with the exception of the statement that $\mathcal{F}_p \otimes L$ is stable. It suffices to prove that $\mathcal{F}_p$ is stable. Note that $\mu(\mathcal{F}_p) = 1/2$. If L is a sub-line bundle of $\mathcal{F}_p$ such that the induced map $L \to \mathcal{O}_C(p)$ is nonzero, then $\deg L \leq \deg \mathcal{O}_C(p) = 1$, and $\deg L = 1$ if and only if $L \cong \mathcal{O}_C(p)$. But in this last case the extension is split, contradicting the definition of $\mathcal{F}_p$. Thus, $\deg L \leq 0 < \mu(\mathcal{F}_p) = 1/2$. If the induced map $L \to \mathcal{O}_C(p)$ is zero, then $L \subseteq \mathcal{O}_C$. In this case $\deg L \leq 0 < \mu(\mathcal{F}_p)$ again. Thus, $\mathcal{F}_p$ is stable. $\square$

For curves of genus at least 2, we have the following existence result:

Theorem 10. *Let* C *be a curve of genus at least 2.*

(i) *If* L *is a line bundle on* C *with* $\deg L = -1$, *then there exist stable bundles* V *over* C *of the form*

$$0 \to L \to V \to L^{-1} \to 0.$$

(ii) **If M is a line bundle on C with $\deg M = 1$, then there exist stable bundles V over C of the form**

$$0 \to \mathcal{O}_C \to V \to M \to 0;$$

in fact, every such extension which is nonsplit is stable.

Proof. (i) For a curve C of genus g and a line bundle L on C of degree -1, the set of extensions V of the form

$$0 \to L \to V \to L^{-1} \to 0$$

is classified by $H^1(C; L^{\otimes 2})$. Since $\deg L^{\otimes 2} = -2 < 0$, we have $h^1(C; L^{\otimes 2}) = g + 1 \geq 3$ by Riemann-Roch. In particular nonsplit extensions always exist.

Let V be such a nonsplit extension. When is V stable? Since $\mu(V) = 0$, we would like to show that, for all sub-line bundles F of V, $\deg F < 0$. If the composite map $F \to L^{-1}$ is zero, then the image of F is contained in L, so that $F = L \otimes \mathcal{O}_C(D)$, where D is effective. In this case $\deg F \leq -1$. Otherwise, the map $F \to L^{-1}$ is nonzero. Thus, $F^{-1} \otimes L^{-1}$ has a section and so $\deg F \leq 1$. If $\deg F = 1$, then $F = L^{-1}$ and the sequence is split, contrary to hypothesis. If $\deg F < 0$, then F is not destabilizing. So the only problem is when $\deg F = 0$. In this case $F = L^{-1} \otimes \mathcal{O}_C(-p)$ and the map $F \to L^{-1}$ is, up to a nonzero scalar, the natural map $L^{-1} \otimes \mathcal{O}_C(-p) \to L^{-1}$. Thus, by construction, we are in the following situation: V is an extension such that the natural map $L^{-1} \otimes \mathcal{O}_C(-p) \to L^{-1}$ lifts to a map $L^{-1} \otimes \mathcal{O}_C(-p) \to V$. To see when this is possible, consider the exact sequence

$$\operatorname{Hom}(L^{-1} \otimes \mathcal{O}_C(-p), V) \to \operatorname{Hom}(L^{-1} \otimes \mathcal{O}_C(-p), L^{-1})$$
$$\to H^1(C; (L^{-1} \otimes \mathcal{O}_C(-p))^{-1} \otimes L).$$

Since $g(C) \geq 2$, $\dim H^0(C; \mathcal{O}_C(p)) = 1$, and the obstruction to lifting the essentially unique element of $\operatorname{Hom}(L^{-1} \otimes \mathcal{O}_C(-p), L^{-1}) = H^0(C; \mathcal{O}_C(p))$ to an element of $\operatorname{Hom}(L^{-1} \otimes \mathcal{O}_C(-p), V)$ lies in $H^1(C; L^{\otimes 2} \otimes \mathcal{O}_C(p))$. Now we have a commutative diagram

$$
\begin{array}{ccc}
H^0(C; \mathcal{O}_C) & \overset{\partial}{\longrightarrow} & H^1(C; L^{\otimes 2}) \\
\downarrow \cong & & \downarrow \\
H^0(C; \mathcal{O}_C(p)) & \longrightarrow & H^1(C; L^{\otimes 2} \otimes \mathcal{O}_C(p)),
\end{array}
$$

where ∂ is the coboundary map in the appropriate long exact cohomology sequence. Moreover, the image of $\partial(1)$ is just the extension class. So a nonzero element of $\operatorname{Hom}(L^{-1} \otimes \mathcal{O}_C(-p), L^{-1})$ can be lifted to a homomorphism $L^{-1} \otimes \mathcal{O}_C(-p) \to V$ exactly when the extension class is in the kernel of the natural map

$$H^1(C; L^{\otimes 2}) \to H^1(C; L^{\otimes 2} \otimes \mathcal{O}_C(p)).$$

For a fixed $p \in C$, we have an exact sequence

$$0 \to L^{\otimes 2} \to L^{\otimes 2} \otimes \mathcal{O}_C(p) \to \mathbb{C}_p \to 0,$$

where $\mathbb{C}_p$ is a skyscraper sheaf at p with stalk $\mathbb{C}$. Since $\deg L^{\otimes 2} \otimes \mathcal{O}_C(p) = -1$, there is an inclusion $\mathbb{C} = H^0(\mathbb{C}_p) \subset H^1(C; L^{\otimes 2})$. In fact, $H^1(C; L^{\otimes 2})$ is the dual of $H^0(C; K_C \otimes L^{\otimes -2})$, and the linear system defined by $K_C \otimes L^{\otimes -2}$ is base point free (since $\deg K_C \otimes L^{\otimes -2} = 2g$). Via Serre duality, the line $H^0(\mathbb{C}_p) \subset H^1(C; L^{\otimes 2})$ defines a point in $\mathbb{P}(H^0(C; K_C \otimes L^{\otimes -2})^\vee)$ which is the hyperplane of all sections vanishing at p, and this point is then the image of p under the morphism defined by $K_C \otimes L^{\otimes -2}$. We see that V is stable if and only if the extension class corresponding to V, viewed as an element in $\mathbb{P}H^1(C; L^{\otimes 2})$, does not lie on the image of C. Since $\mathbb{P}H^1(C; L^{\otimes 2})$ is a projective space of dimension at least 2, the general such extension class will indeed correspond to a stable bundle.

(ii) Let us first show that there exist nonsplit extensions of M by $\mathcal{O}_C$. The set of all such extensions is equal to $H^1(C; M^{-1})$. Since $\deg M^{-1} = -1$, the Riemann-Roch theorem gives $\dim H^1(C; M^{-1}) = g \geq 2$. In particular, there exist nontrivial extensions V. If V is a nonsplit extension, then V is stable by an argument identical to that given in the proof of Theorem 9 for $\mathcal{F}_p$. □

More generally, along these lines we shall show (Exercise 3) that, for every line bundle L on C of negative degree, there exist extensions of L^{-1} by L which are stable. Moreover, if $\deg L > -(g-1)/2$, then the "generic" extension of L^{-1} by L is a stable bundle V such that $\operatorname{Hom}(L', V) = 0$ for all L' such that $\deg L' \geq \deg L$ and $L \neq L'$. A theorem stated classically by Corrado Segre [135] (and rediscovered by Nagata [110]) asserts that, for every rank 2 vector bundle V on C with $\det V = 0$, there exists a line bundle L of degree at least $-(g-1)/2$ such that $\operatorname{Hom}(L, V) \neq 0$. Thus, there is an exact sequence

$$0 \to L' \to V \to (L')^{-1} \to 0,$$

where $\deg L' \geq -(g-1)/2$. Moreover, for the "generic" stable bundle V, if g is odd, then there are exactly 2^g line bundles L of degree $-(g-1)/2$ such that $\operatorname{Hom}(L, V) \neq 0$, and for every other line bundle M such that $\operatorname{Hom}(M, V) \neq 0$, we have $\deg M < -(g-1)/2$.

Some examples of stable bundles on $\mathbb{P}^2$

In this section, we shall give some of the elementary properties of stable rank 2 bundles on $\mathbb{P}^2$. In particular we shall show that a rank 2 bundle on $\mathbb{P}^2$ is stable if and only if it is simple. Many of the following arguments work equally well for any surface X with $\operatorname{Pic} X = \mathbb{Z}$.

Throughout this section, V denotes a rank 2 bundle on $\mathbb{P}^2$. By stability we mean stability with respect to $\mathcal{O}_{\mathbb{P}^2}(1)$. In fact as $\operatorname{Pic}\mathbb{P}^2 \cong \mathbb{Z}$, there is a unique notion of stability for $\mathbb{P}^2$.

Lemma 11. *For a rank 2 vector bundle V on $\mathbb{P}^2$, there exists an integer k_V and a sub-line bundle $\mathcal{O}_{\mathbb{P}^2}(k_V)$ of V such that, for every integer k, if there is a nonzero map $\mathcal{O}_{\mathbb{P}^2}(k) \to V$, then $k \le k_V$. Moreover, the quotient of V by a sub-line bundle of V isomorphic to $\mathcal{O}_{\mathbb{P}^2}(k_V)$ is torsion free.*

Proof. For $\ell \gg 0$, $V \otimes \mathcal{O}_{\mathbb{P}^2}(\ell)$ is generated by its global sections, and thus for $\ell \ll 0$ there exists a nonzero map $\mathcal{O}_{\mathbb{P}^2}(\ell) \to V$. Using Proposition 5 of Chapter 2, there exists an effective divisor D on $\mathbb{P}^2$ such that the map $\mathcal{O}_{\mathbb{P}^2}(\ell) \to V$ factors through the inclusion $\mathcal{O}_{\mathbb{P}^2}(\ell) \to \mathcal{O}_{\mathbb{P}^2}(\ell) \otimes \mathcal{O}_{\mathbb{P}^2}(D)$ and the quotient is of the form $L' \otimes I_Z$. Now $\mathcal{O}_{\mathbb{P}^2}(D) = \mathcal{O}_{\mathbb{P}^2}(m)$ for some integer m and $L' = \mathcal{O}_{\mathbb{P}^2}(k')$ for some integer k'. Thus, there exists an integer k and an exact sequence

$$0 \to \mathcal{O}_{\mathbb{P}^2}(k) \to V \to \mathcal{O}_{\mathbb{P}^2}(k') \otimes I_Z \to 0.$$

It follows that, if $n > \max\{k, k'\}$, then

$$\operatorname{Hom}(\mathcal{O}_{\mathbb{P}^2}(n), V) = H^0(V \otimes \mathcal{O}_{\mathbb{P}^2}(-n)) = 0.$$

Thus, there is a largest integer k_V such that $\operatorname{Hom}(\mathcal{O}_{\mathbb{P}^2}(k_V), V) \ne 0$; indeed $k_V \le \max\{k, k'\}$ in the above notation. Clearly, for a sub-line bundle of V isomorphic to $\mathcal{O}_{\mathbb{P}^2}(k_V)$, the quotient must be torsion free, for otherwise as before the map would factor through $\mathcal{O}_{\mathbb{P}^2}(k_V) \otimes \mathcal{O}_{\mathbb{P}^2}(D) = \mathcal{O}_{\mathbb{P}^2}(k_V + m)$ for some positive integer m, contradicting the maximality of k_V. $\square$

Note that we do not claim in the above lemma that there is a unique sub-line bundle $\mathcal{O}_{\mathbb{P}^2}(k_V)$.

Lemma 12. *V is stable if and only if $2k_V < d$, where $\det V = \mathcal{O}_{\mathbb{P}^2}(d)$, and V is strictly semistable if and only if $2k_V = d$.*

Proof. V is stable if and only if for all nonzero maps $\mathcal{O}_{\mathbb{P}^2}(k) \to V$, we have $k < \frac{1}{2}d$ if and only if $k_V < \frac{1}{2}d$. The proof of the second statement is similar. $\square$

Corollary 13. *A rank 2 bundle V over $\mathbb{P}^2$ is stable if and only if it is simple.*

Proof. If V is stable, then it is simple by Corollary 8. Conversely, suppose that V is not stable, and let $\det V = \mathcal{O}_{\mathbb{P}^2}(d)$. Then there exists an exact sequence

$$0 \to \mathcal{O}_{\mathbb{P}^2}(k) \to V \to \mathcal{O}_{\mathbb{P}^2}(k') \otimes I_Z \to 0$$

with $k + k' = d$ and $2k \geq d$. Thus, $k' \leq k$, and there is an inclusion $\mathcal{O}_{\mathbb{P}^2}(k') \otimes I_Z \subseteq \mathcal{O}_{\mathbb{P}^2}(k') \subseteq \mathcal{O}_{\mathbb{P}^2}(k)$. Hence there is a nonzero map $V \to \mathcal{O}_{\mathbb{P}^2}(k') \otimes I_Z \to \mathcal{O}_{\mathbb{P}^2}(k) \to V$ which is not multiplication by a scalar. Therefore V is not simple. $\square$

Remark. If V is not stable and not of the form $\mathcal{O}_{\mathbb{P}^2}(k) \oplus \mathcal{O}_{\mathbb{P}^2}(k)$, then it is easy to check that $\mathrm{Hom}(\mathcal{O}_{\mathbb{P}^2}(k_V), V)$ has dimension 1, i.e., that the extension

$$0 \to \mathcal{O}_{\mathbb{P}^2}(k_V) \to V \to \mathcal{O}_{\mathbb{P}^2}(k') \otimes I_Z \to 0$$

is canonically determined by V. Moreover, since $k_V - k' \geq 0$, we have $H^2(\mathcal{O}_{\mathbb{P}^2}(-k') \otimes \mathcal{O}_{\mathbb{P}^2}(k_V)) = 0$, so that we may in fact construct all such extensions via Theorem 12 of Chapter 2. This result goes back to Schwarzenberger, who analyzed nonsimple bundles ("almost decomposable bundles" in his terminology) and used the method of double covers to show the existence of simple bundles.

Next we give some examples of stable bundles on $\mathbb{P}^2$. For example, consider the bundle V defined in Exercise 6 of Chapter 2. This bundle has $c_1(V) = \mathcal{O}_{\mathbb{P}^2}(-1)$ and $k_V \leq -1$ since $H^0(V) = 0$. Thus, V is stable by Lemma 12. Another example of a stable bundle is the tangent bundle $\Theta_{\mathbb{P}^2}$ (Exercise 6). In fact, these bundles agree up to a twist by a line bundle. As an exercise, we ask the reader to show that the bundles $V_{a,b}$ constructed in the example after the proof of Proposition 28 in Chapter 2 are stable as long as $a \neq b$ or $b \pm 1$.

We conclude this section by studying stable rank 2 bundles on $\mathbb{P}^2$ with $c_1(V) = 0$ and $c_2(V)$ small. Let us begin with the following:

Proposition 14. *Let V be a rank 2 bundle on $\mathbb{P}^2$ with $c_1(V) = 0$, and let $c = c_2(V)$.*

(i) *V is stable if and only if $k_V < 0$ if and only if $h^0(V) = 0$.*
(ii) *If V is stable and $c \leq 5$, then there exists an extension*

(*) $$0 \to \mathcal{O}_{\mathbb{P}^2}(-1) \to V \to \mathcal{O}_{\mathbb{P}^2}(1) \otimes I_Z \to 0.$$

(iii) *Conversely, if V is given by an extension (*) as above, then $c = \ell(Z) - 1$ and V is stable if and only if Z is not contained in a line.*
(iv) *If V is stable, then $c \geq 2$.*
(v) *Let V be given by an extension (*). Then $h^0(V \otimes \mathcal{O}_{\mathbb{P}^2}(1))$ is equal to $\dim h^0(\mathcal{O}_{\mathbb{P}^2}(2) \otimes I_Z) + 1$, i.e., to 2+ the dimension of the space of conics containing Z (or 1 if there are no conics containing Z).*

Proof. (i) Using Lemma 12, since $\det V = 0$, V is stable if and only if every sub-line bundle of V has strictly negative degree, which is clearly the case if and only if V has no sections.

(ii) Applying the Riemann-Roch formula to $V \otimes \mathcal{O}_{\mathbb{P}^2}(1)$, we obtain

$$\chi(V \otimes \mathcal{O}_{\mathbb{P}^2}(1)) = 6 - c.$$

Hence, if $c \leq 5$, then either $h^0(V \otimes \mathcal{O}_{\mathbb{P}^2}(1)) \geq 1$ or $h^2(V \otimes \mathcal{O}_{\mathbb{P}^2}(1)) = h^0(V \otimes \mathcal{O}_{\mathbb{P}^2}(-1) \otimes K_{\mathbb{P}^2}) \geq 1$. Since a section of $V \otimes \mathcal{O}_{\mathbb{P}^2}(-1) \otimes K_{\mathbb{P}^2}$ is equivalent to a nonzero map $\mathcal{O}_{\mathbb{P}^2}(4) \to V$, it follows from (ii) that since $k_V < 0$ this case cannot occur for stable bundles V. Thus, we must have $h^0(V \otimes \mathcal{O}_{\mathbb{P}^2}(1)) \neq 0$. Hence there exists a nonzero map $\mathcal{O}_{\mathbb{P}^2}(-1) \to V$. The cokernel of this map must be torsion free, else there would exist a nonzero map from $\mathcal{O}_{\mathbb{P}^2}(k)$ to V with $k \geq 0$, contradicting (i). Thus, we obtain the exact sequence $(*)$, proving (ii).

(iii) Suppose that V is given by $(*)$. By (2.9) of Chapter 2 we have $c = \ell(Z) - 1$. By (ii) above, V is stable if and only if $h^0(V) = 0$. Since $H^1(\mathbb{P}^2, \mathcal{O}_{\mathbb{P}^2}(-1)) = 0$, $h^0(V) = h^0(\mathcal{O}_{\mathbb{P}^2}(1) \otimes I_Z)$, and this is nonzero if and only if there exists a section of $\mathcal{O}_{\mathbb{P}^2}(1)$ vanishing on Z, i.e., if and only if Z is contained in a line.

(iv) We may clearly assume that $c \leq 5$, so that V is given by an extension $(*)$. Since every 0-dimensional subscheme of $\mathbb{P}^2$ of length at most 2 is contained in a line, if V is stable, then necessarily $\ell(Z) \geq 3$ and so $c \geq 2$.

(v) This follows from the exact sequence

$$0 \to \mathcal{O}_{\mathbb{P}^2} \to V \otimes \mathcal{O}_{\mathbb{P}^2}(1) \to \mathcal{O}_{\mathbb{P}^2}(2) \otimes I_Z \to 0. \qquad \square$$

We shall next describe the bundles V corresponding to the first few choices of c.

Proposition 15. *Suppose that V is a stable rank 2 bundle on $\mathbb{P}^2$ with $c_1(V) = 0$ and $c_2(V) = 2$. Then there is an exact sequence*

$$0 \to \mathcal{O}_{\mathbb{P}^2}(-2) \oplus \mathcal{O}_{\mathbb{P}^2}(-2) \to (\mathcal{O}_{\mathbb{P}^2}(-1))^4 \to V \to 0.$$

Moreover, two subbundles W_1 and W_2 of $(\mathcal{O}_{\mathbb{P}^2}(-1))^4$, both isomorphic to $\mathcal{O}_{\mathbb{P}^2}(-2) \oplus \mathcal{O}_{\mathbb{P}^2}(-2)$, give isomorphic quotients V if and only if there exists a bundle automorphism of $(\mathcal{O}_{\mathbb{P}^2}(-1))^4$ taking W_1 to W_2.

Proof. Using (ii) and (iii) of Proposition 14 we may write V as an extension $(*)$ with $\ell(Z) = 3$ and Z is not contained in a line. It follows, for example, by [61, Prop. 4.1, p. 396] that Z imposes independent conditions on conics, so that $h^0((\mathcal{O}_{\mathbb{P}^2}(2) \otimes I_Z) = 3$. Next we claim that $\mathcal{O}_{\mathbb{P}^2}(2) \otimes I_Z$ is generated by its global sections, so that the natural map $(\mathcal{O}_{\mathbb{P}^2}(-1))^3 \to \mathcal{O}_{\mathbb{P}^2}(1) \otimes I_Z$ is surjective. One way to see this is to note that, again by [61, Prop. 4.1, p. 396], there exists a smooth conic C containing Z corresponding to a section of $\mathcal{O}_{\mathbb{P}^2}(2) \otimes I_Z$. The quotient of $\mathcal{O}_{\mathbb{P}^2}(2) \otimes I_Z$ by $\mathcal{O}_{\mathbb{P}^2}$ is then $(\mathcal{O}_{\mathbb{P}^2}(C)/\mathcal{O}_{\mathbb{P}^2}) \otimes I_Z$, which is a line bundle of degree 1 on C and so is generated by its global sections. Thus, $\mathcal{O}_{\mathbb{P}^2}(2) \otimes I_Z$, too, must be generated by its global sections.

Putting this together, there is a natural surjective map $(\mathcal{O}_{\mathbb{P}^2}(-1))^4 \to V$, which is more invariantly given by the natural map $H^0(V \otimes \mathcal{O}_{\mathbb{P}^2}(1)) \otimes \mathcal{O}_{\mathbb{P}^2}(-1) \to V$. Let W be the kernel. We must show that $W \cong \mathcal{O}_{\mathbb{P}^2}(-2) \oplus \mathcal{O}_{\mathbb{P}^2}(-2)$, or equivalently that $W \otimes \mathcal{O}_{\mathbb{P}^2}(2) \cong \mathcal{O}_{\mathbb{P}^2} \oplus \mathcal{O}_{\mathbb{P}^2}$. By the Whitney product formula, $c_1(W \otimes \mathcal{O}_{\mathbb{P}^2}(2)) = c_2(W \otimes \mathcal{O}_{\mathbb{P}^2}(2)) = 0$. From (iv) of Proposition 14, $W \otimes \mathcal{O}_{\mathbb{P}^2}(2)$ cannot be stable. Let $k \geq 0$ be the largest integer such that there exists a nonzero map $\mathcal{O}_{\mathbb{P}^2}(k) \to W \otimes \mathcal{O}_{\mathbb{P}^2}(2)$. If $k = 0$, then $W \otimes \mathcal{O}_{\mathbb{P}^2}(2)$ is an extension of $\mathcal{O}_{\mathbb{P}^2}$ by $\mathcal{O}_{\mathbb{P}^2}$ which must necessarily split, and we are done in this case. To rule out the possibility that $k > 0$, note that $W \otimes \mathcal{O}_{\mathbb{P}^2}(2)$ is a subbundle of $(\mathcal{O}_{\mathbb{P}^2}(-1))^4 \otimes \mathcal{O}_{\mathbb{P}^2}(2) = (\mathcal{O}_{\mathbb{P}^2}(1))^4$. Hence any nonzero map $\mathcal{O}_{\mathbb{P}^2}(k) \to W \otimes \mathcal{O}_{\mathbb{P}^2}(2)$ induces a nonzero map $\mathcal{O}_{\mathbb{P}^2}(k) \to (\mathcal{O}_{\mathbb{P}^2}(1))^4$. Thus, if $k > 0$, then $k = 1$ and the map $\mathcal{O}_{\mathbb{P}^2}(k) \to W \otimes \mathcal{O}_{\mathbb{P}^2}(2)$ has no zeros. But then $W \otimes \mathcal{O}_{\mathbb{P}^2}(2)$ is an extension of $\mathcal{O}_{\mathbb{P}^2}(-1)$ by $\mathcal{O}_{\mathbb{P}^2}(1)$ and $c_2(W \otimes \mathcal{O}_{\mathbb{P}^2}(2)) = 1 \neq 0$.

We have proved all of Proposition 15 except the last statement, which follows from the more general observation that any isomorphism $V_1 \to V_2$ between two stable bundles as in the theorem induces an isomorphism $H^0(V_1 \otimes \mathcal{O}_{\mathbb{P}^2}(1)) \otimes \mathcal{O}_{\mathbb{P}^2}(-1) \to H^0(V_2 \otimes \mathcal{O}_{\mathbb{P}^2}(1)) \otimes \mathcal{O}_{\mathbb{P}^2}(-1)$ which is compatible with the natural maps to V_1, V_2. $\square$

In the exercises we shall also show that every stable rank 2 bundle V on $\mathbb{P}^2$ with $c_1(V) = 0$ and $c_2(V) = 2$ is of the form $V_{2,-1}$ for a unique double cover $f: Q \to \mathbb{P}^2$, in the notation of Exercise 14 of Chapter 2. Each double cover $f: Q \to \mathbb{P}^2$ of the plane by a quadric is specified completely by its branch locus, which is a smooth conic in $\mathbb{P}^2$. In this way, we can identify the moduli space of stable rank 2 bundles V on $\mathbb{P}^2$ with $c_1(V) = 0$ and $c_2(V) = 2$ with the space of conics in $\mathbb{P}^2$, which is an open subset of the projective space $|\mathcal{O}_{\mathbb{P}^2}(2)| = \mathbb{P}^5$.

Next we turn to the case $c = 3$.

Proposition 16. *Suppose that V is a stable rank 2 bundle on $\mathbb{P}^2$ with $c_1(V) = 0$ and $c_2(V) = 3$, and write V as an extension* (∗). *If no three of the points in Z are collinear, then there is a natural exact sequence*

$$0 \to \mathcal{O}_{\mathbb{P}^2}(-3) \to (\mathcal{O}_{\mathbb{P}^2}(-1))^3 \to V \to 0.$$

The set of all such V is then identified with an open subset of the space of nets of conics in $\mathbb{P}^2$ which is isomorphic to the Grassmannian $G(3, 6)$.

Proof. Using the exact sequence

$$0 \to \mathcal{O}_{\mathbb{P}^2} \to V \otimes \mathcal{O}_{\mathbb{P}^2}(1) \to \mathcal{O}_{\mathbb{P}^2}(2)I_Z \to 0,$$

and the fact that there is a smooth conic through four points Z (possibly infinitely near) as long as no three are collinear, we easily check as above that with assumptions on Z as above, the natural map $(\mathcal{O}_{\mathbb{P}^2}(-1))^3 = H^0(V \otimes \mathcal{O}_{\mathbb{P}^2}(1)) \otimes \mathcal{O}_{\mathbb{P}^2}(-1) \to V$ is surjective. The kernel, which is a line bundle, is necessarily $\mathcal{O}_{\mathbb{P}^2}(-3)$, and all rank 2 bundles V so obtained are stable. Any bundle map $\mathcal{O}_{\mathbb{P}^2}(-3) \to (\mathcal{O}_{\mathbb{P}^2}(-1))^3$ is determined by a generic element in $H^0((\mathcal{O}_{\mathbb{P}^2}(2))^3)$, or more invariantly, by a generic net of conics. $\square$

The proof of the following for the cases $c = 4, 5$ is similar to the above proofs and so will be omitted.

Proposition 17.

(i) *Suppose that V is a stable rank 2 bundle on $\mathbb{P}^2$ with $c_1(V) = 0$ and $c_2(V) = 4$, and write V as an extension $(*)$. Then V may be written as an extension*

$$0 \to \mathcal{O}_{\mathbb{P}^2}(-1) \oplus \mathcal{O}_{\mathbb{P}^2}(-1) \to V \to (I_Z/I_C) \otimes \mathcal{O}_{\mathbb{P}^2}(1) \to 0,$$

where C is a conic containing Z. This extension is canonical if there is a unique conic containing Z, or equivalently by Proposition 14 if $h^0(V \otimes \mathcal{O}_{\mathbb{P}^2}(1)) = 2$.

(ii) *Suppose that V is a stable rank 2 bundle on $\mathbb{P}^2$ with $c_1(V) = 0$ and $c_2(V) = 5$. Then V may be uniquely written as an extension $(*)$ if and only if Z does not lie on a conic.* $\square$

We shall deal with the case $c_1(V) = \mathcal{O}_{\mathbb{P}^2}(1)$ for small values of $c_2(V)$ in the exercises.

Gieseker stability

For constructing compact moduli spaces, another notion of stability due to Gieseker has proved to be extremely important.

Definition 18. Let $p_1(n)$ and $p_2(n)$ be two real valued functions with domain the natural numbers. Then $p_1 \prec p_2$ (resp., $\preceq$) if, for all $n \gg 0$ we have $p_1(n) < p_2(n)$ (resp., $\leq$). Let V be a torsion free sheaf on X of rank r and H an ample line bundle. Define the *normalized Hilbert polynomial* $p_{H,V}(n) = (1/r)\chi(V \otimes H^{\otimes n})$. (It would amount to the same thing if we had used the function $(1/r)h^0(V \otimes H^{\otimes n})$ instead.) Then V is *Gieseker stable* (resp., *semistable*) if for all coherent subsheaves W of V with $0 < \operatorname{rank} W < \operatorname{rank} V$, we have $p_{H,W} \prec p_{L,V}$ (resp., $\preceq$).

If C is a curve, then for V a vector bundle of rank r and degree d on C, it follows from the Riemann-Roch theorem that

$$p_{H,V}(n) = hn + \frac{d}{r} + 1 - g(C),$$

where $h = \deg H$. Thus, V is Gieseker stable if and only if it is Mumford stable, and similarly for Gieseker semistability. For a surface X and a vector bundle V of rank r on X, we have by a slightly tedious calculation using (2.6) and Theorem 2 (the Riemann-Roch theorem) of Chapter 2,

$$p_{H,V}(n) = \frac{H^2 n^2}{2} + \left[\frac{(c_1(V) \cdot H)}{r} - \frac{(K_X \cdot H)}{2} \right] n$$
$$+ \frac{1}{r} \left(\frac{c_1(V)^2 - (c_1(V) \cdot K_X)}{2} - c_2(V) \right) + \chi(\mathcal{O}_X).$$

The constant term is just $(\chi(V))/r$. It follows that V is Gieseker stable if and only if, for all rank s subsheaves W of V with $0 < s < r$, either $\mu(W) < \mu(V)$ or $\mu(W) = \mu(V)$ and $\chi(W)/s < \chi(V)/r$. Thus, we have proved the following lemma in the surface case for vector bundles (although it is true in general):

Lemma 19. *If V is Mumford stable, it is Gieseker stable. If V is Gieseker semistable, it is Mumford semistable.* $\square$

The normalized Hilbert polynomial has the same convexity properties as the normalized degree (Exercise 9). Thus, Gieseker stable bundles have some properties which are similar to those enjoyed by stable bundles. For example, a Gieseker stable bundle is simple. On the other hand, Gieseker stability differs in many ways from ordinary stability. For example, it is possible for V to be Gieseker stable but for $V \otimes F$ to be Gieseker unstable, where F is a line bundle on X.

Unstable and semistable sheaves

Why is stability a good definition? I don't really know the answer to this question. But for many questions either Mumford or Gieseker stability is exactly what is needed. One partial answer to the question above is that unstable and strictly semistable bundles are much simpler to understand than stable ones. In fact, we can canonically construct an unstable bundle out of semistable torsion free sheaves of lower rank by successive extensions (this is called the *Harder-Narasimhan filtration*). Likewise, a strictly semistable bundle V with $\mu(V) = \mu$ is a successive extension of stable torsion free sheaves of smaller rank, all with normalized degree equal to μ, although in this case the corresponding filtration on V is not necessarily canonical. We shall just make this explicit in the case of a rank 2 bundle on a surface. In any case, the above says that the stable bundles are the interesting ones, since they are the ones we don't know how to describe canonically! From this point of view, stability is a nondegeneracy condition. Another partial answer to the above question is provided by the discussion in the next section of the complex differential geometry of stable bundles.

Let us now prove the above statements in the rank 2 case, leaving the general cases as a series of exercises.

Proposition 20. *Suppose that V is an unstable rank 2 bundle. Then there exists a unique sub-line bundle F of V with torsion free quotient such that $\mu(F) > \mu(V)$. Indeed, if L is a sub-line bundle of V such that $\mu(L) \geq \mu(V)$, then L is a subsheaf of F and $\mu(L) \leq \mu(F)$, with equality if and only if $L = F$.*

Proof. If V is unstable, then there exists some sub-line bundle F of V with torsion free quotient such that $\mu(F) > \mu(V)$ (Lemma 5). Thus, there is an exact sequence

$$0 \to F \to V \to F' \otimes I_Z \to 0,$$

where $\mu(F') < \mu(V)$. Now let L be a sub-line bundle of V such that $\mu(L) \geq \mu(V)$. We claim that the composite map $L \to F' \otimes I_Z$ is zero. For otherwise there is a nonzero map $L \to F'$, so that, by Lemma 3, $\mu(L) \leq \mu(F') < \mu(V)$, a contradiction. Thus, the map $L \to V$ factors through F, and so, if the quotient is torsion free, then $L = F$. □

Proposition 21. *Let V be a semistable but not stable rank 2 bundle. Then exactly one of the following holds:*

(i) *There is a unique sub-line bundle F of V with $\mu(F) = \mu(V)$. The quotient V/F is necessarily torsion free, and V is given canonically as an extension*

$$0 \to F \to V \to F' \otimes I_Z \to 0.$$

(ii) *There are exactly two distinct sub-line bundles F and G of V with $\mu(F) = \mu(G) = \mu(V)$. In this case $V = F \oplus G$.*

(iii) *$V = F \oplus F$, and there are infinitely many sub-line bundles with normalized degree $\mu(V)$, exactly corresponding to the choice of a line in $H^0(V \otimes F^{-1})$.*

More precisely, the following holds: Suppose that V is an arbitrary rank 2 vector bundle which is given given as an extension

$$0 \to F \to V \to F' \otimes I_Z \to 0,$$

and such that $\mu(F) = \mu(V)$. Then V is H-semistable and either F is the unique destabilizing sub-line bundle with torsion free quotient or $Z = \emptyset$ and $V = F \oplus F'$, i.e., the extension splits.

Proof. We shall just prove the last statement, leaving the remaining ones as an exercise. By Lemma 2, as $\mu(F) = \mu(V)$, we also have $\mu(F') = H \cdot F' = \mu(V)$. Thus, by Lemma 6, V is H-semistable. Let M be a sub-line bundle of V such that $H \cdot M \geq \mu(V)$. If the map $M \to V$ factors through F, then

$M = F \otimes \mathcal{O}_X(-D)$, where D is effective. Thus, $H \cdot M = H \cdot F - H \cdot D \le H \cdot F$, with equality holding if and only if $D = \emptyset$ and $M = F$. So in this case M is destabilizing only when $M = F$. Otherwise, the induced map $M \to F' \otimes I_Z$ is nonzero. Thus, $F' = M \otimes \mathcal{O}_X(D)$ for an effective divisor D, where $D = \emptyset$ only when $Z = \emptyset$ and $M = F'$. In this last case the extension

$$0 \to F \to V \to F' \to 0$$

clearly splits. Assuming that the extension does not split, we have $H \cdot M = H \cdot F' - H \cdot D \le H \cdot F' - 1$ and so $\mu(M) = H \cdot M \le \mu(V) - 1 < \mu(V)$. Thus, M is not destabilizing. $\square$

Change of polarization

Suppose that H_1 and H_2 are two ample divisors (although the proofs go through in case the H_i are just assumed to be nef and big). When does there exist a vector bundle which is H_1-stable but is not H_2-stable? We shall only look at this question for rank 2 vector bundles on a surface X (it can be analyzed for higher rank also, but the analysis requires Bogomolov's inequality and will be discussed in the exercises to Chapter 9). Most of these results and further developments can be found in [126]. Fix a rank 2 vector bundle V with $\det V = \mathcal{O}_X(\Delta)$ for some divisor Δ. Instead of using the classes $c_1(V) = \Delta$ and $c_2(V)$, we will use the classes $w_2(V) = \Delta$ mod $2 \in \operatorname{Num} X / 2 \operatorname{Num} X$ and $p_1(\operatorname{ad} V) = c_1(V)^2 - 4c_2(V)$. (Here $\operatorname{ad} V$ is the kernel of the trace map $\operatorname{Hom}(V, V) \to \mathcal{O}_X$, and $c_2(\operatorname{ad} V) = 4c_2(V) - c_1(V)^2 = -p_1(\operatorname{ad} V)$.) The advantage of this choice is that the classes $w_2(V)$ and $p_1(\operatorname{ad} V)$ are unchanged if we replace V by $V \otimes F$, where F is a line bundle. For V fixed, we set $w = w_2(V)$ and $p = p_1(\operatorname{ad} V)$.

Proposition 22. *In the above notation, an H_1-stable bundle V is not H_2-stable if and only if there exists a sub-line bundle $\mathcal{O}_X(D)$ of V with torsion free quotient such that*

$$H_1 \cdot (2D - \Delta) < 0 \le H_2 \cdot (2D - \Delta),$$

and such that

$$p \le (2D - \Delta)^2 < 0.$$

Moreover, $\mathcal{O}_X(D)$ is the unique sub-line bundle of V with torsion free quotient with the above properties. Finally, V is strictly semistable with respect to an ample divisor which is a convex combination of H_1 and H_2.

Proof. V is not H_2-stable if and only if there exists a sub-line bundle $\mathcal{O}_X(D)$ of V with torsion free quotient such that

$$H_2 \cdot D = \mu_{H_2}(\mathcal{O}_X(D)) \ge \mu_{H_2}(V) = H_2 \cdot \Delta/2.$$

Thus, $H_2 \cdot (2D - \Delta) \geq 0$. Since V is H_1-stable, $H_1 \cdot (2D - \Delta) < 0$. Now $2D - \Delta$ is orthogonal to a convex combination of H_1 and H_2, so by the Hodge index theorem $(2D - \Delta)^2 \leq 0$, with equality only if $2D - \Delta$ is numerically trivial. However, $(2D - \Delta) \cdot H_1 < 0$, so that this last case does not occur. There is an exact sequence

$$0 \to \mathcal{O}_X(D) \to V \to \mathcal{O}_X(\Delta - D) \otimes I_Z \to 0$$

for some 0-dimensional subscheme Z. Thus,

$$c_2(V) = -D^2 + D \cdot \Delta + \ell(Z) \geq -D^2 + D \cdot \Delta.$$

So

$$(2D - \Delta)^2 = 4D^2 - 4D \cdot \Delta + \Delta^2 \geq -4c_2(V) + c_1(V)^2 = p.$$

The uniqueness of $\mathcal{O}_X(D)$ follows from Propositions 20 and 21: either V is H_2-unstable and $\mathcal{O}_X(D)$ is the unique destabilizing sub-line bundle, or V is H_2-semistable. However, as V is H_1-stable, it cannot be a direct sum of line bundles, so that only the first case of Proposition 21 can occur. Finally, if we choose H to be a convex linear combination of H_1 and H_2 which is orthogonal to $2D - \Delta$, then H is ample and $\mu_H(\mathcal{O}_X(D)) = \mu_H(\mathcal{O}_X(\Delta - D))$. Thus, by Lemma 6, V is strictly semistable. $\square$

The meaning of the conditions of Proposition 22 is as follows. Given $\zeta \in \operatorname{Num} X$, ζ is a *class of type* (w, p) if the mod 2 reduction of ζ is w and $p \leq \zeta^2 < 0$. For such ζ, we define the *wall* $W^\zeta = \zeta^\perp \subset \mathcal{A}(X)$, where $\mathcal{A}(X)$ is the ample cone of X, provided that $W^\zeta \neq \emptyset$ and $W^\zeta \neq \mathcal{A}(X)$, i.e., provided that there exists an ample divisor orthogonal to ζ and provided that ζ is not numerically trivial. In fact, ζ determines an *oriented wall*, meaning that $\mathcal{A}(X) - W^\zeta$ has two connected components and on one of them the linear from $(\cdot \zeta)$ is positive. Notice that ζ is not uniquely determined by the (oriented) wall, but that any two classes defining the same oriented wall are positive rational multiples of each other. In particular, since $\zeta^2 \geq p$, only finitely many ζ define the same oriented wall. There is also the following standard fact, for whose proof we refer to [38]:

Proposition 23. *For fixed w and p, the set*

$$\{W^\zeta : \zeta \text{ is a class of type } (w, p)\}$$

is locally finite in $\mathcal{A}(X)$. $\square$

We shall refer to the set of walls described in Proposition 23 as the *walls of type* (w, p). The connected components of the complement of the set of walls of type (w, p) are called the *chambers of type* (w, p). It follows that the definition of an L-stable rank 2 vector bundle V with $w_2(V) = w$ and $p_1(\operatorname{ad} V) = p$ only depends on the chamber containing L, provided that L lies in a chamber, in other words, does not lie on a wall of type (w, p). The chambers in $\mathcal{A}(X)$ for the walls of type (w, p) are the algebro-geometric analogue of the chambers needed to define the Donaldson invariants for a 4-manifold with $b_2^+ = 1$. We shall return to this point in Chapter 8. Finally, let us note that we can in some sense reverse the analysis of Proposition 22 to find bundles which are H_1-stable but H_2-unstable:

Proposition 24. *Suppose that H_1 and H_2 are two ample divisors, and that W^ζ is the unique wall of type (w, p) separating H_1 and H_2, and assume further that $\zeta \cdot H_1 < 0 < \zeta \cdot H_2$. Let V be given by a nonsplit exact sequence*

$$0 \to \mathcal{O}_X(D) \to V \to \mathcal{O}_X(\Delta - D) \otimes I_Z \to 0,$$

where $\zeta = 2D - \Delta$ and $w_2(V) = w$, $p_1(\operatorname{ad} V) = p$. Then V is H_1-stable and H_2-unstable.

Proof. Clearly, if $\zeta \cdot H_2 > 0$, then V is H_2-unstable. We claim that V is H_1-stable. Suppose not, i.e., suppose that there exists a sub-line bundle $\mathcal{O}_X(F)$ of V with torsion free quotient and such that $\mu_{H_1}(\mathcal{O}_X(F)) \geq \mu_{H_1}(V)$. Since $\zeta \cdot H_1 < 0$, $\mathcal{O}_X(F) \neq \mathcal{O}_X(D)$ and in fact $\mathcal{O}_X(F)$ is not a subsheaf of $\mathcal{O}_X(D)$. Since $\mathcal{O}_X(F)$ is not a subsheaf of $\mathcal{O}_X(D)$, it follows from Proposition 20 applied to the H_2-unstable bundle V that $\mu_{H_2}(\mathcal{O}_X(F)) < \mu_{H_2}(V)$. Hence $\eta \cdot H_1 > 0 > \eta \cdot H_2$, where η is the class $2F - \Delta$. Clearly, η is a class of type (w, p) separating H_1 and H_2, so that η is proportional to ζ. Now choose an ample $H \in W^\zeta = W^\eta$ (for example, a suitable convex combination of H_1 and H_2). By Lemma 6 V is H-semistable. By applying Proposition 21 to V, since $\mathcal{O}_X(F) \neq \mathcal{O}_X(D)$, we must have $V = \mathcal{O}_X(D) \oplus \mathcal{O}_X(F)$, contradicting the assumption that V was not split. $\square$

Note that, if the above extension is split, then $V = \mathcal{O}_X(D) \oplus \mathcal{O}_X(\Delta - D)$ is unstable for every H_1 not on the wall W^ζ. Finally, for $\ell(Z) \gg 0$, we can use the discussion of Chapter 2 to find bundles V corresponding to nonsplit extensions for which V is H_1-stable and H_2-unstable.

The differential geometry of stable vector bundles

In this section, we assume that the reader knows a little complex differential geometry (which can be found, for example, in [55]) and describe some of the special properties of stable vector bundles on Kähler manifolds. We can interpret these results as giving another answer to the question raised at the beginning of the last section: Why is stability a good definition for a holomorphic vector bundle?

Let M be a manifold and let E be a C^∞ complex vector bundle on M of rank r. Recall that a *connection* on E is a $\mathbb{C}$-linear map D from C^∞ sections of E to sections of $A^1(E) = E \otimes A^1(M)$, where here $A^1(M)$ is the bundle of C^∞ 1-forms on M, satisfying the Leibniz rule: for all sections s of E and C^∞ functions f on M,

$$D(fs) = fDs + s \otimes df.$$

It follows that the difference of two connections is a C^∞ 1-form with coefficients in End E, and in fact the space of all connections is an affine space for $A^1(\text{End } E)$. There is a natural extension of D to an operator from $A^p(E)$ to $A^{p+1}(E)$, where $A^p(E)$ is the vector bundle of p-forms with coefficients in E, by requiring the graded Leibniz rule

$$D(\phi \otimes s) = d\phi \otimes s + (-1)^p \phi \otimes Ds.$$

The *curvature* D^2 of the connection D is a C^∞ section of $A^2(\text{End } E) = A^2(M) \otimes \text{End } E$. Choosing a local basis $s_1, \ldots, s_r$ of C^∞ sections, we can identify a section s with a vector of functions and we can write $Ds = ds + As$, where A is a matrix of 1-forms, called the *connection matrix*. We frequently use the letter A to denote the connection D as well. In this case the curvature D^2 is locally given by the matrix $F_A = dA + A \wedge A$, which transforms as a section of $A^2(\text{End } E)$. The vector bundle E (or more precisely the pair (E, D)) is *flat* if $D^2 = 0$, in which case we say that D is *integrable*. As a corollary of the Frobenius theorem, if E is flat and M is simply connected, then E is trivialized by global sections $s_1, \ldots, s_r$ such that $Ds_i = 0$ for all i. More generally, for an arbitrary manifold M, flat vector bundles E correspond to representations of $\pi_1(M, *)$ into $GL(r, \mathbb{C})$. Given bundles E_1 and E_2 and connections D_i on E_i, there is a naturally induced connection $D = D_1 \oplus D_2$ on $E_1 \oplus E_2$, and $D^2 = D_1^2 \oplus D_2^2$ in the obvious sense. Likewise, there is an induced connection $D_1 \otimes \text{Id} + \text{Id} \otimes D_2$ on $E_1 \otimes E_2$, and its curvature is equal to $D_1^2 \otimes \text{Id} + \text{Id} \otimes D_2^2$. A connection D is *reducible* if E is a direct sum $E = E_1 \oplus E_2$, where both E_1 and E_2 have positive rank, and $D = D_1 \oplus D_2$ for some connections D_1 and D_2 on E_1 and E_2, respectively. The connection E is *irreducible* if it is not reducible.

In the cases of interest, E will have a Hermitian metric $\langle \cdot, \cdot \rangle$, and D will be compatible with the metric in the sense that

$$\langle Ds_1, s_2 \rangle + \langle s_1, Ds_2 \rangle = d\langle s_1, s_2 \rangle.$$

Thus, if s_i is an orthonormal basis with respect to the inner product, then the connection matrix A is skew-Hermitian, or in other words it lies in the Lie algebra $\mathfrak{u}(r)$ of the unitary group $U(r)$. We say that the connection A is *unitary* or *Hermitian*. In this case, the curvature, computed in a local orthonormal frame, is a skew-Hermitian matrix of 2-forms. The flat vector bundles E whose connections are compatible with a Hermitian metric essentially correspond to representations of $\pi_1(M, *)$ into $U(r)$. For unitary connections, we will take reducible to mean that $E = E_1 \oplus E_2$ is an orthogonal direct sum of (nonzero) vector bundles E_i, and D_i is a unitary connection on E_i, such that $D = D_1 + D_2$.

If E is a Hermitian vector bundle and D is a connection which is compatible with the metric on E, then we can consider the characteristic polynomial

$$\det\left(\frac{i}{2\pi}D^2 + t\operatorname{Id}\right) = \sum_{k=0}^{r} c_k(E)t^{r-k}.$$

Here the coefficients $c_k(E)$ turn out to be closed forms of degree $2k$ representing the Chern classes of the vector bundle E. Thus, for example, $c_1(E) = (i/2\pi)\operatorname{trace} D^2$. Note that, if D is flat, then $c_i(E) = 0$ for all $i > 0$.

Now suppose that M is a complex manifold, so that $d = \partial + \bar\partial$. Let $\Omega^{p,q}(M)$ be the vector bundle of forms of type (p, q), and, for a complex vector bundle E, define $\Omega^{p,q}(E)$ similarly. If E is holomorphic, then $\bar\partial$ is well defined on C^∞ sections of E, and we say that the connection D is *compatible with the complex structure* if $\pi^{0,1}(D) = \bar\partial$, where $\pi^{0,1}\colon A^1(E) \to \Omega^{0,1}(E)$ is the projection induced from the projection of the 1-forms on M to the $(0, 1)$-forms. In this case $\pi^{0,2}(D^2) = 0$, in other words, the curvature has no component of type $(0, 2)$. Conversely, if E is a C^∞ vector bundle and D is a connection on E such that $\pi^{0,2}(D^2) = 0$, then there exists a unique holomorphic structure on E for which D is a compatible connection (this is an easier special case of the Newlander-Nirenberg theorem on integrable almost complex structures). For example, it is easy to see that a flat complex vector bundle on M has a natural holomorphic structure. Every holomorphic vector bundle E with a Hermitian metric has a unique unitary connection D which is compatible with the complex structure [55, p. 73]. We shall refer to D as the *compatible unitary connection* associated to the metric. In this case, since $\bar\partial^2 = 0$, D^2 has no component of type $(0, 2)$, and since it is skew-Hermitian, it has no $(2, 0)$-component either. Thus, the curvature D^2 lives in $\Omega^{1,1}(E)$ (and is skew-Hermitian). It follows that the Chern classes $c_k(E)$ are represented by real forms of type (k, k). If $D = D_1 + D_2$ is a reducible connection on E, corresponding to a direct sum decomposition $E = E_1 \oplus E_2$, then it is easy to check that the bundles E_i are again holomorphic and the D_i are compatible with the complex structures.

Suppose in addition that M is a Kähler manifold with Kähler metric ω. If E is a holomorphic vector bundle on M with a Hermitian metric and D is a unitary connection on E which is compatible with the complex structure, then D^2 is a $(1, 1)$-form with coefficients in $\operatorname{End} E$, and so $D^2 \wedge \omega^{n-1}$ is a form of type (n, n) with coefficients in $\operatorname{End} E$. Thus, we can write $D^2 \wedge \omega^{n-1} = \hat{F} \cdot \omega^n$, where $\hat{F}$ is a C^∞ section of $\operatorname{End} E$. Note that, up to a positive scalar, $\hat{F}$ is the same as the contraction of D^2 with ω (since ω^n is $n!$ times the volume form on M).

Definition 25. Let M be a compact Kähler manifold with Kähler metric ω and let E be a holomorphic vector bundle on M with a Hermitian metric. If D is a unitary connection on E which is compatible with the complex structure, then D is a *Hermitian-Einstein* connection if, in the above notation, $\hat{F} = \lambda \operatorname{Id}$ for some constant λ.

If $\hat{F} = \lambda \operatorname{Id}$ for some constant λ, then $D^2 \wedge \omega^{n-1} = \lambda \operatorname{Id} \cdot \omega^n$ and so $\operatorname{trace}(D^2) \wedge \omega^{n-1} = r\lambda \omega^n$. Thus, integrating over M, we find that

$$-2\pi i \int_M c_1(E) \wedge \omega^{n-1} = r\lambda \int_M \omega^n.$$

Here $\int_M \omega^n = n! \operatorname{vol}(M)$ and the formula says that, up to universal positive constants, $\lambda = -i\mu_\omega(E)$, where we define μ_ω by analogy with the normalized degree to be

$$\mu_\omega(E) = \frac{1}{r} \int_M c_1(E) \wedge \omega^{n-1}.$$

Thus, if $E = E_1 \oplus E_2$ and each E_i is a vector bundle of positive rank with a Hermitian-Einstein connection D_i, then $D = D_1 + D_2$ is a Hermitian-Einstein connection on E if and only if $\mu_\omega(E_1) = \mu_\omega(E_2)$. On the other hand, if E_i has a Hermitian-Einstein connection D_i for $i = 1, 2$, then the connection $D_1 \otimes \operatorname{Id} + \operatorname{Id} \otimes D_2$ is always a Hermitian-Einstein connection on $E_1 \otimes E_2$.

We turn now to the study of Hermitian-Einstein connections in special cases. If E is a holomorphic line bundle, a connection D is a Hermitian-Einstein connection on E if and only if $D^2 \wedge \omega^{n-1}$ is a constant multiple of ω^n. Starting with a fixed metric $|\cdot|^2$ on E, there is an associated compatible unitary connection D_0. If we replace $|\cdot|^2$ by $e^h |\cdot|^2$, then D_0 is replaced by $D = D_0 + \partial h$ and D_0^2 by $D_0^2 + \bar{\partial}\partial h$ (compare [55, p. 73]). By the $\partial\bar{\partial}$-lemma ([55, p. 149]), there exists a choice of h so that $D_0^2 + \bar{\partial}\partial h$ is a harmonic $(1, 1)$-form. Every such form can be written as $\lambda\omega + \psi$, where λ is constant and ψ is a harmonic $(1, 1)$-form satisfying $\int_M \psi \wedge \omega^{n-1} = 0$, since ω is harmonic and ω^n is nonzero in cohomology. Now since M is a Kähler manifold and ψ is harmonic, the form $\psi \wedge \omega^{n-1}$ is a harmonic (n, n)-form and is thus a constant multiple of the volume form, so it must be identically zero, since its integral over M is 0. Hence ψ is pointwise orthogonal to ω^{n-1} and so $(\lambda\omega + \psi) \wedge \omega^{n-1} = \lambda\omega^n$. Thus, we see that, if D^2 is harmonic, then D is a Hermitian-Einstein connection. In fact, there is a unique choice of a metric on E, up to a constant factor, so that the corresponding compatible connection D is Hermitian-Einstein. It suffices to show that, for a real-valued C^∞ function h, if $\bar{\partial}\partial h$ is pointwise orthogonal to ω^{n-1}, then h is constant. If we set $\xi = \bar{\partial}\partial h$, then ξ is an imaginary $(1, 1)$-form pointwise orthogonal to ω^{n-1}, and $\int_M \xi \wedge \xi \wedge \omega^{n-2} = 0$ by Stokes' theorem, since $\xi = d\partial h$. On the other hand, a slight generalization of the proof of the Hodge index theorem in [55, p. 125], shows that $\xi \wedge \xi \wedge \omega^{n-2}$ is a nonpositive multiple of ω^n, which is 0 if and only if $\xi = 0$. Thus, $\bar{\partial}\partial h = 0$. It is easy to see that the maximum principle holds for such functions h. In fact, using the Kähler identities, it is easy to check that in our case $\bar{\partial}\partial h = 0$ implies that h is harmonic. Since M is compact, h must be constant. We conclude that there is a C^∞ real valued function h, unique up to a constant, such that the compatible connection with respect to $e^h |\cdot|^2$ is Hermitian-Einstein. There is also an equivalent formulation in terms of bundle automorphisms of a fixed unitary bundle which we shall not describe here.

Next we consider the meaning of Hermitian-Einstein connections on a vector bundle E with $c_1(E) = 0$ for compact complex manifolds of dimension 1 or 2. If M is a compact complex curve and $c_1(E) = 0$ as a real cohomology class, then D is a Hermitian-Einstein connection on M if and only if $D^2 = 0$, in other words if and only if D is a flat unitary connection.

Suppose that M is a compact Kähler surface with Kähler form ω. Then the Hodge $*$-operator acts on 2-forms, and $*\omega = \omega$. In fact, for a general 4-manifold M, the bundle of C^∞ 2-forms $A^2(M)$ splits into the $(+1)$ and (-1) eigenspaces for $*$: $A^2(M) = \Omega_+^2(M) \oplus \Omega_-^2(M)$, and likewise for $A^2(E)$. Thus, given a connection D on E, we can split its curvature into two parts, D_+^2 and D_-^2. We say that D is *self-dual* if $D_-^2 = 0$ and *anti-self-dual* if $D_+^2 = 0$. In case M is a Kähler surface, it is easy to check that the complexification of $\Omega_+^2(M)$ is just $\Omega^{2,0}(M) \oplus \Omega^{0,2}(M) \oplus A^0(M)_{\mathbb{C}} \cdot \omega$, where $A^0(M)_{\mathbb{C}}$ is the bundle of complex-valued C^∞ functions on M, and that the complexification of $\Omega_-^2(M)$ is the orthogonal complement to ω in $\Omega^{1,1}(M)$. Thus, a connection D on E is anti-self-dual if and only if the curvature D^2 is of type $(1,1)$ and orthogonal to ω. Hence, if E is a holomorphic Hermitian vector bundle with $c_1(E) = 0$ and D is the unitary connection on E compatible with the complex structure, then D is anti-self-dual if and only if D is a Hermitian-Einstein connection.

Note that the definition of the normalized degree μ_ω enables us to define ω-stability for a general Kähler metric on a compact Kähler manifold M, by copying Definition 4. In case ω is a Hodge metric associated to an ample divisor H, then ω-stability is the same as H-stability. The following is the main result concerning Hermitian-Einstein connections:

Theorem 26. *Let M be a compact Kähler manifold, with Kähler form ω, and let E be a holomorphic vector bundle on M. If there exists a Hermitian metric on E whose associated compatible unitary connection is an irreducible Hermitian-Einstein connection, then E is ω-stable. Conversely, if E is ω-stable, then there is a Hermitian metric on E whose associated compatible unitary connection is an irreducible Hermitian-Einstein connection on E, and this connection is unique up to C^∞ bundle automorphisms of E.*

The case $\dim M = 1$ was proved by Narasimhan and Seshadri [112]. In this case, if $c_1(E) = 0$, the theorem essentially asserts that a stable bundle

V on M such that $\deg \det V = 0$ is equivalent to an irreducible unitary representation of $\pi_1(M, *)$. The uniqueness part of the general statement was proved by Kobayashi [70] and Lübke [84]. Donaldson [23] showed the existence for an algebraic surface, and then Uhlenbeck and Yau [146] proved the result for a general Kähler manifold. (See also [25] for a proof for smooth projective varieties.)

We shall prove one very special case of a small part of Theorem 26. Suppose that E is a rank 2 holomorphic vector bundle on M with a Hermitian-Einstein connection D. We claim that either E is ω-stable, in other words that, for every holomorphic sub-line bundle L of E, $\mu_\omega(L) < \mu_\omega(E)$, or that there exists a holomorphic sub-line bundle L of E with $\mu_\omega(L) = \mu_\omega(E)$ and $E \cong L \oplus L'$ for some holomorphic line bundle L' such that $\mu_\omega(L') = \mu_\omega(L) = \mu_\omega(E)$ (in which case E is strictly ω-semistable). To see this, suppose that L is a holomorphic sub-line bundle of E. Note that L has a Hermitian-Einstein connection since it is a line bundle, and so we can consider the induced Hermitian-Einstein connection on $E \otimes L^{-1}$. Now $\mu_\omega(E \otimes L^{-1}) = \mu_\omega(E) - \mu_\omega(L)$. Moreover, $E \otimes L^{-1}$ has a holomorphic section and $\mu_\omega(L) \geq \mu_\omega(E)$ if and only $\mu_\omega(E \otimes L^{-1}) \leq 0$. Thus, after replacing E by $E \otimes L^{-1}$, it suffices to show: if E is a holomorphic rank 2 vector bundle with $\mu_\omega(E) \leq 0$ and a Hermitian-Einstein connection, then E has a nonzero holomorphic section if and only if $\mu_\omega(E) = 0$ and in this case $E = \mathcal{O}_M \oplus L'$ for some holomorphic line bundle L' with $\mu_\omega(L') = 0$.

To see this last statement, we begin by proving the identity, for an arbitrary compatible unitary connection D,

$$2\bar{\partial}^*\bar{\partial} = D^*D - c_n i\hat{F},$$

where c_n is a positive real number and $*$ represents the formal adjoint of the appropriate differential operator. Indeed, if Λ denotes the operator which is given by contraction with ω, then we have the Kähler identities [55, p. 111]:

$$\bar{\partial}^* = i[\partial, \Lambda]; \qquad \partial^* = -i[\bar{\partial}, \Lambda],$$

and similar identities hold for the $(1,0)$ and $(0,1)$ parts of D, which we can write as $D^{1,0}$ and $\bar{\partial}$. Thus, for $(0,1)$- or $(1,0)$-forms, we have

$$\bar{\partial}^* = -i\Lambda D^{1,0}; \qquad (D^{1,0})^* = i\Lambda \bar{\partial},$$

and so

$$\begin{aligned} D^*D &= ((D^{1,0})^* + \bar{\partial}^*)(D^{1,0} + \bar{\partial}) \\ &= (i\Lambda\bar{\partial} - i\Lambda D^{1,0})(D^{1,0} + \bar{\partial}) \\ &= -i\Lambda(D^{1,0}\bar{\partial} - \bar{\partial}D^{1,0}), \end{aligned}$$

since the $(2,0)$ part of D^2 is 0. Now D^2 is equal to its $(1,1)$ component, namely $D^{1,0}\bar{\partial} + \bar{\partial}D^{1,0}$, and so $i\Lambda D^2 = i\Lambda(D^{1,0}\bar{\partial} + \bar{\partial}D^{1,0})$. Thus,

$$D^*D = i\Lambda D^2 - 2i\Lambda D^{1,0}\bar{\partial}$$

$$= i\Lambda D^2 + 2\bar{\partial}^*\bar{\partial},$$

where we have used the Kähler identities again. Since $\Lambda D^2 = \hat{F}$ up to a positive factor, we are done.

In the Hermitian-Einstein case, up to a positive scalar, $\hat{F} = -i\mu_\omega(E)$. Thus, after replacing c_n by another suitable positive constant, we find, for a holomorphic section s of E, that

$$D^*Ds - c_n\mu_\omega(E)s = 0.$$

Taking the inner product of this expression with s itself and integrating over M gives

$$\|Ds\|^2 - c_n\mu_\omega(E)\int_M |s|^2 = 0.$$

Since $s \neq 0$ and $\mu_\omega(E) \leq 0$, the only way that this expression can be 0 is if $\mu_\omega(E) = 0$ and $Ds = 0$, in other words s is a covariant constant section of E. In this case, it follows that s does not vanish at any point of E, and so defines an inclusion of $\mathcal{O}_M$ as a holomorphic subbundle. Let L' be the orthogonal complement to the corresponding C^∞ subbundle of E. Then L' is a C^∞ complex line bundle, and since $Ds = 0$ it is easy to see that D induces a connection on L'. Thus, $E = \mathcal{O}_M \oplus L'$ as C^∞ bundles in such a way that the connection $D = D_1 \oplus D_2$, where D_1 is the trivial connection. But then D_2 defines a holomorphic structure on L'. Finally, $0 = \mu_\omega(E) = \mu_\omega(L')$, as desired. $\square$

Our final result is the analytic proof of Bogomolov's inequality, which we only state in the rank 2 case:

Theorem 27. *Let E be a rank 2 holomorphic vector bundle on M and let D be a Hermitian-Einstein connection on E. Then $\int_M(4c_2(E) - c_1^2(E)) \wedge \omega^{n-2} \geq 0$, or in other words $\int_M p_1(\mathrm{ad}\,E) \wedge \omega^{n-2} \leq 0$. Moreover, if $\int_M(4c_2(E) - c_1^2(E)) \wedge \omega^{n-2} \geq 0$ and $c_1(E) = 0$, then D is a flat connection.*

Proof. In a local orthonormal frame, the curvature matrix D^2 looks like

$$\begin{pmatrix} \lambda\omega + i\alpha & \varphi \\ -\bar{\varphi} & \lambda\omega + i\beta \end{pmatrix},$$

where α, β, φ are pointwise orthogonal to ω^{n-1} and α, β are real. Thus, $(4c_2(E) - c_1^2(E)) \wedge \omega^{n-2}$ is represented by

$$-\frac{1}{4\pi^2}\left[4\lambda^2\omega \wedge \omega - 4\alpha \wedge \beta + 4\varphi \wedge \bar{\varphi} - (2\lambda\omega + i(\alpha + \beta))^2\right] \wedge \omega^{n-2},$$

where we have used $\alpha \wedge \omega^{n-1} = \beta \wedge \omega^{n-1} = 0$. Expanding this out, we are left with

$$-\frac{1}{4\pi^2}\left[(\alpha - \beta)^2 + 4\varphi \wedge \bar{\varphi}\right] \wedge \omega^{n-2}.$$

Again by looking at the analytic proof of the Hodge index theorem, we see that $-(1/4\pi^2)\left[(\alpha - \beta)^2 + 4\varphi \wedge \bar{\varphi}\right] \wedge \omega^{n-2}$ is a pointwise nonnegative multiple of the volume form on M, and thus $\int_M (4c_2(E) - c_1^2(E)) \wedge \omega^{n-2} \geq 0$. If equality holds, we must have $\alpha = \beta$ and $\varphi = 0$. If in addition $c_1(E) = 0$, then $\lambda = 0$ and $\alpha + \beta = 2\alpha$ is a real $(1,1)$-form pointwise orthogonal to ω^{n-1} which is exact, since it represents $c_1(E)$. Thus, $\int_M \alpha \wedge \alpha \wedge \omega^{n-2} = 0$. It follows as in the discussion of Hermitian-Einstein metrics on line bundles that $\alpha = 0$. Thus, $D^2 = 0$ and so D is flat. $\square$

If $\int_M (4c_2(E) - c_1^2(E)) \wedge \omega^{n-2} = 0$ but $c_1(E) \neq 0$, then E need not be flat. For example, if E is of the form $E' \otimes L$, where E' is flat and L is a holomorphic line bundle with $c_1(L) \neq 0$, then $\int_M (4c_2(E) - c_1^2(E)) \wedge \omega^{n-2} = 0$. However, not all ω-stable rank 2 bundles E satisfying $\int_M (4c_2(E) - c_1^2(E)) \wedge \omega^{n-2} = 0$ are of the form $E' \otimes L$, where E' is flat and L is a holomorphic line bundle (see Exercise 3 in Chapter 6).

Exercises

1. Let X be a scheme, proper over a field k, and let $\mathcal{F}$ be a coherent sheaf on X. Then an injective map φ from $\mathcal{F}$ to itself is an isomorphism. (Since $\mathrm{Hom}(\mathcal{F}, \mathcal{F})$ is finite-dimensional, φ satsifies a polynomial equation, and since φ is injective we can assume that this polynomial has a nonzero constant term. Clearly, φ satisfies such a polynomial equation on each fiber $\mathcal{F}/\mathfrak{m}_x\mathcal{F}$, which is a finite-dimensional vector space. Thus, on each fiber φ is injective and therefore surjective, so it is surjective by Nakayama's lemma.)

2. Let V be a semistable rank 2 bundle of degree d on a curve C of genus g. Show that, for every line bundle L on C, if $\deg L \geq 2g - [d/2] - 1$, then $H^1(V \otimes L) = 0$, and if $\deg L \geq 2g - [d/2]$, then $V \otimes L$ is generated by its global sections. (If $H^1(V \otimes L) \neq 0$, then $H^0(V^\vee \otimes L^{-1} \otimes K_C) \neq 0$, so that there is a nonzero map from V to a line bundle of degree $[d/2] - 1$. Likewise, if $\deg L \geq 2g - [d/2]$, then $H^1(V \otimes L \otimes \mathcal{O}_C(p)) = 0$ for every $p \in C$.)

3. Let C be a smooth curve of genus at least 2 and let e be a positive integer. Let L be a line bundle of degree $-e \geq -(g-1)/2$. Show that, for the generic extension

$$0 \to L \to V \to L^{-1} \to 0,$$

V is stable and there does not exist a nonzero map from a line bundle L' to V unless $L = L'$ or $\deg L' < \deg L$. (Imitate the proof of Theorem 10 as follows: suppose that V is an extension corresponding to $\xi \in H^1(L^2)$. By Riemann-Roch, $\dim H^1(L^2) = 2e + g - 1$. Thus, the set of all extensions is parametrized by $\mathbb{P}^{2e+g-2}$. Suppose that L' is another line bundle of degree $-d$, where d is allowed to be negative. If $\deg L' \geq \deg L$ and $L' \neq L$, then $\mathrm{Hom}(L', L) = 0$, and so there must exist an element of $\mathrm{Hom}(L', L^{-1})$ which lifts to an element of $\mathrm{Hom}(L', V)$. Now

$L' = L^{-1} \otimes \mathcal{O}_C(-D)$, where D is an effective divisor of degree $d + e$. The method of proof of Theorem 10 shows that ξ is in the kernel of the map $H^1(L^2) \to H^1(L^2 \otimes \mathcal{O}_C(D))$. Show that this kernel is a linear space of dimension $d + e$, and so we need ξ not in the union, over all divisors D of degree $d + e$, of a linear space of dimension $d + e - 1$ in $\mathbb{P}^{2e+g-2}$. Compare $2e + 2d - 1$ with $\dim \mathbb{P}^{2e+g-2} = 2e + g - 2$ and use $d \leq e \leq (g-1)/2$. What happens for $e = (g-1)/2$?)

4. If V is a rank 2 bundle on a surface X, use the splitting principle and the Riemann-Roch theorem for vector bundles to show that

$$\chi(Hom(V, V)) = c_1^2(V) - 4c_2(V) + 4\chi(\mathcal{O}_X).$$

5. If V is a rank 2 stable bundle on $\mathbb{P}^2$, then $c_1^2(V) < 4c_2(V)$. (As V is simple, $h^0(Hom(V, V)) = 1$. By Serre duality, $h^2(Hom(V, V)) = h^0(Hom(V, V) \otimes K_{\mathbb{P}^2}) = 0$ since $K_{\mathbb{P}^2} = \mathcal{O}_{\mathbb{P}^2}(-3) \subset \mathcal{O}_{\mathbb{P}^2}$. Now apply the previous exercise.) Show that if V is a rank 2 strictly semistable bundle on $\mathbb{P}^2$, then $c_1^2(V) \leq 4c_2(V)$.

6. Suppose that $0 \leq a \leq b \leq c$ are three integers. Show that there exist vector bundles V which fit into an exact sequence

$$0 \to \mathcal{O}_{\mathbb{P}^2} \to \mathcal{O}_{\mathbb{P}^2}(a) \oplus \mathcal{O}_{\mathbb{P}^2}(b) \oplus \mathcal{O}_{\mathbb{P}^2}(c) \to V \to 0.$$

What is k_V in this case and when is V stable or semistable? Use this to show that the tangent bundle of $\mathbb{P}^2$ is stable, via the Euler exact sequence

$$0 \to \mathcal{O}_{\mathbb{P}^2} \to \bigoplus_0^2 \mathcal{O}_{\mathbb{P}^2}(1) \to \Theta_{\mathbb{P}^2} \to 0.$$

7. Let V be a rank 2 bundle on $\mathbb{P}^2$ with $\det V = \mathcal{O}_{\mathbb{P}^2}(1)$.
   (a) Show that if V is given by an extension

$$0 \to \mathcal{O}_{\mathbb{P}^1} \to V \to \mathcal{O}_{\mathbb{P}^1}(1) \otimes I_Z \to 0,$$

then V is stable if and only if $Z \neq \emptyset$.
   (b) Conversely, show that, if $c_2(V) \leq 4$ and V is stable, then V is given by an exact sequence as in (a), with $c \geq 1$.
   (c) When does a locally free extension as in (a) exist, and how many are there?
   (d) Show that there is a unique stable bundle W, up to isomorphism, such that $\det W = \mathcal{O}_{\mathbb{P}^2}(1)$ and $c_2(W) = 1$. In particular, $\Theta_{\mathbb{P}^2}(-1)$ and $V(1)$, where V is the bundle of Exercise 6 of Chapter 2, are both isomorphic to W.
   (One approach is as follows: Writing W as an extension as in (a), W is specified up to isomorphism by $Z = \{p\}$. So it suffices to show that there exists one bundle W on $\mathbb{P}^2$, say $W = \Theta_{\mathbb{P}^2}(-1)$, such that for every $p \in \mathbb{P}^2$, there exists a section of W vanishing at p. Now use the description $W = \bigoplus_1^3 \mathcal{O}_{\mathbb{P}^2}/\mathcal{O}_{\mathbb{P}^2}(-1)$.)

For another approach, suppose that W and W' are two isomorphic stable bundles with the correct Chern classes. Use the Riemann-Roch theorem and the splitting principle to conclude that $H^0(Hom(W, W')) \neq 0$, and finish by Proposition 7.)

8. Deduce (1)–(3) of Proposition 21 from the last statement.

9. Prove the analogue of Lemmas 2 and 3 for $p_{H,V}$, restricting yourself to the case of vector bundles on a surface. Deduce that Lemma 6, Proposition 7, and Corollary 8 hold for Gieseker stable bundles.

10. In the notation of the example after the proof of Proposition 28 of Chapter 2, show that $V_{a,b}$ is stable if $a \neq b, b \pm 1$. (If $f: Q \to \mathbb{P}^2$ is the double cover map and $V_{a,b} = f_* \mathcal{O}_Q(af_1 + bf_2)$, let $\mathcal{O}_{\mathbb{P}^2}(k)$ be a sub-line bundle of $V_{a,b}$. Then there is a nonzero map $f^* \mathcal{O}_{\mathbb{P}^2}(k) \to \mathcal{O}_Q(af_1 + bf_2)$.)

11. Let V be a stable rank 2 bundle on $\mathbb{P}^2$ with $c_1(V) = 0$ and $c_2(V) = 2$. Then there is a unique conic C in $\mathbb{P}^2$ such that, if $f: Q \to \mathbb{P}^2$ is the associated double cover, then $V = f_* \mathcal{O}_Q(2f_1 - f_2)$. To see this, argue as follows:

(a) For the existence of such a double cover, first note by the previous exercise that for a fixed conic C and associated double cover $f: Q \to \mathbb{P}^2$, the bundle $V_C = f_* \mathcal{O}_Q(2f_1 - f_2)$ is stable. Next show that, if $g \in \operatorname{Aut} \mathbb{P}^2 = PGL(3, \mathbb{C})$, then $g^* V_C = V_{g^{-1}(C)}$. Thus, it suffices to show that $PGL(3, \mathbb{C})$ acts transitively on the set of stable bundles V with $c_1(V) = 0$ and $c_2(V) = 2$. To see this, note by (2) of Proposition 14 that V is given by an exact sequence

$$0 \to \mathcal{O}_{\mathbb{P}^2}(-1) \to V \to \mathcal{O}_{\mathbb{P}^2}(1) \otimes I_Z \to 0,$$

where $\ell(Z) = 3$. First, we can assume that Z consists of three distinct noncollinear points; the proof of this is deferred to (b). Now $PGL(3, \mathbb{C})$ acts transitively on the set of such points. Finally, the connected component of the stabilizer of a set of three noncollinear points in $\mathbb{P}^2$ is $\mathbb{C}^* \times \mathbb{C}^*$. Show that this group acts transitively on the set of locally free extensions inside

$$\mathbb{P} \operatorname{Ext}^1(\mathcal{O}_{\mathbb{P}^2}(1) \otimes I_Z, \mathcal{O}_{\mathbb{P}^2}(-1)) = \mathbb{P} H^0(\mathcal{O}_Z) = \mathbb{P}^2;$$

here you will need to identify the term $\mathcal{O}_Z$ more intrinsically as $\det N_{Z/\mathbb{P}^2} \otimes \mathcal{O}_{\mathbb{P}^2}(-2)$, where $N_{Z/\mathbb{P}^2}$ is the normal bundle to Z in $\mathbb{P}^2$.

(b) Show that we can find a map $\mathcal{O}_{\mathbb{P}^2}(-1) \to V$ such that the cokernel is $\mathcal{O}_{\mathbb{P}^2}(1) \otimes I_Z$, where Z consists of three distinct points (necessarily not collinear), as follows: we have the exact sequence

$$0 \to \mathcal{O}_{\mathbb{P}^2} \to V \otimes \mathcal{O}_{\mathbb{P}^2}(1) \to \mathcal{O}_{\mathbb{P}^2}(2) \otimes I_Z \to 0,$$

where $\mathcal{O}_{\mathbb{P}^2}(2) \otimes I_Z$ is generated by its global sections. Show more generally that, if W is a rank 2 bundle on a smooth surface X which fits into an exact sequence

$$0 \to \mathcal{O}_X \to W \to L \otimes I_Z \to 0,$$

$L \otimes I_Z$ is generated by its global sections, and the map $H^0(X;W) \to H^0(X; L \otimes I_Z)$ is surjective, then there is a nonzero $\mathcal{O}_X \to W$ such that the cokernel is of the form $L \otimes I_{Z'}$ with Z' smooth. (This is a local question: If locally the map is of the form $0 \to R \to R \oplus R \to I_Z \to 0$, with $1 \mapsto (f,g)$, $(a,b) \mapsto -ag+bf$, and given $t_1, t_2 \in \mathbb{C}$, the hypotheses say that we can find a section of W which has the local form (F,G), where

$$F = t_1(fh_1 + 1) + m_1 + f, G = t_2(gh_2 + 1) + m_2 + g,$$

where h_1, h_2 are arbitrary and $m_1, m_2 \in \mathfrak{m} \cdot I_Z$. For generic small values of t_1 and t_2, $F = G = 0$ will consist of smooth points.)

(c) To see that C is uniquely associated to the bundle V_C, show that, for a line $\ell \subset \mathbb{P}^2$, $V_C|\ell = \mathcal{O}_\ell \oplus \mathcal{O}_\ell$ if ℓ is not tangent to C, and $V_C|\ell = \mathcal{O}_\ell(1) \oplus \mathcal{O}_\ell(-1)$ if ℓ is tangent to C. (If ℓ is not tangent to C, then $V_C|\ell = (f_\ell)_* \mathcal{O}_{\mathbb{P}^1}(1)$, where f_ℓ is the restriction of f to $f^{-1}(\ell)$ which is a hyperplane section of $Q \subset \mathbb{P}^3$. Thus, $c_1(V_C|\ell) = 0$, $h^0(V_C|\ell) = 2$, and $h^0(V_C|\ell \otimes \mathcal{O}_\ell(-1)) = 0$. If ℓ is tangent to C, then $F = f^{-1}(\ell) = f_1 \cup f_2$, where f_1 and f_2 are two fibers of different rulings of Q. Tensoring the exact sequence

$$0 \to f^* \mathcal{O}_{\mathbb{P}^2}(-\ell) \to \mathcal{O}_Q \to \mathcal{O}_F \to 0$$

by $\mathcal{O}_Q(2f_1 - f_2)$, there is an exact sequence

$$0 \to \mathcal{O}_Q(2f_1 - f_2)f^* \mathcal{O}_{\mathbb{P}^2}(-\ell) \to \mathcal{O}_Q(2f_1 - f_2) \to \mathcal{O}_F(2f_1 - f_2) \to 0.$$

Applying f_*, it follows that $V_C|\ell = f_* \mathcal{O}_F(2f_1 - f_2)$. Now use the exact sequence

$$0 \to \mathcal{O}_F(2f_1 - f_2) \to \mathcal{O}_{f_1}(-1) \oplus \mathcal{O}_{f_2}(2) \to \mathbb{C}_p \to 0$$

to conclude that $c_1(V_C|\ell) = 0$, $h^0(V_C|\ell \otimes \mathcal{O}_\ell(-1)) = 1$ and $h^0(V_C|\ell \otimes \mathcal{O}_\ell(-2)) = 0$.)

12. Let V be a torsion free sheaf of rank r on the smooth projective variety X. We define $\det V$ to be the sheaf $(\bigwedge^r V)^\vee$. Thus, $\det V$ is a reflexive rank 1 sheaf on X and hence is a line bundle. Show that $c_1(V) = \det V$ and that the proof of Lemma 3 carries over to the case where V and W are torsion free.

13. Give a proof of Lemma 5 in general.

14. Let V be a torsion free sheaf of rank r on the smooth projective variety X, and let H be a fixed ample divisor. We shall always abbreviate μ_H by μ. Show that the set

$$\{\mu(W) : W \text{ is a subsheaf of } V, 0 < \operatorname{rank} W < r\}$$

is bounded above, in the following steps. First show that there is a filtration of V by subsheaves $\{0\} = F^0 \subset F^1 \subset \cdots \subset F^r = V$ such that the successive quotients F^i/F^{i-1} are torsion free of rank 1, and thus of the form $L_i \otimes I_Z$, where L is a line bundle and Z is a subscheme (possibly empty) of codimension at least 2. Now, if rank $W = 1$, show that $c_1(W) = L_i - [E]$ for some effective divisor E, and thus that $\mu(W) \le \mu(L_i)$ for some i. In the general case where rank $W = k$, find a similar filtration for $\bigwedge^k V$ and compare $\mu(\bigwedge^k W)$ with $\mu(W)$ by the splitting principle.

15. Continuing Exercise 14, let μ_0 be the maximum value of $\mu(W)$, for W a subsheaf of V with $0 < \operatorname{rank} W < r$. Show that, if W_1 and W_2 are two subsheaves of V with $\mu(W_1) = \mu(W_2) = \mu_0$, then $\mu(W_1 + W_2) = \mu_0$. Conclude that there is a maximal W with $\mu(W) = \mu_0$, that V/W is torsion free, that every subsheaf W' of W has $\mu(W') \le \mu(W)$, and that every subsheaf W'' of V/W has $\mu(W'') < \mu_0$. Finally, argue by induction that there is a canonical filtration (the *Harder-Narasimhan filtration*) $\{0\} = F^0 \subset F^1 \subset \cdots \subset F^k = V$ such that F^i/F^{i-1} is torsion free and semistable for every i and such that $\mu(F^{i+1}/F^i) < \mu(F^i/F^{i-1})$ for every i.

16. Suppose that V is a semistable torsion free sheaf with $\mu(V) = \mu$. Show that there is a filtration $\{0\} = F^0 \subset F^1 \subset \cdots \subset F^k = V$ such that F^i/F^{i-1} is torsion free and stable for every i and $\mu(F^i/F^{i-1}) = \mu$ for all i. Such a filtration (which is not in general canonical) is called a *Jordan-Hölder filtration* of V. One can also show that the associated graded sheaf $\bigoplus_i (F^i/F^{i-1})$ is independent of the choice of the filtration, i.e., is canonically associated to V.

5

Some Examples of Surfaces

Rational ruled surfaces

In this chapter, we give a leisurely tour of some of the important classes of surfaces: certain blowups of $\mathbb{P}^2$, ruled surfaces, and $K3$ surfaces. Our plan will be to begin with certain linear systems with assigned base points on $\mathbb{P}^2$ and then to see where this study leads. Linear systems of plane curves of small degree with assigned base points have been extensively studied and lead to a rich source of surfaces with interesting projective geometry. We will try to touch upon some of these examples where appropriate.

The simplest example is that of lines. Let H be a line in $\mathbb{P}^2$. The complete linear system $|H|$ gives the "embedding" $\mathbb{P}^2 \cong \mathbb{P}^2$. If we choose a base point $p \in \mathbb{P}^2$, then $|H - p|$ defines a base point free linear series on the blowup of $\mathbb{P}^2$ at p (Exercise 2 of Chapter 3). The fibers of the induced morphism to $\mathbb{P}^1$ are the proper transforms of the lines passing through p, and so all of the fibers are $\cong \mathbb{P}^1$. The exceptional curve E meets each fiber H' transversally at one point, and is thus a section with self-intersection -1. Thus, we have exhibited the blowup of $\mathbb{P}^2$ at one point as the rational ruled surface $\mathbb{F}_1$, and we will return to this example shortly. If we consider the linear system $|H - p - q|$, then on the two point blowup of $\mathbb{P}^2$ the linear system consists of a single curve, necessarily a fixed component, and for general choices of three points p, q, r the linear system $|H - p - q - r|$ is empty since there are no lines passing through three general points. Thus, we shall have to move up a degree, to consider linear systems of conics.

If we take the complete linear system of conics $|2H|$, we get an embedding of $\mathbb{P}^2$ in a projective space of dimension $\binom{2+2}{2} - 1 = 5$. The image of $\mathbb{P}^2$ in $\mathbb{P}^5$ is a surface of degree 4, the *Veronese surface*. Imposing one assigned base point p gives a very ample linear system on the blowup $\mathbb{F}_1$ of $\mathbb{P}^2$ at p. Here the base point free linear system on $\mathbb{F}_1$ corresponding to $|2H - p|$ is $|2\pi^* H - E|$. The image of $\mathbb{F}_1$ in $\mathbb{P}^4$ has degree $(2\pi^* H - E)^2 = 3$. Note that $(2\pi^* H - E) \cdot (\pi^* H - E) = 1$, so the images of the proper transforms of the lines through p are embedded in $\mathbb{P}^3$ as (disjoint) lines. For this reason the image of $\mathbb{F}_1$ in $\mathbb{P}^4$ is called a (cubic) *scroll*. As $(2\pi^*H - E) \cdot E = 1$, the image of E is also a line in $\mathbb{P}^4$.

Now suppose that we consider $|2H - p_1 - p_2|$, where p_1 and p_2 are two distinct points on $\mathbb{P}^2$. It is easy to verify directly that the corresponding linear system on the blowup X of $\mathbb{P}^2$ at the points p_1 and p_2 has no base points and defines a morphism from X to $\mathbb{P}^3$, such that the images of the exceptional curves E_i corresponding to p_i are (disjoint) lines. But if ℓ is the line joining p_1 and p_2, then its proper transform on X is $\pi^*\ell - E_1 - E_2 = \ell'$, and $(2\pi^*H - E_1 - E_2) \cdot \ell' = 0$. Thus, ℓ' is contracted under the morphism $X \to \mathbb{P}^3$. The image of X has degree 2 and thus is a smooth quadric Q in $\mathbb{P}^3$. As is well known, a smooth quadric in $\mathbb{P}^3$ is isomorphic to the surface $\mathbb{P}^1 \times \mathbb{P}^1 = \mathbb{F}_0$. The quadric Q has two distinct families of lines f_1 and f_2, with $f_i^2 = 0$ and $f_1 \cdot f_2 = 1$, and there are two different morphisms $Q \to \mathbb{P}^1$ corresponding to the two projections. In fact $f_i = \pi^*H - E_i$, where $f_1 + f_2 = 2\pi^*H - E_1 - E_2$. Thus, X is the blowup of the quadric Q at a point q, with exceptional divisor ℓ', and the proper transforms of the two lines f_1 and f_2 passing through q, namely $\pi^*H - E_1 - (\pi^*H - E_1 - E_2) = E_2$ and $\pi^*H - E_2 - (\pi^*H - E_1 - E_2) = E_1$, are exceptional curves. Blowing these down gives back $\mathbb{P}^2$. Finally, the blowup of $\mathbb{P}^2$ at two points, or in other words the blowup of $\mathbb{F}_1$ at a point not lying on the exceptional curve, is isomorphic to the blowup of $\mathbb{F}_0$ at a point.

We may now define the *rational ruled surfaces* $\mathbb{F}_n$, for $n \geq 0$, inductively: we suppose by induction that we have constructed the surface $\mathbb{F}_n$, with the following properties:

1. There exists a morphism $\pi \colon \mathbb{F}_n \to \mathbb{P}^1$ such that all fibers of π are isomorphic to $\mathbb{P}^1$;
2. There exists a curve $\sigma \subset \mathbb{F}_n$, the *negative section*, such that $\sigma \cdot f = 1$ for every fiber f of π and $\sigma^2 = -n$.

Then we construct $\mathbb{F}_{n+1}$ as follows: blow up a point $p \in \sigma$. Let $\widetilde{\mathbb{F}_n}$ be the blowup and let E the exceptional curve. There is an induced morphism from $\widetilde{\mathbb{F}_n}$ to $\mathbb{P}^1$. The proper transform of σ is a new curve σ' with $(\sigma')^2 = -n-1$. Since $f \cdot \sigma = 1$, there is a unique fiber (which we shall again denote by f) passing through p, which necessarily meets σ transversally, and its proper transform f' on the blowup is a curve disjoint from σ' and such that $(f')^2 = -1$. Clearly, $E + f'$ is a fiber of the morphism from $\widetilde{\mathbb{F}_n}$ to $\mathbb{P}^1$. Thus, if we contract f' via the Castelnuovo criterion we obtain a new surface $\mathbb{F}_{n+1}$, together with a morphism $\mathbb{F}_{n+1} \to \mathbb{P}^1$, all of whose fibers are isomorphic to $\mathbb{P}^1$. Moreover, the image of σ' on $\mathbb{F}_{n+1}$ is a curve satisfying $(\sigma')^2 = -n - 1$ and $\sigma' \cdot f = 1$ for every fiber of the new morphism to $\mathbb{P}^1$. This completes the inductive construction of the $\mathbb{F}_n$. The procedure of passing from $\mathbb{F}_n$ to $\mathbb{F}_{n+1}$ is called an *elementary transformation*. Note that if instead we blow up a point of $\mathbb{F}_n$ not on σ and then contract the fiber through this point, the result is $\mathbb{F}_{n-1}$ and we have performed the inverse of the elementary modification which starts with $\mathbb{F}_{n-1}$ and yields $\mathbb{F}_n$.

There are at least three reasons why the $\mathbb{F}_n$ are important:

1. They are exactly the (geometrically) ruled surfaces over a rational base curve;
2. They provide almost all of the examples of the nondegenerate surfaces in $\mathbb{P}^N$ of smallest degree;
3. Along with $\mathbb{P}^2$ itself, and excluding $\mathbb{F}_1$ which blows down to $\mathbb{P}^2$, they are a complete set of isomorphism classes of minimal models for $\mathbb{P}^2$.

Before we discuss these properties more fully, let us describe the numerical invariants of $\operatorname{Pic} \mathbb{F}_n$ and the nef and ample cones of $\mathbb{F}_n$.

Lemma 1. $q(\mathbb{F}_n) = P_m(\mathbb{F}_n) = 0$ for all $n \geq 0$ and all $m > 0$. Furthermore $c_1^2(\mathbb{F}_n) = 8$.

Proof. These statements follow from Corollary 5 and Proposition 6 of Chapter 3, using the standard results for $\mathbb{P}^2$. □

Lemma 2. Let f be a fiber of the map $\mathbb{F}_n \to \mathbb{P}^1$ and let σ be the negative section. Then $\operatorname{Pic} \mathbb{F}_n \cong \operatorname{Num} \mathbb{F}_n \cong H^2(\mathbb{F}_n; \mathbb{Z}) \cong \mathbb{Z} \cdot [f] \oplus \mathbb{Z} \cdot \sigma$. Thus, $c_2(\mathbb{F}_n) = 4$.

Proof. Clearly, $\operatorname{Pic} \mathbb{F}_1 \cong \mathbb{Z} \cdot [f] \oplus \mathbb{Z} \cdot \sigma$, and the general case follows by comparing $\operatorname{Pic} \mathbb{F}_n$ with $\operatorname{Pic} \mathbb{F}_{n+1}$. Since the intersection pairing on $\mathbb{Z} \cdot [f] \oplus \mathbb{Z} \cdot \sigma$ is nondegenerate, $\operatorname{Pic} \mathbb{F}_n \cong \operatorname{Num} \mathbb{F}_n$. The statement about H^2, and thus the calculation of $c_2(\mathbb{F}_n)$, follows from the exponential sheaf sequence, or is a standard comparison of H^2 of a blowup with H^2 of the original surface, which follows easily from the Mayer-Vietoris sequence and is left to the reader. □

Proposition 3. Let f and σ denote the fiber and negative section of $\mathbb{F}_n$. The following are equivalent:

(i) The divisor $a\sigma + bf$ is nef;
(ii) $a \geq 0$ and $b \geq an$;
(iii) $|a\sigma + bf|$ is base point free;
(iv) $|a\sigma + bf|$ has no fixed components.

Proof. (i) implies (ii): If $a\sigma + bf$ is nef, then $(a\sigma + bf) \cdot f = a \geq 0$ and $(a\sigma + bf) \cdot \sigma = -an + b \geq 0$.

(ii) implies (iii): Write $a\sigma + bf = a(\sigma + nf) + (b - an)f = a(\sigma + nf) + cf$. It suffices to show that $\sigma + nf$ is base point free and that cf is base point free provided that $c \geq 0$. The second statement is clear. To see the first, suppose that $k \geq n$ and consider the exact sequence

$$0 \to \mathcal{O}_{\mathbb{F}_n}(kf) \to \mathcal{O}_{\mathbb{F}_n}(\sigma + kf) \to \mathcal{O}_\sigma(k - n) \to 0.$$

Claim 4. *The induced map $H^0(\mathcal{O}_{\mathbb{F}_n}(\sigma + kf)) \to H^0(\mathcal{O}_\sigma(k - n))$ is onto.*

Proof. The cokernel of this map lives in $H^1(\mathcal{O}_{\mathbb{F}_n}(kf))$. We claim that, for all $k \geq 0$, we have $H^1(\mathcal{O}_{\mathbb{F}_n}(kf)) = 0$. For $k = 0$ this is the statement that $q(\mathbb{F}_n) = 0$. The general statement follows by induction on k, using the exact sequence

$$0 \to \mathcal{O}_{\mathbb{F}_n}(kf) \to \mathcal{O}_{\mathbb{F}_n}((k+1)f) \to \mathcal{O}_f \to 0,$$

and the associated long exact cohomology sequence. $\square$

Returning to the proof of Proposition 3, we see that, for the case $k = n$, we can lift the nonzero section of $\mathcal{O}_\sigma$ to a section of $H^0(\mathcal{O}_{\mathbb{F}_n}(\sigma + nf))$, and thus $|\sigma + nf|$ has no base points along σ. Since it also contains the base point free series $|nf|$, it has no base points elsewhere either.

(iii) implies (iv): Trivial.

(iv) implies (i): Trivial. $\square$

Lemma 5.

(i) $a\sigma + bf$ *is effective if and only if* $a, b \geq 0$.
(ii) $a\sigma + bf$ *is ample if and only if* $a > 0$ *and* $b > an$.

Proof. Clearly, if $a, b \geq 0$, then $a\sigma + bf$ is effective. Conversely, if $a\sigma + bf$ is effective, then using the fact that $|f|$ and $|\sigma + nf|$ are base point free, we have $a = (a\sigma + bf) \cdot f \geq 0$ and $b = (a\sigma + bf) \cdot (\sigma + nf) \geq 0$. If $a\sigma + bf$ is ample, then $a > 0$ and $(a\sigma + bf) \cdot \sigma = -an + b > 0$. Conversely, if $a > 0$ and $b > an$, then by applying (i) we see that $(a\sigma + bf) \cdot C > 0$ for every effective divisor, and moreover $(a\sigma + bf)^2 = 2ab - a^2n = a(2b - an) > 0$. So $a\sigma + bf$ is ample by the Nakai-Moishezon criterion. $\square$

Of course, it is easy to give a direct argument that $a\sigma + bf$ is ample if $a > 0$ and $b > an$, along the lines of the proof of Proposition 3.

The next result describes the morphism defined by $\sigma + kf$ for $k \geq n$. We recall that a *rational normal curve* of degree n is the image of $\mathbb{P}^1$ in $\mathbb{P}^n$ under the embedding defined by the complete linear system associated to $\mathcal{O}_{\mathbb{P}^1}(n)$.

Proposition 6. *If $k > n$, then the linear system $|\sigma + kf|$ is very ample. In this case, $\dim |\sigma + kf| = 2k - n + 1$, and if φ is the corresponding morphism $\mathbb{F}_n \to \mathbb{P}^{2k-n+1}$, then $\varphi(\mathbb{F}_n)$ is a surface of degree $2k - n$. Moreover:*

(i) $\varphi(f)$ *is a line for every fiber f;*
(ii) $\varphi(\sigma) = C_1$ *is a rational normal curve of degree $k - n$ inside a $\mathbb{P}^{k-n} \subset \mathbb{P}^{2k-n+1}$;*

(iii) *For every choice of a smooth section $\sigma' \in |\sigma + nf|$, necessarily disjoint from σ, $\varphi(\sigma') = C_2$ is a rational normal curve of degree k contained in a $\mathbb{P}^k \subset \mathbb{P}^{2k-n+1}$, disjoint from the $\mathbb{P}^{k-n}$.*

Moreover, $\varphi(\mathbb{F}_n)$ is obtained as follows: start with C_1 and C_2. For each point $p \in C_1$, there is a unique point $q \in C_2$ such that $\pi(p) = \pi(q)$, where $\pi: \mathbb{F}_n \to \mathbb{P}^1$ is the natural map. Then $\varphi(\mathbb{F}_n)$ is the union over all $p \in C_1$ of the lines $\overline{pq}$.

For $k = n$, the morphism φ corresponding to $|\sigma + nf|$ contracts σ and $\varphi(\mathbb{F}_n)$ is the cone over a rational normal curve in $\mathbb{P}^n$.

Proof. We leave these statements for the reader to work out, using Claim 4. $\square$

It is a standard fact that, if $X \subset \mathbb{P}^N$ is an irreducible nondegenerate surface (i.e., is not contained in a hyperplane), then $\deg X \geq N - 1$. In fact, if $\deg X = N - 1$, then either $X = \varphi(\mathbb{F}_n)$ for some n and some φ corresponding to $|\sigma + kf|$, $k \geq n$, or X is the Veronese surface in $\mathbb{P}^5$.

General ruled surfaces

Definition 7. A smooth surface X is a (geometrically) *ruled* surface if there exists a morphism $\pi: X \to C$, where C is a smooth curve, such that all fibers of π are isomorphic to $\mathbb{P}^1$. The surface X is *ruled* if there exists a morphism $\pi: X \to C$ such that at least one fiber of π is isomorphic to $\mathbb{P}^1$.

Next we shall show that every ruled surface is obtained by blowing up a geometrically ruled surface:

Lemma 8. *If $\pi: X \to C$ is a ruled surface, then the morphism π factors through a smooth blowdown $X \to \bar{X}$, where $\bar{X}$ is a geometrically ruled surface.*

Proof. If f is an irreducible fiber of π, then $f^2 = 0$ and $f \cong \mathbb{P}^1$. Thus, $K_X \cdot f = -2$. Now suppose that $f = \sum_i n_i C_i$ is a reducible fiber $\pi^* t$ of π (here t is a point of C and as a reduced curve $\pi^{-1}(t) = \bigcup_i C_i$). Since f^2 and $f \cdot K_X$ are independent of the choice of f, we still have $f^2 = 0$ and $K_X \cdot f = -2$. By the connectedness theorem, $\bigcup_i C_i$ is connected. Thus, for every j there exists a k such that $C_j \cdot C_k > 0$. It then follows from $0 = C_j \cdot (\sum_i n_i C_i)$ that $C_j^2 < 0$ for every j. Moreover, $K_X \cdot (\sum_i n_i C_i) = -2$, so that there exists a j such that $C_j^2 < 0$ and $C_j \cdot K_X < 0$. Thus, C_j is exceptional and we can contract it. Continuing in this way, we eventually reach a stage where this is no longer possible, i.e., all fibers are irreducible. At this point we claim that the resulting surface $\bar{X}$ is geometrically ruled.

Indeed all fibers are irreducible, but there might exist a multiple fiber of π, i.e., a point $t \in C$ and a positive integer m such that, as divisors, $\pi^* t = mC$ with $m > 1$. However, $(mC)^2 = f^2 = 0$ and $K_X \cdot (mC) = -2$. Thus, $C^2 = 0$ and $K_X \cdot C$ is even and divides -2. It follows that $K_X \cdot C = -2$ and $m = 1$. $\square$

Note that if X is not minimal, the blowdown $X \to \bar{X}$ is never unique. In fact, at the second to last stage we must have started with a geometrically ruled surface and blown up a point p. The proper transform of the fiber through p is then an exceptional curve and we may contract it. The resulting birational map from $\bar{X}$ to a new geometrically ruled surface is called an *elementary transformation*, just as for the rational ruled surfaces. If the genus $g(C)$ of the base curve is at least 1, then it is not too difficult to show that every birational map from X to X', where X and X' are two geometrically ruled surfaces over C, is obtained as a sequence of elementary transformations. (See also Exercise 2 of this chapter and Exercise 9 of Chapter 2.) In particular a ruled surface over C is birational to $C \times \mathbb{P}^1$, which we can also see directly from Theorem 9 below. Finally, again if $g(C) \geq 1$ and X is a (not necessarily geometrically) ruled surface over C, then the morphism $\pi \colon X \to C$ exhibiting X as a ruled surface is unique. Indeed, if $\pi' \colon X \to C'$ is another morphism to a smooth curve C' such that a fiber f' of π' is isomorphic to $\mathbb{P}^1$, then the restriction of π to f' is a morphism from $\mathbb{P}^1$ to C. Since every morphism from $\mathbb{P}^1$ to C is constant, it follows that the fibers of π' are also fibers of π and thus that $\pi = \pi'$. More intrinsically π is given by the Albanese map of X (which we will define in Chapter 10).

From now on, unless otherwise specified, we shall take ruled to mean geometrically ruled. There are natural examples of geometrically ruled surfaces: take a rank 2 vector bundle V over the curve C and consider the surface $\mathbb{P}(V)$ together with the obvious morphism $\pi \colon \mathbb{P}(V) \to C$. (Note that our conventions are opposite to those of EGA [59] or [61], so that the fiber of π over $x \in C$ is the projective space associated to the fiber of V over x. Hence, if $\mathcal{O}_{\mathbb{P}(V)}(1)$ is the tautological line bundle on $\mathbb{P}(V)$, then $\pi_* \mathcal{O}_{\mathbb{P}(V)}(1) = V^\vee$.) The morphism π gives $\mathbb{P}(V)$ the structure of a geometrically ruled surface over C, and the next result says that all geometrically ruled surfaces arise in this way:

Theorem 9. *Let* $\pi \colon X \to C$ *be a (geometrically) ruled surface.*

(i) *There exists a section* σ *on* X, *i.e., an irreducible curve* σ *such that* $\sigma \cdot f = 1$ *for every fiber* f *of* π.
(ii) X *is isomorphic to* $\mathbb{P}(V)$, *where* V *is the dual of the rank 2 vector bundle* $R^0 \pi_* \mathcal{O}_X(\sigma)$.
(iii) $\mathbb{P}(V) \cong \mathbb{P}(V')$ *if and only if there is a holomorphic line bundle* L *such that* $V' \cong V \otimes L$.

Proof. All fibers of π are smooth curves, in the sense of divisors. Thus, π is a smooth morphism, and locally on X in the analytic topology π is a product. In particular there exist local analytic sections of π around every point of C.

Next, let T be a smooth complex curve, not necessarily compact, and let $\pi\colon Y \to T$ be a smooth and proper holomorphic map from the complex surface Y to T such that all fibers of π are isomorphic to $\mathbb{P}^1$. Suppose that σ is a section of Y, in other words a curve in Y such that $f \cdot \sigma = 1$ for all fibers f of π, and consider the coherent sheaf $R^0\pi_*\mathcal{O}_Y(\sigma)$. Since $H^0(f; \mathcal{O}_Y(\sigma)|f) = H^0(\mathbb{P}^1; \mathcal{O}_{\mathbb{P}^1}(1))$ has dimension 2 for every fiber f, $R^0\pi_*\mathcal{O}_Y(\sigma)$ is a holomorphic vector bundle of rank 2. If moreover $R^0\pi_*\mathcal{O}_Y(\sigma)$ is holomorphically trivial, choose two generating sections s_0, s_1 of $R^0\pi_*\mathcal{O}_Y(\sigma)$. The pulled back sections of $\pi^*R^0\pi_*\mathcal{O}_Y(\sigma)$ define an isomorphism $Y \to \mathbb{P}(R^0\pi_*\mathcal{O}_Y(\sigma))^\vee = T \times \mathbb{P}^1$. It is easy to check that the isomorphism $Y \to \mathbb{P}(R^0\pi_*\mathcal{O}_Y(\sigma))^\vee$ is canonical, i.e., does not depend on the choice of the sections. Thus, for arbitrary Y such that there exists a section σ as above, there is a global isomorphism $Y \to \mathbb{P}(R^0\pi_*\mathcal{O}_Y(\sigma)^\vee)$. In particular (ii) of the theorem follows from (i). Conversely, suppose that $X = \mathbb{P}(V)$, and let s be a meromorphic section of V. Then s defines a section σ of X, so that (i) follows from (ii). We will show that every X is of the form $\mathbb{P}(V)$, thereby proving (i) and (ii).

Choose an open cover of C, say $\{U_\alpha\}$, for which we can find holomorphic local sections σ_α of π for which $R^0\pi_*\mathcal{O}_{\pi^{-1}(U_\alpha)}(\sigma_\alpha)$ is trivial. Fix an isomorphism $\psi_\alpha\colon R^0\pi_*\mathcal{O}_{\pi^{-1}(U_\alpha)}(\sigma_\alpha)^\vee \to \mathcal{O}_{U_\alpha}^2$. There is an induced isomorphism $\tilde{\psi}_\alpha\colon \pi^{-1}(U_\alpha) \to U_\alpha \times \mathbb{P}^1$. On the overlaps $U_\alpha \cap U_\beta$, let $A_{\alpha\beta} = \psi_\alpha \circ \psi_\beta^{-1}$ be the 2×2 matrix with coefficients in $\mathcal{O}_C(U_\alpha \cap U_\beta)$ and let $\bar{A}_{\alpha\beta}$ be the induced element of $PGL(2, \mathcal{O}_C(U_\alpha \cap U_\beta))$. Now it is easy to check directly from the construction that, over $U_\alpha \cap U_\beta \cap U_\gamma$, $\bar{A}_{\alpha\beta}\bar{A}_{\beta\gamma} = \bar{A}_{\alpha\gamma}$, in other words, that $\bar{A}_{\alpha\beta}$ is a 1-cocycle for the cover $\{U_\alpha\}$ with values in the nonabelian sheaf $PGL(2, \mathcal{O}_C)$. On the other hand, it need not be the case that the lifted cochain $A_{\alpha\beta}$ is a 1-cocycle for the sheaf $GL(2, \mathcal{O}_C)$. Indeed, all we can say is that, over $U_\alpha \cap U_\beta \cap U_\gamma$, there exists an element $f_{\alpha\beta\gamma} \in \mathcal{O}_C^*(U_\alpha \cap U_\beta \cap U_\gamma)$ such that $A_{\alpha\beta}A_{\beta\gamma} = f_{\alpha\beta\gamma}A_{\alpha\gamma}$.

Claim. The 2-cochain $f_{\alpha\beta\gamma}$ is a 2-cocycle for $\{U_\alpha\}$ with values in $\mathcal{O}_C^*$.

Proof of the Claim. We must show that, over $U_\alpha \cap U_\beta \cap U_\gamma \cap U_\delta$,

$$f_{\beta\gamma\delta}f_{\alpha\gamma\delta}^{-1}f_{\alpha\beta\delta}f_{\alpha\beta\gamma}^{-1} = 1,$$

or equivalently that $f_{\beta\gamma\delta}f_{\alpha\gamma\delta}^{-1}f_{\alpha\beta\gamma}^{-1}f_{\alpha\beta\delta} = 1$. By definition, it suffices to show that

$$(A_{\beta\gamma}A_{\gamma\delta}A_{\beta\delta}^{-1})(A_{\alpha\delta}A_{\gamma\delta}^{-1}A_{\alpha\gamma}^{-1})(A_{\alpha\gamma}A_{\beta\gamma}^{-1}A_{\alpha\beta}^{-1})(A_{\alpha\beta}A_{\beta\delta}A_{\alpha\delta}^{-1}) = \mathrm{Id}.$$

First rewrite this expression as

$$(A_{\beta\gamma}A_{\gamma\delta}A_{\beta\delta}^{-1})(A_{\alpha\delta}A_{\gamma\delta}^{-1}A_{\beta\gamma}^{-1}A_{\beta\delta}A_{\alpha\delta}^{-1}),$$

and then use the fact that $(A_{\beta\gamma}A_{\gamma\delta}A_{\beta\delta}^{-1})$ lies in the center to further rewrite it as

$$A_{\alpha\delta}A_{\gamma\delta}^{-1}A_{\beta\gamma}^{-1}(A_{\beta\gamma}A_{\gamma\delta}A_{\beta\delta}^{-1})A_{\beta\delta}A_{\alpha\delta}^{-1} = \mathrm{Id},$$

as claimed. □

Claim. $H^2(C; \mathcal{O}_C^*) = 0.$

Proof of the Claim. This follows from the exponential sheaf sequence on C and the fact that $H^2(C; \mathcal{O}_C) = H^3(C; \mathbb{Z}) = 0.$ □

Thus, possibly after passing to a refinement of the cover $\{U_\alpha\}$, $f_{\alpha\beta\gamma}$ is a Čech coboundary of $v_{\alpha\beta}$, say. In other words, $f_{\alpha\beta\gamma} = v_{\alpha\beta}v_{\beta\gamma}v_{\alpha\gamma}^{-1}$. Consider the isomorphisms $A'_{\alpha\beta} \colon \mathcal{O}_{U_\alpha \cap U_\beta}^2 \to \mathcal{O}_{U_\alpha \cap U_\beta}^2$ defined by $A'_{\alpha\beta} = v_{\alpha\beta}^{-1}A_{\alpha\beta}$. We leave to the reader the check that $A'_{\alpha\beta}A'_{\beta\gamma} = A'_{\alpha\gamma}$. Thus, the $A'_{\alpha\beta}$ are the transition functions of a rank 2 vector bundle V on C. Moreover, it is easy to check from the construction that $X \cong \mathbb{P}(V)$, compatibly with the projection to C. For instance, this is clear over U_α since both ruled surfaces are products over U_α, and over the intersections $U_\alpha \cap U_\beta$ the transition functions $A_{\alpha\beta}$ and $A'_{\alpha\beta}$ agree as elements of $PGL(2, \mathcal{O}_C(U_\alpha \cap U_\beta))$.

Finally, suppose that $\mathbb{P}(V) = \mathbb{P}(V')$. Let $\mathcal{L} = \mathcal{O}_{\mathbb{P}(V)}(1)$ be the tautological line bundle over $\mathbb{P}(V)$ and let $\mathcal{L}' = \mathcal{O}_{\mathbb{P}(V')}(1)$ the tautological line bundle over $\mathbb{P}(V') = \mathbb{P}(V)$. Thus, canonically $\pi_*\mathcal{L} = V^\vee$ and $\pi_*\mathcal{L}' = (V')^\vee$. On the other hand, $\mathcal{L} \otimes (\mathcal{L}')^{-1}$ has degree 0 on the fibers, and so is trivial since the fibers are all $\mathbb{P}^1$. By base change $\pi_*(\mathcal{L} \otimes (\mathcal{L}')^{-1}) = L$ is a line bundle on C. Moreover, the natural map $\pi^*\pi_*(\mathcal{L} \otimes (\mathcal{L}')^{-1}) \to \mathcal{L} \otimes (\mathcal{L}')^{-1}$ is an isomorphism, and hence $\mathcal{L} \otimes (\mathcal{L}')^{-1} = \pi^*L$. Thus,

$$V^\vee = \pi_*\mathcal{L} = \pi_*(\mathcal{L}' \otimes (\mathcal{L} \otimes (\mathcal{L}')^{-1})) = \pi_*(\mathcal{L}' \otimes \pi^*L) = \pi_*\mathcal{L}' \otimes L = (V')^\vee \otimes L.$$

It follows that $L \otimes (V')^\vee \cong V^\vee$, and thus $V' \cong V \otimes L$. □

The calculations of the above proof can be summarized as follows: there is an exact sequence of sheaves

$$0 \to \mathcal{O}_C^* \to GL(2, \mathcal{O}_C) \to PGL(2, \mathcal{O}_C) \to 0,$$

with $\mathcal{O}_C^*$ in the center of $GL(2, \mathcal{O}_C)$. Thus, there is an exact sequence (of pointed sets)

$$H^1(\mathcal{O}_C^*) \to H^1(GL(2, \mathcal{O}_C)) \to H^1(PGL(2, \mathcal{O}_C)) \to H^2(\mathcal{O}_C^*) = 0.$$

Every $\mathbb{P}^1$-bundle over C defines a unique element in $H^1(C; PGL(2, \mathcal{O}_C))$. By the exactness of the above sequence, this element can be lifted to an element of $H^1(C; GL(2, \mathcal{O}_C))$, corresponding to a rank 2 vector bundle over

C, and two such lifts differ by an element in $H^1(\mathcal{O}_C^*)$, in other words the two possible vector bundles differ by twisting by a line bundle on C.

In general, as is customary, we will call any curve σ on a geometrically ruled surface X such that $\sigma \cdot f = 1$ a *section*. Thus, π defines a bijection between σ and C, which is an isomorphism since C is smooth. We now analyze the geometry of ruled surfaces in more detail.

Lemma 10. *Let X be a geometrically ruled surface. There is a one-to-one correspondence between sections σ of X and rank 2 vector bundles V such that $X = \mathbb{P}(V)$, together with a choice of a nowhere vanishing section of $V^\vee$. More precisely, given a section σ, the bundle V is defined by $V^\vee = R^0\pi_*\mathcal{O}_X(\sigma)$. Moreover, with $V^\vee = R^0\pi_*\mathcal{O}_X(\sigma)$, we have $\mathcal{O}_{\mathbb{P}(V)}(1) = \mathcal{O}_X(\sigma)$. There is an exact sequence*

(5.1) $$0 \to \mathcal{O}_C \to V^\vee \to L \to 0,$$

where L is a line bundle on C with $\deg L = \sigma^2$, and hence $c_1(V) = -\sigma^2$. Finally, if τ is another section of X, then there exists a unique line bundle λ on C such that $\mathcal{O}_X(\tau) = \mathcal{O}_X(\sigma) \otimes \pi^*\lambda$.

Proof. The correspondence between sections of X and quotients of $V^\vee$ is a general fact [61, II, 7.12]. In our case, we can see this follows: fix a section σ and let $V^\vee = R^0\pi_*\mathcal{O}_X(\sigma)$. By the proof of Theorem 9, $X = \mathbb{P}(V)$. Moreover, the surjection $\pi^*\pi_*\mathcal{O}_X(\sigma) = \pi^*V^\vee \to \mathcal{O}_X(\sigma)$ gives a dual inclusion $\mathcal{O}_X(-\sigma) \to \pi^*V$ which identifies $\mathcal{O}_X(-\sigma)$ with $\mathcal{O}_{\mathbb{P}(V)}(-1)$ and thus $\mathcal{O}_X(\sigma)$ with $\mathcal{O}_{\mathbb{P}(V)}(1)$. Consider the exact sequence

$$0 \to \mathcal{O}_X \to \mathcal{O}_X(\sigma) \to \mathcal{O}_\sigma(\sigma) \to 0.$$

Apply $R^0\pi_*$ to this sequence. The direct image $R^1\pi_*\mathcal{O}_X = 0$ since $H^1(f; \mathcal{O}_f) = 0$, and $R^0\pi_*\mathcal{O}_X = \mathcal{O}_C$. Thus, there is an exact sequence

$$0 \to \mathcal{O}_C \to V^\vee \to \mathcal{O}_C(\sigma) \to 0,$$

where we identify σ with C via π and thus $\mathcal{O}_\sigma(\sigma)$ with a line bundle on C, denoted by $\mathcal{O}_C(\sigma) = L$. Moreover, $\deg \mathcal{O}_C(\sigma) = \sigma^2$.

To go the other way, suppose that V is a rank 2 bundle and that there exists an exact sequence (4.1). Then, for every $p \in C$, we let σ_p be the point in $\mathbb{P}(V_p)$ defined by the vanishing of the linear form on V_p corresponding to the image of a nonzero element of $H^0(\mathcal{O}_C)$ in $(V^\vee)_p$. It is easy to check that these are inverse constructions.

Now suppose that τ is another section. As $\mathcal{O}_X(\tau - \sigma)$ is a line bundle whose restriction to each fiber f has degree 0, $R^1\pi_*\mathcal{O}_X(\tau - \sigma) = 0$, and $R^0\pi_*\mathcal{O}_X(\tau - \sigma)$ is a line bundle λ on C such that the natural map $\pi^*R^0\pi_*\mathcal{O}_X(\tau - \sigma) \to \mathcal{O}_X(\tau - \sigma)$ is an isomorphism. Thus, $\mathcal{O}_X(\tau) = \mathcal{O}_X(\sigma) \otimes \pi^*\lambda$. Moreover, λ is determined by this property since $\lambda = R^0\pi_*\pi^*\lambda$, by the projection formula, and so $\lambda = R^0\pi_*\mathcal{O}_X(\tau - \sigma)$. □

Proposition 11. $\operatorname{Pic} X \cong \operatorname{Pic} C \oplus \mathbb{Z}$. *More precisely, there is an exact sequence*

$$0 \to \operatorname{Pic} C \xrightarrow{\pi^*} \operatorname{Pic} X \to \mathbb{Z} \to 0,$$

where the second map is given by $L \mapsto \deg(L|f)$, *and the choice of a section splits this sequence.*
Thus, $\operatorname{Num} X \cong \mathbb{Z} \cdot f \oplus \mathbb{Z} \cdot \sigma$.

Proof. Clearly, the map $L \mapsto \deg(L|f)$ is a surjection from $\operatorname{Pic} X$ to $\mathbb{Z}$ which is split by the map $1 \mapsto \mathcal{O}_X(\sigma)$. We must identify the kernel with $\operatorname{Pic} C$ via π^*. If $L \in \operatorname{Pic} X$ and $\deg(L|f) = 0$, by base change $\pi_* L = \lambda$ is a line bundle on C and the map $\pi^* \pi_* L \to L$ is an isomorphism. Thus, every L such that $\deg(L|f) = 0$ is in $\pi^* \operatorname{Pic} C$, and the arguments of Lemma 10 show that $\pi^* \colon \operatorname{Pic} C \to \operatorname{Pic} X$ is an isomorphism onto its image, with inverse π_*. $\square$

We next want to find a generalization of the negative section of $\mathbb{F}_n$. For a rank 2 bundle V over C, we could try to consider

$$\max\{\deg L : \text{ there exists a nonzero map } L \to V\}.$$

However, this number changes when we replace V by $V \otimes \lambda$ for a line bundle λ on C of nonzero degree, and instead we consider

$$e(V) = \max\{2 \deg L - \deg V : \text{ there exists a nonzero map } L \to V\}.$$

It is easy to see that $e(V) < \infty$. For example, if there is an exact sequence $0 \to L_1 \to V \to L_2 \to 0$, then $e(V) \le |\deg L_1 - \deg L_2|$.

Proposition 12.

  (i) $e(V) = e(V \otimes \lambda)$ *for all line bundles* λ *on* C.
 (ii) $e(V) = e(V^\vee)$.
(iii) *Suppose that* $V = L_1 \oplus L_2$, *with* $\deg L_1 = d_1 \ge \deg L_2 = d_2$. *Then* $e(V) = d_1 - d_2 \ge 0$.
 (iv) $-e(V) = \min\{\sigma^2 : \sigma \text{ is a section of } V\}$.
  (v) V *is stable if and only if* $e < 0$ *and semistable if and only if* $e \le 0$.

Proof. Part (i) is clear, and (ii) follows from (i) and the fact that $V^\vee = V \otimes \det V^{-1}$. For (iii), in that notation it is clear that the maximum possible degree of a line bundle L for which there is a nonzero map $L \to V$ is d_1. Thus, $e(V) = 2d_1 - (d_1 + d_2) = d_1 - d_2$. For (iv), first suppose that there is an exact sequence

$$0 \to L \to V \to \mathcal{O}_C \to 0.$$

Then $2 \deg L - \deg V = \deg L$. Thus, if L is chosen so that $2 \deg L - \deg V$ is maximal, after replacing V by $V \otimes \lambda$ for some line bundle λ we can assume that $V/L = \mathcal{O}_C$ (note that for such L, V/L is necessarily torsion free). In this case there is the dual exact sequence

$$0 \to \mathcal{O}_C \to V^\vee \to L^{-1} \to 0,$$

and by Lemma 10 there is a section σ of X with $\sigma^2 = -\deg L = -e(V)$. Conversely, if σ is a section, then by Lemma 10 there is a line bundle λ and an exact sequence

$$0 \to \mathcal{O}_C \to V^\vee \otimes \lambda \to M \to 0,$$

with $\deg M = \sigma^2$. Thus, $e(V) = e(V^\vee \otimes \lambda) \geq 0 - \deg M = -\sigma^2$. It follows that $\sigma^2 \geq -e(V)$ for every possible choice of σ. Thus, we obtain (iv). Finally, (v) follows from the definitions.   □

It follows that $e(V)$ only depends on the ruled surface $X = \mathbb{P}(V)$, and we will also use the notation $e(X)$. Note that $e(V) > 0$ if and only if V is unstable, and in this case there is a unique maximal destabilizing sub-line bundle. There is thus a unique section of X of negative self-intersection. See also Exercise 1 for a generalization.

Example. Over $\mathbb{P}^1$, every rank 2 vector bundle is a direct sum of line bundles. After twisting, we see that every ruled surface over $\mathbb{P}^1$ is of the form $\mathbb{P}(\mathcal{O}_{\mathbb{P}^1} \oplus \mathcal{O}_{\mathbb{P}^1}(n))$. Clearly, this ruled surface is just $\mathbb{F}_n$, and $e(\mathbb{F}_n) = n$.
Over an elliptic curve C, every ruled surface is either $\mathbb{P}(\mathcal{O}_C \oplus L)$ for a line bundle L with $\deg L = d \geq 0$, or $\mathbb{P}(\mathcal{E})$, where $\mathcal{E}$ is the nontrivial extension of $\mathcal{O}_C$ by $\mathcal{O}_C$ (Theorem 6 of Chapter 2) or $\mathbb{P}(\mathcal{F}_p)$. The possibilities for e are: d in the first case, 0 in the second, and -1 in the last case.

By the Segre-Nagata theorem, we always have the inequality $e(V) \geq -g$, where g is the genus $g(C)$. (See the remarks after Theorem 10 in Chapter 4.) For example, in case $c_1(V) = 0$, there always exists a line subbundle L of V with $\deg L \geq -(g-1)/2$ and thus $e(V) \geq -(g-1)$.
Next let us determine the numerical invariants and the canonical bundle of X:

Lemma 13. *If X is a ruled surface, then $q(X) = g(C)$ and $p_g(X) = 0$.*

Proof. By base change $R^0\pi_*\mathcal{O}_X = \mathcal{O}_C$ and $R^i\pi_*\mathcal{O}_X = 0$ for $i > 0$. Thus, the Leray spectral sequence gives $H^1(X; \mathcal{O}_X) = H^1(C; \mathcal{O}_C)$, which has dimension $g(C)$, and $H^2(X; \mathcal{O}_X) = H^2(C; \mathcal{O}_C) = 0$.   □

Lemma 14. *Let σ be a section of X and write $\mathcal{O}_\sigma(\sigma) = \pi^*\mathcal{O}_C(\mathbf{d})|\sigma$ for some divisor $\mathbf{d}$ on C. Then*

$$K_X = -2\sigma + \pi^*(K_C + \mathbf{d}).$$

Thus, if $\deg \mathbf{d} = d$, *then the numerical equivalence class of* K_X *is* $-2\sigma + (2g - 2 + d)f$ *and* $K_X^2 = -8(g - 1)$.

Proof. We may write $K_X = a\sigma + \pi^*\mathbf{b}$ for an integer a and a divisor class $\mathbf{b}$ on C, by Proposition 11. By adjunction $K_X \cdot f = a = -2$. Next, applying adjunction to $\sigma \cong C$, we see that

$$K_\sigma = K_X|\sigma \otimes \mathcal{O}_\sigma(\sigma)$$

giving (as divisor classes) $K_C = -2\mathbf{d} + \mathbf{b} + \mathbf{d} = \mathbf{b} - \mathbf{d}$. Thus, $\mathbf{b} = K_C + \mathbf{d}$. The remaining statements are then clear. $\square$

Another way to see Lemma 14 is as follows: Since π is smooth, the relative canonical line bundle $K_{X/C}$ is an invariantly defined line bundle on X satisfying $K_X = K_{X/C} \otimes \pi^* K_C$. On the other hand, there is the Euler exact sequence

$$0 \rightarrow \mathcal{O}_X \rightarrow \pi^* V \otimes \mathcal{O}_{\mathbb{P}(V)}(1) \rightarrow T_{X/C} \rightarrow 0,$$

where $T_{X/C} = K_{X/C}^{-1}$ is the relative tangent bundle. Thus, $K_{X/C} = \pi^*(\det V)^{-1} \otimes \mathcal{O}_{\mathbb{P}(V)}(-2)$. In case σ is a section of V, so that $\det V$ corresponds to the divisor $\mathcal{O}_\sigma(-\sigma)$ and $\mathcal{O}_{\mathbb{P}(V)}(-2) = \mathcal{O}_X(-2\sigma)$, we recover the statement of Lemma 14. (See also [61, III, ex. 8.4.])

Finally, let us determine the ample cone of a ruled surface X. Let f be the numerical equivalence class of a fiber and let σ be a section with $\sigma^2 = -e$.

Proposition 15. *If* $e \geq 0$, *then* $a\sigma + bf$ *is ample if and only if* $a > 0$ *and* $b > ae$. *If* $e < 0$, *then* $a\sigma + bf$ *is ample if and only if* $a > 0$ *and* $b > \frac{1}{2}ae$.

Proof. We shall just prove the statement when $e \geq 0$, referring to [61, V 2.21] for the case $e < 0$. If $e \geq 0$ and $a\sigma + bf$ is ample, then $(a\sigma + bf) \cdot f = a$ and $(a\sigma + bf) \cdot \sigma = b - ae$. Thus, $a > 0$ and $b > ae$. Conversely, suppose that $a > 0$ and $b > ae$. Then $(a\sigma + bf) \cdot f = a > 0$ and $(a\sigma + bf) \cdot \sigma = b - ae > 0$. Moreover, $(a\sigma + bf)^2 = -a^2 e + 2ab = a(2b - ae) > ab \geq 0$. Now suppose that C is an irreducible curve on X with $C \neq f, \sigma$. Write $C = n\sigma + mf$. Then $f \cdot C = n > 0$ since f moves in a base point free system and C is not contained in a fiber and $\sigma \cdot C = -ne + m \geq 0$ since $C \neq \sigma$. Thus,

$$(a\sigma + bf) \cdot C = -ane + am + bn > -ane + a(ne) + (ae)n = ane > 0.$$

The Nakai-Moishezon criterion then implies that $a\sigma + bf$ is ample. $\square$

Linear systems of cubics

We turn now to the study of linear systems of cubics in $\mathbb{P}^2$. While the study of linear systems of conics led us to general ruled surfaces, here we will be led to elliptic surfaces and "$K3$-like" surfaces.

Let $p_1, \ldots, p_n \in \mathbb{P}^2$ be distinct points and let X be the blowup of $\mathbb{P}^2$ at the points p_i, with E_i the exceptional divisors. By analogy with the case of conics, we want to determine when $|3H - \sum_i p_i|$ induces an *ample* divisor on X.

Theorem 16. *Let X be the blowup of $\mathbb{P}^2$ at the distinct points $p_1, \ldots, p_n$, with $n \leq 8$, and let $D = 3\pi^* H - \sum_i E_i$, where $\pi \colon X \to \mathbb{P}^2$ is the blowup map. Then D is ample if and only if no three of the points p_i are collinear, no six lie on a conic, and, if $n = 8$, the p_i are not all contained in a plane cubic such that one of the p_i is a singular point. In this case, D is very ample if $n \leq 6$.*

Proof. We begin with the following lemma on distinct points $p_1, \ldots, p_n \in \mathbb{P}^2$.

Lemma 17. *Suppose that the points $p_1, \ldots, p_n$ are contained in an irreducible cubic D_0, and that none of the points is a singular point of D_0.*

(i) *If $n \leq 8$, the proper transform D of D_0 on the blowup X of $\mathbb{P}^2$ at the p_i is nef and big, and $D = -K_X$.*
(ii) *If C is an irreducible curve on X with $D \cdot C = 0$, then $C^2 = -2$ and C is a smooth rational curve.*
(iii) *The linear system $|D|$ is base point free if $n \leq 7$, whereas there is a unique base point of $|D|$ if $n = 8$, and in this case $|2D|$ is base point free.*
(iv) *If $n \leq 6$, the morphism defined by $|D|$ is birational to its image and exactly contracts the smooth rational curves C on X with $C^2 = -2$. If $n = 7$, the morphism defined by $|D|$ has degree 2 onto $\mathbb{P}^2$.*

Proof. Note that, since none of the points p_i is a singular point of D_0, $D = 3\pi^* H - \sum_i E_i$, where $\pi \colon X \to \mathbb{P}^2$ is the blowup map. Thus, $D^2 = 9 - n \geq 1$ and D is big. As $K_{\mathbb{P}^2} = -3H$, we see by Proposition 4 of Chapter 3 that $K_X = -3\pi^* H + \sum_i E_i = -D$. Since D is irreducible, it follows that $D \cdot C \geq 0$ for every irreducible curve $C \neq D$ and $D \cdot D > 0$ by the above, so that D is nef. If C is a curve such that $C \cdot D = 0$, then by the Hodge index theorem $C^2 < 0$. Moreover, as $D = -K_X$, we have $2p_a(C) - 2 = K_X \cdot C + C^2 = C^2 < 0$. Thus, $p_a(C) = 0$, C is rational, and $C^2 = -2$.

To determine the base points of $|D|$, since there is a section of $\mathcal{O}_X(D)$ vanishing exactly along D, the base points of $|D|$, if any, are contained in

D. Consider the exact sequence

$$0 \to \mathcal{O}_X \to \mathcal{O}_X(D) \to \mathcal{O}_X(D)|D \to 0.$$

Since X is regular, $H^1(\mathcal{O}_X) = 0$ and so every section of $H^0(\mathcal{O}_X(D)|D)$ lifts to a section of $H^0(\mathcal{O}_X(D))$. The line bundle $\mathcal{O}_X(D)|D$ is a line bundle of degree $D^2 = 9 - n$ on the curve D. Moreover, $p_a(D) = 1$, which follows from the corresponding fact for D_0 or by adjunction for D: $2p_a(D) - 2 = K_X \cdot D + D^2 = -D^2 + D^2 = 0$. Now it is well known that for a smooth elliptic curve D, a line bundle of degree at least 3 is very ample, a line bundle of degree 2 has no base points and the corresponding morphism maps D to $\mathbb{P}^1$ with degree 2, and that every line bundle of degree 1 is of the form $\mathcal{O}_D(p)$ for a unique point $p \in D$, which is thus a base point. Similar results hold for an irreducible plane curve of arithmetic genus 1, with essentially the same proofs. Thus, $\mathcal{O}_X(D)|D$ has no base points if $D^2 \geq 2$, i.e., $n \leq 7$, and is birational to its image as long as $D^2 \geq 3$, i.e., $n \leq 6$, and the same must hold for $\mathcal{O}_X(D)$. Clearly, an irreducible curve C contracted by $|D|$ satisfies $C \cdot D = 0$ and thus C is a smooth rational curve with $C^2 = -2$. We see that we have proved all of the statements of the lemma, except the one about $2D$ in case $n = 8$. In this case, use instead the sequence

$$0 \to \mathcal{O}_X(D) \to \mathcal{O}_X(2D) \to \mathcal{O}_X(2D)|D \to 0.$$

The above sequence for $\mathcal{O}_X(D)$ and the fact that $H^1(\mathcal{O}_X(D)|D) = 0$ as long as $D^2 \geq 1$ shows that $H^1(\mathcal{O}_X(D)) = 0$, and we can argue as before, using the fact that $\mathcal{O}_X(2D)|D$ has no base points. □

Note that the above lemma shows that eight points $p_1, \ldots, p_8$ lying on an irreducible plane cubic, and not singular points of the cubic, determine a ninth point p_9 lying on the cubic, and every other plane cubic passing through $p_1, \ldots, p_8$ also passes through p_9. In other words, $p_1, \ldots, p_9$ have the Cayley-Bacharach property with respect to $\mathcal{O}_{\mathbb{P}^2}(3)$ (Definition 14 in Chapter 2).

Our next concern will be to find out when a set of at most eight points $\{p_1, \ldots, p_n\}$ in $\mathbb{P}^2$ is contained in an irreducible cubic, and when there exist smooth rational curves C on the blowup X with $C^2 = -2$.

Lemma 18. *Suppose that $n \leq 9$, and that no three of the p_i are collinear and no six lie on a conic. Then there exists an irreducible cubic containing $p_1, \ldots, p_n$. If $n \leq 7$, we can assume in addition that the cubic is smooth.*

Proof. The dimension of the linear system of cubics in $\mathbb{P}^2$ is 9. Thus, any set of no more than nine points is contained in a cubic. Given $p_1, \ldots, p_n$ with $n < 9$ with no three of the p_i collinear and no six on a conic, we can always find a p_{n+1} which does not lie on any line through two of the p_i or any conic through five of the p_i. Thus, $p_1, \ldots, p_{n+1}$ also satisy: no three of the p_i are collinear and no six lie on a conic. So we may assume that $n = 9$. Suppose that $p_1, \ldots, p_9$ is contained in a reducible cubic, necessarily the union of a line and a (possibly reducible) conic. At most two of the points can lie on a given line. Thus, the remaining seven lie on a (possibly singular) conic, contradicting the hypothesis.

Now the above argument implies that, for $n \leq 8$, the linear system containing $p_1, \ldots, p_n$ has no fixed curve. The only possible singularities of a general member of the linear system, by Bertini's theorem, are at the base points. To prove the last statement, it suffices to find an element of the linear system $|3H - \sum_i p_i|$ which is smooth at the p_i. We may assume that $n = 7$. Consider a line ℓ containing p_1 and p_2 and a conic c containing the remaining five points. If c were reducible, then it would be a union of two lines, and thus three of the p_i would have to be collinear. Thus, c is irreducible and therefore smooth. The singular locus of $\ell + c$ is then exactly $\ell \cap c$. If p_i lies on a singular point of $\ell + c$, then $p_i \in \ell \cap c$ and so either ℓ contains three of the p_i or c contains six, contrary to hypothesis. Thus, the general cubic containing $p_1, \ldots, p_n$ is smooth at $p_1, \ldots, p_n$.   □

Next, we must determine when there exist smooth rational curves C on X with $C^2 = -2$. Such a curve must be the proper transform of a plane curve of degree d. Thus, $C = d\pi^* H + \sum_i a_i E_i$, with $a_i \geq 0$ and

$$\left(d\pi^* H + \sum_i a_i E_i \right)^2 = d^2 - \sum_i a_i^2 = -2;$$

$$\left(d\pi^* H + \sum_i a_i E_i \right) \cdot \left(3\pi^* H - \sum_i E_i \right) = 3d - \sum_i a_i = 0.$$

Solving for $d = (\sum_i a_i)/3$, this becomes

$$\frac{1}{9}\left(\left(\sum_i a_i \right)^2 - 9 \sum_i a_i^2 \right) = -2.$$

By the Cauchy-Schwarz inequality, $(\sum_i a_i)^2 \leq r \sum_i a_i^2$, where r is the number of the a_i which are nonzero. Thus, we must estimate $\frac{1}{9}(r - 9)(\sum_i a_i^2)$. First suppose that $a_i = 0$ or 1 for all i. Then $3d = r$ and $d^2 = r - 2$. Thus, $d^2 - 3d + 2 = 0$, so that $(d - 1)(d - 2) = 0$ and $d = 1$ or 2. So if C is effective, there is either a line or a conic containing all the points p_i for which $a_i = 1$. The only way to have $C^2 = -2$ in this case is for three of the points to lie on a line or six on a conic.

In the remaining case, at least one of the a_i is ≥ 2. Thus,

$$\tfrac{1}{9}(r - 9)\left(\sum_i a_i^2 \right) \leq \tfrac{1}{9}(r - 9)(r + 3).$$

Direct inspection shows that this is always < -2 unless $r = 8$. So there is no solution unless $r = 8$ and at least one of the a_i is greater than 1. An easy if somewhat tedious calculation rules out the possibility that $r = 8$

and two of the a_i are greater than 1 or one of them is greater than 2. We are left with $3d = 7 + 2 = 9$, so that $d = 3$, and C is the proper transform of a cubic passing through seven points with multiplicity 1 and a remaining point with multiplicity 2. Thus, if the points p_i satisfy the hypotheses of Theorem 16, then D is ample. Conversely, if the points fail to satisfy all of the conditions of Theorem 16, then we can use the discussion above to locate a curve C on X with $D \cdot C \leq 0$, so that D cannot be ample.

To finish the proof of Theorem 16, we note that, under the hypotheses of Theorem 16, $D^2 > 0$ and $D \cdot C > 0$ for every irreducible curve C, so that D is ample by the Nakai-Moishezon criterion. To see that D is very ample when $n \leq 6$, note that we have shown that $|D|$ defines a morphism from X to $\mathbb{P}^{9-n}$ which is birational to its image. One argument that $|D|$ separates points and tangent directions on X proceeds along the lines of Lemma 17; see [61, V.4.6] for details. Another argument, which can be generalized to the case where D is not necessarily ample, goes as follows: since D is ample, there exists a $k > 0$ such that kD is very ample. For this k, we claim that the natural morphism $H^0(\mathcal{O}_{\mathbb{P}^d}(k)) \to H^0(\mathcal{O}_X(kD))$ is surjective (here $d = 9 - n$). It follows that the morphism $X \to \mathbb{P}^N$ defined by kD factors as follows: consider the morphism $X \to \mathbb{P}^d$ defined by D, followed by the Veronese embedding $\mathbb{P}^d \to \mathbb{P}^M$ defined by $|\mathcal{O}_{\mathbb{P}^d}(k)|$. Then the image of X is contained in a linear subspace $\mathbb{P}^N \subseteq \mathbb{P}^M$ and this is exactly the embedding of X in $\mathbb{P}^N$ via kD. Thus, the original morphism $X \to \mathbb{P}^d$ must have been an embedding as well.

So we must show that $H^0(\mathcal{O}_{\mathbb{P}^d}(k)) \to H^0(\mathcal{O}_X(kD))$ is surjective for all $k \geq 0$. We argue by induction, the case $k = 0$ being clear. For $k > 0$, choose a smooth curve in $|D|$, which we will also denote by D. Then the image of D in $\mathbb{P}^d$ is an elliptic normal curve in $\mathbb{P}^{d-1}$, in other words the image of the smooth elliptic curve D in $\mathbb{P}^{d-1}$ under a complete linear system of degree $d \geq 3$. By [61, IV, Ex. 4.2], D is projectively normal, i.e., the map $H^0(\mathcal{O}_{\mathbb{P}^{d-1}}(k)) \to H^0(\mathcal{O}_D(k))$ is surjective for every k. Now consider the diagram with exact rows

$$
\begin{array}{ccccccc}
H^0(\mathcal{O}_{\mathbb{P}^d}(k-1)) & \longrightarrow & H^0(\mathcal{O}_{\mathbb{P}^d}(k)) & \longrightarrow & H^0(\mathcal{O}_{\mathbb{P}^{d-1}}(k)) & \longrightarrow & 0 \\
\downarrow & & \downarrow & & \downarrow & & \\
H^0(\mathcal{O}_X((k-1)D)) & \longrightarrow & H^0(\mathcal{O}_X(kD)) & \longrightarrow & H^0(\mathcal{O}_D(k)). & &
\end{array}
$$

The right-hand vertical arrow is surjective for all k, and the left-hand vertical arrow is surjective by induction. Thus, the middle arrow is surjective as well, and so D is very ample. $\square$

Arguments similar to those above show that, in case $n \leq 6$ and D is nef but contracts some curves C with $C^2 = -2$, the image of X under the linear system D is normal and has just rational double point singularities (see also the remarks after Definition 24 of Chapter 1). The surfaces X above with $-K_X$ ample are called *del Pezzo surfaces*. We should also add

$\mathbb{P}^1 \times \mathbb{P}^1$ to this list, since $-K_X = 2f_1 + 2f_2$, where the f_i are the fibers of the projections of $\mathbb{P}^1 \times \mathbb{P}^1$ to the two factors. The del Pezzo surfaces have the following property: if $D = -K_X$ is very ample and embeds X in $\mathbb{P}^N$, then the lines on $X \subset \mathbb{P}^N$ are exactly the exceptional curves on X. Indeed, if E is an exceptional curve, then $E \cdot D = -E \cdot K_X = 1$, so that E has degree 1 as a subvariety of $\mathbb{P}^N$ and is thus a line. Conversely, if $\ell \subset X \subset \mathbb{P}^N$ is a line, then $\ell \cdot D = 1$, so that $\ell \cdot K_X = -1$ and ℓ is a smooth rational curve. Thus, ℓ is exceptional. In case $n = 7$ and D is ample, the morphism defined by D is a finite morphism from X to $\mathbb{P}^2$ and the branch curve is a smooth quartic plane curve. In this case, we could also analyze when D is nef but not ample in terms of the branch curve.

The most famous example of a del Pezzo surface is of course the cubic surface in $\mathbb{P}^3$. It is a fact which we shall not prove here (see [55], [61], etc.) that every smooth cubic surface in $\mathbb{P}^3$ is in fact the image of the blowup of $\mathbb{P}^2$ at six points in general position as above. The lines on the cubic are: the six disjoint exceptional curves, the proper transforms of the $\binom{6}{2} = 15$ lines through two of the p_i, and the proper transforms of the $\binom{6}{5} = 6$ conics passing through five of the p_i, for a grand total of 27. Similarly, we could work out the total number of exceptional curves on any of the X with $-K_X$ ample. The configuration of the lines and the projective geometry of the surfaces X in $\mathbb{P}^N$ are a source of a vast amount of really intricate geometry. There is also a deep connection with sub-root systems of the root systems E_6, E_7, E_8. One reason is the following: for a del Pezzo surface X of degree d, the lattice $[K_X]^\perp \subset H^2(X; \mathbb{Z})$ is, up to sign, a root lattice of type E_{9-d}, at least for $d = 3, 2, 1$. In fact, viewing X as the blowup of $\mathbb{P}^2$ at $r = 9 - d$ points, let H be the pullback of the hyperplane class and let $E_1, \dots, E_r$ be the classes of the exceptional curves. Then $[K_X]^\perp$ is spanned by the classes $E_1 - E_2, \dots, E_{r-1} - E_r, H - E_1 - E_2 - E_3$, which are elements of square -2 whose associated Dynkin diagram is that of type E_r. There are also connections with the possible rational double point singularities which can lie on a generalized del Pezzo surface, for which we refer to [20].

One reason for the importance of the del Pezzo surfaces is that they are surfaces of "almost minimal" degree. If $-K_X$ is very ample on X, then it embeds X as a surface of degree $9 - n$ in $\mathbb{P}^{9-n}$. Thus, the del Pezzo surfaces are examples of surfaces of degree d in $\mathbb{P}^d$, for $3 \le d \le 9$. Conversely, it is a beautiful theorem of classical algebraic geometry that a nondegenerate smooth surface of degree d in $\mathbb{P}^d$ is either a del Pezzo surface or the embedding of $\mathbb{P}^1 \times \mathbb{P}^1$ via $|2f_1 + 2f_2|$. In particular it follows that $d \le 9$. The general idea of the proof is as follows. One first shows that if X is a smooth surface in $\mathbb{P}^d$ of degree d and $d \ge 4$, then X is an intersection of quadrics in $\mathbb{P}^d$—this follows by looking at the hyperplane sections, which must be elliptic curves embedded by a complete linear system of degree d, and using classical facts about them. From this it follows easily that there can be only finitely many lines lying on X. Projecting X to $\mathbb{P}^{d-1}$ from a point of X not lying on any line, we obtain a new surface X' in $\mathbb{P}^{d-1}$ of degree $d-1$. We can repeat this process until we reach a cubic surface in $\mathbb{P}^3$. But each time we do so, we introduce a new line, disjoint from the ones previously introduced by projection. Moreover, the cubic surface in $\mathbb{P}^3$ has 27 lines, but a maximal set of mutually disjoint lines can have at most six elements. Thus, $d \le 3 + 6 = 9$, and X is obtained from a cubic surface by blowing down exceptional curves. From this it is easy to show that our examples of such surfaces exhaust the possibilities.

Next let us determine the ample cone of a del Pezzo surface. More generally, we make the following definition [38]:

Definition 19. A rational surface X is a *good generic surface* if $K_X = -D$ for some irreducible curve D and there are no smooth rational curves C on X with $C^2 = -2$.

In fact, it follows from classification theory that a good generic surface is automatically rational. (See also Exercise 3.) As we have seen, if $D^2 > 0$, a good generic surface is a del Pezzo surface. There exist good generic surfaces for all possible choices of $D^2 \le 0$ as well; in fact, we leave it as an exercise to show that the blowup of n generic points on D_0, where D_0 is a smooth cubic in $\mathbb{P}^2$, is a good generic surface for every $n \ge 0$. However, for $n \ge 9$, we do not have an explicit description of what exactly the genericity condition is.

Proposition 20. *If X is a good generic surface and D is an irreducible curve on X such that $D = -K_X$, then the irreducible curves of negative self-intersection on X are exactly the exceptional curves and possibly D, if $D^2 < 0$.*

Proof. If C is an irreducible curve with $C^2 < 0$ and $C \ne D$, then $C \cdot D \ge 0$. If $C \cdot D > 0$, then $C \cdot K_X = -C \cdot D < 0$ and C is exceptional by Lemma 11 of Chapter 3. Otherwise, $C \cdot K_X = 0$ and $C^2 < 0$, so that $C^2 = -2$ and C is a smooth rational curve. But we have assumed that no such curves exist on X. $\square$

In particular, we see that if $D^2 \ge 0$, the walls of the ample cone $\mathcal{A}(X)$ are exactly the classes of the exceptional curves, whereas if $D^2 < 0$, the class $D = -K_X$ defines an additional wall. Thus, for example, for the cubic surface X, a divisor D is ample if and only if $D \cdot \ell > 0$ for each of the 27 lines ℓ on X.

Next we shall give a characterization of the exceptional classes on X, in case $D^2 \ge 0$.

Proposition 21. *Let X be a good generic surface with $K_X^2 \geq 0$ and let F be a divisor on X with $F^2 = F \cdot K_X = -1$. Then there exists an exceptional curve E with $E = F$ as divisor classes.*

Proof. Of course, the main point here is that F is just some divisor, not necessarily the divisor associated to an irreducible curve, and we need to show that F is effective. As before, let D be an irreducible curve with $D = -K_X$ as divisor classes. First note that, since $D^2 \geq 0$ by hypothesis, $D \cdot C \geq 0$ for every effective divisor C. Furthermore, if $D \cdot C = 0$, then $C^2 \leq 0$, by the Hodge index theorem, and the case $C \neq D$, $C^2 < 0$ is impossible since then C would be a smooth rational curve of self-intersection -2. If $C^2 = 0$, then C and D must span a 1-dimensional lattice in Num X, by the Hodge index theorem. Thus, C is a positive rational multiple of D.

By the Riemann-Roch formula,

$$h^0(\mathcal{O}_X(F)) + h^2(\mathcal{O}_X(F)) \geq \tfrac{1}{2}(F^2 - F \cdot K_X) + 1 = 1.$$

Thus, either $h^0(\mathcal{O}_X(F)) \neq 0$, so that F is effective, or $h^2(\mathcal{O}_X(F)) \neq 0$, so that by Serre duality $K_X - F = -D - F$ is effective. Suppose that $-D - F$ is effective. Now $D \cdot (-D - F) = -D^2 - D \cdot F \leq -1$, which is impossible. Thus, F is effective. Moreover, $D \cdot F = 1$. If $F = \sum_i n_i C_i$, where the C_i are irreducible and $n_i > 0$, then $D \cdot C_i = 1$ for exactly one C_i, with $n_i = 1$, and $D \cdot C_j = 0$ for $i \neq j$. By the remarks at the beginning of the proof, $C_i = E$ is irreducible and C_j is a positive rational multiple of D for all $i \neq j$. Thus, there exists an irreducible curve E with $K_X \cdot E = -D \cdot E = -1$ such that $F = E$ if $D^2 > 0$, and $F = E + rD$ with $r \geq 0$ if $D^2 = 0$. However, in this last case we have $F^2 = -1 = E^2 + 2r$. Thus, $E^2 = -1 - 2r < 0$, so that E is an exceptional curve, $E^2 = -1$ and $r = 0$.   □

Let us consider the case $n = 9$ above, i.e., $D^2 = 0$. Then we have the following example of Kodaira:

Proposition 22. *If X is a good generic surface with $K_X^2 = 0$, then choosing a fixed exceptional curve E_0, there is a bijection from the set of all exceptional curves E on X to the lattice $(K_X)^\perp / K_X \cong \mathbb{Z}^8$. In particular there are infinitely many exceptional curves on X.*

Proof. Given an exceptional curve E on X, map it to the image of $E - E_0 \in (K_X)^\perp / K_X$, noting that $K_X \cdot E = K_X \cdot E_0$ and thus $E - E_0 \perp K_X$. Conversely, given a class $e \in (K_X)^\perp / K_X$, choose a lift of e to a class $F \in K_X^\perp$ and consider $E_0 + F$. We are free to modify this by a multiple of K_X, say $m K_X$. Now, if $a = E_0 \cdot F$, then

$$(E_0 + F + m K_X)^2 = -1 + 2a + 2m + F^2,$$

where F^2 is even since $F \perp K_X$. Thus, for a unique choice of m we have $(E_0 + F + m K_X)^2 = -1$. Since $(E_0 + F + m K_X) \cdot K_X = -1$ as well,

$E_0 + F + mK_X = E$ is the class of an exceptional curve, by Proposition 21, and we have constructed the desired bijection. $\square$

To find an example satisfying the hypotheses of Proposition 22, it follows from Exercise 8 that the blowup of nine points in $\mathbb{P}^2$ in general position is a good generic surface. In this case, there is a unique (smooth) plane cubic passing through all nine. However, if the points are in somewhat special position, a whole new phenomenon occurs, and we see our first examples of *elliptic surfaces*, which we shall study more systematically in Chapter 7. Let us note here, however, that we can indeed find a morphism to $\mathbb{P}^1$ from certain blowups of $\mathbb{P}^2$ at nine points. Let $p_1, \ldots, p_9$ be contained in a smooth cubic and let D be the proper transform of the cubic. Consider the exact sequence

$$0 \to \mathcal{O}_X \to \mathcal{O}_X(D) \to \mathcal{O}_D(D) \to 0.$$

Since $D^2 = 0$, $\mathcal{O}_D(D)$ is a line bundle of degree 0 on D. If it is in fact trivial, then since $H^1(\mathcal{O}_X) = 0$ we can lift a nonvanishing section to a section of $\mathcal{O}_X(D)$. It follows that $h^0(\mathcal{O}_X(D)) = 2$ and that $|D|$ defines a morphism to $\mathbb{P}^1$, one of whose fibers is D. Note that every exceptional curve E satisfies $E \cdot D = 1$ and so the exceptional curves E are sections of the map $X \to \mathbb{P}^1$. Conversely, every section E is isomorphic to $\mathbb{P}^1$ and $E \cdot K_X = -E \cdot D = -1$, so that E is exceptional.

A similar construction works if $\mathcal{O}_D(D)$ is a line bundle of finite order on D. In this case, if m is its order, then $|mD|$ defines a morphism to $\mathbb{P}^1$ whose general fiber is a smooth elliptic curve, and such that one fiber is mD. We leave this as an exercise.

An introduction to $K3$ surfaces

By definition a $K3$ *surface* X is a surface such that $K_X = 0$ and $q(X) = 0$. Thus, $K3$ surfaces are one natural generalization of an elliptic curve. The other possible generalization is an *abelian surface* or complex torus of dimension 2. Here a complex torus of dimension 2 is a compact complex surface of the form $\mathbb{C}^2/\Lambda$, where $\Lambda \cong \mathbb{Z}^4$ is a discrete subgroup of $\mathbb{C}^2$. The torus is an abelian surface if in addition it is a smooth projective surface, or equivalently carries a Hodge metric. An abelian surface X also has $K_X = 0$, but $q(X) = 2$, and these are the only two classes of surfaces with $K_X = 0$, or in other words with a nowhere vanishing holomorphic 2-form. Despite the apparent similarities between complex tori and $K3$ surfaces, it is the $K3$ surfaces which have received by far the lion's share of attention in surface theory, partly because many statements which are elementary to prove for abelian surfaces have very difficult proofs in the case of $K3$ surfaces.

Here are some of the basic examples of $K3$ surfaces:

1. A smooth surface of degree 4 in $\mathbb{P}^3$;

2. A smooth complete intersection of a quadric and a cubic surface in $\mathbb{P}^4$;
3. A smooth complete intersection of three quadrics in $\mathbb{P}^5$;
4. A double cover of $\mathbb{P}^2$ branched along a smooth sextic curve;
5. (Kummer surfaces.) Let A be an abelian surface. Then there is an involution $\iota\colon A \to A$ defined by $\iota(a) = -a$. This involution has 16 fixed points, the points of order 2 on A. The quotient A/ι has 16 singular points, which are all ordinary double points. The minimal resolution of the quotient is then a $K3$ surface with 16 disjoint curves C with $C^2 = -2$.

Since the canonical bundle of a $K3$ surface X is trivial, $p_g(X) = 1$ and in fact $P_n(X) = 1$ for all n, and moreover $c_1(X)^2 = 0$. By definition $q(X) = 0$. Thus, $\chi(\mathcal{O}_X) = 2$ and so $c_2(X) = 24$. By the Hodge index theorem $b_2^+(X) = 3$. Since K_X is trivial and thus divisible by 2, the intersection form on X is of Type II, and in fact the intersection form on X is $I \oplus I \oplus I \oplus (-E_8) \oplus (-E_8)$, where I is the hyperbolic plane and $-E_8$ is the unique unimodular negative definite intersection form of rank 8 and Type II.

Most of the deeper properties of $K3$ surfaces require looking at nonalgebraic surfaces as well and studying the period map. For example, Kodaira showed that all $K3$ surfaces fit into one 20-dimensional family of complex surfaces, and in particular they are all diffeomorphic. Thus, since by the Lefschetz theorem on hyperplane sections a smooth surface in $\mathbb{P}^3$ is simply connected, all $K3$ surfaces are simply connected. Using the global Torelli theorem, one can show that, for every $n \geq 1$, the set of isomorphism classes of pairs (X, H), where X is a $K3$ surface and H is a nef divisor on X with $H^2 = 2n$ which is primitive (not the multiple of another divisor), forms an irreducible 19-dimensional family $\mathcal{F}_{2n}$. We have listed the first few cases above.

The structure of ample and nef divisors on X is a topic for which we do not need the period map, and we want to say something about it here.

Lemma 23. *Let X be a $K3$ surface and let C be an irreducible curve on X. Then $\mathcal{O}_C(C) = \omega_C$. Thus, $C^2 \geq -2$, $C^2 = -2$ if and only if C is a smooth rational curve, and $C^2 = 0$ if and only if $p_a(C) = 1$. In all other cases, C is a nef and big divisor.*

Proof. It follows from adjunction that $\mathcal{O}_C(C) = \omega_C$, and thus $C^2 = 2p_a(C) - 2$. The other statements are clear. $\square$

By Lemma 23, the closure of the ample cone $\mathcal{A}(X)$ of X is exactly the set of x in the positive cone such that $x \cdot C \geq 0$ for every smooth rational curve C on X, or equivalently every curve of square -2. The set of walls defined by such curves is locally finite, and the interior $\mathcal{A}(X)$ is exactly the set of x in the positive cone of $\operatorname{Num} X$ with $x \cdot C > 0$ for all C irreducible of square -2.

Lemma 24. *If C is an irreducible curve on X with $p_a(C) \geq 1$, then the linear system $|C|$ is base point free.*

Proof. From the exact sequence

$$0 \to \mathcal{O}_X \to \mathcal{O}_X(C) \to \omega_C \to 0,$$

and the fact that $H^1(\mathcal{O}_X) = 0$, it follows that the map $H^0(\mathcal{O}_X(C)) \to H^0(\omega_C)$ is surjective. By Proposition 6 of Chapter 1, ω_C has no base locus if $p_a(C) \geq 1$. Thus, $|C|$ has no base locus either. $\square$

By Bertini's theorem, if C is an irreducible curve on X with $C^2 > 0$, then the linear system $|C|$ contains smooth curves of genus equal to $p_a(C) \geq 2$. So we may assume that C is itself smooth. Then $\mathcal{O}_X(C)|C = K_C$. Now either C is hyperelliptic or it is not. If it is not hyperelliptic, then it is not difficult to show that show that the morphism φ defined by C has degree 1 onto its image, and the image $\varphi(X)$ is a normal surface of degree $2g - 2$ in $\mathbb{P}^g$, all of whose smooth hyperplane sections are curves of genus g embedded by the canonical bundle (these are referred to as canonical curves). The argument here is similar to the last part of the proof of Theorem 16, but using the properties of canonical curves instead of elliptic normal curves. The singularities of $\varphi(X)$ correspond to the curves D such that $C \cdot D = 0$. By the Hodge index theorem, for such a curve D we have $D^2 < 0$, and thus $D^2 = -2$ and D is smooth rational. Hence $\varphi(X)$ has just rational double point singularities (or no singularities, if C is ample). Conversely, any surface in $\mathbb{P}^g$ with at worst rational double point singularities, all of whose hyperplane sections are canonical curves, is a $K3$ surface. (See Exercise 12.)

Lemma 25. *Let D be a divisor on X with $D^2 \geq -2$. Then either D or $-D$ is effective.*

Proof. Apply Riemann-Roch to $\mathcal{O}_X(D)$: we have

$$h^0(\mathcal{O}_X(D)) + h^2(\mathcal{O}_X(D)) \geq \frac{D^2}{2} + \chi(\mathcal{O}_X) \geq -1 + 2 \geq 1.$$

Thus, either $h^0(\mathcal{O}_X(D)) \neq 0$, in which case D is effective, or $h^2(\mathcal{O}_X(D)) \neq 0$. By Serre duality, $h^2(\mathcal{O}_X(D)) = h^0(\mathcal{O}_X(-D))$ since K_X is trivial. Thus, either D or $-D$ is effective. $\square$

Lemma 26. *If D is a nef and big divisor on the K3 surface X, then*

$$h^0(\mathcal{O}_X(D)) = 2 + D^2/2.$$

Proof. Since D is nef, $H \cdot D \geq 0$ for all ample divisors H, and $H \cdot D = 0$ implies that $D^2 \leq 0$, contradicting the assumption that D is big. Thus,

$H \cdot D > 0$ and $-D$ cannot be effective. Thus, $h^2(\mathcal{O}_X(D)) = 0$. Since D is nef and big, $h^1(\mathcal{O}_X(D)) = 0$ as well by the Mumford vanishing theorem (Theorem 26 of Chapter 1). So $\chi(\mathcal{O}_X(D)) = h^0(\mathcal{O}_X(D))$, and by Riemann-Roch, using the fact that $K_X = 0$, this is $D^2/2 + 2$. $\square$

The following result of Mayer [88] gives a complete description of nef and big divisors on X:

Theorem 27. *Let D be a nef and big divisor on the K3 surface X. Then $|D|$ has a base point if and only if $|D|$ has a fixed curve if and only if $D = kE + R$, where E is a smooth elliptic curve, R is a smooth rational curve, $R \cdot E = 1$, and $k \geq 2$. In this last case $2D$ has no base points. Thus, every nef and big divisor on X is eventually base point free, and defines a morphism from X onto a normal surface with only rational double point singularities.*

Proof. By Lemma 25, D is effective. Write $D = D_f + D_m$, where D_f is the fixed curve of D and $D_m = D - D_f$ has no fixed curves. Clearly, by construction $h^0(\mathcal{O}_X(D_f)) = 1$, $h^0(\mathcal{O}_X(D)) = h^0(\mathcal{O}_X(D_m))$, and D_m is nef.

Case I. $(D_m)^2 > 0$.

By Lemma 26, $h^0(\mathcal{O}_X(D_m)) = 2 + (D_m)^2/2$ since D_m is nef and big. Likewise, $h^0((\mathcal{O}_X(D)) = 2 + D^2/2$. From the assumption that $h^0((\mathcal{O}_X(D)) = h^0((\mathcal{O}_X(D_m))$ we see that $D_m^2 = D^2 = (D_f + D_m)^2 = (D_f)^2 + 2D_f \cdot D_m + (D_m)^2$. Thus, $(D_f)^2 + 2D_f \cdot D_m = 0$. Rewrite this as $D_f \cdot (D_f + D_m) + D_f \cdot D_m = 0$. As $D_f + D_m = D$, we have $D_f \cdot D + D_f \cdot D_m = 0$. Since both D and D_m are nef, $D_f \cdot D \geq 0$ and $D_f \cdot D_m \geq 0$, and hence $D_f \cdot D = D_f \cdot D_m = 0$. Thus, $0 = D_f \cdot D = D_f \cdot (D_f + D_m) = (D_f)^2$, since $D_f \cdot D_m = 0$. If $D_f \neq 0$, then $-D_f$ is not effective, and so $h^2(\mathcal{O}_X(D_f)) = 0$. But then by Riemann-Roch $h^0(\mathcal{O}_X(D_f)) \geq 2 + (D_f)^2/2 = 2$, contradicting $h^0(\mathcal{O}_X(D_f)) = 1$. It follows that D has no fixed curves. Using Exercise 11, the general element of $|D|$ is of the form $\sum_i D_i$, where all the D_i are numerically equivalent irreducible curves. Since D is big, each D_i is an irreducible curve of positive square, and thus $|D_i|$ is base point free by Lemma 24. It follows that $|D|$ is base point free as well.

Case II. $(D_m)^2 = 0$.

Since $D_m^2 = 0$, either the general element of D_m is irreducible, in which case a general element of D_m equals E for a smooth elliptic curve E, or D_m is composite with a pencil by Bertini's theorem [61, p. 280, Ex. 11.3]. In this last case $D_m = kE$ again, where E is a smooth elliptic curve. If $D_m = E$, then $h^0(\mathcal{O}_X(D)) = h^0(\mathcal{O}_X(D_m)) = 2 + D^2/2$. From the exact sequence

$$0 \to \mathcal{O}_X \to \mathcal{O}_X(E) \to \mathcal{O}_E \to 0,$$

and using $h^1(\mathcal{O}_X) = 0$, there is an exact sequence

$$0 \to H^1(\mathcal{O}_X(E)) \to H^1(\mathcal{O}_E) \to H^2(\mathcal{O}_X) \to H^2(\mathcal{O}_X(E)).$$

Since E is effective and nonzero, $h^2(\mathcal{O}_X(E)) = h^0(\mathcal{O}_X(-E)) = 0$. Thus, $H^1(\mathcal{O}_E) \to H^2(\mathcal{O}_X)$ is surjective and so, since both spaces have dimension 1, an isomorphism. Hence $H^1(\mathcal{O}_X(E)) = 0$ and so $h^0(\mathcal{O}_X(D_m)) = 2 + E^2/2 = 2 = h^0(\mathcal{O}_X(D))$. In this case $D^2 = 0$ as well, contradicting the fact that D was big.

So we may assume that $k \geq 2$. There exists a component R of D_f such that $E \cdot R \neq 0$, since otherwise $E \cdot D_f = 0$ and so $D_m \cdot D_f = 0$, and thus $D_f^2 = (D_m + D_f)^2 = D^2 > 0$. But then by Riemann-Roch we would have $h^0(D_f) \geq 2$, a contradiction. So we can find such an R, necessarily a smooth rational curve with $R^2 = -2$. Next we claim that $E \cdot R = 1$: If $E \cdot R \geq 2$, then $(E + R) \cdot R \geq 0$, so that $E + R$ is nef, and $(E + R)^2 \geq 2$. Thus, $E + R$ is big, and $h^0(\mathcal{O}_X(E + R)) = 2 + (E + R)^2/2 > 2$. It follows that $|E + R|$ is strictly bigger than E, so that R is not a fixed component of $|E + R|$. But then $D_m + R = (k - 1)E + (E + R)$ does not have a fixed component either, contradicting the choice of $R \subseteq \operatorname{Supp} E_f$. So $E \cdot R = 1$.

Now let $D_1 = kE + R$ and $D_2 = D_f - R$. Thus, D_1 is still nef and D_2 is effective, and $(kE + R)^2 = 2k - 2 > 0$, so that $kE + R$ is big. An argument identical to the previous argument with the decomposition $D = D_m + D_f$, but applied to the decomposition $D = D_1 + D_2$, shows that if $D_2 \neq 0$, then $D_1 \cdot D_2 = D_2^2 = 0$. Thus, $\dim |D_2| \geq 1$ and $|D_2| \subseteq |D_f|$ which has dimension 0, a contradiction. It follows that $D_2 = 0$ and $D = kE + R$. We leave it to the reader to check that $h^0(\mathcal{O}_X(kE + R)) = k + 1$, and thus $\dim |kE + R| = \dim |kE|$, so that R is indeed a fixed curve of $kE + R$.

Finally, we must show that, if $D = kE + R$, then $|2D|$ has no base locus. It suffices to show that $|2D|$ has no fixed components. The only possible fixed component is R. But consider the exact sequence

$$0 \to \mathcal{O}_X(2kE + R) \to \mathcal{O}_X(2kE + 2R) \to \mathcal{O}_R(2kE + 2R) \to 0.$$

Since $k \geq 2$, $2k \geq 4$, and $2kE + R$ is nef and big. Thus, $H^1(\mathcal{O}_X(2kE + R)) = 0$ (we could also show this directly). So the map $H^0(\mathcal{O}_X(2kE + 2R)) \to H^0(\mathcal{O}_R(2kE + 2R))$ is surjective. As $\mathcal{O}_R(2kE + 2R)$ is a line bundle of nonnegative degree on the smooth rational curve R, it has a nonzero section which lifts to $H^0(\mathcal{O}_X(2kE + 2R))$. Thus, R is not a fixed curve of $|2D|$. □

Using arguments similar to the proof of Theorem 16, and in particular the classical fact that a canonical curve is projectively normal, one can show the following: if D is nef, big, and base point free, with $D^2 = 2g - 2 > 0$, then either every smooth curve in $|D|$ is nonhyperelliptic, and the morphism from X to $\mathbb{P}^g$ is birational to its image, which is a normal surface with at worst rational double point singularities, or every smooth curve in $|D|$ is hyperelliptic and the image of X is a surface of minimal degree $g-1$ in $\mathbb{P}^g$, and is thus a rational normal scroll (the image of $\mathbb{F}_n$ under a linear system of the form $|\sigma + kf|$, $k \geq n$, or the Veronese surface in $\mathbb{P}^5$).

Exercises

1. Show that, for $n > 0$, the negative section σ on $\mathbb{F}_n$ is the unique irreducible curve with negative self-intersection. Likewise, for X a geometrically ruled surface with $e(X) > 0$, there is a unique irreducible curve with negative self-intersection.

2. Show that elementary transformations of a ruled surface $X = \mathbb{P}(V)$ in the sense of this chapter correspond to elementary modifications of the vector bundle V over C, in the sense of Chapter 2.

3. For which ruled surfaces X is $-K_X$ effective? Show that, if X is a ruled surface such that $K_X = -D$, where D is irreducible, then X is a rational ruled surface. (Show that $p_a(D) = 1$, so that either X is rational or X is ruled over an elliptic curve C, D is smooth, and the morphism $D \to C$ is a covering space. Thus, $H^1(\mathcal{O}_X) \cong H^1(\mathcal{O}_C) \cong H^1(\mathcal{O}_D)$. Use the long exact sequence associated to

$$0 \to \mathcal{O}_X(-D) \to \mathcal{O}_X \to \mathcal{O}_D \to 0$$

to show that $H^1(\mathcal{O}_X(-D)) = H^1(K_X) = 0$, and derive a contradiction.)

4. Let σ and σ' be two sections of a (geometrically) ruled surface. Show that $\sigma^2 \equiv (\sigma')^2 \mod 2$. Show also that, for every rank 2 vector bundle V over the curve C, we have $c_1(\mathcal{O}_{\mathbb{P}(V)}(1))^2 = -c_1(V)$. (We have seen this for the case where $V = R^0\pi_*\mathcal{O}_X(\sigma)^\vee$. Reduce the general case to this case.)

5. Let A be an abelian surface, i.e., a complex torus of dimension 2. Thus, $K_A = 0$. Show that there is no effective curve C on A with $C^2 < 0$. Show moreover that if C is irreducible and $C^2 > 0$, then C is ample. More generally, use the Riemann-Roch theorem to show that if D is a divisor on X with $D^2 > 0$, then either D or $-D$ is effective and in addition ample. What can you say about the case $C^2 = 0$, C irreducible?

6. Let $X = \operatorname{Sym}^2 E$, the second symmetric product of an elliptic curve. Thus, X is the quotient of the abelian surface $E \times E$ by the involution $(a,b) \mapsto (b,a)$. Using the group law, there is a natural map $X \to E$ defined by $(a,b) \mapsto a + b$. Show that the fibers of this map are $\mathbb{P}^1$ so that X is geometrically ruled over E. Which ruled surface over E is it (bearing in mind our classification in the example after the proof of Proposition 12)? (Here is one approach. Note that the curves $E \times \{p\}$, for $p \in E$, give sections σ_p of X. Also, $\sigma_p^2 = \sigma_p \cdot \sigma_q = 1$ for all p, q. Now use the last two exercises.)

7. Let $\pi\colon X \to C$ be a nonminimal ruled surface. Let $t \in C$ and let $\pi^{-1}(t) = \bigcup_{i=1}^{k} C_i$. Let $I \subset \{1, \ldots, k\}$ be a proper set of indices such that $\bigcup_{i \in I} C_i$ is connected. Show that the lattice spanned by the classes of the C_i, $i \in I$, is negative definite and that $\bigcup_{i \in I} C_i$ contracts to a rational singularity. (Note that, for a general fiber f of π, $H^1(nf; \mathcal{O}_{nf}) = 0$ for every $n > 0$ and thus the same must hold for all fibers.)

   Can you locate a D_4 rational double point configuration (the dual graph is D_4 and all curves have self-intersection -2) this way?

8. Let X be a blowup of $\mathbb{P}^2$ at distinct points $p_1, \ldots, p_n$ lying on a smooth cubic D_0. Show that if there exists a smooth curve C disjoint from the proper transform D of D_0 on X, then there is a nontrivial relation $3n_0 h + \sum_i n_i p_i \equiv 0$ in the divisor class group of D_0, where the n_i are integers and h is the divisor class of degree three on D_0 which corresponds to the restriction of $\mathcal{O}_{\mathbb{P}^2}(1)$ to D_0. Deduce that for general choices of the p_i there is no such curve C.

9. Let X be the blowup of $\mathbb{P}^2$ at nine points lying on a smooth cubic, and let D be the proper transform of the cubic. Suppose that $\mathcal{O}_X(D)|D$ has order exactly m. Show that $|nD|$ consists of one point for $0 < n < m$ and that $|mD|$ defines a morphism to $\mathbb{P}^1$ whose general fiber is a smooth elliptic curve and such that mD is a fiber.

10. There exist elliptic surfaces which are the blowup of $\mathbb{P}^2$ at nine points which are the base locus of a pencil of cubics, such that all fibers are irreducible and such that the generic fiber has only a single ordinary double point, i.e., is a nodal cubic. Show at least that there exist such surfaces with all fibers irreducible as follows. The space of all cubics is a projective space of dimension 9. What is the dimension of the space of reducible cubics? Thus, show that a general line in the space of all cubics misses the set of reducible ones.

    We can show moreover that the general irreducible singular cubic C is nodal as follows. If C is not nodal, then as $p_a(C) = 1$, C has a cusp. Show that all cuspidal cubics are projectively equivalent (since all rational curves with a cusp are isomorphic and the automorphism group acts transitively on the set of degree three line bundles on C) and that the cubic $y^2 z = x^3$ has a 1-parameter family of projective automorphisms. The upshot is that the dimension of the space of cuspidal plane cubic curves is $7 = \dim PGL(3) - 1$, whereas the dimension of the space of nodal plane cubic curves is $8 = \dim PGL(3)$. Thus, the general singular cubic is a nodal curve.

    Show that if X is a good generic surface and $X \to X'$ is the contraction of an exceptional curve, then X' is again a good generic surface. Thus, again we see directly that good generic surfaces exist which are blowups of $\mathbb{P}^2$ at $n \leq 9$ points.

11. Let $|D|$ be a linear system without fixed curves on an algebraic surface X. Then the general element of $|D|$ is of the form $\sum_i D_i$, where all

of the D_i are numerically equivalent. (If $|D|$ has no base locus and the general element of $|D|$ is reducible, then by [61, p. 280, Ex. 11.3], $|D|$ is composite with a pencil, i.e., the corresponding morphism to $\mathbb{P}^n$ has image a curve C_0. If $X \to C \to C_0$ is the Stein factorization, then each D_i is a fiber of $X \to C$ and thus they are all numerically equivalent. If the general element of $|D|$ is reducible but connected, it must have a singular point. Such a point must be in the base locus, by Bertini's theorem. After blowing up the base points of a general pencil, we obtain a linear system without base points whose general element is not connected, and we can reduce this case to the previous case.)

12. Let $X \subset \mathbb{P}^g$ be a smooth surface such that all hyperplane sections are canonical curves. Then X is a $K3$ surface. (Note that $K_X \cdot H = 0$ for every hyperplane section H. Choose a general pencil in $|H|$, such that every member is irreducible, and blow up the base locus which consists of distinct points p_i. Let $\rho \colon \tilde{X} \to X$ be the blowup, with exceptional curves E_i, and let $\pi \colon \tilde{X} \to \mathbb{P}^1$ be the morphism defined by the pencil. Let C be a smooth fiber of π. From

$$K_C = K_{\tilde{X}} \otimes \mathcal{O}_X(C)|C = K_{\tilde{X}}|C = \rho^* K_X \otimes \mathcal{O}_{\tilde{X}}\left(\sum_i E_i\right)|C,$$

conclude that $\rho^* K_X|C$ is trivial for all smooth C. For every fiber C, smooth or not, use semicontinuity and the fact that $\rho^* K_X|C$ and $(\rho^* K_X|C)^{-1}$ have sections for general C to conclude that $\rho^* K_X|C$ is trivial. Thus, $\rho^* K_X = \pi^* \pi_* \rho^* K_X$ must be pulled back from $\mathbb{P}^1$, so is $\mathcal{O}_{\tilde{X}}(nC)$ for some n. Using $\rho^* K_X \cdot E_i = 0$, conclude that $n = 0$ and thus that K_X is trivial. Finally, argue that $H^1(\mathcal{O}_X) = 0$ by using the vanishing of $H^1(\mathcal{O}_X(H))$ for an ample divisor H.)

6

Vector Bundles over Ruled Surfaces

Suitable ample divisors

In this chapter and in Chapter 8 we will discuss the structure of stable bundles on special classes of surfaces, ruled and elliptic surfaces. In general, it seems a little difficult to obtain detailed information about moduli spaces when stability is defined with respect to an arbitrary ample divisor. We will need instead to consider ample divisors adapted to the geometry of the surfaces at hand. The examples we have in mind are all given as fibrations over a base curve, and the fibration itself will be the most interesting geometric feature of the surface. Thus, we will try to consider ample divisors which reflect the fibration. The class of a fiber is not ample; it lives at the boundary of the ample cone. Instead we shall consider ample divisors which are sufficiently close to the class of a fiber, where how close will depend on the particular choice of Chern classes of the problem we want to study. We will further study bundles on the surface X by studying their restrictions to the fibers of the fibration. This method of studying bundles on a surface by looking at their restrictions to curves on the surface is one which has been successfully applied in a wide variety of contexts.

We begin by discussing the general strategy for analyzing the meaning of stability for a fibration $X \to C$, and then specialize to the case of ruled surfaces. After an expository section describing the global and local structure of moduli spaces, we will find a Zariski open and dense subset of the moduli space of bundles over a ruled surface. In the exercises, we shall sketch how these ideas may be applied to prove that the moduli space of rank 2 bundles over $\mathbb{P}^2$ is irreducible.

Fix the following notation for this section: X is a surface and $\pi\colon X \to C$ is a morphism from X to the smooth curve C. We denote by f the numerical equivalence class of a general fiber. Next we recall the following terminology from Chapter 4: For a rank 2 vector bundle V, we shall denote the mod 2 reduction of $c_1(V)$ by $w(V) = w$ and the integer $c_1(V)^2 - 4c_2(V)$ by $p_1(\mathrm{ad}\,V) = p$. If w is a numerical equivalence class mod 2 and p is an integer, a class of type (w, p) is a class ζ such that $p \leq \zeta^2 < 0$ and such that the mod 2 reduction of ζ equals w. A wall of type (w, p) is the intersection W^ζ of the hyperplane $\zeta^\perp$ with $\mathcal{A}(X)$. The chambers of type (w, p) are the connected components of the complement in $\mathcal{A}(X)$ of the set of walls of type (w, p).

Definition 1. Let Δ be a divisor on X, let c be an integer, and set $w = \Delta \bmod 2$ and $p = \Delta^2 - 4c$. An ample divisor H is (w, p)-*suitable* or (Δ, c)-*suitable* if H does not lie on a wall of type (w, p) and, for all ζ of type (w, p) such that $\zeta \cdot f \neq 0$, we have sign $f \cdot \zeta = $ sign $H \cdot \zeta$.

Clearly, H is (w, p)-suitable if and only if H is not separated from f by any wall, if and only if f lies in the closure of the chamber containing H. It is easy to see that, if H_1 and H_2 are both (w, p)-suitable, then the only walls of type (w, p) separating H_1 and H_2 must contain f in their closure. If H is (w, p)-suitable, then clearly $H + tf \in \operatorname{Num} X \otimes \mathbb{Q}$ is (w, p)-suitable for all $t \in \mathbb{Q}^+$, and by Proposition 22 of Chapter 4, a rank 2 bundle V is H-semistable if and only if it is $H + tf$-semistable. Finally, note the following useful lemma:

Lemma 2. If H is an ample divisor lying on no wall of type (w, p) and V is a strictly H-semistable bundle with $c_1(V) = \Delta$ and $p_1(\operatorname{ad} V) = p$, then there exists an exact sequence

$$0 \to \mathcal{O}_X(D) \to V \to \mathcal{O}_X(\Delta - D) \otimes I_Z \to 0$$

with $2D - \Delta$ numerically equivalent to zero.

Proof. By Proposition 21 of Chapter 4, there exists such an exact sequence, with $\mathcal{O}_X(D)$ a destabilizing sub-line bundle. Thus, $H \cdot (2D - \Delta) = 0$. By the Hodge index theorem $(2D - \Delta)^2 \leq 0$, with equality if and only if $2D - \Delta$ is numerically trivial. On the other hand,

$$p = p_1(\operatorname{ad} V) = \Delta^2 - 4D \cdot (\Delta - D) - 4\ell(Z) = (2D - \Delta)^2 - 4\ell(Z),$$

so that $(2D - \Delta)^2 \geq p$. Thus, either $2D - \Delta$ defines a wall of type (w, p) or $2D - \Delta$ is numerically equivalent to zero. The first case cannot arise since we have assumed that H lies on no wall of type (w, p). So $2D - \Delta$ is numerically equivalent to zero. $\square$

Lemma 3. For every w and p, (w, p)-suitable ample divisors exist.

Proof. Let H_0 be an ample divisor. We may assume that H_0 lies on no wall of type (w, p). For $n \geq 0$, let $H_n = H_0 \otimes \mathcal{O}_S(nf)$. It follows from the Nakai-Moishezon criterion that H_n is ample as well. We claim that if $n > -p(H_0 \cdot f)/2$, then H_n is (w, p)-suitable.

  To see this, let $\zeta = 2F - \Delta$ be a class of type (w, p) with $p \leq \zeta^2 < 0$ and $\zeta \cdot f \neq 0$. We may assume that $a = \zeta \cdot f > 0$, and must show that $\zeta \cdot H_n > 0$

as well. The class $aH_0 - (H_0 \cdot f)\zeta$ is perpendicular to f. Since $f^2 = 0$, we may apply the Hodge index theorem to conclude:

$$0 \geq (aH_0 - (H_0 \cdot f)\zeta)^2 = a^2 H_0^2 - 2a(H_0 \cdot f)(H_0 \cdot \zeta) + (H_0 \cdot f)^2\zeta^2.$$

Using the fact that $\zeta^2 \geq \Delta^2 - 4c = p$, we find that

$$H_0 \cdot \zeta \geq \frac{a(H_0^2)}{2(H_0 \cdot f)} + \frac{\zeta^2}{2a}(H_0 \cdot f) > \frac{p}{2a}(H_0 \cdot f).$$

Thus,

$$H_n \cdot \zeta = (H_0 \cdot \zeta) + n(\zeta \cdot f) > \frac{p}{2a}(H_0 \cdot f) - \frac{pa}{2}(H_0 \cdot f)$$

$$= -\frac{p}{2}(H_0 \cdot f)\left(a - \frac{1}{a}\right) \geq 0.$$

Thus, $H_n \cdot \zeta > 0$, and to see that H_n is (w,p)-suitable, it will suffice to show that H_n does not lie on a wall of type (w,p). We have just seen above that, if ζ is a wall of type (w,p) such that $f \cdot \zeta \neq 0$, then $H_n \cdot \zeta \neq 0$ as well. On the other hand, if ζ is a wall of type (w,p) such that $f \cdot \zeta = 0$, then $H_n \cdot \zeta = H_0 \cdot \zeta \neq 0$ by assumption. In all cases, H_n does not lie on a wall of type (w,p), and so it is (w,p)-suitable.   □

For example, suppose that X is a geometrically ruled surface with section σ and fiber f, and that the invariant of X is e. Since stability is unchanged by taking positive multiples and only depends on the numerical equivalence class of a divisor, we may as well work with elements in $\operatorname{Num} X \otimes \mathbb{Q}$. Thus, using the description of the ample cone in the last chapter, we may assume that $H = \sigma + rf$, where $r \in \mathbb{Q}$ satisfies $r > e$ if $e \geq 0$ and $r > e/2$ if $e < 0$. Suppose that $\mathbf{d}$ is a divisor class on C of degree d and that c is a positive integer, and consider rank 2 vector bundles V over X with $c_1(V) = \pi^*(\mathbf{d})$ and $c_2(V) = c$. (If $c \leq 0$, then there are no walls of type (w,p).) Thus, $w = df \mod 2$ and $p = -4c$. Although we shall not need the precise description of suitability, we include it for the reader's edification:

Lemma 4. *In the above notation, suppose that $e + c \equiv d \mod 2$. Then $\sigma + rf$ is (w,p)-suitable if and only if $r > \max\{e, (e+c)/2\}$ if $e \geq 0$ and $r > (e+c)/2$ if $e < 0$.*

Proof. The condition $r > e$ if $e \geq 0$ is simply to insure that $\sigma + rf$ is ample, which is automatic in case $e < 0$. Note that, if $\zeta_0 = 2\sigma + (e-c)f$, then $\zeta_0^2 = -4e + 4(e-c) = -4c$ and $\zeta_0 \equiv df \mod 2$, so that ζ_0 is a class of type (w,p). Moreover, $\zeta_0 \cdot f = 2 > 0$ and $\zeta_0 \cdot (\sigma + rf) = -2e + (e-c) + 2r = 2r - e - c$. Thus, if

$$\operatorname{sign} f \cdot \zeta_0 = \operatorname{sign}(\sigma + rf) \cdot \zeta_0,$$

then $r > (e+c)/2$.

Conversely, suppose that $r > (e + c)/2$ and let $\zeta = 2a\sigma + (d + 2b)f$ be a wall of type (w, p). We may assume that $\zeta \cdot f = a > 0$. Now

$$-4c \leq \zeta^2 = -4a^2 e + 4a(d + 2b) < 0,$$

and so, since $a > 0$, we have $0 < a(ae - (d + 2b)) \leq c$. Thus,

$$(\sigma + rf) \cdot \zeta = -2ae + (d + 2b) + 2ar$$
$$= 2a(r - e) + (d + 2b) > 2a(c - e)/2 + (d + 2b)$$
$$= ac - ae + (d + 2b) \geq ac - c/a = c(a - 1/a) \geq 0.$$

Thus, sign $f \cdot \zeta = \text{sign}(\sigma + rf) \cdot \zeta$. $\quad\square$

Next we discuss the geometric meaning of suitability:

Theorem 5. *Let $\pi \colon X \to C$ be a morphism from X to the smooth curve C. For given w and p, let H be a (w, p)-suitable divisor on X. Let V be a rank 2 vector bundle on X with $w_2(V) = w$ and $p_1(\text{ad} V) = p$.*

(i) *If V is H-semistable, then its restriction to almost all fibers f is semistable.*

(ii) *If there exists a smooth fiber f such that the restriction of V to f is stable, then V is H-stable.*

(iii) *If there exists a smooth fiber f such that the restriction of V to f is semistable and V is not H-stable, then exactly one of the following holds:*

    (a) *There exists a sub-line bundle $\mathcal{O}_X(D)$ such that $f \cdot (2D - \Delta) = 0$ and $2D - \Delta$ is a wall passing through f such that $H \cdot (2D - \Delta) > 0$. In this case V is H-unstable.*

    (b) *There exists a sub-line bundle $\mathcal{O}_X(D)$ such that $f \cdot (2D - \Delta) = 0$ and $2D - \Delta$ is numerically equivalent to a rational multiple rf of f, with $r \geq 0$. Moreover, V is H-semistable if $r = 0$ and H-unstable if $r > 0$.*

Part of the proof. We begin with the easy part, the proofs of the second and third statements. We shall prove the first statement in Chapter 9. Suppose that there exists an f with $V|f$ semistable. If V is not H-stable, then there exists a divisor D on X such that $\mathcal{O}_X(D)$ is a sub-line bundle of V and such that $\mu(L) \geq \mu(V)$. In other words, $H \cdot (2D - \Delta) \geq 0$. We claim that $f \cdot (2D - \Delta) \geq 0$ as well. As we may assume that $V/\mathcal{O}_X(D)$ is torsion free, there is the usual exact sequence

$$0 \to \mathcal{O}_X(D) \to V \to \mathcal{O}_X(\Delta - D) \otimes I_Z \to 0,$$

and thus as we saw in the proof of Lemma 2, $(2D - \Delta)^2 \geq p$. If $(2D - \Delta)^2 \geq 0$, then since $H \cdot (2D - \Delta) \geq 0$, it follows from Lemma 19 of Chapter 1 that $f \cdot (2D - \Delta) \geq 0$ as well. Since $V|f$ is semistable, $f \cdot (2D - \Delta) = 0$, and we see that $V|f$ is strictly semistable. In particular, if conversely $V|f$ is stable, then V is H-stable. Now suppose that V is not H-stable and that $V|f$ is semistable, so that we are in Case (iii). By the Hodge index theorem $(2D-\Delta)^2 \leq 0$, with equality if and only if $2D-\Delta$ is numerically equivalent to a rational multiple of f. If $(2D-\Delta)^2 < 0$, $2D-\Delta$ is a wall of type (w,p). By the definition of suitability, $H \cdot (2D - \Delta) \neq 0$ and $H \cdot (2D - \Delta) > 0$, and so we are in Case (a) of (iii) in Theorem 5.

In the remaining case, $(2D - \Delta)^2 = 0$, so that $2D - \Delta$ is numerically equivalent to a rational multiple rf of f. If $r = 0$, then V is H-semistable by Lemma 6 of Chapter 4, and V is clearly unstable if $r > 0$. Finally, Proposition 20 in Chapter 2, which says that the maximal destabilizing sub-line bundle of V is essentially unique, shows that Cases (a) and (b) are mutually exclusive.

The proof of Part (i) is harder, and I don't know a proof which does not involve descent theory or something equivalent. We will discuss the main idea in the general case and then give the proof in case the fibers have genus 0 or 1. Let us start with the general case. Suppose that $V|f$ is unstable for infinitely many fibers f. General properties of stability imply that in fact $V|f$ is unstable for all smooth f. Thus, for every t such that $f_t = \pi^{-1}(t)$ is smooth, there is a unique maximal destabilizing sub-line bundle L_t of $V|f_t$. (Here the only reason we restrict ourselves to smooth fibers is because we have not bothered as yet to try and define stability for a singular curve.) What we would like to say is the following:

Claim. There exists a sub-line bundle $\mathcal{O}_X(D)$ of V on X whose restriction to f_t is L_t for almost all t.

Assuming the claim, let us finish the argument. We may assume that $V/\mathcal{O}_X(D)$ is torsion free. Thus, we have the usual exact sequence

$$0 \to \mathcal{O}_X(D) \to V \to \mathcal{O}_X(\Delta - D) \otimes I_Z \to 0,$$

and by a familiar calculation $(2D - \Delta)^2 \geq p$. Moreover, since $\mathcal{O}_X(D)|f$ is destabilizing, $\deg((2D - \Delta)|f) = f \cdot (2D - \Delta) > 0$. If $(2D - \Delta)^2 \geq 0$, then by the Hodge index theorem $H \cdot (2D - \Delta) \geq 0$ as well, and $H \cdot (2D - \Delta) = 0$ only if $2D - \Delta$ is numerically equivalent to 0, contradicting the fact that $f \cdot (2D - \Delta) > 0$. Thus, $H \cdot (2D - \Delta) > 0$, contradicting the assumption that V is H-semistable.

The remaining case is $(2D - \Delta)^2 < 0$, so that $2D - \Delta$ is a wall of type (w,p). By the definition of suitability, $H \cdot (2D - \Delta) > 0$ in this case as well. Hence in all cases V is H-unstable, contradicting the assumption on V.

How do we find $\mathcal{O}_X(D)$? One method is via base change and descent theory. However, in the case of ruled and elliptic surfaces we can find $\mathcal{O}_X(D)$ directly, and this is what we shall proceed to do.

First assume that all of the fibers have genus 0, so that X is geometrically ruled and there exists a section σ of the ruling. In this case, we are allowed to twist by $\mathcal{O}_X(k\sigma)$ for any k, so that we may assume that $V|f$ has degree either 0 or 1 for all f.

Lemma 6. *Suppose that V is a rank 2 vector bundle on the ruled surface X with $\deg(V|f) = 0$. Then there is a nonnegative integer a and a nonempty Zariski open subset U of C such that $V|f = \mathcal{O}_{\mathbb{P}^1}(a) \oplus \mathcal{O}_{\mathbb{P}^1}(-a)$ for all fibers f lying over a point of U. A similar conclusion holds if $\deg(V|f) = 1$, in which case $V|f = \mathcal{O}_{\mathbb{P}^1}(a) \oplus \mathcal{O}_{\mathbb{P}^1}(1-a)$ for all fibers f lying over a point of U.*

Proof. First assume that $\deg V|f = 0$. For every fiber $f_t = \pi^{-1}(t)$ we can write $V|f_t = \mathcal{O}_{\mathbb{P}^1}(a_t) \oplus \mathcal{O}_{\mathbb{P}^1}(-a_t)$ for a uniquely specified nonnegative integer a_t, and we need to show that the function a_t is constant on a Zariski open set. It suffices to show that the function a_t is upper semicontinuous in the Zariski topology. In fact, consider the upper semicontinuous function

$$h^0(V \otimes \mathcal{O}_X(-\sigma)|f_t) = H^0(\mathcal{O}_{\mathbb{P}^1}(a_t - 1) \oplus \mathcal{O}_{\mathbb{P}^1}(-a_t - 1)) = a_t.$$

If the minimum value of $h^0(V \otimes \mathcal{O}_X(-\sigma)|f_t)$ is h, then for a nonempty Zariski open subset U of C, $h = a_t$. This proves the lemma in case $\deg(V|f) = 0$ and a very similar argument takes care of the case $\deg(V|f) = 1$. $\square$

We return to the proof of Theorem 5 for the ruled surface X. In case $\deg(V|f) = 0$, the assertion is that if V is semistable, then the nonnegative integer a of Lemma 6 is 0.

Lemma 7. *In the situation of Lemma 6, suppose that $\deg(V|f) = 0$ and that $a > 0$. Then there exists a unique sub-line bundle of V of the form $\mathcal{O}_X(a\sigma + \pi^*\mathbf{b})$, where $\mathbf{b}$ is a divisor on C, and such that the quotient is torsion free. If $a = 0$, there is also a sub-line bundle $\mathcal{O}_X(\pi^*\mathbf{b})$ of V, not unique, with torsion free quotient. A similar conclusion holds if $\deg(V|f) = 1$, where in this case the sub-line bundle with torsion free quotient is always unique.*

Proof. First suppose either that $\deg(V|f) = 0$ and $a > 0$ or that $\deg(V|f) = 1$, and consider the bundle $V \otimes \mathcal{O}_X(-a\sigma)$. For a nonempty Zariski open subset U of C we have $h^0(\mathcal{O}_X(-a\sigma) \otimes V|f_t) = 1$ for all $t \in U$. Thus, the torsion free sheaf $R^0\pi_*(\mathcal{O}_X(-a\sigma) \otimes V) = L$ has rank 1 on C and so is a line bundle, which we can write as $\mathcal{O}_C(\mathbf{b})$. Thus,

$$L \otimes L^{-1} = R^0\pi_*(\mathcal{O}_X(-a\sigma) \otimes \pi^*L^{-1} \otimes V) = R^0\pi_*(\mathcal{O}_X(-a\sigma - \pi^*\mathbf{b}) \otimes V)$$

is trivial. Since $H^0(X; \mathcal{O}_X(-a\sigma - \pi^*\mathbf{b}) \otimes V)) = H^0(C; R^0\pi_*(\mathcal{O}_X(-a\sigma - \pi^*\mathbf{b}) \otimes V))$, there is a section of $H^0(X; \mathcal{O}_X(-a\sigma - \pi^*\mathbf{b}) \otimes V)$, or in other words a homomorphism from $\mathcal{O}_X(a\sigma + \pi^*\mathbf{b})$ to V. This homomorphism factors through the inclusion of $\mathcal{O}_X(a\sigma + \pi^*\mathbf{b})$ in $\mathcal{O}_X(a\sigma + \pi^*\mathbf{b} + D)$, where D is an effective divisor, necessarily of the form $n\sigma + \pi^*\mathbf{b}'$ for $n \geq 0$. But clearly $n = 0$, for otherwise we would get a nonzero map $\mathcal{O}_{\mathbb{P}^1}(a+n) \to \mathcal{O}_{\mathbb{P}^1}(a) \oplus \mathcal{O}_{\mathbb{P}^1}(-a)$ for almost all fibers f, with $n > 0$, which is clearly impossible. Thus, after changing notation we can assume that there is a nonzero map

$$\mathcal{O}_X(a\sigma + \pi^*\mathbf{b}) \to V$$

with torsion free quotient. We leave the uniqueness as an exercise. If $\deg(V|f) = 0$ and $a = 0$, choose a subbundle of rank 1 of the rank 2 bundle $R^0\pi_*V = W$ and argue similarly. $\square$

We have thus proved Theorem 5 in the case of a geometrically ruled surface X. Note that, in case $w \cdot f \neq 0$, or equivalently $\Delta \cdot f$ is odd, the theorem says that there are no H-semistable vector bundles V with $c_1(V) = \Delta$ if H is (w,p)-suitable, since there are no semistable rank 2 vector bundles on $\mathbb{P}^1$ of odd degree.

Next consider an elliptic fibration $\pi\colon X \to C$. We shall just give an outline of the proof in this case. The main point is the following: if W is an unstable rank 2 bundle on an elliptic curve C, then W splits into a direct sum of line bundles $L_1 \oplus L_2$ of unequal degree, say $\deg L_1 = d_1 < \deg L_2 = d_2$. Thus, $Hom(W,W) = \mathcal{O}_C \oplus \mathcal{O}_C \oplus (L_2^{-1} \otimes L_1) \oplus (L_1^{-1} \otimes L_2)$. Here the two factors $\mathcal{O}_C$ correspond to diagonal maps from W to itself. Since $\deg(L_2^{-1} \otimes L_1) < 0$, the line bundle $L_2^{-1} \otimes L_1$ has no sections, whereas $\deg(L_1^{-1} \otimes L_2) > 0$ so that $L_1^{-1} \otimes L_2$ does have sections. It follows that every section of $Hom(W,W)$ is strictly upper triangular, and that there exist sections where the off-diagonal entry is nonzero.

Let V be a rank 2 vector bundle on X such that $V|f$ is unstable for infinitely many smooth fibers f. Consider the bundle $R^0\pi_*Hom(V,V)$. We have the functions det and Trace from $Hom(V,V)$ to $\mathbb{C}$. From the map $\pi^*R^0\pi_*Hom(V,V) \to Hom(V,V)$, there are induced functions det and Trace on $\mathbb{V}R^0\pi_*Hom(V,V)$, where $\mathbb{V}R^0\pi_*Hom(V,V)$ is the total space of the vector bundle $R^0\pi_*Hom(V,V)$. Now for the space of 2×2 upper triangular matrices, the vanishing of det and Trace defines the linear subspace of strictly upper triangular matrices. For infinitely many fibers $f_t = \pi^{-1}(t)$ such that $V|f_t$ is unstable, the map $R^0\pi_*Hom(V,V)_t \to H^0(Hom(V|f_t,V|f_t))$ is onto, and, for such t, the rank d of $R^0\pi_*Hom(V,V)$ is equal to $h^0(Hom(V|f_t,V|f_t))$. We consider the (homogeneous) subvariety $\mathbb{Y}$ of $\mathbb{V}R^0\pi_*Hom(V,V)$ defined by the vanishing of det and Trace, with its reduced structure. For infinitely many t corresponding to unstable V_t as above, the fiber of $\mathbb{Y}$ over t is a linear subspace of the fiber of $\mathbb{V}R^0\pi_*Hom(V,V)$ over t, of dimension $d - 2$. The same must hold for a general fiber of $\mathbb{Y}$. In this way we can find a subsheaf $\mathcal{S}$ of $R^0\pi_*Hom(V,V)$ such that the image of $\pi^*\mathcal{S}$ in $Hom(V,V)$ via the natural map $\pi^*\mathcal{S} \to \pi^* R^0 \pi_* Hom(V,V) \to Hom(V,V)$ consists of homomorphisms from V to itself which are generically of rank 1. The image of $\pi^*\mathcal{S}$ is thus a rank 1 subsheaf of $Hom(V,V)$. A local generator σ of the image defines a sub-line bundle $\mathcal{O}_X(D) = \text{Ker}\,\sigma$ of V, which by construction restricts to the destabilizing line bundle L_2 on a fiber f such that $V|f = L_1 \oplus L_2$ with $\deg L_2 > \deg L_1$. This proves the claim in the case of a fibration whose fibers have genus 1. $\square$

We now give the general outline for attacking the study of H-stable bundles on fibrations $\pi\colon X \to C$ as above, when H is a (w,p)-suitable ample divisor. Begin with a vector bundle V such that $w_1(V) = w$ and $p_1(\text{ad}\,V) = p$. By Theorem 5, if V is H-semistable, then $V|f$ is semistable for almost all f. Now suppose that f is a smooth fiber such that $V|f$ is **not** semistable. Let $j\colon f \to X$ be the inclusion. Then there is a maximal destabilizing sub-line bundle L of $V|f$, necessarily a subbundle since f is a curve, and thus an exact sequence

$$0 \to L \to V|f \to L' \to 0.$$

Here $2 \deg L' < \deg V$, and moreover L' is the unique quotient line bundle of $V|f$ with this property. Now we can define the elementary modification of V with respect to the surjection $V \to j_* L'$:

$$0 \to V' \to V \to j_* L' \to 0.$$

Let us make a definition formalizing the kinds of elementary modifications that arise:

Definition 8. Let V be a rank 2 vector bundle on X and $j\colon f \to X$ the inclusion of a fiber f in X such that $V|f$ is unstable. There is a unique maximal destabilizing subbundle L of $V|f$. Let L' be the line bundle quotient $(V|f)/L$; thus $\mu(L') < \mu(V|f)$. We call the corresponding elementary modification V' of V defined above *allowable*.

Lemma 9. *If V' is an allowable elementary modification of V, then $p_1(\text{ad}\,V') \geq p_1(\text{ad}\,V) + 2$.*

Proof. By the formulas of Lemma 16 of Chapter 2,

$$
\begin{aligned}
p_1(\text{ad}\,V') &= c_1(V')^2 - 4c_2(V') \\
&= (c_1(V) - f)^2 - 4(c_2(V) - c_1(V) \cdot f) + \deg L') \\
&= c_1(V)^2 - 4c_2(V) + 2c_1(V) \cdot f - 4 \deg L' \\
&= p_1(\text{ad}\,V) + 4(\mu(V|f) - \mu(L')) \\
&\geq p_1(\text{ad}\,V) + 2,
\end{aligned}
$$

where the last line follows since $\mu(V|f) - \mu(L')$ is a positive element of $\frac{1}{2}\mathbb{Z}$. $\square$

This then is the general strategy: beginning with V, make all possible allowable elementary modifications. At each stage $p_1(\text{ad } V)$ strictly increases. There is an absolute bound on $p_1(\text{ad } V)$ by Bogomolov's inequality (we will also see this directly in the cases described in this book), and so this procedure must stop. Of course, this idea is a little incomplete since we have not done anything with the singular fibers. In Chapter 8, we shall give an analysis of stability for singular fibers f which are irreducible curves with just one singular point which is an ordinary double point. For a ruled surface, of course, we do not have to worry about singular fibers. So we will reach a stage where $V|f$ is semistable for all fibers f, smooth or not. At this stage, of course, the work really begins. But in certain cases, namely ruled surfaces or elliptic surfaces where $\deg V|f$ is odd, we will be able to use this preliminary reduction to give a complete classification of stable bundles.

Ruled surfaces

From now on in this chapter, X is a geometrically ruled surface. We shall describe H-stable vector bundles V on X for suitable ample divisors H. After normalizing, we can assume that $H = \sigma + rf$ for some positive rational number r. We need only consider the case where $\deg(V|f) = 0$ or 1. By Theorem 5, there are no H-stable vector bundles V when $\deg(V|f) = 1$. Thus, we may assume that $\deg(V|f) = 0$, or in other words that the determinant of V is pulled back from C. Fix a divisor $\mathbf{d}$ on C and an integer c. We consider vector bundles V on X with $c_1(V) = \pi^*(\mathbf{d})$, and let w and $p = -4c$ be the corresponding classes. Thus, $c_1(V)^2 = 0$. First let us note that we can give a direct proof of Bogomolov's inequality $c_1(V)^2 \leq 4c_2(V)$ in our case.

Theorem 10. *Let H be an arbitrary ample divisor on X. If V is H-semistable, then $c_2(V) \geq 0$, and $c_2(V) = 0$ if and only if V is given as the pullback of a semistable bundle W on C.*

Proof. We may assume that $c_2(V) \leq 0$. Thus, there are no walls of type $(\pi^*\mathbf{d}, c)$, so that every ample H is suitable. By Theorem 5, the restriction of V to almost every fiber f is $\mathcal{O}_f \oplus \mathcal{O}_f$. Using Lemma 7, there is an exact sequence

$$0 \to \mathcal{O}_X(\pi^*\mathbf{b}) \to V \to \mathcal{O}_X(-\pi^*\mathbf{b} + \pi^*\mathbf{d}) \otimes I_Z \to 0.$$

Thus, $c_2(V) = \ell(Z) \geq 0$, and $c_2(V) = 0$ if and only if $\ell(Z) = 0$. In this case V is given as an extension of $\mathcal{O}_X(-\pi^*\mathbf{b} + \pi^*\mathbf{d})$ by $\mathcal{O}_X(\pi^*\mathbf{b})$, and such extensions are classified by $H^1(\mathcal{O}_X(2\pi^*\mathbf{b} - \pi^*\mathbf{d})$. An application of the Leray spectral sequence shows that $H^1(\pi^*L) = H^1(R^0\pi_*\pi^*L) = H^1(L)$ for every line bundle L on C. Thus, every extension of $\mathcal{O}_X(-\pi^*\mathbf{b} + \pi^*\mathbf{d})$ by $\mathcal{O}_X(\pi^*\mathbf{b})$ is pulled back from an extension of $\mathcal{O}_C(-\mathbf{b}+\mathbf{d})$ by $\mathcal{O}_C(\mathbf{b})$ on C. Thus, $V = \pi^*W$ for some W. Finally, we must decide when $V = \pi^*W$ is H-semistable. Assuming that $H = \sigma + rf$, $\mu(V) = \frac{1}{2}d = \frac{1}{2}\deg W$. If L is a sub-line bundle of W, then $(\sigma + rf) \cdot \pi^*L = \deg L$, from which it follows that, if π^*W is H-semistable, then W is semistable. Conversely, suppose that W is semistable. Since $c_1(V)^2 = 0$ and $c = 0$, every ample divisor is $(\pi^*\mathbf{d}, 0)$-suitable. The ample cone is a line segment and the walls are just certain points. In particular, there is no wall passing through the class of f. By (iii) of Theorem 5, the only possible destabilizing sub-line bundles are thus of the form $\mathcal{O}_X(D)$ where D is numerically equivalent to a rational multiple of f. It follows that $\mathcal{O}_X(D) = \pi^*L$ for some line bundle L on X, by the description of $\operatorname{Pic} X$ given in Chapter 5, Proposition 11. Since $\operatorname{Hom}(\pi^*L, \pi^*W) = \operatorname{Hom}(L, W)$ and $\mu(\pi^*L) = \deg L$, the above arguments show that, if W is semistable, then π^*W is semistable as well. □

Note that the argument that $c_2(V) \geq 0$ only used the fact that $V|f = \mathcal{O}_f \oplus \mathcal{O}_f$ for almost all fibers f.

We now study the case where $c > 0$. By Theorem 5, we know that if H is a (w, p)-suitable divisor on X, then an H-semistable bundle on X restricts to $\mathcal{O}_{\mathbb{P}^1} \oplus \mathcal{O}_{\mathbb{P}^1}$ on almost all fibers. So we shall study such bundles without at first assuming that they are semistable. The following result is due to Brosius [14]:

Proposition 11. Let V be a rank 2 bundle on X such that $V|f = \mathcal{O}_f \oplus \mathcal{O}_f$ for almost all fibers f. Then there is a natural exact sequence

$$0 \to \pi^*W \to V \to Q \to 0,$$

where $W = R^0\pi_*V$, the map $\pi^*W \to V$ is the natural map $\pi^*R^0\pi_*V \to V$, and Q is a torsion sheaf supported in a union of fibers of π. Finally, we have $H^0(Q) = 0$.

Proof. By assumption $\pi_*V = W$ is a torsion free rank 2 sheaf on C, hence is a rank 2 vector bundle. Moreover, the natural map $\pi^*W = \pi^*\pi_*V \to V$ is an isomorphism over almost all fibers and hence it is injective. Set $Q = V/\pi^*W$.

To see that $H^0(Q) = 0$, note that the projection formula gives

$$R^i\pi_*\pi^*W = R^i\pi_*\pi^*\pi_*V = \pi_*V \otimes R^i\pi_*\mathcal{O}_X$$

and thus $\pi_*\pi^*W = W = \pi_*V$ and $R^i\pi_*\pi^*W = 0$ for $i > 0$. Applying $R^i\pi_*$ to the exact sequence

$$0 \to \pi^*W \to V \to Q \to 0,$$

we thus obtain

$$0 \to \pi_* V \cong \pi_* V \to \pi_* Q \to 0.$$

Thus, $\pi_* Q = 0$, and therefore $H^0(Q) = H^0(\pi_* Q) = 0$. $\square$

Less canonically, we have the following description of V and W:

Lemma 12. *With V as in Proposition 11, there exists a divisor $\mathbf{b}$ on C and an exact sequence*

$$0 \to \mathcal{O}_X(\pi^* \mathbf{b}) \to V \to \mathcal{O}_X(\pi^*(\mathbf{d} - \mathbf{b})) \otimes I_Z \to 0.$$

Thus, $c_2(V) = \ell(Z) \geq 0$. Finally, a fiber f contains a point of the support of Z if and only if the restriction $V|f = \mathcal{O}_{\mathbb{P}^1}(a) \oplus \mathcal{O}_{\mathbb{P}^1}(-a)$ with $a > 0$.

Proof. The existence of the exact sequence is a consequence of Lemma 7, in the case $a = 0$. Restricting the exact sequence to a fiber f and noting that $\mathcal{O}_X(\pi^*(\mathbf{d} - \mathbf{b}))|f$ is trivial gives a surjection $V|f \to I_Z \otimes \mathcal{O}_f \to I_Z \mathcal{O}_f$. Now $I_Z \mathcal{O}_f$ is an ideal sheaf on f, which is equal to $\mathcal{O}_f$ if and only if $\operatorname{Supp} Z \cap f = \emptyset$, and thus there is a surjection $V|f \to \mathcal{O}_{\mathbb{P}^1}(-a)$, where $a \geq 0$ and $a = 0$ if and only if f contains no points of the support of Z. Since the kernel of the surjection must be $\mathcal{O}_{\mathbb{P}^1}(a)$, there is an exact sequence

$$0 \to \mathcal{O}_{\mathbb{P}^1}(a) \to V|f \to \mathcal{O}_{\mathbb{P}^1}(-a) \to 0,$$

which is split since $H^1(\mathcal{O}_{\mathbb{P}^1}(2a)) = 0$ if $a \geq 0$. Thus, $V|f = \mathcal{O}_{\mathbb{P}^1}(a) \oplus \mathcal{O}_{\mathbb{P}^1}(-a)$ where $a = 0$ if f does not contain any point in the support of Z and $a > 0$ otherwise. $\square$

We now describe more concretely how to obtain V from W. Suppose that f_0 is a fiber such that $V|f_0 = \mathcal{O}_{\mathbb{P}^1}(a) \oplus \mathcal{O}_{\mathbb{P}^1}(-a)$ with $a > 0$. The surjection $V|f_0 \to \mathcal{O}_{\mathbb{P}^1}(-a)$ is then unique mod scalars. Let $j \colon f_0 \to X$ be the inclusion. Then we can make the allowable elementary modification

$$0 \to V' \to V \to j_* \mathcal{O}_{\mathbb{P}^1}(-a) \to 0.$$

By Lemma 16 of Chapter 2, $c_1(V') = c_1(V) - f_0$, and in fact $\det V' = \pi^*(\mathbf{b} - p_0)$, where $\det V = \mathbf{b}$ and $p_0 \in C$ is the point lying under f_0. Moreover, $c_2(V') = c_2(V) - a$, again by Lemma 16 of Chapter 2. Thus, $c_2(V') < c_2(V)$. Clearly, the restriction of V' to a general fiber f is $\mathcal{O}_f \oplus \mathcal{O}_f$. The restriction of V' to f_0 fits into an exact sequence

$$0 \to \mathcal{O}_{\mathbb{P}^1}(-a) \to V'|f_0 \to \mathcal{O}_{\mathbb{P}^1}(a) \to 0.$$

Thus, it is easy to check that $V'|f_0 = \mathcal{O}_{\mathbb{P}^1}(a') \oplus \mathcal{O}_{\mathbb{P}^1}(-a')$ with $0 \leq a' \leq a$. Applying $R^0 \pi_*$ to the defining exact sequence for V' and using $R^0 \pi_* j_* \mathcal{O}_{\mathbb{P}^1}(-a) = 0$ since $H^0(\mathcal{O}_{\mathbb{P}^1}(-a)) = 0$, we see that $R^0 \pi_* V' = R^0 \pi_* V$. So the bundle $\pi^* W$ is still a subbundle of V'.

If $V'|f = \mathcal{O}_f \oplus \mathcal{O}_f$ for all f, then $V' = W$ by base change. Otherwise, we may repeat the process with another allowable elementary modification. This process must terminate, for example, because c_2 drops each time and $c_2 \geq 0$ for a bundle with trivial restriction to almost all fibers. (We could also compare c_1 at each stage with $\pi^* c_1(W)$.)

Summarizing, then:

Theorem 13. *Let V be a rank 2 bundle on X such that $V|f = \mathcal{O}_f \oplus \mathcal{O}_f$ for almost all fibers f. Then $R^0 \pi_* V = W$ is a rank 2 vector bundle on C, and $\pi^* W$ is obtained from V by a finite series of allowable modifications. Conversely, V is obtained from $\pi^* W$, up to a twist by a line bundle of the form $\pi^* L$, where L is a line bundle on C, by a sequence of elementary modifications which are the duals of allowable modifications.* $\square$

We must now analyze when such bundles are stable. We will have to consider Case (iii) of Theorem 5, where the possible destabilizing sub-line bundles are of the form $\mathcal{O}_X(rf)$, and semistability on the general fiber does not necessarily imply that V is actually stable.

Theorem 14. *Let V be a rank 2 bundle on X with $\det V = \mathcal{O}_X(\pi^* \mathbf{d})$ for a divisor $\mathbf{d}$ on C of degree d and with $c_2(V) = c$ and such that $V|f = \mathcal{O}_f \oplus \mathcal{O}_f$ for almost all f. Let $W = R^0 \pi_* V$ be the corresponding rank 2 bundle on C. Suppose that $H = \sigma + rf$ is a (w, p)-suitable ample $\mathbb{Q}$-divisor, where w is the mod 2 reduction of df and $p = -4c$. Then V is H-stable if and only if, for all sub-line bundles L of W, $\deg L < d/2$.*

Proof. First suppose that V is H-stable. If L is a sub-line bundle of W, then $\pi^* L$ is a sub-line bundle of $\pi^* W$ and hence of V. Thus, $H \cdot \pi^* L < d/2$. But $H \cdot \pi^* L = \deg L$. Thus, $\deg L < d/2$.

Conversely, suppose that W satisfies the condition that $\deg L < d/2$ for all sub-line bundles L of W. We must show that V is H-stable. Since there are no walls passing through f, the argument of (iii) of Theorem 5 shows that it suffices to check stability for all sub-line bundles $\mathcal{O}_X(D)$, where $2D - \pi^* \mathbf{d}$ is numerically equivalent to a rational multiple of f. This says that $D = \pi^* \mathbf{b}$ for some divisor $\mathbf{b}$ on C. Applying $R^0 \pi_*$ to the nonzero map

$$\mathcal{O}_X(D) = \mathcal{O}_X(\pi^* \mathbf{b}) = \pi^* \mathcal{O}_C(\mathbf{b}) \to V$$

gives a nonzero map $\mathcal{O}_C(\mathbf{b}) \to R^0 \pi_* V = W$. Thus, by assumption $b = \deg \mathbf{b} < d/2$. But $H \cdot \pi^* \mathbf{b} = (\sigma + rf) \cdot bf = b < d/2$ as well. Thus, V is H-stable. $\square$

Remarks. (1) The sheaf Q of Proposition 11 can be shown to be I_Z / I_E, where Z is a 0-dimensional local complete intersection subscheme on X contained in $E = \sum_i n_i f_i$, an effective divisor which is a sum of fibers. In a certain sense Q records all of the elementary modifications simultaneously.

However, the sheaf Q can be quite complicated, as a result of performing several elementary modifications on the same fiber.

(2) In Theorem 14, the bundle W has degree strictly less than d if $c_2(V) > 0$. In fact, if we need to perform $k \leq c_2(V)$ allowable elementary modifications on V to get π^*W, then $\deg W = d - k$. Thus, the condition that every sub-line bundle of W have degree less than $d/2$ is always weaker than assuming that W is semistable. However, one can show that the generic such bundle W, in an appropriate sense, is stable. One consequence of Theorem 14 is that, if V is H-stable and V' is obtained from V by an allowable elementary modification, then V' is not necessarily H-semistable.

(3) In case $g(C) = 0$ and $d = 0$, a bundle W on $\mathbb{P}^1$ such that every subbundle of W has degree less than 0 is of the form $\mathcal{O}_{\mathbb{P}^1}(a) \oplus \mathcal{O}_{\mathbb{P}^1}(b)$ with both $a, b < 0$. To get from π^*W to V by elementary modifications along fibers, if we want to have $\det V = 0$ we must make exactly $-(a + b)$ elementary modifications which are the duals of allowable elementary modifications. Each such modification will increase c_2. Thus, if V is H-stable for a suitable H, then $c_2(V) \geq 2$, which is also easy to see directly. Likewise, for an H-stable rank 2 bundle V on a surface X ruled over an elliptic curve with $\det V = 0$, one can show that $c_2(V) \geq 1$. In all other cases every value of $c_2(V) \geq 0$ is possible, and in fact Theorem 10 shows that every such bundle V with $c_2(V) = 0$ is the pullback π^*W of a stable bundle W on C.

A brief introduction to local and global moduli

The abstract point of view on moduli problems is that every moduli space arises from a functor from the category of schemes or analytic spaces to the category of sets. For example, the moduli space of stable vector bundles V over X with fixed Chern classes arises from the following functor $\mathbf{M}$ from the category of schemes (over $\mathbb{C}$, say) to the category of sets: Given a scheme T, let $\mathbf{M}(T)$ be the set of all equivalence classes of vector bundles $\mathcal{V}$ over $X \times T$ such that the restriction of $\mathcal{V}$ to each slice $X \times \{t\}$ is a stable rank 2 bundle V over $X \cong X \times \{t\}$ with $\det V = \Delta$ and $c_2(V) = c$. Here two bundles $\mathcal{V}$ and $\mathcal{V}'$ are *equivalent* if there exists a line bundle $\mathcal{L}$ on T such that $\mathcal{V}' \cong \mathcal{V} \otimes \pi_2^*\mathcal{L}$. Ideally the moduli functor would be *representable*. In other words, there would be a universal parameter space $\mathfrak{M}$ and a universal bundle $\mathcal{U}$ over $X \times \mathfrak{M}$ such that given a bundle $\mathcal{V}$ over $X \times T$ as above, there would exist a unique morphism $f \colon T \to \mathfrak{M}$ such that $\mathcal{V} = (\mathrm{Id} \times f)^*\mathcal{U}$ up to a twist by a line bundle $\pi_2^*\mathcal{L}$. Unfortunately, the moduli functor is not in general representable, for the simple reason that equivalence as defined above is too weak to force universal bundles defined locally on an open cover of the moduli space to agree on overlaps. What is unique is the space

$\mathfrak{M}$ and the morphism $f\colon T \to \mathfrak{M}$. More precisely, we call $\mathfrak{M}$ a *coarse moduli space* for $\mathbf{M}(T)$ if:

1. The points of $\mathfrak{M}$ are in 1–1 correspondence with stable rank 2 vector bundles V over X with $\det V = \Delta$ and $c_2(V) = c$;
2. For every scheme T and bundle $\mathcal{V}$ over $X \times T$ corresponding to an element of $\mathbf{M}(T)$, there is a morphism $f\colon T \to \mathfrak{M}$, functorial under pullback;
3. Conversely, suppose that N is a scheme with the property that, for every scheme T and bundle $\mathcal{V}$ over $X \times T$ corresponding to an element of $\mathbf{M}(T)$, there is a morphism $g\colon T \to N$, functorial under pullback. Then there is a unique morphism $h\colon \mathfrak{M} \to N$ so that $g = h \circ f$.

Then it is a theorem in geometric invariant theory due to Gieseker in the case of surfaces [47] that there exists a quasiprojective variety $\mathfrak{M}_H(\Delta, c) = \mathfrak{M}$ parametrizing H-semistable rank 2 vector bundles V on X with $c_1(V) = \Delta$ and $c_2(V) = c$. (For the much simpler case of curves, this theorem is due to Mumford. For proofs in this case, see, for example, [48], [113], [140], and [40].) The space $\mathfrak{M}$ is a coarse moduli space, but the universal bundle $\mathcal{U}$ only exists locally in the classical (or étale) topology on $\mathfrak{M}$. Moreover, $\mathfrak{M}$ is almost never compact. To compactify it, we have to consider equivalence classes of Gieseker semistable torsion free rank 2 sheaves. Here, if V is a Gieseker semistable torsion free sheaf, then there exists a (noncanonical) filtration on V, the Jordan-Hölder filtration, defined in Chapter 4, Exercise 16 (see also Proposition 21 in Chapter 4 for the case of a rank 2 bundle). Given such a filtration, there is the associated graded sheaf $\operatorname{gr} V$, which is canonical and is also Gieseker semistable. We say that V_1 and V_2 are *S-equivalent* if $\operatorname{gr} V_1 = \operatorname{gr} V_2$. The set of S-equivalence classes of Gieseker semistable torsion free rank 2 sheaves on X with $c_1(V) = \Delta$ and $c_2(V) = c$ can then be given the structure of a projective variety which is a compactification $\overline{\mathfrak{M}}_H(\Delta, c)$ of $\mathfrak{M}_H(\Delta, c)$. This theorem was proved by Gieseker in the case of a surface X and generalized to higher dimension by Maruyama [85]. (See also Simpson [141] for a new approach to these questions.)

Let us describe why geometric invariant theory plays a role in the construction of the moduli space. First, fix Δ and c. Standard arguments show that, if V is a Gieseker semistable torsion free sheaf or rank r with the given invariants, then there is an integer N, depending only on r, Δ, c such that $h^i(V \otimes \mathcal{O}_X(NH)) = 0$, $i > 0$, and $V \otimes \mathcal{O}_X(NH)$ is generated by its global sections. In particular $h^0(V \otimes \mathcal{O}_X(NH)) = K$ can be calculated from Riemann-Roch. Since $V \otimes \mathcal{O}_X(NH)$ is generated by its global sections, it is a quotient of $\mathcal{O}_X^K$, and since an isomorphism from V_1 to V_2 induces an isomorphism of global sections, given two surjections

$$\Phi_i\colon \mathcal{O}_X^K \to V_i \otimes \mathcal{O}_X(NH), \qquad i = 1, 2,$$

inducing isomorphisms on global sections, $V_1 \cong V_2$ if and only if there is an automorphism A of $\mathcal{O}_X^K$ such that $\Phi_2 = \Phi_1 \circ A$. Now by Grothendieck's theory of the Quot scheme (which is a generalization of the Hilbert scheme), there is a scheme Q parametrizing quotient maps $\Phi\colon \mathcal{O}_X^K \to V \otimes \mathcal{O}_X(NH)$ as above, and $\mathrm{Aut}(\mathcal{O}_X^K) = GL(K, \mathbb{C})$ acts on Q. In general, however, the action of $GL(K, \mathbb{C})$ on Q is very bad, and to get a reasonable quotient we have to restrict ourselves to the open subset Q^0 corresponding to quotients $V \otimes \mathcal{O}_X(NH)$ such that V is Gieseker semistable. Even here, because of the existence of strictly Gieseker semistable sheaves, if we want to take a quotient by the action of $GL(K, \mathbb{C})$ it is necessary to identify two S-equivalent sheaves. However, aside from this technical problem, geometric invariant theory constructs a quotient which is in fact a projective variety.

What is the local structure of $\mathfrak{M}$? In the crudest sense, we could ask for some very basic information about $\mathfrak{M}$ at a point x corresponding to V: What is the dimension of $\mathfrak{M}$ at x? Is $\mathfrak{M}$ smooth at x? It is these kinds of questions that deformation theory can sometimes answer. We run into some trouble where there are strictly semistable torsion free sheaves, because the notion of equivalence is not just that of isomorphism. So we will only worry about this problem for V stable; this will usually take care of a Zariski open and dense subset of $\mathfrak{M}$. Deformation theory works in ideal circumstances when the moduli space $\mathfrak{M}$ is a *fine moduli space*. This means that there is a universal sheaf $\mathcal{V}$ over $X \times \mathfrak{M}$ such that $\mathcal{V}|X \times \{x\} = V$ under the isomorphism $X \times \{x\} \cong X$, where the point $x \in \mathfrak{M}$ corresponds to the sheaf V. In general $\mathfrak{M}$ is not a fine moduli space, but a universal sheaf does exist locally in the classical (or étale) topology around every point corresponding to a stable sheaf V, and this is enough for our purposes. Related to this issue is the fact that stable torsion free sheaves are simple, by Corollary 8 (and Exercise 9) of Chapter 4. Thus, the automorphisms of a stable sheaf cannot suddenly jump at a point. By contrast, given a strictly semistable sheaf, it is S-equivalent to its associated graded object which always has at least an extra $\mathbb{C}^*$ of automorphisms, and these can create singularities in the moduli space.

Instead of starting with a moduli space $\mathfrak{M}$, deformation theory takes a completely local point of view: start with a family of, say vector bundles over some parameter space T, or in other words a single vector bundle $\mathcal{V}$ over $X \times T$. We assume that T has a distinguished point t_0 and that we are given a fixed isomorphism from the restriction of $\mathcal{V}$ to the slice $X \times \{t_0\}$ to V. To start, we will also assume that T is smooth of dimension 1, with coordinate t. How do we describe $\mathcal{V}$? Like all bundles, $\mathcal{V}$ is given by transition functions. Let A_{ij} be transition functions for V with respect to some open cover $\{U_i\}$ of X. After possibly shrinking T, we may assume that we have trivialized $\mathcal{V}$ on the open cover $\{U_i \times T\}$ of $X \times T$. The transition functions for $\mathcal{V}$ can then be taken to be of the form

$$A_{ij}(t) = A_{ij} + B_{ij} \cdot t + O(t^2).$$

The main idea of deformation theory is then to look at the linear term $B_{ij} \cdot t$, or more naively to take the derivative $\frac{d}{dt} A_{ij}(t)\Big|_{t=0}$. Since $A_{ij}(t)$ is a 1-cocycle, i.e., $A_{ij}(t) \cdot A_{jk}(t) = A_{ik}(t)$ on $U_i \cap U_j \cap U_k$, the B_{ij} satisfy the following condition: on $U_i \cap U_j \cap U_k$, we have

$$B_{ij} A_{jk} + A_{ij} B_{jk} = B_{ik}.$$

Using the cocycle condition for A_{ij}, with a little manipulation we can rewrite this as

$$B_{ij} A_{ij}^{-1} + A_{ij} (B_{jk} A_{jk}^{-1}) A_{ij}^{-1} = B_{ik} A_{ik}^{-1}.$$

This condition says exactly that $B_{ij} A_{ij}^{-1}$ transforms like a 1-cocycle for $Hom(V, V)$. It is not hard to show that different choices lead to a 1-coboundary for $Hom(V, V)$, so that we have intrinsically defined an element in $H^1(X; Hom(V, V))$. This element is the *Kodaira-Spencer class* of the family $\mathcal{V} \to X \times T$. We can partially reverse this procedure: given an element $C_{ij} \in H^1(X; Hom(V, V))$, we can define first-order terms in a power series expansion for $A_{ij}(t)$, by the rule $B_{ij} = C_{ij} A_{ij}$.

More intrinsically we can start with any family $\mathcal{V} \to X \times T$, where T can be any scheme or complex space; for example, T could be the dual numbers $\operatorname{Spec} \mathbb{C}[t]/(t^2)$. Then this construction leads to a map from the Zariski tangent space of T at t_0 to $H^1(X; Hom(V, V))$. Thus, we can think of $H^1(X; Hom(V, V))$ as the space of all first-order deformations of V.

Given the first-order terms in a potential power series expansion for $A_{ij}(t)$, what is the obstruction to extending this to second order? We would like to find terms B'_{ij} so that $A_{ij} + B_{ij} \cdot t + B'_{ij} \cdot t^2$ is a 1-cocycle mod t^3. So we must compute the t^2 term in

$$(A_{ij} + B_{ij} \cdot t + B'_{ij} \cdot t^2)(A_{jk} + B_{jk} \cdot t + B'_{jk} \cdot t^2)(A_{ik} + B_{ik} \cdot t + B'_{ik} \cdot t^2)^{-1}.$$

Let us first work out this product in case $B'_{ij} = 0$. In this case, since

$$\begin{aligned}(A_{ik} + B_{ik} \cdot t)^{-1} &= (\operatorname{Id} + A_{ik}^{-1} B_{ik} \cdot t) A_{ik}^{-1} \\ &= (\operatorname{Id} - A_{ik}^{-1} B_{ik} \cdot t + (A_{ik}^{-1} B_{ik})^2 \cdot t^2 + \cdots) A_{ik}^{-1},\end{aligned}$$

a rather tedious calculation using the cocycle rules shows that the product

$$(A_{ij} + B_{ij} \cdot t)(A_{jk} + B_{jk} \cdot t)(A_{ik} + B_{ik} \cdot t)^{-1}$$

is equal to $\operatorname{Id} + \omega_{ijk}$, where

$$\omega_{ijk} = B_{ij} B_{jk} A_{ik}^{-1} = (B_{ij} A_{ij}^{-1}) A_{ij} (B_{jk} A_{jk}^{-1}) A_{ij}^{-1},$$

which is the product of the cocycle $\theta = B_{ij} A_{ij}^{-1}$ with itself under the natural cup product map from $H^1(X; Hom(V, V))$ to $H^2(X; Hom(V, V))$ induced by multiplication on $Hom(V, V)$. We can also write this as $\frac{1}{2}[\theta, \theta]$, where $[\cdot, \cdot]$ is the Lie bracket on $Hom(V, V)$, which also induces a cup product map. (Since the Lie bracket is anticommutative, the associated cup product map from H^1 to H^2 is commutative. Thus, $[\theta, \theta]$ will not in general be 0.) If the element $\frac{1}{2}[\theta, \theta]$ is 0 in cohomology, in other words is a 2-coboundary, then this will solve the problem of finding B'_{ij} so that $A_{ij} + B_{ij} \cdot t + B'_{ij} \cdot t^2$ is a 1-cocycle mod t^3, and conversely; we leave these calculations to the patient reader.

The general principle is the same, although the formulas get progressively messier. At each stage, assuming that we have a 1-cocycle to order n, we try to lift to order $n+1$ and find an obstruction that lives in $H^2(X; Hom(V, V))$. Thus, for example if $H^2(X; Hom(V, V)) = 0$, we can always lift to order $n+1$. The general formalism of deformation theory is just some technical glue to put these calculations into more respectable form.

For our applications, we shall just be interested in the case of deformations of V where the determinant is held fixed, equal to Δ, say. In this case the differentiated form of $\det V = $ constant is Trace $= 0$. We can interpret this as follows: there is a natural splitting $Hom(V, V) = \mathrm{ad}\, V \oplus \mathcal{O}_X$, where $\mathrm{ad}\, V$ is the kernel of the trace and we map $\mathcal{O}_X \to Hom(V, V)$ via the map $f \mapsto (1/r)f\,\mathrm{Id}$ (r is the rank of V). In this case the obstruction space is $H^2(X; \mathrm{ad}\, V)$ (in fact this is true in characteristic 0 even for the usual deformation theory of V, where we don't fix the determinant). Globally one can prove the following result about the local structure of $\mathfrak{M}$:

Theorem 15. *Suppose that $x \in \mathfrak{M}$ is a point corresponding to a stable bundle V. If $H^2(X; \mathrm{ad}\, V) = 0$, then $\mathfrak{M}$ is smooth at x of dimension $h^1(\mathrm{ad}\, V)$. In general, there is an analytic neighborhood of x in $\mathfrak{M}$ which is isomorphic to the zero set of h holomorphic functions $f_1, \ldots, f_h$ defined in a neighborhood of the origin in $H^1(X; \mathrm{ad}\, V)$, where $h = \dim H^2(X; \mathrm{ad}\, V)$. Moreover, the f_i have no constant or linear terms and thus the Zariski tangent space of $\mathfrak{M}$ at x may be identified with $H^1(X; \mathrm{ad}\, V)$.* $\square$

We make the following definition:

Definition 16. A vector bundle V on X is *good* if $H^2(X; \mathrm{ad}\, V) = 0$. Equivalently, the moduli space is smooth at x (in the sense of schemes) of dimension equal to $h^1(X; \mathrm{ad}\, V)$.

On a curve C, we can calculate $h^1(\mathrm{ad}\, V)$ for a rank 2 bundle V if V is simple, i.e., $h^0(\mathrm{ad}\, V) = 0$, by Riemann-Roch: the degree of $\mathrm{ad}\, V$ is the same as the degree of $Hom(V, V) = V^\vee \otimes V$, namely $-\deg V + \deg V = 0$, and the rank is 3, so $\chi(\mathrm{ad}\, V) = -h^1(\mathrm{ad}\, V) = 3(1 - g)$. Thus, $\dim H^1(\mathrm{ad}\, V) = 3g - 3$. Likewise, for a surface X and a rank 2 vector bundle V on X, if $h^2(\mathrm{ad}\, V) = 0$, then $\chi(\mathrm{ad}\, V) = -h^1(\mathrm{ad}\, V)$ again, and by Riemann-Roch and the splitting principle we find that

$$h^1(\mathrm{ad}\, V) = 4c_2(V) - c_1^2(V) - 3\chi(\mathcal{O}_X) = -p_1(\mathrm{ad}\, V) - 3\chi(\mathcal{O}_X).$$

In general, if $h^2(\operatorname{ad} V) \neq 0$, the expectation is that all of the possible $h = h^2(\operatorname{ad} V)$ equations to define the moduli space locally really are there. Thus, the moduli space always has an expected dimension $h^1(\operatorname{ad} V) - h^2(\operatorname{ad} V)$, and we say that the moduli space has the *expected dimension* at x if indeed its dimension is $h^1(\operatorname{ad} V) - h^2(\operatorname{ad} V)$. The Riemann-Roch calculations above show that the expected dimension for a surface is just $-p_1(\operatorname{ad} V) - 3\chi(\mathcal{O}_X)$. We also see readily that the dimension of the moduli space at a point x is always at least its expected dimension, and, if it is equal to the expected dimension, then the moduli space is a local complete intersection at x, or in other words is defined by exactly as many equations as the codimension of its local embedding in the smooth space $H^1(\operatorname{ad} V)$.

Next we shall give some criteria for the moduli space to be smooth:

Proposition 17.

(i) *For a curve C of genus g, every vector bundle is good. If V is a simple bundle of rank 2, then the moduli space is always smooth of dimension $3g - 3$.*

(ii) *If X is a surface and $K_X = \mathcal{O}_X(-D)$, where D is effective, then a simple vector bundle is good.*

(iii) *(Maruyama.) Let X be a surface and H an ample divisor on X such that $H \cdot K_X < 0$. Then every H-semistable rank 2 vector bundle on X is good.*

Proof. Part (i) follows since, by dimension reasons, $H^2(C; \operatorname{ad} V) = 0$. For (ii), let V be simple. By Serre duality $H^2(X; \operatorname{ad} V)$ is dual to $H^0(X; (\operatorname{ad} V)^\vee \otimes K_X)$. Now $Hom(V, V)^\vee = (V^\vee \otimes V)^\vee \cong V \otimes V^\vee \cong Hom(V, V)$, and similarly $(\operatorname{ad} V)^\vee \cong \operatorname{ad} V$. (In fact, Trace defines a nondegenerate pairing on $Hom(V, V)$ for which $\operatorname{ad} V$ is the perpendicular space to $\mathcal{O}_X \cdot \mathrm{Id}$.) Thus, we must show that $H^0(X; \operatorname{ad} V \otimes K_X) = 0$. But since $K_X = \mathcal{O}_X(-D)$, there is an inclusion $K_X \subseteq \mathcal{O}_X$ and therefore an inclusion $H^0(X; \operatorname{ad} V \otimes K_X) \subseteq H^0(X; \operatorname{ad} V)$. By assumption $H^0(X; \operatorname{ad} V) = 0$, and so $H^0(X; \operatorname{ad} V \otimes K_X) = 0$ as well.

Finally, let us prove (iii). First note that $H^0(X; nK_X) = 0$ for all $n > 0$, since $H \cdot (nK_X) < 0$. Let V be an H-semistable bundle and let $\varphi \in H^0(X; V \otimes K_X)$. Then $\det \varphi$ is a section of $2K_X$ and so is 0. If $\varphi \neq 0$, it follows that φ is a rank 1 map from V to $V \otimes K_X$. Thus, the kernel of φ is a torsion free rank 1 subsheaf of V and the cokernel of φ is a torsion free rank 1 subsheaf of $V \otimes K_X$, necessarily of the form $\mathcal{O}_X(F') \otimes I_Z$, where $\mathcal{O}_X(F')$ is a line bundle. There is a surjection $V \to \mathcal{O}_X(F') \otimes I_Z$ and an inclusion $\mathcal{O}_X(F') \otimes I_Z \to V \otimes K_X$. Now $V \otimes K_X$ is H-semistable since V is H-semistable, and $c_1(V \otimes K_X) = c_1(V) + 2K_X$. So

$$H \cdot c_1(V) \leq 2H \cdot F' \leq H \cdot (c_1(V) + 2K_X).$$

Thus, $2H \cdot K_X \geq 0$, contrary to hypothesis, and therefore $\varphi = 0$. $\square$

Corollary 18. *If X is a ruled surface and H is a (w,p)-suitable ample divisor, then every H-semistable rank 2 vector bundle V on X with $w_2(V) = w$ and $p_1(\operatorname{ad} V) = p$ is good. Thus, the moduli space $\mathfrak{M}_H(w,p)$ is everywhere smooth of dimension $-p - 3\chi(\mathcal{O}_X) = -p + 3g - 3$.*

Proof. Recall from Lemma 14 of Chapter 5 that $K_X = -2\sigma + \pi^*(K_C - \mathbf{e})$ for some divisor $\mathbf{e}$ on C, where $\mathcal{O}_\sigma(\sigma) = \mathcal{O}_C(-\mathbf{e})$ and $\deg \mathbf{e} = e$. If H is numerically equivalent to $\sigma + rf$, then

$$H \cdot K_X = 2e + (2g - 2 - e - 2r) = 2g - 2 + e - 2r.$$

Thus, if $r > g - 1 + e/2$, then $H \cdot K_X < 0$. However, if H is an arbitrary (w,p)-suitable divisor, then $H + tf$ is also suitable if $t > 0$ and if V is H-semistable, then it is $(H + tf)$-semistable. Thus, we may assume that V is semistable with respect to a divisor of the form $\sigma + rf$, with $r > g - 1 + e/2$. Applying (iii) of the previous proposition, we see that V is good. $\square$

A Zariski open subset of the moduli space

Finally, we shall give a description of a Zariski open and dense subset of the moduli space of H-semistable bundles on a ruled surface. What should the generic bundle V look like? It is natural to expect that most bundles V have the following form: let $c = c_2(V)$. Then $V|f = \mathcal{O}_f \oplus \mathcal{O}_f$ for almost all fibers f, and for the remaining fibers $V|f \cong \mathcal{O}_{\mathbb{P}^1}(1) \oplus \mathcal{O}_{\mathbb{P}^1}(-1)$, the smallest possible deviation from being trivial. If π^*W is obtained from V by making k elementary modifications, then by the formula for $c_2(V)$ given in Lemma 16 of Chapter 2, $c_2(V) = c_2(\pi^*W) + k = k$ and thus $k = c$. We obtain π^*W canonically from V as follows: there are c distinct fibers $f_1, \ldots, f_c$ such that $V|f_i \cong \mathcal{O}_{\mathbb{P}^1}(1) \oplus \mathcal{O}_{\mathbb{P}^1}(-1)$. For each such fiber f_i there is a canonical map from $V|f_i$ to $\mathcal{O}_{\mathbb{P}^1}(-1)$. Hence we have the exact sequence (where we leave off the notation for the inclusion of f_i in X):

$$0 \to \pi^*W \to V \to \bigoplus_{i=1}^{c} \mathcal{O}_{f_i}(-1) \to 0.$$

Taking the dual elementary modification, we see that we can recover V as follows:

$$0 \to V^\vee \to \pi^*W^\vee \to \bigoplus_{i=1}^{c} \mathcal{O}_{f_i}(1) \to 0.$$

Now if $\det V = \pi^*\mathcal{O}_C(\mathbf{d})$, then $\det W = \mathbf{d} - \sum_i p_i$, which has degree $d - c$, and $\det W^\vee = \sum_i p_i - \mathbf{d}$. If for the sake of neatness we wish to twist W by an appropriate line bundle so as to wind up with V at the end, we can use the fact that $V = V^\vee \otimes \det V$ and take the elementary modifications of $\pi^*W^\vee \otimes \det V = \pi^*(W^\vee \otimes \mathcal{O}_C(\mathbf{d}))$ corresponding to the ones for $\pi^*W^\vee$ given above. If we set $W_0 = W^\vee \otimes \mathcal{O}_C(\mathbf{d})$, then $\det W_0 = \det(W^\vee \otimes \mathcal{O}_C(\mathbf{d})) = \mathbf{d} + \sum_i p_i$ and V is obtained from π^*W_0 by a canonical sequence of elementary modifications.

Reversing this procedure, start with c points $p_1, \ldots, p_c \in C$. For such a choice, choose W_0 a rank 2 vector bundle on C with $\det W_0 = \mathbf{d} + \sum_i p_i$. We also need to impose a weak stability condition on W_0, coming from the condition on W. Without writing this condition down, we expect that the generic W_0 will actually be stable. Finally, we need to choose surjective maps $\pi^*W_0 \to \mathcal{O}_{f_i}(1)$ for $i = 1, \ldots, c$. Since $\pi^*W_0|f_i = \mathcal{O}_{f_i} \oplus \mathcal{O}_{f_i}$, a map from π^*W_0 to $\mathcal{O}_{f_i}(1)$ is the same as the choice of two sections of $\mathcal{O}_{f_i}(1)$ and the generic choice of sections corresponds to a surjection from π^*W_0 to $\mathcal{O}_{f_i}(1)$. Thus, the set of surjections from π^*W_0 to $\mathcal{O}_{f_i}(1)$ is a Zariski open subset of $\operatorname{Hom}(\pi^*W_0, \mathcal{O}_{f_i}(1)) \cong \mathbb{C}^4$. Now given a map $\pi^*W_0 \to \bigoplus_{i=1}^c \mathcal{O}_{f_i}(1)$, the kernel will be unchanged if we compose with an automorphism of $\bigoplus_{i=1}^c \mathcal{O}_{f_i}(1)$, namely an element of $(\mathbb{C}^*)^c$. So we should divide out by the set of such automorphisms. Of course, we should also divide out by $\operatorname{Aut} W_0$. But if W_0 is simple, then $\operatorname{Aut} W_0 = \mathbb{C}^*$, and we may identify this $\mathbb{C}^*$ with the diagonal of $(\mathbb{C}^*)^c$. So we don't have to factor out by this additional $\mathbb{C}^*$.

Let us now count the number of moduli involved in this construction. For simplicity we will just consider here the case $g \geq 2$, and will say a few words about the cases $g = 0, 1$ later. First we choose c points $p_1, \ldots, p_c$ of C, using c moduli. Next, given the p_i, we choose W_0 a stable bundle over C with determinant equal to $\mathbf{d} + \sum_i p_i$. This choice involves $3g - 3$ moduli. Finally, we need to choose a surjection $\pi^*W_0 \to \mathcal{O}_{f_i}(1)$, modulo $\mathbb{C}^*$, for every i; this choice involves $c(4 - 1) = 3c$ parameters. Tallying these all up, we arrive at

$$c + 3g - 3 + 3c = 4c + 3g - 3 = -p - 3\chi(\mathcal{O}_X)$$

parameters, which is the expected dimension of the moduli space. Moreover, we have the following birational picture of the moduli space $\mathfrak{M}$: There is a rational map $\mathfrak{M} \dashrightarrow \operatorname{Sym}^c C$, whose fibers rationally look like a bundle over $\mathfrak{M}(C)$, the moduli space of rank 2 bundles of a given fixed determinant over C, and the fiber of the bundle over $\mathfrak{M}(C)$ looks like a Zariski open subset of $(\mathbb{C}^4)^c$, modulo an action of $(\mathbb{C}^*)^c$. The moduli space $\mathfrak{M}(C)$ is known to be irreducible and unirational [113], [140]. (A variety Z is *unirational* if there is a dominant rational map $\mathbb{P}^N \dashrightarrow Z$; see Definition 32 in Chapter 10.) Thus, the moduli space $\mathfrak{M}$ exhibits behavior very much like that of X itself: it fibers over $\operatorname{Sym}^c C$, and the fibers are unirational varieties.

How do we go about making the above heuristic discussion more precise, and showing that we really have described an open dense subset of the moduli space in this way? The basic idea is to analyze the functorial properties of the moduli space, and especially the fact that coarse moduli spaces have a weak universal property. To identify an open subset of $\mathfrak{M}$, we start by finding a space M and a bundle $\mathcal{U}'$ over $X \times M$ with the following property: the restriction of $\mathcal{U}'$ to every slice $X \times \{p\}$ is stable. The bundle $\mathcal{U}'$ induces a morphism $f \colon M \to \mathfrak{M}$. For a good construction of M and $\mathcal{U}'$, f will be an embedding onto an open subset of $\mathfrak{M}$. If $\mathfrak{M}$ is singular, this may be rather difficult to show, in the sense that it may be very hard to determine the exact scheme structure on $\mathfrak{M}$ or to check that a given morphism is an isomorphism. However, if we know, for example from deformation theory, that $\mathfrak{M}$ is smooth we can often just use the easy case of Zariski's main theorem (Theorem 8 of Chapter 3) to conclude that the map $M \to \mathfrak{M}$ is an open embedding. We will not work through the technical details of these constructions, but will just give an informal approach to the description of the moduli space.

Having found a Zariski open subset of the moduli space, how do we then conclude that it is dense? The problem, as should already be apparent in the discussion for ruled surfaces above, is that it is rare to have a uniform description of *all* stable bundles. Instead we tend to have a stratified description of the moduli space, with a description of a set of presumably "generic bundle" as well as more specialized descriptions of all of the others. One way to prove that the "generic" description really does give a Zariski open and dense subset of the moduli space is to show that there exist a finite number of schemes T_i and bundles $\mathcal{V}_i$ over $X \times T_i$ parametrizing all of the "other" stable bundles, in other words such that, if the stable bundle V is not in the image of $f \colon M \to \mathfrak{M}$, then V is isomorphic to the restriction of $\mathcal{V}_i$ to some slice $X \times \{t\}$ for some i and for some $t \in T_i$. Suppose further that, for all i, $\dim T_i$ is less than the expected dimension of the moduli space. If so, letting $f_i \colon T_i \to \mathfrak{M}$ be the morphism induced by $\mathcal{V}_i$, the closure of $\bigcup_i f_i(T_i)$ will be contained in a subvariety of $\mathfrak{M}$ of dimension strictly less than $\dim \mathfrak{M}$. In particular the remaining bundles, in other words the image of M, must be Zariski dense.

To see how this works in the case of a ruled surface, we shall make an informal parameter count. To be nongeneric, we should take W_0 to be an unstable bundle, or allow some of the points p_i to coincide, or make an elementary modification using $\mathcal{O}_{f_i}(a)$ with $a > 1$. Let us argue that all of these will lower the number of parameters in the construction. For the case where W_0 is not semistable, W_0 is uniquely described by an extension using the maximal destablizing subsheaf. We leave it as an exercise (Exercise 4) to show that all such extensions are parametrized by a parameter space of dimension at most $2g - 2$. If $g \geq 2$, then $2g - 2 < 3g - 3$, so we are done in this case. Clearly, if two or more points coincide, then the points p_i depend on fewer moduli. Finally, what happens if we do an elementary modification using $\mathcal{O}_{f_i}(a_i)$ with $a_i > 1$, possibly at a point where we have already done an elementary modification? We shall just analyze what happens as we go from V to a vector bundle V' by a single elementary modification as above. As we noted in the discussion after Lemma 12, if $V|f = \mathcal{O}_{\mathbb{P}^1}(a) \oplus \mathcal{O}_{\mathbb{P}^1}(-a)$ with $a > 0$, then $V'|f = \mathcal{O}_{\mathbb{P}^1}(a') \oplus \mathcal{O}_{\mathbb{P}^1}(-a')$ with $0 \leq a' \leq a$. Conversely, to go from $(V')^\vee$ to $V^\vee$ we must do an elementary modification coming from a surjection from $\mathcal{O}_{\mathbb{P}^1}(a') \oplus \mathcal{O}_{\mathbb{P}^1}(-a')$ to $\mathcal{O}_{\mathbb{P}^1}(a)$. Now

$$\dim \operatorname{Hom}(\mathcal{O}_{\mathbb{P}^1}(a') \oplus \mathcal{O}_{\mathbb{P}^1}(-a'), \mathcal{O}_{\mathbb{P}^1}(a)) = h^0(\mathcal{O}_{\mathbb{P}^1}(a - a') \oplus \mathcal{O}_{\mathbb{P}^1}(a + a'))$$
$$= a - a' + 1 + a + a' + 1 = 2a + 2.$$

So the total number of parameters needed to make an elementary modification dual to an allowable elementary modification is $2a + 2$. On the other hand, the total change in the number of expected moduli is the amount by which $4c_2$ increases from V' to V, and by the discussion after Lemma 12, this change is $4a$. Now $4a \geq 2a + 2$ as long as $a > 0$, and equality holds if and only if $a = 1$. So the only way we can get the full number $4c_2 + 3g - 3$ parameters is to make the generic construction described above. Summarizing:

Theorem 19. *Let X be a ruled surface over a curve C of genus $g \geq 2$, and let $\mathbf{d}$ be a divisor on C. Let H be a $(\pi^*\mathbf{d}, c)$-suitable divisor on X. Then the moduli space of H-stable rank 2 vector bundles V on X with $\det V = \pi^*\mathbf{d}$ and $c_2(V) = c$ is nonempty for all $c \geq 0$, and it is irreducible and smooth of dimension $4c + 3g - 3$. A Zariski open and dense subset fibers over $\operatorname{Sym}^c C$ and the fibers are unirational varieties.*

Variations of the above arguments prove similar results for $g(C) = 0, 1$. For example, in case $g = 0$, there is just one semistable bundle $\mathcal{O}_{\mathbb{P}^1}(n) \oplus \mathcal{O}_{\mathbb{P}^1}(n)$, but there are infinitely many unstable bundles $\mathcal{O}_{\mathbb{P}^1}(n) \oplus \mathcal{O}_{\mathbb{P}^1}(m)$ for $m \neq n$, and we cannot simply argue that they have fewer moduli than the semistable ones. In this case, however, $\operatorname{Hom}(\pi^*W_0, \bigoplus_i \mathcal{O}_{f_i}(1))$ is acted on not only by $\operatorname{Aut}(\bigoplus_i \mathcal{O}_{f_i}(1)) = (\mathbb{C}^*)^c$, but also by $\operatorname{Aut}(\pi^*W_0) = \operatorname{Aut}(W_0)$. If $W_0 = \mathcal{O}_{\mathbb{P}^1}(n) \oplus \mathcal{O}_{\mathbb{P}^1}(n)$, then $\dim \operatorname{Aut} W_0 = \dim GL(2, \mathbb{C}) = 4$, but if $W_0 = \mathcal{O}_{\mathbb{P}^1}(m_1) \oplus \mathcal{O}_{\mathbb{P}^1}(m_2)$ for $m_1 \neq m_2$, $m_1 + m_2 = 2n$, then

$$\dim \operatorname{Aut} W_0 = \dim \operatorname{End} W_0$$
$$= h^0(\mathcal{O}_{\mathbb{P}^1}(m_1 - m_2) \oplus \mathcal{O}_{\mathbb{P}^1} \oplus \mathcal{O}_{\mathbb{P}^1} \oplus \mathcal{O}_{\mathbb{P}^1}(m_2 - m_1))$$
$$= |m_1 - m_2| + 3 > 4.$$

So we again get away with a smaller number of moduli to describe the nongeneric bundles, those where W_0 is not semistable. Using this circle of ideas, one can then prove:

Theorem 20. *Let X be a rational ruled surface. Let H be an $(\pi^*\mathcal{O}_{\mathbb{P}^1}(d), c)$-suitable divisor on X. Then the moduli space of H-stable rank 2 vector bundles V on X with $\det V = \pi^*\mathcal{O}_{\mathbb{P}^1}(d)$ and $c_2(V) = c$ is nonempty for $c \geq 2$ if d is even, and for $c \geq 1$ if d is odd. It is irreducible, smooth, and unirational of dimension $4c - 3$.*

These ideas may then be applied to the study of bundles over $\mathbb{P}^2$. For example, Barth, Maruyama, and Hulek have proved that the space of stable bundles over $\mathbb{P}^2$ is irreducible and unirational, and indeed rational if the determinant is a line bundle of odd degree. Their proof uses the study of bundles of $\mathbb{P}^2$ via monads. (See [117] for a discussion of these methods.) Using Theorem 20, Qin [125] gave another proof, starting with the rational ruled surface $\mathbb{F}_1$ and using the study of the way the moduli space changes as we cross walls in the ample cone coming from chambers of type (w, p), to pass from the chamber containing (w, p)-suitable divisors to one more suited to studying blowups. We outline this approach in the exercises.

Exercises

1. Let X be an arbitrary surface and H an ample divisor on X. Show that a strictly H-semistable rank 2 vector bundle V always satisfies Bogomolov's inequality $p_1(\operatorname{ad} V) \leq 0$.

2. Show that, if V is a semistable rank 2 bundle (for *some* divisor H) on a ruled surface X with $\deg(V|f) = 1$, then V satisfies Bogomolov's inequality. (Use the fact that there are no stable bundles for a suitable ample divisor, so that every stable bundle is actually strictly semistable for some ample divisor.)

3. Let $X = C_1 \times C_2$ be a product of two curves, let $\pi_i \colon X \to C_i$ denote the projection, and let W be a stable rank 2 vector bundle on C_1. Show that, for a suitable ample divisor on X, $\pi_1^* W$ is a stable vector bundle on X with $c_1^2 = c_2 = 0$. Moreover, if W has odd degree (which is only possible if $g(C_1) > 0$), then $\pi_1^* W$ is not of the form $V \otimes L$, where V is a flat rank 2 vector bundle on X and L is a holomorphic line bundle.

4. Let C be a curve of genus $g \geq 2$. Show that the set of rank 2 vector bundles V of fixed determinant such that there exists a nonsplit exact sequence

$$0 \to L \to V \to L' \to 0$$

with $\deg L > \deg V/2$ may be parametrized by a space of dimension at most $2g - 2$. (If L is fixed, use Clifford's theorem and Riemann-Roch to estimate $H^1(C; (L')^{-1} \otimes L)$. Then use the fact that the space of all L is parametrized by the Jacobian $J(C)$, which has dimension g.) What about an estimate for the space of strictly semistable bundles?

5. Our discussion of ruled surfaces concentrated on the case of bundles with determinant the pullback of a line bundle on C, or equivalently bundles V such that $\deg V|f$ is even. In the case where $\deg V|f$ is odd, say $\det V = \sigma$, there are no H-stable bundles if H is a (w, p)-suitable ample line bundle. Show that for a divisor H lying in the chamber $\mathcal{C}_1$ in $\mathcal{A}(X)$ next to the chamber $\mathcal{C}_0$ containing (σ, c)-suitable divisors, we can construct stable bundles via extensions corresponding to the wall separating $\mathcal{C}_1$ from $\mathcal{C}_0$ as follows. We suppose that $X = \mathbb{F}_e$ is a rational ruled surface. Show that the unique wall separating $\mathcal{C}_0$ from $\mathcal{C}_1$ is given by W^ζ, where $\zeta = \sigma - 2cf$ (argue as in the proof of Lemma

4). Show that for every $H \in C_1$, every H-stable rank 2 vector bundle V with $\det V = \sigma$ and $c_2(V) = c$ is given as an extension

$$0 \to \mathcal{O}_{\mathbb{F}_e}(\sigma - cf) \to V \to \mathcal{O}_{\mathbb{F}_e}(cf) \to 0$$

(since V must be C_0-unstable, and necessarily a destabilizing sub-line bundle $\mathcal{O}_X(D)$ satsifies $2D - \sigma = \zeta$), and conversely by Proposition 24 of Chapter 4 such a bundle V is H-stable. Conclude that the moduli space is a projective space $\mathbb{P}^N$, and show directly that $N = -p - 3$. If X is not rational, argue informally that the moduli space is a $\mathbb{P}^N$-bundle over $\mathrm{Pic}^0 X$, where $N = 4c + e + 2g - 3 = -p + 2g - 3$, and again recover the fact that the moduli space is smooth of dimension $-p - 3\chi(\mathcal{O}_X)$.

6. Let $X = \mathbb{F}_e$ be a rational ruled surface, and let $\mathfrak{M}_0$ be the moduli space of stable bundles V with $c_1(V) = 0$ and $c_2(V) = c$ for the chamber C_0 corresponding to $(0, c)$-suitable ample divisors. In this exercise, we outline a proof that, for every chamber $\mathcal{C}$ in $\mathcal{A}(X)$, the moduli space $\mathfrak{M}$ of stable bundles V for $\mathcal{C}$ is birational to $\mathfrak{M}_0$. In particular, it is irreducible and unirational. Using Theorem 20 and Proposition 24 of Chapter 4, it is enough to show that, for every ζ a wall of type $(0, c)$ and for every local complete intersection subscheme Z such that $-\zeta^2 + 4\ell(Z) = -p = 4c$, for D the divisor such that $\zeta = 2D$, the set of extensions of the form

$$0 \to \mathcal{O}_X(D) \to V \to \mathcal{O}_X(-D) \otimes I_Z \to 0$$

may be parametrized by a scheme of dimension less that $4c - 3$. Argue informally that $Z \in \mathrm{Hilb}^{\ell(Z)}(X)$ depends on $2\ell(Z)$ moduli and that, given Z, the set of all extensions depends on $\dim \mathrm{Ext}^1(\mathcal{O}_X(-D) \otimes I_Z, \mathcal{O}_X(D))$ moduli. Using the fact that $2D$ is orthogonal to an ample divisor and that $K_X = -E$ for an effective divisor E, show that $h^0(2D) = h^2(2D) = 0$ and thus that $\dim \mathrm{Ext}^1(\mathcal{O}_X(-D) \otimes I_Z, \mathcal{O}_X(D)) = \ell(Z) + h^1(2D)$. Thus, the total number of moduli of such an extension is $h^1(2D) + 3\ell(Z) - 1$, since if two given extension classes differ by a nonzero scalar, the corresponding bundles are isomorphic.

Now since $h^0(-2D) = h^2(-2D) = 0$ as well, we have

$$h^1(2D) \leq h^1(2D) + h^1(-2D) = -2\frac{(2D)^2}{2} - 2$$

(Riemann-Roch) and so $h^1(2D) \leq -\zeta^2 - 2$, with equality only if $h^1(-2D) = 0$. Thus,

$$h^1(2D) + 3\ell(Z) - 1 \leq -\zeta^2 + 3\ell(Z) - 3 = 4c - 3 - \ell(Z) \leq 4c - 3,$$

with equality if and only if $h^1(-2D) = 0$ and $\ell(Z) = 0$. In this case, if $D = a\sigma + bf$, show that $a(ae - 2b) = c$ and that (Riemann-Roch and the canonical bundle formula) $ae - 2b = 2a + 2c - 2$ and derive a contradiction, since $c \geq 1$. What if $\det V = df$ instead?

Show similarly that for $\det V = \Delta$ with $\Delta \cdot f = 1$, then for every chamber $\mathcal{C}$, the corresponding moduli space $\mathfrak{M}$ is empty if $\mathcal{C}$ is the chamber containing (Δ, c)-suitable ample divisors, and (using the previous exercise) $\mathfrak{M}$ is birational to a projective space in all other cases.

For an explicit description of the blowups and blowdowns required to get $\mathfrak{M}$ from $\mathfrak{M}_0$, see [29], [44], and [86].

7. Let $\rho\colon \mathbb{F}_1 \to \mathbb{P}^2$ be the blowup of $\mathbb{P}^2$ at a point. If V is a stable rank 2 vector bundle on $\mathbb{P}^2$, show that $\rho^* V$ is H-stable for all ample line bundles of the form $\rho^* \mathcal{O}_{\mathbb{P}^2}(N) \otimes \mathcal{O}_{\mathbb{F}_1}(-E)$, where E is the exceptional curve and $N \gg 0$, and that ρ^* defines an open embedding from $\mathfrak{M}_{\mathbb{P}^2}(0, c)$ to $\mathfrak{M}$, where $\mathfrak{M}$ is the moduli space of corresponding vector bundles for an appropriate chamber of $\mathbb{F}_1$. Conclude that $\mathfrak{M}_{\mathbb{P}^2}(0, c)$ is irreducible and unirational. Establish a similar result for rank 2 stable vector bundles on $\mathbb{P}^2$ with determinant $\mathcal{O}_{\mathbb{P}^2}(1)$, where the moduli space is in fact rational. (If, say, V is a stable bundle on $\mathbb{P}^2$ with $c_1(V) = 0$ and $\mathcal{O}_{\mathbb{F}_1}(D)$ is a sub-line bundle of $\rho^* V$, use the fact that $\rho_* \rho^* V = V$ to show that $\mathcal{O}_{\mathbb{F}_1}(D) = \rho^* \mathcal{O}_{\mathbb{P}^2}(k) \otimes \mathcal{O}_{\mathbb{F}_1}(aE)$ with $k < 0$. To see that ρ^* defines an open immersion, look at Zariski tangent spaces.)

For more detailed analysis of stable bundles on blown up surfaces, see Theorem 17 in Chapter 9 as well as [38] and [15].

7

An Introduction to Elliptic Surfaces

The goal of this chapter is to survey the classification of elliptic surfaces. We begin with a preliminary section on general fibrations. Next we give Kodaira's classification of singular fibers of elliptic fibrations, and describe the canonical bundle and other basic invariants of an elliptic surface. We give a complete discussion of elliptic surfaces with a section and discuss, without proofs, the mechanism for obtaining all elliptic surfaces starting with those which have a section. Finally, we describe some of the basic topological invariants of an elliptic surface.

Singular fibers

Let $\pi\colon X \to C$ be a (proper) morphism from a smooth surface X to a smooth curve C, such that the general fiber of π is connected. We shall refer to such a map as a *fibration*. Let f be a smooth fiber, and let g be the genus of f. We shall primarily be concerned with the case where $g = 1$, but let us record here some general facts about such fibrations. For much of this study, we will look at the local case where $C = \Delta$ is a disk in $\mathbb{C}$ and π is smooth except possibly at $0 \in \Delta$. In this local setting, if $C_1, \ldots, C_r$ are the components of $\pi^{-1}(0)$, then standard arguments show that X deformation retracts onto $\pi^{-1}(0) = \bigcup_i C_i$. In particular $H^2(X; \mathbb{Z}) = \bigoplus_i \mathbb{Z} \cdot [C_i]$. Intersection pairing on $H^2(X; \mathbb{Z})$ may then be defined in the usual way, by first defining $C_i \cdot C_j = \deg \mathcal{O}_X(C_i)|C_j$ and extending by linearity. Note that the pairing will never be nondegenerate, since if $\pi^*0 = \sum_i n_i C_i$, then $\mathcal{O}_X(\sum_i n_i C_i)$ is a trivial line bundle on X and so $\sum_i n_i C_i$ is in the radical of the intersection pairing. However, up to a rational multiple, this is the only element in the radical of the lattice spanned by the C_i:

Lemma 1. *In the above situation, let $\pi^{-1}(0) = \bigcup_i C_i$ and let $\pi^*0 = \sum_i n_i C_i$. Then $\bigcup_i C_i$ is connected, $n_i > 0$ for all i, and the lattice spanned by the classes of the C_i is negative semidefinite with a radical of rank 1. Moreover, the radical is spanned over $\mathbb{Z}$ by an element $\sum_i a_i C_i = e$ with $a_i > 0$ for all i and $\gcd a_i = 1$ such that $f = me$ for some integer $m > 0$.*

Proof. That $\bigcup_i C_i$ is connected follows from the connectedness theorem, and clearly $n_i > 0$ for all i. That the lattice spanned by the C_i is negative semidefinite with a rank 1 radical spanned over $\mathbb{Q}$ by $f = \sum_i n_i C_i$ follows, in case X is projective, from the Hodge index theorem and the fact that $C_i \cdot f = 0$ for all i. Let us show that this conclusion holds in the local case as well, where $C = \Delta$ and the fiber over 0 is the unique singular fiber.

First note that $f = \sum_i n_i C_i$ with $f \cdot C_i = 0$ for all i. If $i = 1$, i.e., if the reduction of $\pi^{-1}(0)$ is connected, then $f = nC_1$ and the statement about the lattice is clear. Otherwise, by connectedness, for every i there exists a $j \neq i$ such that $C_i \cap C_j \neq 0$. Thus, we have

$$0 = f \cdot C_i = n_i(C_i^2) + \sum_{j \neq i} n_j(C_j \cdot C_i) > n_i(C_i^2).$$

So $C_i^2 < 0$ for all i. Let $v_i = n_i C_i$, so that $v_i^2 < 0$ for all i as well. For every i there exists a j such that $v_i \cdot v_j > 0$. Finally, $(\sum_j v_j) \cdot v_i = 0$ for all i. A minor adaptation of the arguments of Theorem 21 in Chapter 3 shows that $(\sum_i \lambda_i v_i)^2 \leq 0$ for all choices of $\lambda_i \in \mathbb{R}$, and that $(\sum_i \lambda_i v_i)^2 = 0$ if and ony if $\lambda_i = \lambda_j$ for all i, j. Thus, the lattice is negative semidefinite and its radical is generated over $\mathbb{Q}$ by $\sum_i v_i = \sum_i n_i C_i$. If $e = \sum_i a_i C_i$ is the primitive generator of the radical over $\mathbb{Z}$ such that f is a positive multiple of e, then $a_i > 0$ for all i, $\gcd a_i = 1$, and $f = me$ for some positive integer m. $\square$

Corollary 2. *In the above situation, if $\{C_i : i \in A\}$ is a proper subset of the components of $\pi^*\{0\}$, then $\{C_i : i \in A\}$ spans a negative definite lattice.* $\square$

Next we show that in the above situation $h^0(\mathcal{O}_e) = 1$:

Lemma 3. *Let $\pi\colon X \to \Delta$ be a fibration with fiber f, and let $e = \sum_i a_i C_i$ be the divisor over 0 which is the primitive generator for the radical of the pairing. Then $h^0(\mathcal{O}_e) = 1$. In other words, the only holomorphic functions on e are the constants.*

Proof. First we claim that we can find e via a modified computation sequence as in Exercise 9 in Chapter 3 as follows. Let $Z_1 = C_i$ be any component of the fiber over 0. Thus, Z_1 is contained in the support of e, i.e., $Z_1 \leq e$. Inductively suppose that we have found Z_i and $Z_i \leq e$. If $Z_i \cdot C_j \leq 0$ for all j, then in fact we must have $Z_i \cdot C_j = 0$ for all j, for otherwise the proof of Theorem 21 in Chapter 3 shows that the C_j span a negative definite lattice (cf. Exercise 9 in Chapter 3), contradicting the fact that the lattice has a rank 1 radical. So Z_i lies in the radical and is thus a positive multiple of e. As $Z_i \leq e$, $Z_i = e$. Conversely, if $Z_i \neq e$, then there exists a j such that $Z_i \cdot C_j > 0$. Set $Z_{i+1} = Z_i + C_j$. Since $(e - Z_i) \cdot C_j = -(Z_i \cdot C_j) < 0$, C_j is contained in the support of the effective divisor $e - Z_i$. Thus, $e - (Z_i + C_j)$ is effective, so that $Z_{i+1} \le e$ as well. This procedure must eventually terminate with $Z_k = e$.

As in Exercise 10 of Chapter 3, using the exact sequence

$$0 \to \mathcal{O}_{C_j}(-Z_i) \to \mathcal{O}_{Z_{i+1}} \to \mathcal{O}_{Z_i} \to 0$$

and induction, we see that $h^0(\mathcal{O}_{Z_{i+1}}) = 1$ for all i. In particular $h^0(\mathcal{O}_e) = 1$. $\square$

Definition 4. If $f = me$, where e is the primitive generator of the radical of intersection pairing and $m > 1$, then we call $f = me$ a *multiple fiber* of multiplicity m.

In fact, multiple fibers can exist only if $\bigcup_i C_i$ is not simply connected:

Lemma 5. *Suppose that $f = me$ is a multiple fiber of the fibration $\pi\colon X \to C$. Then the reduction of f has a covering space of order m. Moreover, the normal bundle $\mathcal{O}_X(e)|e$ is a torsion line bundle of order exactly m on the possibly nonreduced scheme e.*

Proof. It is enough to consider the local case where $C = \Delta$. Let t be a coordinate on Δ. Locally on X, there exist coordinates x, y such that $\pi^*t = g(x, y)$, where $g = h^m$ is an mth power (and h is a local defining equation for e). We have the map $\Delta \to \Delta$ defined by $t = s^m$. Let X' be the pullback; thus X' is the subset of $X \times \Delta$ locally given by

$$\{(x, y, s) : s^m = h^m(x, y)\}.$$

Let $\tilde{X}$ be the normalization of X'. Since we can factor

$$s^m - h^m(x, y) = \prod_{\zeta^m = 1} (s - \zeta h(x, y)),$$

we see that locally $\tilde{X}$ is the union of m smooth pieces. Clearly, $\tilde{X} \to X$ is a covering space. Since the general fiber of $\tilde{X} \to \Delta$ is connected, the fiber $\tilde{e}$ over 0 is connected as well. Thus, the fiber over 0 of X has a covering space of order m (in fact the cover is Galois with group $\mathbb{Z}/m\mathbb{Z}$). Note that, if we write $\tilde{e}$ as a sum of irreducible curves with positive coefficients, then the coefficients of the components of $\tilde{e}$ are the same as the coefficients of the components of e. In particular the gcd of the coefficients of the components of $\tilde{e}$ is 1. Thus, $\tilde{e}$ is a primitive generator of the radical of intersection pairing for $\tilde{X}$, and it is not a multiple fiber.

We turn now to the statement concerning the normal bundle. Let N be the normal bundle of e in X and let $\tilde{N}$ be the normal bundle of $\tilde{e}$ in $\tilde{X}$. Since $\tilde{X} \to X$ is a covering space, $\tilde{N}$ is the pullback of N to $\tilde{e}$. Now $\tilde{N}$ is trivial, and indeed an everywhere generating section is given by s. The group of covering transformations $\mathbb{Z}/m\mathbb{Z}$ acts on $\tilde{X}$ fixing $\tilde{e}$, and thus acts on $\tilde{N}$ and on its sections, as well as on various tensor powers of $\tilde{N}$.

Clearly, $\mathbb{Z}/m\mathbb{Z}$ acts on the section s by the roots of unity. Since $\tilde{N}$ is trivial, Lemma 3 applied to $\tilde{X}$ shows that $h^0(\tilde{N}^{\otimes k}) = 1$, and that a generator for the $\mathbb{Z}/m\mathbb{Z}$-action acts on a nonzero element of $H^0(\tilde{N}^{\otimes k})$ as ζ^k, where ζ is a primitive m^{th} root of unity. Now the line bundle N on e has order dividing m since $\mathcal{O}_e(e)^{\otimes m} = \mathcal{O}_e(f) = \mathcal{O}_e$. A section of $N^{\otimes k}$ would give a section of $\tilde{N}^{\otimes k}$ invariant under the $\mathbb{Z}/m\mathbb{Z}$-action. Clearly, this is only possible if m divides k. Thus, the order of N is exactly m. □

Corollary 6. *For every fiber f of $\pi\colon X \to C$, $h^0(\mathcal{O}_f) = 1$ and $h^1(\mathcal{O}_f) = g$.*

Proof. Let f be a fiber, and suppose that $f = me$ where the gcd of the coefficients of e is 1. It follows from Lemma 3 that $h^0(\mathcal{O}_e) = 1$. For $k \geq 1$, we have the exact sequence

$$0 \to \mathcal{O}_e(-ke) \to \mathcal{O}_{(k+1)e} \to \mathcal{O}_{ke} \to 0.$$

If $0 < k < m$, the line bundle $\mathcal{O}_e(-ke)$ is a nontrivial torsion line bundle on e. In particular its restriction to each component of e has degree 0 and is either trivial or has no sections. It follows that a nonzero section φ of $\mathcal{O}_e(-ke)$ restricts on each component C of e to a section of $\mathcal{O}_e(-ke)|C$ which is either identically 0 or nowhere 0. Now consider the computation sequence used in the proof of Lemma 3, which we can start at an arbitrary component C of e: we have

$$0 \to \mathcal{O}_{C_j}(-Z_i - ke) \to \mathcal{O}_{Z_{i+1}}(-ke) \to \mathcal{O}_{Z_i}(-ke) \to 0,$$

where $\deg \mathcal{O}_{C_j}(-Z_i - ke) = -(C_j \cdot Z_i) - k(C_j \cdot e) = -(C_j \cdot Z_i) < 0$ and $Z_1 = C$. It follows that, for all i, we have $H^0(\mathcal{O}_{Z_{i+1}}(-ke)) \subseteq H^0(C; \mathcal{O}_C(-ke))$. Choosing in particular $Z_{i+1} = e$, we see that, for every component C of e, $H^0(\mathcal{O}_e(-ke)) \subseteq H^0(C; \mathcal{O}_C(-ke))$. Thus, if $H^0(\mathcal{O}_e(-ke))$ is nonzero, then a nonzero section φ of $H^0(\mathcal{O}_e(-ke))$ restricts to a nonzero section of $H^0(C; \mathcal{O}_C(-ke))$ for every component C of e, which must be everywhere generating since $\deg \mathcal{O}_C(-ke) = 0$. In particular $\mathcal{O}_C(-ke) \cong \mathcal{O}_C$ for every component C of e. But then the map $\mathcal{O}_e \to \mathcal{O}_e(-ke)$ defined by φ generates $\mathcal{O}_e(-ke)$ mod the maximal ideal at every point. Thus, $\mathcal{O}_e \to \mathcal{O}_e(-ke)$ is surjective and hence is an isomorphism. This contradicts the fact that $\mathcal{O}_e(-ke)$ is not trivial. It follows that $\mathcal{O}_e(-ke)$ has no sections and by induction $h^0(\mathcal{O}_{(k+1)e}) = h^0(\mathcal{O}_{ke}) = 1$ for all $k < m$. In particular, taking $k = m - 1$ we see that $h^0(\mathcal{O}_f) = 1$. Next we claim π is a flat map: this follows from the local criterion of flatness [61, p. 276, Ex. 10.9], since X is a smooth surface over a smooth base curve and all fibers of π have dimension 1. Thus, $\chi(\mathcal{O}_f)$ is independent of f. Since

$$\chi(\mathcal{O}_f) = h^0(\mathcal{O}_f) - h^1(\mathcal{O}_f) = 1 - h^1(\mathcal{O}_f),$$

it follows that $h^1(\mathcal{O}_f)$ is also independent of f, and is equal to its value on a smooth fiber, namely g. □

We may define relatively minimal models for fibrations by analogy with Definition 16 of Chapter 3. We shall show that there is a unique relatively minimal model, as long as the genus g of a smooth fiber is at least 1.

Definition 7. Let $\pi\colon X \to C$ be a fibration from the smooth surface X to the smooth curve C. A *relatively minimal model* X' of X is a smooth surface X' obtained by contracting exceptional curves lying in fibers of π, and such that X' has no exceptional curves lying in fibers of the induced map $\pi'\colon X' \to C$. Equivalently, there is a commutative diagram

$$
\begin{array}{ccc}
X & \longrightarrow & X' \\
\Big\downarrow{\scriptstyle \pi} & & {\scriptstyle \pi'}\Big\downarrow \\
C & =\!=\!= & C,
\end{array}
$$

such that the morphism $X \to X'$ is birational and such that X' has no exceptional curves contained in fibers of π'. Strong relatively minimal models are similarly defined.

We then have the analogue of Theorem 19 of Chapter 3:

Theorem 8. *With notation as above, relatively minimal models always exist. If $g \geq 1$, then every relatively minimal model is a strong relatively minimal model.*

Proof. Arguing as in Lemma 17 of Chapter 3, we can keep contracting exceptional curves in fibers until there are none left, and so we reach a relatively minimal model. If it is not a strong relatively minimal model, then the proof of Theorem 19 in Chapter 3 shows that there is a fibration $\tilde{X} \to C$, birational to X over C, such that the fiber over a point contains exceptional curves F and E with $F \cdot E = n \geq 1$. Since the intersection matrix determined by F and E is negative semidefinite, $1 - n \geq 0$. Thus, $n \leq 1$ and so $n = 1$. Hence $E + F$ has square 0, so it generates the radical of the lattice spanned by the components of the fiber over $\mathbb{Q}$. There can thus be no other components of the fiber, which is then a multiple of $E + F$. Since $E + F$ is simply connected, the fiber must then be $E + F$, by Lemma 5. Since $E + F$ has arithmetic genus 0, $g = 0$. Conversely, if the genus of a general fiber is at least 1, relatively minimal models are strong relatively minimal models. $\square$

One way to characterize a (strong) minimal model in case $g \geq 1$ is by nef properties of the canonical bundle:

Definition 9. Let $\pi\colon X \to C$ be a proper morphism from the smooth surface X to a smooth (not necessarily compact) curve C. A line bundle L

on X is *nef relative to* π or π-*nef* if $L \cdot D \geq 0$ for every irreducible curve D contained in a fiber of π.

Proposition 10. *Let* $\pi\colon X \to C$ *be a fibration and let* g *be the genus of a general fiber. If* $g \geq 1$, *then* X *is relatively minimal if and only if* K_X *is* π-*nef*.

Proof. If X is not relatively minimal, then there exists an exceptional curve E contained in a fiber of π, and $K_X \cdot E = -1$. Thus, K_X is not π-nef. Conversely, suppose that K_X is not π-nef, and let D be an irreducible curve contained in a fiber such that $K_X \cdot D < 0$. Now $D^2 \leq 0$, and $D^2 = 0$ if and only if D generates the radical of the intersection pairing on the components of the fiber containing D. We have $2p_a(D) - 2 = K_X \cdot D + D^2 < 0$, so that D is a smooth rational curve. If $D^2 = 0$, since D is in the radical of the pairing, there can be no other component of the fiber. Since D is simply connected, the fiber is not multiple, by Lemma 5. Thus, D is a smooth fiber and $g = 0$. In the remaining case $D^2 < 0$ and D is exceptional. In this case X is not relatively minimal. $\square$

Singular fibers of elliptic fibrations

From now on we shall only consider the case where the genus of all smooth fibers is 1. We shall refer to such an X as an *elliptic fibration*. We may as well only consider the case where X is a (strong) relatively minimal model as well. Our first goal will be to classify the possible singular fibers.

Lemma 11. *Let* $\pi\colon X \to C$ *be a relatively minimal elliptic fibration, and let* D *be an irreducible component of a fiber of* π. *Then* $K_X \cdot D = 0$ *and either* $p_a(D) = 1$ *and* D *is the only component of the fiber, or* D *is a smooth rational curve and* $D^2 = -2$.

Proof. By Proposition 10, $K_X \cdot D \geq 0$. If f is a smooth fiber of π, then $K_X \cdot f + f^2 = 0$, and so $K_X \cdot f = 0$. If $D = D_1, \ldots, D_r$ are the components of the fiber containing D, then f is algebraically equivalent to $\sum_i n_i D_i$ for some positive integers n_i. Thus, $\sum_i n_i(K_X \cdot D_i) = 0$, and since $K_X \cdot D_i \geq 0$ for all i, we must have $K_X \cdot D_i = 0$ for all i. Hence $K_X \cdot D = 0$. If $D^2 = 0$, then, by an argument as in the proof of Proposition 10, D is the unique component of the fiber in which it lies. In this case $p_a(D) = 1$. Otherwise, $D^2 < 0$, and so $2p_a(D) - 2 < 0$. Hence $p_a(D) = 0$, D is a smooth rational curve, and $D^2 = -2$. $\square$

The irreducible curves D of arithmetic genus 1 lying on a smooth surface are easy to classify, using Exercise 4 of Chapter 1: either D is a smooth elliptic curve or the normalization of D is a smooth rational curve and D

has exactly one singular point, which is either an ordinary double point or a cusp. Moreover, D is simply connected if and only if it has a cusp, so this case cannot correspond to a multiple fiber.

To handle the case of a reducible fiber, we define the *dual graph* Γ of the fiber as follows: if the fiber is of the form $\sum_i n_i D_i$, consider the graph Γ whose vertices v_i correspond to the components D_i. We join v_i to v_j by $m(i,j)$ edges, where $m(i,j) = D_i \cdot D_j$. There is also the lattice spanned by the D_i with intersection pairing. More generally, if Λ is a free $\mathbb{Z}$-module with a bilinear form and $v_1,\dots,v_n \in \Lambda$, we can define the dual graph Γ associated to $v_1,\dots,v_n$ in the analogous way. We then have a purely algebraic lemma, whose proof is left to the exercises:

Lemma 12. *Let Λ be a lattice spanned by $v_1,\dots,v_n$ such that $v_i^2 = -2$ for all i. Then Λ is negative semidefinite with a 1-dimensional radical and every sublattice spanned by a proper subset of the v_i is negative definite if and only if the dual graph Γ determined by the v_i is the Dynkin diagram of an extended root system $\tilde{A}_n, \tilde{D}_n, \tilde{E}_6, \tilde{E}_7, \tilde{E}_8$.* $\square$

Here the Dynkin diagrams of $\tilde{A}_n, \tilde{D}_n, \tilde{E}_6, \tilde{E}_7, \tilde{E}_8$ are depicted in Figure 3 on the next page, together with the coefficients of the primitive positive generator of the radical. (In the notation, the number of vertices is $n+1$ and $\tilde{D}_n$ is defined only if $n \geq 4$.) Note that $\tilde{E}_6 = T_{3,3,3}$, $\tilde{E}_7 = T_{2,4,4}$, and $\tilde{E}_6 = T_{2,3,6}$.

We make the following remarks in the case where Λ is the lattice spanned by the curves D_i in a reducible fiber of an elliptic fibration:

1. It follows from the description of the lattices (or by an easy direct argument) that the D_i meet transversally in at most one point and that no three pass through a single point except in the following cases: In case the lattice is $\tilde{A}_1$, there are two components D_1 and D_2 and either D_1 and D_2 meet transversally in two points or they meet at one point x and $D_1 \cdot_x D_2 = 2$ (the curves are simply tangent). In case the lattice is $\tilde{A}_2$, there are three components D_1, D_2, D_3, and each D_i meets the other two curves transversally in one point. However, it is possible for all three curves to pass through the same point, as in three concurrent lines in $\mathbb{P}^2$.

2. The graphs $\tilde{D}_n, \tilde{E}_6, \tilde{E}_7, \tilde{E}_8$ are simply connected, and so (since the components are smooth rational curves) the fibers are also simply connected in this case. The graphs $\tilde{A}_n$ are not simply connected. The fibers are also not simply connected, except in the exceptional cases where there are two components which are simply tangent at one point or three components which all pass through the same point.

3. Direct inspection of the radical for the lattices corresponding to $\tilde{A}_n, \tilde{D}_n, \tilde{E}_6, \tilde{E}_7, \tilde{E}_8$, shows that the radical is generated by a combination of the basis vectors corresponding to the vertices with at least one coefficient

equal to 1. Assume that the fiber is not multiple; for example, this is always the case if the lattice is not of type $\tilde{A}_n$. Then the fiber is of the form $\sum_i n_i D_i$ with at least one $n_i = 1$. There is always a local holomorphic section passing through such a component. Thus, if the fiber is not multiple, there always exist local holomorphic cross sections (and cross sections in the étale topology in the algebraic case). Of course, for a multiple fiber of multiplicity m, the best we can hope for is a local m-section, in other words a local holomorphic curve $\Xi \subset X$ such that $\Xi \cdot f = m$, and in fact we can always find such an m-section locally (either in the classical or étale topology).

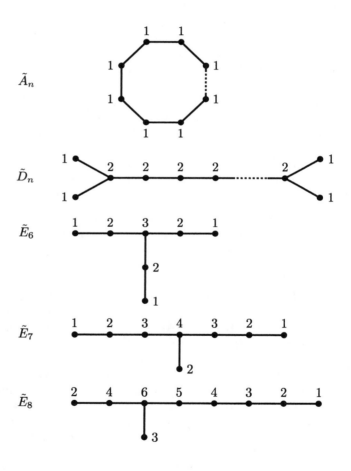

Figure 3

We conclude by giving Kodaira's notation for the types of singular fibers we have described:

$_mI_0$: A multiple fiber of multiplicity m, with smooth reduction. In case $m = 1$, we simply write I_0 for this fiber.

$_mI_n$, $n \geq 1$: A multiple fiber of multiplicity m, whose reduction is a cycle of smooth rational curves meeting transversally. Here a "cycle" of length 1 is an irreducible fiber with an ordinary double point and a cycle of length 2 consists of two smooth curves meeting transversally at two points. The corresponding Dynkin diagram is $\tilde{A}_{n-1}$. As before, in the case $m = 1$ we simply write I_n for this case.

II: An irreducible fiber with a cusp (whose normalization is then rational).

III: Two smooth rational curves meeting at one point with local intersection number two. (The dual graph is $\tilde{A}_1$.)

IV: Three smooth rational curves meeting at one point, with each pair meeting transversally. (The dual graph is $\tilde{A}_2$.)

I_n^*, $n \geq 0$: This corresponds to $\tilde{D}_{n+4}$.

II^*:   This corresponds to $\tilde{E}_8$.

III^*: This corresponds to $\tilde{E}_7$.

IV^*:   This corresponds to $\tilde{E}_6$.

We shall not show here that all of these types are realized, nor explain the reason for the notation $*$ or the local form of monodromy. General references which deal with these questions are [7], [71], [72], and [94].

The list of singular fibers is rather complicated looking. However, by a theorem of Moishezon, given an elliptic fibration $\pi \colon X \to C$, after a small perturbation of the complex structure of X and C, the only singular fibers are either (nonmultiple) rational curves with an ordinary double point or multiple fibers with smooth reduction [99], [40]. On the other hand, these singular fibers cannot disappear under a deformation of complex structure, and thus are called *stable*. Thus, in contrast to the case of ruled surfaces singular fibers play an essential role in the theory of elliptic surfaces.

Invariants and the canonical bundle formula

In this section, we consider the global situation: $\pi \colon X \to C$ is an elliptic fibration, where X and C are compact. Consider the sheaf $R^1\pi_*\mathcal{O}_X$ on C. Since $h^1(\mathcal{O}_f) = 1$ for every fiber of π, smooth or not, standard base change results imply that $R^1\pi_*\mathcal{O}_X$ is a line bundle on C. We denote the dual line bundle by L and set $d = \deg L$. We will see shortly why it is more convenient to work with L instead of its dual $R^1\pi_*\mathcal{O}_X$. The line bundle L will reappear in a slightly different guise in the next section.

One basic fact about L which we shall prove in Corollary 17 is:

Lemma 13. $\deg L \geq 0$. $\square$

In fact, if $\deg L = 0$, then L is a torsion line bundle on C of degree $1, 2, 3, 4$, or 6.

Lemma 14. *Let $d = \deg L$ and let $g = g(C)$. Then:*

(i) *If L is not trivial, then $q(X) = g$ and $p_g(X) = d + g - 1$.*
(ii) *If L is trivial, then $q(X) = g + 1$ and $p_g(X) = g$.*

In all cases we have $\chi(\mathcal{O}_X) = d$.

Proof. Apply the Leray spectral sequence to calculate $H^i(X; \mathcal{O}_X)$. By dimension reasons, if either p or q is greater than 1, then $H^p(C; R^q \pi_* \mathcal{O}_X) = 0$. It follows that the spectral sequence degenerates at the E_2 term. Thus, there is an exact sequence

$$0 \to H^1(C; \mathcal{O}_C) \to H^1(X; \mathcal{O}_X) \to H^0(C; R^1 \pi_* \mathcal{O}_X) \to 0,$$

and $H^2(X; \mathcal{O}_X) = H^1(C; R^1 \pi_* \mathcal{O}_X)$. Hence

$$q(X) = h^1(X; \mathcal{O}_X) = h^1(C; \mathcal{O}_C) + h^0(C; R^1 \pi_* \mathcal{O}_X).$$

Now $\dim H^1(C; \mathcal{O}_C) = g$. As for $H^0(C; R^1 \pi_* \mathcal{O}_X) = H^0(C; L^{-1})$, since L has nonnegative degree, $H^0(C; L^{-1}) = 0$ unless L is the trivial bundle on C, in which case $H^0(C; L^{-1})$ has dimension 1. Thus, $q(X) = g$ if L is not trivial and $q(X) = g + 1$ if L is trivial. As for $H^2(X; \mathcal{O}_X) = H^1(C; R^1 \pi_* \mathcal{O}_X)$, we must calculate $h^1(C; L^{-1})$. By Riemann-Roch on C,

$$h^1(C; L^{-1}) = h^0(C; L^{-1}) + d + g - 1.$$

Thus, $p_g(X) = h^1(C; L^{-1}) = d + g - 1$ if L is nontrivial, and $p_g(X) = g$ if L is trivial. Since $\chi(\mathcal{O}_X) = 1 - q(X) + p_g(X)$, in all cases $\chi(\mathcal{O}_X) = d$. $\square$

Note that the above proof uses the fact that $\deg L \geq 0$ to calculate $h^i(C; L^{-1})$. However, to calculate $\chi(\mathcal{O}_X)$, it is sufficient to know $\chi(C; L^{-1})$, which follows from Riemann-Roch on C. So the proof that $\chi(\mathcal{O}_X) = d$ does not depend on knowing that $d \geq 0$.

Finally, we have Kodaira's formula for the canonical bundle of an elliptic surface:

Theorem 15. *Let $\pi \colon X \to C$ be a relatively minimal elliptic fibration. Suppose that $F_1, \ldots, F_k$ are the multiple fibers of π and that the multiplicity of F_i is m_i. Then:*

$$K_X = \pi^*(K_C \otimes L) \otimes \mathcal{O}_X \left(\sum_i (m_i - 1) F_i \right).$$

Proof. First note that, for a smooth fiber f, the adjunction formula implies that $K_X|f$ is trivial. Moreover, we have seen that $K_X \cdot D = 0$ for every component of a fiber of π. Now $\pi_* K_X$ is a torsion free rank 1 sheaf on C, and thus is a line bundle λ on C. Moreover, we have a natural map $\pi^* \pi_* K_X \to K_X$, which by general base change results is an isomorphism over the smooth fibers. Thus, this map of line bundles vanishes along an effective divisor supported on the singular fibers of π. We may thus write $K_X = \pi^* \lambda \otimes \mathcal{O}_X(\sum_i a_i D_i)$, where the D_i are components of singular fibers. Moreover, $\sum_i a_i D_i \cdot D = 0$ for every component of a fiber of π. Thus, given a fiber of π, which we may write as mF for some effective divisor F such that the gcd of the coefficients of F is 1, we have that $\sum_{D_i \subseteq F} a_i D_i$ is a positive integral multiple of F. On the other hand, $\mathcal{O}_X(mF) = \pi^* \mathcal{O}_C(p)$, where $p = \pi(F)$. So after absorbing such factors into the term $\pi^* \lambda$, we see that we can write $K_X = \pi^* \lambda' \otimes \mathcal{O}_X(\sum_i a_i F_i)$, where the multiple fibers of π are of the form $m_i F_i$, and $0 \le a_i \le m_i - 1$. By adjunction, $K_X \otimes \mathcal{O}_X(F_i)|F_i$ is the trivial line bundle on F_i. Thus, $\mathcal{O}_{F_i}((a_i+1)F_i)$ is trivial. But as $\mathcal{O}_{F_i}(F_i)$ has order exactly m_i, by Lemma 5, $a_i + 1 = m_i$, so that $a_i = m_i - 1$. Thus, $K_X = \pi^* \lambda' \otimes \mathcal{O}_X(\sum_i (m_i - 1)F_i)$ for some line bundle λ' on C.

Next we must identify the line bundle λ'. Now

$$\pi_* K_X = \lambda' \otimes \pi_* \mathcal{O}_X\left(\sum_i (m_i - 1)F_i\right)$$

by the projection formula. By Exercise 2 below, $\pi_* \mathcal{O}_X(\sum_i (m_i - 1)F_i) = \mathcal{O}_C$. Thus, in fact our new λ' is equal again to $\pi_* K_X = \lambda$. To complete the identification of $\lambda = \pi_* K_X$, we need to recall a few easy facts concerning relative duality (see, for example, [7, p. 98]). First let $K_{X/C} = K_X \otimes (\pi^* K_C)^{-1}$. Thus, $K_X = K_{X/C} \otimes \pi^* K_C$, and so, again by the projection formula, $\pi_* K_X = (\pi_* K_{X/C}) \otimes K_C$. There is a canonical adjunction map $K_{X/C}|F \cong \omega_F$ (if we had not tensored by $(\pi^* K_C)^{-1}$, this would no longer be true). Now for each fiber F of π, Serre duality says that $H^1(\mathcal{O}_F)$ and $H^0(\omega_F)$ are canonically dual. Relative duality is the statement that the line bundles $R^1 \pi_* \mathcal{O}_X$ and $R^0 \pi_* K_{X/C}$ are dual to each other. Thus, $\pi_* K_{X/C} \cong L$, and so $\pi_* K_X = L \otimes K_C$. Putting this together gives the canonical bundle formula. $\square$

Corollary 16. *For a relatively minimal elliptic surface X, $K_X^2 = 0$. Thus, by Noether's formula,*

$$c_2(X) = 12\chi(\mathcal{O}_X) = 12d. \qquad \square$$

Corollary 17. $\deg L \ge 0$, *and* $\deg L = 0$ *if and only if the only singular fibers are multiple fibers with smooth reduction.*

Proof. Using the remark after the proof of Lemma 14, we see that in any case $\chi(\mathcal{O}_X) = d$. Thus, the Euler characteristic of X, which is $c_2(X)$, is equal to $12d$ and we must show that this number is nonnegative. Let $p_1, \ldots, p_k \in C$ be the points lying under singular fibers and let U be the open subset which is a union of sets of the form $U_i = \pi^{-1}(\Delta_i)$, where Δ_i is a disk in C containing p_i and no other singular point. Thus, $H_i(U_i) = H_i(\pi^{-1}(p_i))$ since U_i deformation retracts onto $\pi^{-1}(p_i)$. Let Δ'_i be a smaller disk in C centered at p_i and set $V = \pi^{-1}(C - \bigcup_i \overline{\Delta'_i})$. Then $\{U, V\}$ is an open cover of X and V and $U \cap V$ are fiber bundles with fiber a 2-torus. Thus, the Euler characteristics $\chi(V)$ and $\chi(U \cap V)$ are 0. A Mayer-Vietoris argument shows that

$$\chi(X) = \sum_i \chi(\pi^{-1}(p_i)).$$

An easy argument using the explicit description of the singular fibers shows that $\chi(\pi^{-1}(p_i)) \geq 0$, with equality if and only if $\pi^{-1}(p_i)$ has smooth reduction. (In the remaining cases we leave it as Exercise 4 to calculate $\chi(\pi^{-1}(p_i))$.) Hence $\chi(X) = c_2(X) \geq 0$, with equality if and only if the reduction of every fiber is smooth. $\square$

The method of proof of the above lemma relates the Euler characteristic of X to the singular fibers. For each fiber F, singular or not, we define $e(F)$ to be the Euler characteristic of F. Thus, if F is a smooth fiber $e(F) = 0$. Then

$$c_2(X) = \chi(X) = 12d = \sum_F e(F).$$

For example, if the only singular fibers of X are rational curves with an ordinary double point, then there are exactly $12d$ such singular fibers.

Elliptic surfaces with a section and Weierstrass models

Let $\pi \colon X \to C$ be a relatively minimal fibration with a section σ. First note that since $\sigma \cdot f = 1$, π has no multiple fibers. If F is a reducible fiber of π, then let F_0 be the unique component of F meeting σ. The components of F not meeting σ all have self-intersection -2 and span a negative definite lattice. In fact, by inspection of the possible components of F of multiplicity 1, the dual graph of this lattice is connected. Thus, the components of F not meeting σ can be contracted to rational double points, to obtain a new surface $\bar{\pi} \colon \bar{X} \to C$. Here $\bar{X}$ has just rational double point singularities and $X \to \bar{X}$ is its minimal resolution. The curve σ induces an isomorphic curve in $\bar{X}$, which we shall also denote by σ, and it is an effective Cartier divisor. In fact σ is contained in the smooth locus of $\bar{X}$ and meets each fiber in a smooth point. Since $\bar{X}$ is a normal surface, it is Cohen-Macaulay. Thus, by [61, III.10, Ex. 10.9], $\bar{\pi}$ is flat, and all fibers of $\bar{\pi}$ are reduced irreducible curves of arithmetic genus 1. We have classified such curves in Exercise 4 of Chapter 1: they are either smooth elliptic curves or curves with either an ordinary node or a cusp whose normalization is rational. All such curves are isomorphic to a plane cubic, and hence their dualizing sheaves are the trivial line bundle.

Definition 18. The surface $\bar{X}$ is the *Weierstrass model* of X.

To understand fibrations $\bar{\pi}$ with these properties, we begin by reviewing the theory of embeddings of elliptic curves in the plane. Let D be a reduced irreducible curve of arithmetic genus 1. Choose a smooth point $p \in D$. It follows from Serre duality that $h^1((D; \mathcal{O}_D(np)) = h^0(\omega_D \otimes \mathcal{O}_D(-np)) = 0$ for all $n > 0$. By Riemann-Roch, $h^0(D; \mathcal{O}_D(np)) = n$ for all $n \geq 1$. There is a natural inclusion $H^0(D; \mathcal{O}_D(n_1 p)) \subseteq H^0(D; \mathcal{O}_D(n_2 p))$ for $n_1 \leq n_2$. Under this inclusion $H^0(D; \mathcal{O}_D(p)) = \mathbb{C}$ is just the space of constant meromorphic functions on D. Moreover, there is a nonconstant meromorphic $x \in H^0(D; \mathcal{O}_D(2p))$ such that $\{1, x\}$ is a basis of $H^0(D; \mathcal{O}_D(2p))$. The function x induces a morphism $D \to \mathbb{P}^1$ which is 2–1, for which p is a branch point. Here more invariantly $\mathbb{P}^1 = \mathbb{P}(H^0(D; \mathcal{O}_D(2p))^*)$ and the morphism to $\mathbb{P}^1$ is given by the complete linear system $|2p|$. We note that the usual arguments show that $|2p|$ has no base locus at p, and the inclusion of the constants in $H^0(\mathcal{O}_D(2p))$ shows that $|2p|$ has no base locus elsewhere (including the singular point if D is singular). Now $h^0(D; \mathcal{O}_D(3p)) = 3$ and so there exists an element y of $H^0(D; \mathcal{O}_D(3p))$ having a triple pole at x. The usual arguments show that y^2 is a linear combination of $1, x, x^2, x^3, y, xy$, where the coefficient of x^3 is nonzero. First complete the square for y, by replacing y by $y + cx + d$ for unique $c, d \in \mathbb{C}$. We may then assume that $y^2 = f(x)$, where $f(x)$ is a cubic polynomial in x. Such a choice of y is unique up to a nonzero constant. Completing the cube in x, we see that after replacing x by $ax + b$ for unique $a \in \mathbb{C}^*, b \in \mathbb{C}$, we can assume that

$$(19) \qquad y^2 = 4x^3 - g_2 x - g_3.$$

This choice of x will again be unique up to a nonzero constant. The linear system defined by $3p$ embeds D in the projective space $\mathbb{P}^2 = \mathbb{P}(H^0(D; \mathcal{O}_D(3p))^*)$, and the image of D is equal to the cubic curve defined by the homogeneous polynomial associated to (19). More precisely, if $p \in D$ is a smooth point, the linear system $|3p|$ defines a degree 1 morphism from D to $\mathbb{P}^2$ whose image is a cubic curve D'. Since $p_a(D') = 1 = p_a(D)$, the morphism from D to D' is an isomorphism. We call (19) a *Weierstrass equation* for D. Conversely, every curve in $\mathbb{P}^2$ defined by the affine equation (19) is in fact a reduced irreducible curve of arithmetic genus 1. Note that $p \in D \subset \mathbb{P}^2$ is an inflection point for D, in other words the tangent line through the smooth point p meets D to order 3 at p. Moreover, $y \mapsto -y$ is an involution ι of D, fixing the singular point if D is singular, whose quotient is $\mathbb{P}^1$. Note also that y lies in the (-1)-eigenspace of the $\mathbb{Z}/2\mathbb{Z}$-action on $H^0(D; \mathcal{O}_D(3p))$ defined by ι, and it is the unique nonzero element of the (-1)-eigenspace up to a nonzero constant.

What are the possible choices in the Weierstrass equation? As we have seen, y is unique up to multiplying by a nonzero constant. If we replace y by a nonzero multiple, which for convenience we shall write as $y' = \lambda^3 y$, then we have

$$(y')^2 = \lambda^6 y^2 = 4\lambda^6 x^3 - \lambda^6 g_2 x - \lambda^6 g_3$$
$$= 4(x')^3 - \lambda^4 g_2 x' - \lambda^6 g_3,$$

where we have set $x' = \lambda^2 x$. Thus, x is replaced by $\lambda^2 x$, y by $\lambda^3 y$, g_2 by $\lambda^4 g_2$ and g_3 by $\lambda^6 g_3$. Conversely, making these changes of variable gives us a new Weierstrass equation isomorphic to the original one.

Now let us do the relative version of the above discussion. Let $\pi: \mathcal{D} \to S$ be a proper flat family such that every fiber is a reduced irreducible curve of arithmetic genus 1, and let $\Sigma \subset \mathcal{D}$ be an effective Cartier divisor which is a section of π, i.e., such that the intersection of Σ with every fiber is a reduced point, necessarily contained in the smooth locus of the fiber. Here we can take S to be, for example, a reduced scheme of finite type over $\mathbb{C}$, or a complex analytic space, but any scheme such that 2 and 3 are invertible will do. Since π is flat, standard base change arguments show that $\pi_* \mathcal{O}_D(n\Sigma)$ is a vector bundle $\mathcal{E}_n$ of rank n on S for every $n \geq 1$. Note that $\mathcal{E}_1$ is the trivial bundle, since the natural map $\mathcal{O}_S = \pi_* \mathcal{O}_D \to \mathcal{E}_1$ is an isomorphism. Moreover, there is an induced map $\mathcal{D} \to \mathbb{P}(\mathcal{E}_2^\vee)$ which realizes $\mathcal{D}$ as a double cover of the $\mathbb{P}^1$-bundle $\mathbb{P}(\mathcal{E}_2^\vee)$ over S, and there is an embedding of $\mathcal{D}$ in $\mathbb{P}(\mathcal{E}_3^\vee)$, which is a $\mathbb{P}^2$-bundle over S.

Our goal now will be to identify the bundles $\mathcal{E}_2$ and $\mathcal{E}_3$ explicitly, and to use these identifications to describe the embedding of $\mathcal{D}$ in $\mathbb{P}(\mathcal{E}_3^\vee)$. Begin by setting $\mathcal{L} = \mathcal{O}_D(-\Sigma)|\Sigma$, viewed as a line bundle on $S \cong \Sigma$. From the exact sequence

$$0 \to \mathcal{O}_D(n\Sigma) \to \mathcal{O}_D((n+1)\Sigma) \to \mathcal{O}_D((n+1)\Sigma)|\Sigma \to 0,$$

and the fact that $R^1 \pi_* \mathcal{O}_D(n\Sigma) = 0$ for all $n > 0$, we see that, for all $n > 0$, there is an exact sequence

$$0 \to \mathcal{E}_n \to \mathcal{E}_{n+1} \to \mathcal{L}^{-(n+1)} \to 0.$$

Note further that, applying $R^i \pi_*$ to the above exact sequence for $n = 0$ and using the fact that the map $\pi_* \mathcal{O}_D \to \pi_* \mathcal{O}_D(\Sigma)$ is an isomorphism, we see that

$$\mathcal{L}^{-1} = R^0 \pi_* \mathcal{O}_D(\Sigma)|\Sigma \cong R^1 \pi_* \mathcal{O}_D.$$

In particular, the line bundle $R^0 \pi_* \mathcal{O}_D(\Sigma)|\Sigma = \mathcal{L}^{-1}$ does not depend on the choice of the section Σ.

Cover S with affine open sets U_i such that $\mathcal{L}|U_i$ is trivial and such that the exact sequence $0 \to \mathcal{E}_n \to \mathcal{E}_{n+1} \to \mathcal{L}^{-(n+1)} \to 0$ splits for $n = 1, 2$.

It follows that $\Gamma(U_i, \mathcal{E}_{n+1}) = \Gamma(\pi^{-1}(U_i), \mathcal{O}_D((n+1)\Sigma))$ is a free $H^0(\mathcal{O}_{U_i})$-module for $n = 1, 2$. We may choose sections $x_i \in \Gamma(\pi^{-1}(U_i), \mathcal{O}_D(2\Sigma))$, $y_i \in \Gamma(\pi^{-1}(U_i), \mathcal{O}_D(3\Sigma))$ which restrict on each fiber D to generators of the respective sheaves at $D \cap \Sigma$. Using the case of a single curve and Nakayama's lemma, the sections $1, x_i, (x_i)^2, (x_i)^3, y_i, x_i y_i$ are a basis for $\Gamma(\pi^{-1}(U_i), \mathcal{O}_D(6\Sigma))$ over $H^0(\mathcal{O}_{U_i})$. Thus, we can write $(y_i)^2$ as a combination of these elements, and then complete the square in y_i and the cube in x_i. It follows that there is a local Weierstrass equation

$$y_i^2 = 4x_i^3 - g_{2,i}x_i - g_{3,i},$$

with $g_{2,i}, g_{3,i} \in H^0(\mathcal{O}_{U_i})$. Here the sections x_i, y_i are well defined up to multiplication by a section of $\mathcal{O}_{U_i}^*$. Now compare Weierstrass equations over $U_i \cap U_j$. There must exist invertible functions μ_{ij}, ν_{ij} on $U_i \cap U_j$ such that $x_i = \mu_{ij}x_j$ and $y_i = \nu_{ij}y_j$. Let $\lambda_{ij} = \nu_{ij}/\mu_{ij}$. Note that x_i generates $\mathcal{O}_D(2\Sigma)$ in some neighborhood of $\Sigma \cap \pi^{-1}(U_i) \subset \pi^{-1}(U_i)$, and similarly for y_i. Hence y_i/x_i is a generating section for $\mathcal{O}_D(\Sigma))$ in some neighborhood of $\Sigma \cap \pi^{-1}(U_i)$. It follows that the λ_{ij}, viewed as functions on $\pi^{-1}(U_i \cap U_j)$, are the transition functions for $\mathcal{O}_D(\Sigma))|\Sigma = \mathcal{L}^{-1}$.

Take the local Weierstrass equation $y_i^2 = 4x_i^3 - g_{2,i}x_i - g_{3,i}$ and multiply by ν_{ij}^{-2}. On $U_i \cap U_j$, we then have

$$\nu_{ij}^{-2}(4x_i^3 - g_{2,i}x_i - g_{3,i}) = \nu_{ij}^{-2}y_i^2 = y_j^2 = 4x_j^3 - g_{2,j}x_j - g_{3,j}$$
$$= 4\mu_{ij}^{-3}x_i^3 - \mu_{ij}^{-1}g_{2,j}x_i - g_{3,j}.$$

Thus, $\nu_{ij}^2 = \mu_{ij}^3$ and so, as $\lambda_{ij} = \nu_{ij}/\mu_{ij}$, we see that $\mu_{ij} = \lambda_{ij}^2$ and $\nu_{ij} = \lambda_{ij}^3$. Furthermore $g_{2,i} = \lambda_{ij}^4 g_{2,j}$ and $g_{3,i} = \lambda_{ij}^6 g_{3,j}$. Thus, the $g_{2,i}$ and $g_{3,i}$ fit together to give sections g_2, g_3 of $\mathcal{L}^4$ and $\mathcal{L}^6$, respectively. Clearly, g_2 and g_3 are unique modulo invertible elements of $\Gamma\mathcal{O}_S$ of the form λ^{-4} and λ^{-6}, respectively. Finally, using the basis $1, x_i$ for $\mathcal{E}_2$ in U_i shows that the transition functions for $\mathcal{E}_2$ can be taken to be in the form $\begin{pmatrix} 1 & 0 \\ 0 & \lambda_{ij}^2 \end{pmatrix}$.

Thus, $\mathcal{E}_2 \cong \mathcal{O}_C \oplus \mathcal{L}^{-2}$. Likewise, $\mathcal{E}_3 \cong \mathcal{O}_C \oplus \mathcal{L}^{-2} \oplus \mathcal{L}^{-3}$. It follows that $\mathbb{P}(\mathcal{E}_3^\vee) = \mathbb{P}(\mathcal{O}_C \oplus \mathcal{L}^2 \oplus \mathcal{L}^3)$. Note that, viewing x as a local generator for the factor $\mathcal{L}^3$, y as a local generator for the factor $\mathcal{L}^2$, and z as the generator for the trivial factor $\mathcal{O}_C$, the homogeneous Weierstrass equation for $\mathcal{D}$, namely $y^2z - 4x^3 - g_2xz^2 - g_3z^3$, is then a naturally defined global section of $\mathrm{Sym}^3 \mathcal{E}_3 \otimes \mathcal{L}^6$, which is the relative homogeneous coordinate ring of $\mathbb{P}(\mathcal{E}_3^\vee)$, and the vanishing of this section defines a subscheme of $\mathbb{P}(\mathcal{E}_3^\vee)$ which is isomorphic to $\mathcal{D}$.

With this preliminary discussion, we can state the following result concerning Weierstrass models of elliptic surfaces:

Theorem 20. *Let $\pi: X \to C$ be a relatively minimal elliptic fibration with a section σ. Let $L^{-1} = R^1\pi_*\mathcal{O}_X = \mathcal{O}_X(\sigma)|\sigma$ under the natural identification. Finally, let $\bar{X}$ be the Weierstrass model of X. Then there exist* sections $g_2 \in H^0(C; L^4)$ and $g_3 \in H^0(C; L^6)$ such that $\bar{X}$ is defined by the equation $y^2 z - 4x^3 - g_2 x z^2 - g_3 z^3$ inside $\mathbb{P}(\mathcal{O}_C \oplus L^2 \oplus L^3)$. The sections g_2 and g_3 are unique up to replacing g_2 by $\lambda^2 g_2$ and g_3 by $\lambda^3 g_3$ for $\lambda \in \mathbb{C}^*$. In addition, the g_i satisfy:

1. The section $\Delta = g_2^3 - 27 g_3^2$ of L^{12} is not identically 0.
2. For all $p \in C$, $\min\{3 v_p(g_2), 2 v_p(g_3)\} < 12$, where v_p denotes the order of vanishing of the corresponding section at p.

Conversely, given a line bundle L on C and two sections $g_2 \in H^0(C; L^4)$ and $g_3 \in H^0(C; L^6)$ satisfying (1) and (2) above, the surface $\bar{X}$ defined in $\mathbb{P}(\mathcal{O}_C \oplus L^2 \oplus L^3)$ by the equation $y^2 z - 4x^3 - g_2 x z^2 - g_3 z^3$ is the equation of a surface with at worst rational double points, such that the induced morphism $\bar{X} \to C$ is a flat family of irreducible curves of arithmetic genus 1, and the minimal resolution X is a relatively minimal elliptic fibration with a section. Moreover, every section σ of X satisfies $\sigma^2 = -\deg L$.

Proof. We shall give a proof of Theorem 20 modulo some basic facts about rational double points. First suppose that X is an elliptic surface with a section, and apply the preceding discussion to the family $\bar{X} \to C$. Here $\mathcal{D} = \bar{X}$, $S = C$, and $\Sigma = \sigma$. We see that $\bar{X}$ is indeed defined by a Weierstrass equation. Condition (1) in the theorem is equivalent to saying that the general fiber of $\bar{X}$, or equivalently X, is smooth. Condition (2) is exactly the condition that $\bar{X}$ has rational double points. We will not prove this here, but will simply refer to [67] or [7, III.7 and II.8] for the necessary results on singularities of double covers and simple curve singularities.

Conversely, given a line bundle L on C and two sections $g_2 \in H^0(C; L^4)$ and $g_3 \in H^0(C; L^6)$ satisfying (1) and (2), running the above argument backward constructs a surface $\bar{X}$ with at worst rational double points. Note that $\bar{X}$ has a section σ not passing through the singular points of $\bar{X}$ or of any fiber of $\bar{X}$, by looking at the point at infinity on each fiber (this is the divisor defined locally by $x = z = 0, y = 1$). Let X be the minimal resolution of $\bar{X}$. The section σ then induces a section of X. To see that X is relatively minimal, it suffices to show that the canonical bundle K_X is π-nef. Let $\bar{F}$ be a fiber of $\bar{X}$ and let F be the corresponding fiber of X. We may write $F = F' + \sum_i n_i C_i$, where F' is the proper transform of $\bar{F}$ and the C_i are the components of the resolution of the rational double point singularity. Thus, each C_i is smooth rational with $C_i^2 = -2$, so that $K_X \cdot C_i = 0$. As $K_X \cdot F = 0$, we must have $K_X \cdot F' = 0$, and K_X is π-nef. Finally, note that $L^{-1} = \mathcal{O}_\sigma(\sigma)$ and hence $\sigma^2 = -\deg L$. $\square$

To explain Condition (2) further, note that if there exists a point $p \in C$ such that $\min\{3 v_p(g_2), 2 v_p(g_3)\} \geq 12$, then $v_p(g_2) \geq 4$ and $v_p(g_3) \geq 6$. Thus, g_2 and g_3 induce sections g_2', g_3' of $(L')^4$ and $(L')^6$, respectively, where $L' = L \otimes \mathcal{O}_C(-p)$, and we can make a birational model of the Weierstrass equation which is "better" in some sense by using the sections g_2', g_3'.

Corollary 21. *In the above notation, $\deg L \geq 0$. If $\deg L = 0$, then L has order either $1, 2, 3, 4$, or 6. Finally, L is trivial if and only if X is a product $E \times C$.*

Proof. Since $\Delta = g_2^3 - 27g_3^2$ is not 0, at least one of g_2 and g_3 is nonzero. Thus, either L^4 or L^6 has a nonzero section. It follows that $\deg L \geq 0$, and that if $\deg L = 0$, then either L^4 or L^6 is trivial. Hence the order of L divides either 4 or 6, and so the order of L is as claimed. If L is trivial, then g_2 and g_3 are constant, $\mathbb{P}(\mathcal{O}_C \oplus L^2 \oplus L^3) = \mathbb{P}^2 \times C$, and $X = \bar{X}$ is a product. Conversely, if X is a product surface, then $L^{-1} = R^1\pi_*\mathcal{O}_X = H^1(E; \mathcal{O}_E) \otimes \mathcal{O}_C$, by the Künneth formula, and thus $L^{-1} = \mathcal{O}_C$ and L is trivial. To see this last statement directly, note that we have shown that $R^1\pi_*\mathcal{O}_X = \mathcal{O}_\sigma(\sigma)$, regardless of the choice of the section σ. For a product surface $E \times C$, we can choose the section $\sigma = \{p\} \times C$, and then it is clear that $\mathcal{O}_\sigma(\sigma)$ is trivial. $\square$

One can relate the various types of the singular fibers to the orders of vanishing of g_2, g_3, and Δ at a point p; see, for example, [93].

More general elliptic surfaces

In this section, we wish to describe without proof how to construct all elliptic surfaces starting from an elliptic surface with a section. This procedure is not as elementary as the methods we have used in the preceding sections, and requires either some knowledge of étale cohomology or (over $\mathbb{C}$) working with complex analytic surfaces which are not necessarily algebraic. For a discussion of the first method, see [19] and for the second see [40, Chapter 1].

We begin with the algebraic method. Given $\pi: X \to C$ an elliptic surface over C, let $k = k(C)$ be the function field of C. If we let $\eta = \operatorname{Spec} k$ be the generic point of C, then by base change the generic fiber X_η of X over η is a smooth curve of genus 1 over the field k. Quite generally, let k be an arbitrary field with algebraic closure $\bar{k}$ and let E be a smooth curve of genus 1 defined over k. Then the Jacobian $J(E)$ of divisors of degree 0 on E is again a smooth curve defined over k, $J(E)$ has a k-rational point (the trivial divisor), and there is a k-morphism (given by translation) $\varphi: J(E) \times_k E \to E$ which realizes E as a principal homogeneous space over $J(E)$. Here we may identify E itself with the divisors of degree 1 on E and φ is then the usual sum of divisors. If there is a k-rational point $P \in E$, then $x \mapsto \varphi(x, P) = x + P$ defines a k-isomorphism from $J(E)$ to E, and conversely if E is k-isomorphic to $J(E)$, then it clearly has a k-rational point.

The general mechanism for classifying principal homogeneous spaces E over $J(E)$ is as follows. If E is such a space, then there is a finite extension of k, say K, over which E has a K-rational point P. We may assume that K is Galois over k (in positive characteristic, this amounts to showing that we can assume that K is separable over k). Given an element ρ of the Galois group $G_{K/k}$ of K over k, define $a_\rho = \rho(P) - P$. Thus, a_ρ is a divisor of degree 0 defined over K, in other words an element of $J(E)(K)$, the set of K-rational points of $J(E)$. It is easy to check that a_ρ is a 1-cocycle for $G_{K/k}$ with values in $J(E)(K)$. Changing P to some other point of $J(E)(K)$ replaces a_ρ by a 1-coboundary. Thus, there is a well-defined element of the group cohomology $H^1(G_{K/k}; J(E)(K))$ associated to E, and one can show via descent theory that the principal homogeneous spaces E over $J(E)$ which have a K-rational point are classified by $H^1(G_{K/k}; J(E)(K))$. To deal with all possible extensions K at once, we consider the profinite group $G = \mathrm{Gal}(\bar{k}/k)$, where $\bar{k}$ is an algebraic closure of k (in the case of positive characteristic we would use instead the separable closure) and consider the group cohomology (in the sense of profinite groups) $H^1(G, J(E)(\bar{k}))$, which we shall just denote by $H^1(G, J(E))$.

Returning to the case of $\pi\colon X \to C$, let X_η be the generic fiber and let $J(X_\eta)$ be its Jacobian. We can complete $J(X_\eta)$ to some smooth elliptic surface over C, which has a unique relatively minimal model. We denote this relatively minimal model by $J(X)$ and call it the *Jacobian elliptic surface* associated to X. Clearly, X and $J(X)$ are isomorphic if and only if X has a section. Our goal is to classify those elliptic surfaces X having a fixed Jacobian elliptic surface $B \to C$, using the theory of the previous section to take Jacobian elliptic surfaces as essentially known. Thus, we must compute $H^1(G, B)$ in the above notation.

Now given $p \in C$, we can consider the completion of the local ring $\mathcal{O}_{C,p}$. Call this complete local ring R_p and let its function field be k_p. There are inclusions $\mathrm{Spec}\, k_p \to \mathrm{Spec}\, R_p \to C$, and thus we can consider the restriction of X to $\mathrm{Spec}\, R_p$ and to $\mathrm{Spec}\, k_p$. Here we should think of the restriction to $\mathrm{Spec}\, R_p$ as corresponding to the analytic restriction of X to a small tubular neighborhood of the fiber at p, and the restriction to $\mathrm{Spec}\, k_p$ as corresponding to the tubular neighborhood minus the singular fiber. Hensel's lemma implies that if there is a component of the fiber F_p at p of multiplicity 1 (which is equivalent in the elliptic surface case to saying that F_p is not multiple) then there is a section of the morphism $X \times_C \mathrm{Spec}\, R_p \to \mathrm{Spec}\, R_p$. Now $X \times_C \mathrm{Spec}\, k_p$ is a curve of genus 1 over k_p, and it is a principal homogeneous space over $B \times_C \mathrm{Spec}\, k_p = B_p$. We may then define $H^1(G_p, B_p)$ by using the Galois group G_p of the local field k_p, and there is a homomorphism

$$H^1(G, B) \to \bigoplus_{p \in C} H^1(G_p, B_p),$$

since G_p is the decomposition group of the point p and thus can be identified with a subgroup of G. If ξ is an element of $H^1(G, B)$ corresponding to an elliptic surface $X \to C$ with Jacobian surface B, for each $p \in C$, we let ξ_p

denote the image of ξ in $H^1(G_p, B_p)$. Thus, $\xi_p = 0$ if and only if the curve $X \times_C \operatorname{Spec} k_p$ has a section, which by the above discussion is equivalent to saying that the fiber F_p of X over p is not multiple. This implies that $H^1(G_p, B_p) = 0$ if the fiber of B at p is not either smooth or of Type I_n, $n \geq 1$. More precisely, one can show that (at least in characteristic 0)

$$H^1(G_p, B_p) = \begin{cases} (\mathbb{Q}/\mathbb{Z})^2, & \text{if the fiber is smooth,} \\ \mathbb{Q}/\mathbb{Z}, & \text{if the fiber is of Type } I_n, n \geq 1, \\ 0, & \text{otherwise.} \end{cases}$$

In all cases we can identify ξ_p with an element of finite order in the Jacobian of E_p, the fiber of B over p. It turns out that the order of ξ_p in $H^1(G_p, B_p)$ is exactly equal to the multiplicity of the corresponding multiple fiber of X, and that ξ_p is a certain local invariant of the multiple fiber. The meaning of ξ_p in the analytic case is described below.

Next we may ask when the map $H^1(G, B) \to \bigoplus_{p \in C} H^1(G_p, B_p)$ is surjective. At least over the complex numbers, the answer is as follows: if B is not isomorphic to a product elliptic surface $E \times C$, then the map is always surjective. If B isomorphic to a product elliptic surface $E \times C$, then we can identify all of the fibers E_p with the fixed elliptic curve E, and there is an algebraic elliptic surface with given local invariants ξ_p if and only if $\sum_{p \in C} \xi_p = 0$.

Lastly we may ask about the structure of the kernel of the map $H^1(G, B) \to \bigoplus_{p \in C} H^1(G_p, B_p)$. This kernel is called the *Tate-Shafarevich* group of B and can be analyzed further via Galois cohomology or étale cohomology. See, for example, [132] or [19].

Now let us redo the above discussion in the complex analytic category. Let $\pi \colon X \to C$ be a holomorphic map from the smooth compact complex surface X to the smooth curve C, such that the general fiber of π is a smooth elliptic curve. It is a straightforward argument in complex surface theory that X is algebraic if and only if there exists a holomorphic multisection of π, in other words an irreducible curve $\Xi \subset X$ such that $\Xi \cdot f > 0$, where f is a general fiber of π (or equivalently such that $\pi(\Xi) = C$). The general theory of fibrations and relatively minimal models works equally well in the complex analytic case, and we may therefore assume that π is relatively minimal (no exceptional curves in the fibers). In this case the classification of singular fibers is as in the algebraic case. Our first task is to define the Jacobian elliptic surface $J(X)$ in the complex analytic case. To do so, we introduce three basic invariants of X:

1. The *j-function* $j \colon C \to \mathbb{P}^1$, defined by $j(t) = j(\pi^{-1}(t))$, the j-invariant of the smooth elliptic curve $\pi^{-1}(t)$, whenever $t \in C$ is such that $\pi^{-1}(t)$ is constant. It turns out that j (which is defined on the complement of the points of C lying under the singular fibers) has an extension to a meromorphic function $C \to \mathbb{C}$, or equivalently to a holomorphic function $C \to \mathbb{P}^1$.

2. The *homological invariant* G, defined as follows: let $U \subseteq C$ be the open subset of points lying under smooth fibers. Let π_U denote the restriction of π to $\pi^{-1}(U)$ and let $i: U \to C$ be the inclusion. Then we set G to be the sheaf $i_* R^1(\pi_U)_* \mathbb{Z}$. The restriction $G|U$ is a locally constant sheaf of rank two $\mathbb{Z}$-modules on U. In fact $G = R^1\pi_*\mathbb{Z}$ if there are no multiple fibers, but, if there are multiple fibers, then there is a finite discrepancy between G and $R^1\pi_*\mathbb{Z}$ at the multiple fibers.
3. The line bundle L on C such that $L^{-1} = R^1\pi_*\mathcal{O}_X$.

There are various relations among the above invariants. For example, let $U_0 \subseteq U$ be the open set consisting of smooth fibers whose j-invariants are $\neq 0, 1728$, and assume for simplicity that $U_0 \neq \emptyset$. Then j maps U_0 to $\mathfrak{H}_0/PSL(2,\mathbb{Z})$, where $\mathfrak{H}_0$ is the upper half plane $\mathfrak{H}$, minus the set $PSL(2,\mathbb{Z}) \cdot \{0, 1728\}$, and $PSL(2,\mathbb{Z})$ acts freely on $\mathfrak{H}_0$. Thus, by the theory of covering spaces there is a homomorphism $\pi_1(U_0) \to PSL(2,\mathbb{Z})$, well defined up to conjugation. On the other hand, it is easy to see that the locally constant sheaf $R^1(\pi_U)_*\mathbb{Z}$ has monodromy contained in $SL(2,\mathbb{Z})$, so that it is essentially equivalent to a representation $\pi_1(U) \to SL(2,\mathbb{Z})$. The compatibility between G and j is then that, after conjugation, the homomorphism $\pi_1(U_0) \to PSL(2,\mathbb{Z})$ defined by j is equal to the composition $\pi_1(U_0) \to \pi_1(U) \to SL(2,\mathbb{Z}) \to PSL(2,\mathbb{Z})$. In general, for any curve C, we may consider pairs (j, G) consisting of a holomorphic function $j: C \to \mathbb{P}^1$ and a sheaf G on C of the form $i_* G_0$, where U_0 is the complement of a discrete set of points of C contained in $j^{-1}(\mathbb{P}^1 - \{0, 1728, \infty\}$, and G_0 is a locally constant sheaf on U_0 with fibers $\cong \mathbb{Z}^2$ associated to a representation $\pi_1(U_0) \to SL(2,\mathbb{Z})$. Such a pair (j, G) will be called *compatible* if the homomorphism $\pi_1(U_0) \to PSL(2,\mathbb{Z})$ defined by j is equal up to conjugation to the composition $\pi_1(U_0) \to \pi_1(U) \to SL(2,\mathbb{Z}) \to PSL(2,\mathbb{Z})$. We then have the following theorem of Kodaira [71]:

Proposition 22. *Let C be a curve. Given a compatible pair (j, G), there exists an elliptic surface $\varphi: B \to C$ with a section and with associated invariants j and G, and B is unique up to isomorphism of elliptic surfaces. Moreover, if (j, G) are also associated to the elliptic surface $\pi: X \to C$, then the line bundles $(R^1\pi_*\mathcal{O}_X)^{-1}$ and $(R^1\varphi_*\mathcal{O}_B)^{-1}$ are isomorphic.* $\square$

In the above situation, starting with the elliptic surface X, we call the unique elliptic surface B with a section and with the same invariants (j, G) as X the *Jacobian elliptic surface* $J(X)$ of X. (In Kodaira's terminology B is called the *basic elliptic surface* associated to X.) This definition of $J(X)$ coincides with the usual definition of the Jacobian surface in case X is algebraic.

The next step is to classify those complex elliptic surfaces which have the same Jacobian elliptic surface B and which do not have multiple fibers. Equivalently, these are elliptic surfaces X with $J(X) \cong B$ such that local sections exist above every point $p \in C$. Using the proof of Proposition 22, which is essentially a local argument, one shows the following: the set of pairs (X, ψ), where X is an elliptic surface without multiple fibers and $\psi \colon J(X) \to B$ is an isomorphism of elliptic surfaces from $J(X)$ to B, modulo addition by a section in B, is classified by the sheaf cohomology group $H^1(C; \mathcal{B})$, where $\mathcal{B}$ is the sheaf of abelian groups given by the group of local holomorphic cross-sections of $B \to C$. This group $H^1(C; \mathcal{B})$ is the analytic version of the Tate-Shafarevich group. It can be further analyzed as follows: fix once and for all a holomorphic section Σ of B. Then if Σ' is another section over an open subset U of C, the divisor $\Sigma' - \Sigma$ defines a line bundle $\mathcal{O}_{\pi^{-1}(U)}(\Sigma' - \Sigma)$ on $\pi^{-1}(U)$, of degree 0 on each smooth fiber, and thus it defines an element of $H^1(\pi^{-1}(U); \mathcal{O}^*_{\pi^{-1}(U)})$. Conversely, given a line bundle $\mathcal{L}$ on $\pi^{-1}(U)$ such that $\mathcal{L}$ has degree 0 on each smooth fiber, the line bundle $\mathcal{L} \otimes \mathcal{O}_{\pi^{-1}(U)}(\Sigma)$ has degree 1 on each smooth fiber and so $R^0\pi_* \left(\mathcal{L} \otimes \mathcal{O}_{\pi^{-1}(U)}(\Sigma) \right)$ is a line bundle on U. Moreover, the natural map $R^0\pi_* \left(\mathcal{L} \otimes \mathcal{O}_{\pi^{-1}(U)}(\Sigma) \right) \to \mathcal{L} \otimes \mathcal{O}_{\pi^{-1}(U)}(\Sigma)$ vanishes along a section Σ', and possibly also along the union of some components of reducible fibers. Thus, Σ' can be recovered from $\mathcal{L}$. More precisely, there is a split exact sequence of sheaves

$$0 \to \mathcal{B} \to R^1\pi_*\mathcal{O}^*_B/\mathcal{S} \to \mathbb{Z} \to 0,$$

where $\mathcal{S}$ is the skyscraper subsheaf of $R^1\pi_*\mathcal{O}^*_B$ generated by line bundles of the form $\mathcal{O}_B(D)$, where D is a reducible component of a fiber of π (note that $\mathcal{O}_B(f)$ defines the trivial element of $R^1\pi_*\mathcal{O}^*_B$), $\mathbb{Z}$ is the constant sheaf on C, the map $R^1\pi_*\mathcal{O}^*_B/\mathcal{S} \to \mathbb{Z}$ is given by taking the degree of a line bundle on the general fiber, and the splitting is given by $n \in \mathbb{Z} \mapsto \mathcal{O}_B(n\Sigma)$. From this, the study of $H^1(C; \mathcal{B})$ is essentially reduced to the study of $H^1(C; R^1\pi_*\mathcal{O}^*_B)$. Moreover, this group can be analyzed in detail by applying $R^i\pi_*$ to the exponential exact sequence

$$0 \to \mathbb{Z} \to \mathcal{O}_B \to \mathcal{O}^*_B \to 0$$

and using the Leray spectral sequence. The upshot is that, if π has a singular fiber, then the set of elliptic surfaces classified by $H^1(C; \mathcal{B})$ are all deformation equivalent. For more details see [40] (see also [68] for the case where the base is $\mathbb{P}^1$).

Finally, we must add multiple fibers to complex analytic surfaces. To do so complex analytically, there is the method of logarithmic transformations introduced by Kodaira. For simplicity we shall just consider the case of a multiple fiber with smooth reduction. Let F be such a fiber, of multiplicity m over $p \in C$, and let $\Delta \subset C$ be a small analytic disk centered at p such that $\pi^{-1}(\Delta)$ has F as the only singular fiber. Let z be a coordinate on Δ and make the base change $\tilde{\Delta} \to \Delta$ defined by $z = w^m$. In the following discussion, we shall replace X by $\pi^{-1}(\Delta)$, so that we will assume that X fibers over Δ. Arguing as in the proof of Lemma 5, we see that, if $\tilde{X}$ is the normalization of the fiber product of X with $\tilde{\Delta}$ with coordinate

w, then $\tilde{X}$ is smooth and the map $\tilde{X} \to X$ is a cyclic covering space of order m. Let T be the generator of the covering group corresponding to the automorphism $w \mapsto e^{2\pi i/m}w$ of $\tilde{\Delta}$. The central fiber $\tilde{F}$ of $\tilde{X}$ is a smooth curve, necessarily connected, and the restriction of T to $\tilde{F}$ is a fixed point free morphism with quotient F. Since F is an elliptic curve, $\tilde{F}$ is an elliptic curve and $T|\tilde{F}$ is given by addition by an element $\xi_p \in \tilde{F}$. One can show that this element ξ_p may be identified with the local invariant ξ_p introduced above in the discussion of the algebraic classification of elliptic surfaces. Kodaira's theory then classifies all such group actions T on $\tilde{X}$ and shows how to glue these into elliptic surfaces without multiple fibers to obtain (complex analytic) surfaces with prescribed multiple fibers. Once we have this construction in the complex analytic category, we can then go back and try to determine when such surfaces are actually algebraic. For details on this, we refer to [40] (see also Exercise 11).

The fundamental group

In this final section, we shall briefy describe some of the "classical" topology of an elliptic surface X. The most important invariant is the fundamental group. Before we can describe the fundamental group of an elliptic surface, we need to make the following definition:

Definition 23. Let C be a compact Riemann surface, and let $p_1, \ldots, p_n$ be a set of n distinct points on C. Suppose that for each i we are given an integer $m_i \geq 2$. We call the data of C, the p_i, and the m_i a 2-*orbifold*. We define the *orbifold fundamental group* $\pi_1^{\mathrm{orb}}(C, *)$ as follows: the fundamental group of $C - \{p_1, \ldots, p_n\}$ is generated by the usual generators $\alpha_i, \beta_i, 1 \leq i \leq g$, where $g = g(C)$, together with additional generators $\gamma_1, \ldots, \gamma_n$ corresponding to loops enclosing each p_i simply, not enclosing any $p_j, j \neq i$, and which are homotopic to zero rel $*$ on C. There is also the relation $[\alpha_1, \beta_1] \cdots [\alpha_g, \beta_g]\gamma_1 \cdots \gamma_n = 1$, where the α_i, β_i are the standard generators of $\pi_1(C, *)$ and $[\alpha_i, \beta_i]$ is the commutator of α_i and β_i. Define $\pi_1^{\mathrm{orb}}(C, *)$ to be the quotient of $\pi_1(C - \{p_1, \ldots, p_n\}, *)$ by the smallest normal subgroup containing $\gamma_i^{m_i}$ for all i. Thus, $\pi_1^{\mathrm{orb}}(C, *)$ is freely generated by the elements $\alpha_1, \beta_1, \ldots, \alpha_g, \beta_g, \gamma_1, \ldots, \gamma_n$, subject to the relations $[\alpha_1, \beta_1] \cdots [\alpha_g, \beta_g]\gamma_1 \cdots \gamma_n = 1$ and $\gamma_i^{m_i} = 1$ for all i.

In general, the orbifold fundamental group is quite large. For example, one can show that $\pi_1^{\mathrm{orb}}(C, *)$ determines $g(C)$ and the m_i unless C has genus 0 and there are at most two points p_i, i.e., $n \leq 2$. In case $g(C) = 0$ and there are exactly two multiple fibers of multiplicities m_1 and m_2, $\pi_1^{\mathrm{orb}}(C, *)$ is cyclic of order $\gcd(m_1, m_2)$. In particular it is trivial if and only if m_1 and m_2 are relatively prime. (In case $g(C) = 0$ and $n = 1$, $\pi_1^{\mathrm{orb}}(C, *)$ is always trivial.)

The connection with elliptic surfaces is the following. Let $\pi\colon X \to C$ be an elliptic surface. Then X defines a 2-orbifold structure on C as follows: the points p_i are the points lying under multiple fibers (not necessarily with smooth reduction) and the integers m_i are the corresponding multiplicities. With this said, there is the following result due to Kodaira [74], Moishezon [99], and Dolgachev [22]:

Theorem 24. *Suppose that* $\pi\colon X \to C$ *is a relatively minimal elliptic surface such that* π *has at least one fiber with singular reduction (or equivalently* X *does not have Euler number 0). Then* $\pi_1(X, *) \cong \pi_1^{\mathrm{orb}}(C, *)$. *In particular,* X *is simply connected if and only if* $C \cong \mathbb{P}^1$, *there are at most two multiple fibers, and, if there are two multiple fibers, then their multiplicities are relatively prime.* $\square$

Let us use this to discuss the homotopy type of a simply connected elliptic surface X, always assumed to be relatively minimal. As a 4-manifold, X is then determined up to homotopy type, or equivalently homeomorphism, by the intersection form on $H_2(X; \mathbb{Z})$. Let $p_g = p_g(X)$. Since $c_2(X) = 12(1 + p_g)$, we have

$$b_2(X) = 10 + 12p_g.$$

Moreover, $b_2^+(X) = 2p_g + 1$, by the Hodge index theorem. Thus, we know the rank and signature of $H_2(X; \mathbb{Z})$, and it remains to determine whether the intersection form is even or odd (i.e., of Type II or Type I). The intersection form is even if and only if K_X is divisible by 2, by the Wu formula. Now, by Lemma 14, L is a line bundle over $\mathbb{P}^1$ of degree $d = 1 + p_g$. Thus, since $K_{\mathbb{P}^1} = \mathcal{O}_{\mathbb{P}^1}(-2)$, the canonical bundle formula for an elliptic surface over $\mathbb{P}^1$ with multiple fibers F_1 and F_2 of multiplicities m_1 and m_2 gives

$$K_X = \mathcal{O}_X((p_g - 1)f + (m_1 - 1)F_1 + (m_2 - 1)F_2).$$

Since m_1 and m_2 are relatively prime, there exists a linear combination $am_1 + bm_2 = 1$ with $a, b \in \mathbb{Z}$. Taking $\kappa = bF_1 + aF_2$, we see that

$$m_1\kappa = bm_1F_1 + am_1F_2 = bf + (1 - bm_2)F_2$$
$$= bf + F_2 - bf = F_2.$$

Thus, $m_1\kappa = F_2$, and similarly $m_2\kappa = F_1, m_1m_2\kappa = f$. We also have, by [40, Chap. 2, Prop. 2.7]:

Proposition 25. *Suppose that* $\pi\colon X \to C$ *is a relatively minimal elliptic surface such that* π *has at least one fiber with singular reduction, and let* $m_1, \ldots, m_k$ *be the multiplicities of the multiple fibers. Let* m *be the least common multiple of the* m_i. *Then there exists a class* $x \in H_2(X; \mathbb{Z})$ *such that* $x \cdot f = m$. *Thus, if the* m_i *are pairwise relatively prime, then* m *is the exact order of divisibility of* f *in* $H_2(X; \mathbb{Z})$. $\square$

Applying the above proposition, we see that the class κ described above is a primitive class, in other words it is not the multiple mx of a class $x \in H_2(X; \mathbb{Z})$ with $m > 1$. Thus,

$$K_X = (m_1 m_2(p_g - 1) + m_2(m_1 - 1) + m_1(m_2 - 1))\kappa$$
$$= (m_1 m_2(p_g + 1) - m_1 - m_2)\kappa.$$

If p_g is even,

$$m_1 m_2(p_g + 1) - m_1 - m_2 \equiv m_1 m_2 - m_1 - m_2 = (m_1 - 1)(m_2 - 1) - 1 \bmod 2.$$

Since m_1 and m_2 are relatively prime, at least one of them is always odd, so that $(m_1 - 1)(m_2 - 1)$ is always even. It follows that $m_1 m_2(p_g + 1) - m_1 - m_2 \equiv 1 \bmod 2$ if p_g is even, so that the intersection form on X is always odd. If p_g is odd,

$$m_1 m_2(p_g + 1) - m_1 - m_2 \equiv -m_1 - m_2 \equiv m_1 + m_2 \bmod 2.$$

Thus, if p_g is odd, the intersection form on X is even if and only if $m_1 + m_2 \equiv 0 \bmod 2$.

Summarizing, then:

Proposition 26. *Suppose that X is a minimal simply connected elliptic surface, with two multiple fibers of multiplicities $m_1 \leq m_2$. Then the intersection form on X is of Type I if p_g is even or if p_g is odd and $m_1 + m_2 \equiv 0 \bmod 2$, and is of Type II otherwise. Similar statements hold if X has less than two multiple fibers.* $\square$

Let us conclude by discussing the possible torsion in $H^2(X; \mathbb{Z})$:

Proposition 27. *Suppose that $\pi \colon X \to C$ is a relatively minimal elliptic surface such that π has at least one fiber with singular reduction, and let $m_1, \ldots, m_k$ be the multiplicities of the multiple fibers. Then the torsion subgroup of $H^2(X; \mathbb{Z})$ is isomorphic to the quotient of $\bigoplus_{i=1}^{k} \mathbb{Z}/m_i\mathbb{Z}$ by the image of $\mathbb{Z}$ embedded diagonally, or in other words by the image of the subgroup generated by $(1, \ldots, 1)$.*

Proof. Since the torsion subgroup of $H^2(X; \mathbb{Z})$ is isomorphic to the torsion subgroup of $H_1(X; \mathbb{Z})$, the proof follows easily by working out the abelianization of $\pi_1^{\mathrm{orb}}(C, *)$. $\square$

For example, if the m_i are all pairwise relatively prime, then $H^2(X; \mathbb{Z})$ is torsion free. At the other extreme, if $k = 2$ and $m_1 = m_2 = m$, then the torsion subgroup of $H^2(X; \mathbb{Z})$ is isomorphic to $\mathbb{Z}/m\mathbb{Z}$.

Exercises

1. Generalize the proof of Corollary 6 as follows: recall (Exercise 13 of Chapter 1) that an effective divisor $D = \sum_i n_i C_i$ is numerically connected if, whenever $D = D_1 + D_2$ with each D_i effective and nonzero,

then $D_1 \cdot D_2 \geq 1$. For example, if D is nef and big, then it follows by Exercise 14 of Chapter 1 that D is numerically connected. Show that, if e is the primitive effective divisor supported in a fiber of a fibration which generates the radical of intersection theory, then e is numerically connected. (If $e = D_1 + D_2$, then

$$0 = e \cdot D_1 = (D_1)^2 + (D_1 \cdot D_2),$$

and by assumption $(D_1)^2 < 0$.) What about the fundamental cycle Z_0 of the resolution of an isolated surface singularity? (I don't know the answer to this question.)

Show that, if $D = \sum_i n_i C_i$ is numerically connected and L is a line bundle on D such that $\deg L|C_i \leq 0$, then $h^0(D; L) \leq 1$, and $h^0(D; L) = 1$ if and only if L is trivial (Ramanujam's lemma). In particular $h^0(\mathcal{O}_D) = 1$. (First show that, if C_i is an arbitrary component of D, then there exists a sequence $Z_1 = C_i, Z_2, \ldots, Z_n = D$ with $Z_{i+1} = Z_i + C_j$ for some j such that $Z_i \cdot C_j > 0$, and then argue as in the proof of Corollary 6.)

2. Let $\pi \colon X \to C$ be a fibration, and let F_i be the multiple fibers of π. Suppose that a_i are integers such that $0 \leq a_i \leq m_i - 1$, and that D is a divisor on C. Show that $\pi_*(\pi^*\mathcal{O}_C(D) \otimes \mathcal{O}_X(\sum_i a_i F_i)) = \mathcal{O}_C(D)$. Use this to determine $H^0(X; \mathcal{O}_X(\pi^*D + \sum_i a_i F_i))$.

3. Prove Lemma 12. (Suppose that every sublattice of Λ spanned by a proper subset of the vertices of the dual graph is negative definite. If there is a cycle, show that we are in case $\tilde{A}_n$. If there are no cycles but the dual graph is not a $T_{p,q,r}$ graph, show that we are in case $\tilde{D}_n$. Otherwise, the dual graph is a $T_{p,q,r}$ graph such that the lattices of type $T_{p-1,q,r}, T_{p,q-1,r}$, and $T_{p,q,r-1}$ are all negative definite. Show that, in this case, the only possible choices for (p,q,r) up to permutation are $(3,3,3),(2,4,4),(2,3,6)$.)

4. Let F be a singular fiber of an elliptic fibration. Define the invariant $e(F) = \chi(F)$. Calculate $e(F)$ for the various singular fibers of an elliptic fibration.

5. Generalize the above exercise and Corollary 17 to genus g fibrations: let $\pi \colon X \to C$ be a fibration with general fiber of genus g. Then $\chi(X) = (2-2g)\chi(C) + \sum_F d(F)$, where $d(F) = \chi(F) - 2 + 2g = \chi(F) - 2\chi(\mathcal{O}_F)$. Show $d(F) \geq 0$. (By Corollary 6, $\chi(\mathcal{O}_F) = 1 - g$. First show that $p_a(F) \geq p_a(F_{\text{red}}) = h^1(F_{\text{red}})$, where F_{red} is the reduction of F, and thus $-2\chi(\mathcal{O}_F) \geq -2\chi(\mathcal{O}_{F_{\text{red}}})$. Thus, we may assume that F is reduced. Now $\dim_{\mathbb{C}} H^2(F; \mathbb{C})$ is the number of components of F, which is ≥ 1, so it suffices to show that $2h^1(\mathcal{O}_F) \geq \dim H^1(F; \mathbb{C})$. Do this by considering the normalization $\nu \colon \tilde{F} \to F$ and the commutative

diagram

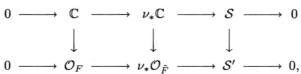

showing that the map $\mathcal{S} \to \mathcal{S}'$ is an injective map of skyscraper sheaves.) Show finally that $d(F) = 0$ implies that F_{red} is smooth.

6. Suppose that $\mathcal{L} \subset |\mathcal{O}_{\mathbb{P}^2}(3)|$ is a pencil of plane cubics containing a smooth member. Show that the resolution of indeterminacy of $\mathcal{L}$ defines an elliptic fibration $\pi\colon X \to \mathbb{P}^1$, where X is the blowup of nine points of $\mathbb{P}^2$, possibly infinitely near. The map π has a section. The general such pencil has 12 singular nodal fibers (see also Exercise 10 of Chapter 5). Without using infinitely near base points, show that we can arrange singular fibers of Types I_n ($n = 1, 2, 3$), II, III, and IV. Using infinitely near base points, show that we can arrange fibers of type $\tilde{D}_4$ (take the pencil defined by a double line plus a reduced line, together with a smooth cubic meeting each line transversally), $\tilde{E}_6$ (a triple line together with a smooth cubic meeting it transversally at three points), $\tilde{E}_7$ (a triple line together with a smooth cubic such that the line is tangent to the cubic at one point), and $\tilde{E}_8$ (a triple line together with a smooth cubic such that the line meets the cubic at an inflection point).

7. The numerical equivalence class of K_X: Show that K_X is numerically equivalent to rf for some $r \in \mathbb{Q}$. Show $r = 0$ if and only if either $g = 1$, $d = 0$, and $m_i = 0$ for all i (X is a complex torus if L is trivial and is a hyperelliptic surface otherwise), $g = 0$, $d = 2$, $m_i = 0$ for all i (X is a $K3$ surface), $g = 0$, $d = 1$, and $m_1 = m_2 = 2$, no other multiple fibers (X is an Enriques surface), or $g = 0$, $d = 0$, and (up to permutation) $(m_1, m_2, \ldots) = (2, 2, 2, 2)$, $(2, 4, 4)$, $(2, 3, 6)$, $(3, 3, 3)$ (X is a hyperelliptic surface). Likewise, $r < 0$ if and only if $g = 0, d = 1$ and there is at most one multiple fiber, or $g = 0$, $d = 0$, there are less than two multiple fibers of arbitrary multiplicity or there are three multiple fibers and the possible multiplicities up to permutation are $(2, 2, m)$, m arbitrary, $(2, 3, 4)$, $(2, 3, 5)$. In all other cases $r \in \mathbb{Q}^+$. (Not all of these cases can arise for an algebraic surface; see Exercise 11.)

8. If, in the notation of Exercise 6, $r = 0$, or equivalently K_X is numerically trivial, then there is a finite unramified cover where K_X is trivial. Thus, K_X is a torsion line bundle, and in fact the order of K_X is either $1, 2, 3, 4$, or 6. In all cases, $12K_X = 0$.

9. An algebraic surface which is an elliptic surface in two different ways satisfies: K_X is numerically equivalent to 0.

10. Let E be an elliptic curve with an automorphism ϕ of order $d = 2, 3, 4, 6$ (here automorphism is understood as automorphism of the group structure, so that $\phi(0) = 0$). Let C be a smooth curve and $\tilde{C}$ a

finite unramified cyclic cover of order d, with $\tau\colon \tilde{C} \to \tilde{C}$ a generator of the automorphisms of $\tilde{C}$ over C. Define the surface X to be the quotient of $E \times \tilde{C}$ by the automorphism group generated by $(e, x) \mapsto (\phi(e), \tau(x))$. Show that X is an elliptic surface over C, and that all fibers of X are smooth elliptic curves isomorphic to C. Show that the sections of $X \to C$ correspond to morphisms $f\colon \tilde{C} \to E$ such that $f(\tau(x)) = \phi(f(x))$.

11. Let L be a line bundle on the curve $\tilde{C}$ such that $L^2 = \mathcal{O}_C$ but L is not trivial. Let $g_2 \in H^0(L^4) \cong \mathbb{C}$ and $g_3 \in H^0(L^6) \cong \mathbb{C}$ be two sections such that the curve with Weierstrass model $y^2 = 4x^3 - g_2 x - g_3$ is a smooth elliptic curve E. (Here we use a fixed trivialization of L^2 to identify L^4 and L^6 with $\mathcal{O}_C$.) Let X be the corresponding Weierstrass model. Show that X is an elliptic surface over C with all fibers smooth, such that the line bundle $(R^1\pi_*\mathcal{O}_X)^{-1}$ is equal to L. If $\tilde{C} \to C$ is the unramified double cover of C corresponding to L, show that the pullback of X to $\tilde{C}$ is the product elliptic surface $E \times \tilde{C}$, and that X is the quotient of $E \times \tilde{C}$ by the involution $(e, x) \mapsto (-e, \tau(x))$, where τ is the involution on $\tilde{C}$ corresponding to the double cover $\tilde{C} \to C$. Do the same for the case where L is a line bundle on C of order $3, 4, 6$ and E is an elliptic curve with an automorphism of order $3, 4, 6$. (Here you will need to use the fact that such a curve always has either g_2 or g_3 equal to 0 in order to define the appropriate Weierstrass model.) Using this, show that, for the elliptic surfaces X constructed in the previous problem, the line bundle L is always nontrivial.

12. In the algebraic case there is a condition on the multiple fibers if L is trivial. Suppose that $\pi\colon X \to C$ is an algebraic surface with L trivial; thus all fibers have smooth reduction. Since X is algebraic, there is a holomorphic multisection Σ of π. Taking the normalization of $X \times_C \Sigma$, there is a finite cover $\pi'\colon X' \to C'$ of $X \to C$ such that X' has no multiple fibers and has a section. We may assume that $C' \to C$ is Galois. By flat base change the line bundle $L' = (R^1(\pi')_*\mathcal{O}_{X'})^{-1}$ is the pullback of L and is therefore trivial. Thus, $X' = E \times C'$ for some smooth elliptic curve E and X is the quotient $(E \times C')/G$, for some finite group of automorphisms G acting faithfully on C' with $C'/G = C$. Show that, after adjusting C', we may assume that G acts faithfully on E. Next, using the previous two problems, show that the assumption that L is trivial forces the image of G to be contained in the translation subgroup of E. Thus, G is a finite subgroup of the translation group E, and is therefore abelian. The multiple fibers of X arise from fixed points for the action of G on C'. For a point $p_i \in C$ which is a ramification point of order m_i for $C' \to C$, a small loop γ_i enclosing p_i simply counterclockwise lifts to a unique element of G (since G is abelian) and thus defines an element ξ_{p_i} of E, corresponding to the local invariant of the multiple fiber. Show that $\sum_i \xi_{p_i} = 0$ in E; this is a formal consequence of the fact that the abelianization of the

fundamental group of $C - \{p_1, \ldots, p_k\}$ is

$$H_1(C;\mathbb{Z}) \oplus \left(\bigoplus_i \mathbb{Z}\gamma_i \right) \Big/ \mathbb{Z}(\gamma_1 + \cdots + \gamma_k).$$

For example, if L is trivial X cannot have just one multiple fiber of multiplicity m, or several multiple fibers whose multiplicities are pairwise relatively prime.

Show by a direct construction that the necessary condition $\sum_i \xi_{p_i} = 0$ is also sufficient. Working a little harder, show that if L has degree 0 but is not trivial, then we can arrange arbitrary preassigned multiplicities for X (this amounts to showing that, if τ is an automorphism of E of order $d = 2, 3, 4, 6$, then $\sum_{i=0}^{d-1} \tau^i(\xi) = 0$ for all $\xi \in E$.) Less trivially, if X has a fiber whose reduction is singular, or equivalently X has positive Euler number, or equivalently $\deg L > 0$, then it follows from work of Shafarevich and Ogg that we can arrange arbitrary multiplicities for the multiple fibers. Likewise, in the complex analytic category, by using logarithmic transforms we can arrange arbitrary multiplicities even when L is trivial.

13. Let $\tilde{C} \to C$ be a double cover branched at points $p_1, \ldots, p_{2k}$ and with corresponding involution $\tau \colon \tilde{C} \to \tilde{C}$. Let E be an elliptic curve. Then the surface X which is the minimal resolution of the quotient of $E \times C$ by the involution $(e, t) \mapsto (-e, \tau(t))$ is an elliptic surface with constant j-invariant and with $2k$ singular fibers of type D_4. Show that the line bundle L satisfies $L^2 = \mathcal{O}_C(p_1 + \cdots + p_{2k})$ and corresponds to the line bundle defining the double cover, and that the sections $g_2 \in H^0(C; \mathcal{O}_C(2p_1 + \cdots + 2p_{2k}))$ and $g_3 \in H^0(C; \mathcal{O}_C(3p_1 + \cdots + 3p_{2k}))$ are constant sections. Consider a cyclic cover $\tilde{C} \to C$ of order 3, 4, or 6, such that τ is a generator for the group of covering transformations and take for E an elliptic curve with the appropriate complex multiplication, denoted by $e \mapsto \mu(e)$. Let X be the minimal resolution of the quotient of $E \times \tilde{C}$ by the group generated by $(e, t) \mapsto (\mu(e), \tau(t))$. Show that at a point of total ramification the singular fiber of $X \to C$ is of type $\tilde{E}_6$ in case the order of μ is 3, $\tilde{E}_7$ in case the order of μ is 4, and $\tilde{E}_8$ in case the order of μ is 6.

14. Suppose that $\pi \colon X \to C$ has a section σ. Using adjunction for σ give an easier proof of the canonical bundle formula in this case.

15. Suppose that X is an elliptic surface with $g = 0$ and $d = 1$ and either no multiple fibers or a single multiple fiber of multiplicity m. Show that K_X is either $-f$ or $-F$, where mF is the multiple fiber. Note that X is a good generic surface if and only if there are no reducible fibers. Show that on every smooth blowdown $\bar{X}$ of X there exists an effective smooth divisor in $|-K_{\bar{X}}|$. Show that if X blows down at all to a rational ruled surface $\mathbb{F}_k$, then it blows down to $\mathbb{P}^2$, and that in this case either f or F is the proper transform of a smooth cubic. Discuss when each case happens. (See Exercise 9 in Chapter 5.)

16. Let X be an elliptic surface with a section. Show that there is a morphism $X \to \mathbb{P}(\mathcal{E}_2^{\vee})$ which realizes X as a double cover. In case X is an elliptic surface over $\mathbb{P}^1$ and $X = \bar{X}$, show that X is a double cover of $\mathbb{F}_{2k}$ with $k = p_g + 1$. Show further that the branch locus of the map $X \to \mathbb{F}_{2k}$ is of the form $\sigma + B'$, where B is a smooth curve in $|3\sigma + 6kf|$ disjoint from σ.

8

Vector Bundles over Elliptic Surfaces

This chapter gives a variety of techniques for understanding stable rank 2 vector bundles V over elliptic surfaces X. Because of constraints of space and time, we do not give complete proofs of all of the results, and this chapter is mainly intended as a sampler of various methods for studying bundles over X. For simplicity, we shall assume that X is simply connected, so that the base curve is $\mathbb{P}^1$, and shall concentrate on the case where X has a section σ and the fibers are generic, i.e., smooth or nodal. After describing allowable elementary modifications for singular fibers, we begin with the technically much simpler case where the vector bundle V has degree 1 on every fiber. In this case, there is a unique stable bundle on each fiber, and correspondingly there is a unique stable rank 2 vector bundle V_0 on X, up to twisting by the pullback of a line bundle on the base curve, which restricts to a stable bundle on each fiber. The bundle V_0 then generates all stable bundles which have degree 1 on every fiber, via elementary modifications. A second approach to the moduli space is via sub-line bundles and extensions. We give a brief description of Donaldson invariants and some methods for computing them, and apply this to calculate the 2-dimensional Donaldson invariants coming from stable bundles of degree 1 on every fiber. Next we turn to the case where the degree of V on every fiber is 0. Here the moduli space of rank 2 semistable bundles of degree 0 on an elliptic curve is a $\mathbb{P}^1$, and so the geometry of the moduli space when the fiber degree is 0 is much more complicated. We have tried to outline most of the important ideas in this case. Finally, we describe the general shape of the Donaldson invariants and how this shape both reflects the geometry of the moduli space and determines the smooth classification of elliptic surfaces.

Stable bundles on singular curves

We begin with more discussion of the program outlined in Chapter 6 of making allowable elementary modifications. Here we will allow singular fibers as well. Let C be an irreducible curve on the smooth surface X. We will be primarily interested in the case where C has only ordinary double point singularities, but will try to do most of the arguments in general. First we define torsion free rank 1 sheaves L on C:

Definition 1. Let C be a reduced curve. A coherent sheaf L is a *torsion free rank 1 sheaf* on C if L has no sections supported at a point and if the restriction of L to the smooth points of C is a line bundle. Equivalently, L is a torsion free rank 1 sheaf if there exists a finite set S on C such that $L|C - S$ is a line bundle and the map $L \to i_* i^* L$ is an inclusion, where $i\colon C - S \to C$ is the natural map.

For example, if $n\colon \tilde{C} \to C$ is the normalization map, then $n_* \mathcal{O}_{\tilde{C}}$ is a torsion free rank 1 sheaf, as is any rank 1 subsheaf of a locally free sheaf. If C is an irreducible curve with a single ordinary double point, then in fact every torsion free rank 1 sheaf on C is either a line bundle or of the form $n_* \tilde{L}$, where $\tilde{L}$ is a line bundle on the normalization $\tilde{C}$ (Exercise 6). Next we define the degree of a torsion free rank 1 sheaf L. Note that if L is a line bundle, then by the Riemann-Roch theorem for a singular curve we have $\deg L = \chi(L) + p_a(C) - 1$. So we simply make this a definition:

Definition 2. Let L be a torsion free rank 1 sheaf on C. Then define $\deg L$ to be the integer $\chi(L) + p_a(C) - 1$.

Lemma 3.

(i) If $0 \to L' \to L \to \tau \to 0$ is an exact sequence of sheaves on C, where L' and L are torsion free rank 1 sheaves on C and τ is supported on a finite number of points, then $\deg L = \deg L' + \ell(\tau)$. Thus, $\deg L' \leq \deg L$, with equality holding if and only if $L' = L$;

(ii) If L' and L are torsion free rank 1 sheaves on C and $\mathrm{Hom}(L', L) \neq 0$, then $\deg L' \leq \deg L$ with equality only if $L = L'$;

(iii) If V is a rank 2 vector bundle and

$$0 \to L' \to V \to L \to 0$$

is exact, where L' and L are torsion free rank 1 sheaves, then $\deg L' + \deg L = \deg V$.

Proof. We leave the proofs of these statements as exercises. □

Thus, for example, $\deg n_* \mathcal{O}_{\tilde{C}} = \delta$, where $n\colon \tilde{C} \to C$ is the normalization and δ is the genus drop, defined in Chapter 1. If $L = \mathfrak{m}_x$ is the ideal of a point on C (not necessarily a smooth point), then $\deg \mathfrak{m}_x = -1$.

Definition 4. Let V be a rank 2 vector bundle on C. Then V is *stable* if, for every torsion free rank 1 subsheaf L of V, $\deg L < \deg V/2$. Semistable and unstable are defined similarly.

Using (iii) and (i) above, we see that V is stable if and only if every torsion free rank 1 quotient M of V satisfies $\deg M > \deg V/2$.

Lemma 5. *If V is a stable bundle on C, then V is simple.*

Proof. Let $\varphi\colon V \to V$ be an endomorphism. As in the case of a smooth variety, either $\det \varphi \neq 0$ and φ is an isomorphism or $\det \varphi$ is identically 0. In this case, assuming that $\varphi \neq 0$, then φ has generic rank 1 and so the image M of φ is a torsion free rank 1 sheaf. But simultaneously $\deg M > \deg V/2$ and $\deg M < \deg V/2$, a contradiction. Thus, every nonzero map from V to itself is an isomorphism, so it follows as in Corollary 8 of Chapter 4 that V is simple. $\square$

Next we consider the situation where V is a rank 2 vector bundle on the smooth surface X. As in Chapter 5, we assume that $V|C$ is unstable for some irreducible curve C on X. Let L be a torsion free rank 1 quotient of $V|C$ with $\deg L < \deg(V|C)/2$. It is easy to see (Exercise 11) that there is a unique such L, although we will not need to use this fact. We then define V' by the elementary modification

$$0 \to V' \to V \to j_*L \to 0,$$

where $j\colon C \to X$ is the inclusion. We then have the following analogue of Lemma 16 in Chapter 2 in the case where L is not necessarily a line bundle.

Proposition 6. *In the above situation, V' is again a vector bundle and*

$$p_1(\operatorname{ad} V') = p_1(\operatorname{ad} V) + 2(\deg(V|C) - 2\deg L) + C^2.$$

Thus, if $\deg L < \deg(V|C)/2$ and $C^2 \geq 0$, then $p_1(\operatorname{ad} V') > p_1(\operatorname{ad} V)$.

Proof. To see that V' is locally free, it suffices by Theorem 17 of Chapter 2 to show that $Ext^i(V', \mathcal{O}_X) = 0$ for all $i > 0$. Using the defining exact sequence for V' and the long exact Ext sequence, we see that it suffices to show that $Ext^i(j_*L, \mathcal{O}_X) = 0$ for all $i > 1$. This follows in general from the observation that $\operatorname{depth} j_*L \geq 1$, by the assumption that L is torsion free, and the Auslander-Buchsbaum dimension formula [139], [87] $\operatorname{proj.dim} j_*L + \operatorname{depth} j_*L = \dim X = 2$, where $\operatorname{proj.dim} j_*L$ is the minimal length of a projective resolution of j_*L, and is equal to the largest i such that $Ext^i(j_*L, \mathcal{O}_X) \neq 0$. Thus, $\operatorname{proj.dim} j_*L = 1$ and so $Ext^i(j_*L, \mathcal{O}_X) = 0$ for all $i > 1$. In case C has just ordinary double points, it is easy to see directly that $\operatorname{proj.dim} j_*L = 1$. In fact, in this case if we set $R = \mathbb{C}\{x, y\}$

and use Exercise 6, j_*L corresponds either to the R-module R/xyR or to $R/xR \oplus R/yR$. In the first case, R/xyR has the short free resolution

$$0 \to R \to R \to R/xyR \to 0,$$

where the first map is multiplication by xy, and, in the second case, there is the resolution

$$0 \to R^2 \to R^2 \to R/xR \oplus R/yR \to 0,$$

where the map $R^2 \to R^2$ corresponds to the matrix $\begin{pmatrix} x & 0 \\ 0 & y \end{pmatrix}$. Thus, in both cases j_*L has projective dimension 1.

Next we must verify the formula for $p_1(\operatorname{ad} V')$. Away from the singularities of C, we have $c_1(V') = c_1(V) - [C]$. Thus, a similar formula holds over all of X. To find $c_2(V')$ we can use Riemann-Roch on X, applied to the formula $\chi(V') = \chi(V) - \chi(L)$. A brief calculation with Riemann-Roch and the adjunction formula gives $c_2(V') = c_2(V) - c_1(V) \cdot C + \deg L$. Thus, $p_1(\operatorname{ad} V') - p_1(\operatorname{ad} V) = 2(\deg(V|C) - 2 \deg L) + C^2$, as claimed. $\square$

Stable bundles of odd fiber degree over elliptic surfaces

For the rest of this chapter, unless otherwise noted, we shall let $\pi\colon X \to \mathbb{P}^1$ denote a regular elliptic surface over $\mathbb{P}^1$ with a section σ (although much of what follows will work under more general circumstances). We shall also assume that all fibers of π are irreducible, with at worst ordinary double points. Let $p_g = p_g(X)$. By the canonical bundle formula $K_X = \mathcal{O}_X((p_g - 1)f)$, where f is a general fiber of π. Thus, $K_X \cdot \sigma = p_g - 1$, so that

$$\sigma^2 = -2 - K_X \cdot \sigma = -p_g - 1.$$

In this section, we shall consider rank 2 vector bundles V on X such that $\deg V|f = 1$, or more precisely such that $\det V = \sigma + kf$. Here since we will make allowable elementary modifications along the fibers we will not try to specify k in advance, since an allowable elementary modification will replace $\det V$ by $\det V - f$. Doing two such modifications replaces $\det V$ by $\det V - 2f = \det(V \otimes \mathcal{O}_X(-f))$, and so it is natural to allow ourselves to identify two bundles which differ by a twist of the form $\mathcal{O}_X(nf)$. Note that $(\sigma + kf)^2 = \sigma^2 + 2k \equiv -p_g - 1 \mod 2$ and $3\chi(\mathcal{O}_X) = 3(p_g + 1) \equiv -p_g - 1 \mod 2$ as well. Thus, since $p_1(\operatorname{ad} V) \equiv \Delta^2 \mod 4$, we see that $-p_1(\operatorname{ad} V) - 3\chi(\mathcal{O}_X) \equiv 0 \mod 2$ for our choices. In particular the expected dimension of the moduli space is always an even integer $2t$.

Since $\det(V|f) = \mathcal{O}_f(p)$, where $p = \sigma \cdot f$, it follows by Theorem 9 of Chapter 4 that there is a unique stable bundle on f with determinant equal to $\mathcal{O}_f(p)$. A similar result holds for the singular fibers:

Lemma 7. *Let C be an irreducible nodal curve of arithmetic genus 1, and let p be a smooth point of C. Then there is a unique stable rank 2 bundle W on C with $\det W = \mathcal{O}_f(p)$.*

Proof. If W is a stable rank 2 bundle on C with $\deg W = 1$, then by the Riemann-Roch theorem there is a section of W, and thus a nonzero map $\mathcal{O}_f \to W$. By stability the cokernel must be torsion free, and thus it is a torsion free rank 1 sheaf L with $\deg L = 1$. A local calculation (Exercise 7) shows that, if $S = \mathbb{C}\{x,y\}/(xy)$, then there are no nontrivial extensions of $\tilde{S}$ by S. It follows that, in the above situation, since W is locally free, L is locally free as well. Thus, L is a line bundle of degree 1 on C, and so by Riemann-Roch $L = \mathcal{O}_f(p)$ for some smooth point p, where $\mathcal{O}_f(p) = \det W$. Hence V is given as an extension of $\mathcal{O}_f(p)$ by $\mathcal{O}_f$. The extension cannot be split if W is stable. Now $H^1(\mathcal{O}_f(-p))$ has dimension 1, so that there is a unique nonsplit extension of $\mathcal{O}_f(p)$ by $\mathcal{O}_f$. An easy argument along the lines of the smooth case (Theorem 9 in Chapter 4) shows that this extension is stable. Thus, there exists a unique stable W with $\det W = \mathcal{O}_f(p)$, as claimed.   □

Now we can begin the analysis of stable bundles on X. Let $\Delta = \sigma + kf$ and let w be the mod 2 reduction of Δ. For an integer p, we consider (w,p)-suitable ample divisors H, and shall always understand stability to mean with respect to such a divisor. By Theorem 5 of Chapter 6 and the fact that there are no strictly semistable bundles on the fibers, a rank 2 vector bundle V is H-stable if and only if its restriction to some fiber f is stable, if and only if its restriction to almost all fibers f is stable. Moreover, the proof of Theorem 5 in Chapter 6 shows that, in case V is not stable, a destabilizing sub-line bundle of V is destabilizing on every fiber. Note finally that there are no strictly semistable bundles in this case.

Our first result says that every H-stable bundle with $\det V = \sigma + kf$ is good.

Lemma 8. *If V is H-stable for a suitable ample divisor H, then V is good. Hence the moduli space corresponding to V is always smooth of dimension equal to $-p_1(\operatorname{ad} V) - 3\chi(\mathcal{O}_X)$.*

Proof. Suppose that $\varphi \in H^0(X; \operatorname{ad} V \otimes K_X)$. Then φ defines a map from V to $V \otimes K_X$ of trace 0. Restricting to a fiber f, and using $K_X|f = \mathcal{O}_f$, $\varphi|f : V|f \to V|f$ is a map of trace 0. Since $V|f$ is simple for almost all f, $\varphi|f = 0$ for almost all f. But then $\varphi = 0$. Thus, $H^0(X; \operatorname{ad} V \otimes K_X) = 0$, so that V is good.   □

Next let us show that there is, up to a twist, a single rank 2 vector bundle V_0 on X such that $V_0|f$ is stable for *every* fiber f. This bundle will turn out to be the ancestor of all stable rank 2 bundles on X.

Proposition 9. *With notation as above, there exists a rank 2 vector bundle V_0 on X such that $V_0|f$ is stable for every fiber f. Moreover, we can construct V_0 as follows: start with an arbitrary rank 2 vector bundle V on X such that there exists a fiber f with $V|f$ stable, and successively perform allowable elementary modifications on V. Then this procedure terminates with a V_0 as desired. Finally, V_0 is unique in the following sense: if V_0' also has the property that $\det V_0' = \sigma + k'f$ and $V_0'|f$ is stable for all f, then $V_0' = V_0 \otimes \mathcal{O}_X(nf)$ for some n.*

Proof. By Lemma 8, there is an absolute bound $p_1(\mathrm{ad}\, V) \leq -3\chi(\mathcal{O}_X)$ for every bundle V such that $V|f$ is stable for almost all f. Thus, if we begin with a fixed V and make allowable elementary modifications, then since this procedure increases p_1 (by Proposition 6) it must terminate. Clearly, the only way this can happen is when we reach a V_0 for which no more allowable elementary modifications are possible, i.e., $V_0|f$ is stable for every fiber f. Thus, it suffices to find a V such that V_f is stable for almost all f, or equivalently for one f.

Fix a fiber f, and let $p = \sigma \cap f$. The unique stable bundle W on f with $\det W = \mathcal{O}_f(p)$ is given as an extension of $\mathcal{O}_f(p)$ by $\mathcal{O}_f$. This suggests that we try to construct V as an extension of $\mathcal{O}_X(\sigma)$ by $\mathcal{O}_X$, such that the restriction of V to f is not split. While this may not always be possible, we are free to modify σ by adding Nf for some unspecified N. The extensions of $\mathcal{O}_X(\sigma + Nf)$ by $\mathcal{O}_X$ are classified by $H^1(X; \mathcal{O}_X(-\sigma - Nf))$ and the extensions of $\mathcal{O}_f(p)$ by $\mathcal{O}_f$ are classified by $H^1(f; \mathcal{O}_f(-p))$. Moreover, the restriction of an extension of $\mathcal{O}_X(\sigma + Nf)$ by $\mathcal{O}_X$ to f corresponds to looking at the restriction map $H^1(X; \mathcal{O}_X(-\sigma - Nf)) \to H^1(f; \mathcal{O}_f(-p))$ which arises from the short exact sequence

$$0 \to \mathcal{O}_X(-\sigma - (N+1)f) \to \mathcal{O}_X(-\sigma - Nf) \to \mathcal{O}_f(-p) \to 0.$$

Thus, the cokernel of the map on the H^1's lies in $H^2(X; \mathcal{O}_X(-\sigma - (N+1)f))$. By Serre duality, $H^2(X; \mathcal{O}_X(-\sigma - (N+1)f))$ is dual to $H^0(X; \mathcal{O}_X(\sigma + (N+1)f + (p_g - 1)f))$. Since f, σ, and p_g are fixed, this group is zero for all $N \ll 0$ (since, for example, we can arrange that the divisor $\sigma + (N+1)f + (p_g-1)f$ has negative intersection with a given ample divisor). Thus, for a suitable N we can find an extension of $\mathcal{O}_X(\sigma + Nf)$ by $\mathcal{O}_X$ such that the induced extension on f is not split, and is therefore stable.

Finally, we must establish the uniqueness property of V_0. If V_0' is another bundle on X with the property that $V_0'|f$ is stable and $\det V_0'|f = \det V_0|f = \mathcal{O}_f(p)$ for all f, then $\dim \mathrm{Hom}(V_0|f, V_0'|f) = 1$ for every f, and every nonzero element of $\mathrm{Hom}(V_0|f, V_0'|f)$ is an isomorphism. Thus, $R^0\pi_* \mathrm{Hom}(V_0, V_0') = L$ is a line bundle on $\mathbb{P}^1$. Moreover, there is a natural section of $R^0\pi_* \mathrm{Hom}(V_0, V_0') \otimes L^{-1} = \mathcal{O}_{\mathbb{P}^1}$, and thus there is a natural section of $\mathrm{Hom}(V_0, V_0' \otimes \pi^* L^{-1})$. The corresponding homomorphism $V_0 \to V_0' \otimes \pi^* L^{-1}$ is an isomorphism on every fiber and thus an isomorphism. Thus, up to a twist by $\pi^* L^{-1} = \mathcal{O}_X(nf)$ for some n, V_0 and V_0' are isomorphic. □

Corollary 10. For $p = -3\chi(\mathcal{O}_X)$ and w such that $w^2 \equiv p \mod 4$, the moduli space of H-stable rank 2 vector bundles V with the invariants w and p consists of a single reduced point. □

As we shall see, Corollary 10 computes a Donaldson invariant: it implies that the 0-dimensional Donaldson invariant corresponding to the choices of w and p above is always 1. We will describe how to compute more interesting invariants below.

A Zariski open subset of the moduli space

Using Proposition 9, we can give a description of a Zariski open subset of the moduli space of H-stable rank 2 bundles V with $p_1(\mathrm{ad}\, V) = -3\chi(\mathcal{O}_X) - 2t$ for a nonnegative integer t. As with ruled surfaces, we first describe the generic behavior of a stable bundle V with given invariants w and p. The bundles on an elliptic curve f with determinant $\mathcal{O}_f(p)$ which are the "least" unstable are those of the form $\mathcal{O}_f(q) \oplus \mathcal{O}_f(p-q)$ for some point $p \in f$, and the maximal destabilizing quotient is $\mathcal{O}_f(p-q)$. Reversing this procedure, we should start with the stable bundle $\mathcal{F}_p$ on f which is the nontrivial extension of $\mathcal{O}_f(p)$ by $\mathcal{O}_f$ and look for surjections $\mathcal{F}_p \to \mathcal{O}_f(q)$.

Lemma 11. For an elliptic curve f and a point $q \in f$, there exists a surjection $\mathcal{F}_p \to \mathcal{O}_f(q)$, and it is unique mod scalars. Moreover, $H^1(f; \mathrm{Hom}(\mathcal{F}_p, \mathcal{O}_f(q))) = 0$. A similar result holds if instead f is an irreducible nodal curve.

Proof. We shall just write out the proof in the case where f is smooth; the case of a singular f is similar. There is an exact sequence

$$0 \to \mathcal{O}_f(q-p) \to \mathrm{Hom}(\mathcal{F}_p, \mathcal{O}_f(q)) \to \mathcal{O}_f(q) \to 0.$$

If $q \neq p$, then $H^0(f; \mathcal{O}_f(q-p)) = H^1(f; \mathcal{O}_f(q-p)) = 0$ and $H^1(f; \mathcal{O}_f(q)) = 0$. Thus, $H^1(f; \mathrm{Hom}(\mathcal{F}_p, \mathcal{O}_f(q))) = 0$ and the unique nonzero section of $H^0(f; \mathcal{O}_f(q))$ mod scalars lifts to give a map $\mathcal{F}_p \to \mathcal{O}_f(q)$. If this map is not surjective, then its image lies in $\mathcal{O}_f(q - \mathbf{d})$, where $\mathbf{d}$ is an effective nonzero divisor. Thus, $\deg(q - \mathbf{d}) \leq 0$, contradicting the stability of $\mathcal{F}_p$. So the map $\mathcal{F}_p \to \mathcal{O}_f(q)$ is surjective. This concludes the proof if $q \neq p$.

In case $q = p$, we are given a surjection $\mathcal{F}_p \to \mathcal{O}_f(p)$ and need to show that it is unique mod scalars. If there were two linearly independent maps from $\mathcal{F}_p$ to $\mathcal{O}_f(p)$, then some linear combination of them would have to vanish at any given point. But as we observed above, by stability every map from $\mathcal{F}_p$ to a line bundle of degree 1 is surjective. Thus,

$\dim \operatorname{Hom}(\mathcal{F}_p, \mathcal{O}_f(p)) = 1$, and then it follows from Riemann-Roch or from the exact cohomology sequence associated to

$$0 \to \mathcal{O}_f \to \operatorname{Hom}(\mathcal{F}_p, \mathcal{O}_f(p)) \to \mathcal{O}_f(p) \to 0$$

that $H^1(f; \operatorname{Hom}(\mathcal{F}_p, \mathcal{O}_f(p))) = 0$. $\square$

Now choose points $x_1, \ldots, x_t \in \mathbb{P}^1$ lying under distinct smooth fibers $f_1, \ldots, f_t$, and choose a line bundle of degree 1 on each f_i. Since a line bundle of degree 1 on f_i can be written in the form $\mathcal{O}_{f_i}(q_i)$ for a unique point $q_i \in f_i$, the choice of the t fibers and t line bundles amounts to a choice of t points $q_1, \ldots, q_t \in X$ such that for different i and j, q_i and q_j lie in different fibers. In this way we find a Zariski open subset of $\operatorname{Sym}^t X$. Let $j_i \colon f_i \to X$ be the inclusion. By Lemma 11, $\operatorname{Hom}(V_0, j_{i*}\mathcal{O}_{f_i}(q_i)) = \mathbb{C}$ and up to a nonzero scalar (corresponding to $\operatorname{Aut} \mathcal{O}_{f_i}(q_i) = \mathbb{C}^*$) there is a unique surjection from V_0 to $\mathcal{O}_{f_i}(q_i)$. Define V by the exact sequence

$$0 \to V \to V_0 \to \bigoplus_i j_{i*}\mathcal{O}_{f_i}(q_i) \to 0.$$

It follows that V is uniquely specified by the choice of points q_i. We see that V is obtained from V_0 by elementary modifications which are dual to allowable elementary modifications, as we would expect.

We still must check that this construction gives a dense open subset of the moduli space, or in other words that the bundles obtained with a repeated choice of fiber, or using destabilizing sub-line bundles on the fiber of higher degree, or using singular fibers, form a subset of the moduli space of dimension less that $-p - 3\chi(\mathcal{O}_X) = 2t$. The argument follows along lines similar to those outlined in Chapter 6 for ruled surfaces, although the details are more complicated (see [36] for details). Summarizing, we have:

Theorem 12. *There exists a Zariski open and dense subset of the moduli space of stable bundles V on X with $c_1(V) = \sigma + kf$ and $p_1(\operatorname{ad} V) = -3\chi(\mathcal{O}_X) - 2t$ which consists of t points in X lying in smooth distinct fibers. Thus, there is a rational fibration $\mathfrak{M} \dashrightarrow \operatorname{Sym}^t \mathbb{P}^1 = \mathbb{P}^t$, and the fibers are a product of t elliptic curves.* $\square$

Once again, we see the structure of X reflected in the birational structure of the moduli space. Note that, instead of describing $\mathfrak{M}$ as a fibration, we could give another description as follows: the choice of t points lying in t distinct fibers is the same as choosing t general points on X. Thus, the moduli space is birational to the Hilbert scheme $\operatorname{Hilb}^t X$.

An overview of Donaldson invariants

We will give a very brief discussion of Donaldson invariants for a smooth oriented 4-manifold M in this section, and discuss how to compute them under special circumstances in case M is an algebraic surface. General references are [26], [27], and [40].

Let X be an algebraic surface and let H an ample divisor on X. Let $\mathfrak{M}$ be the moduli space of H-stable rank 2 vector bundles with invariants (w, p), for a fixed choice of w and p. For our purposes, we will need to make the following very special assumptions:

(i) Every bundle V corresponding to a point of $\mathfrak{M}$ is good. Thus, $\mathfrak{M}$ is smooth of dimension $-p - 3\chi(\mathcal{O}_X) = d$.

(ii) The space $\mathfrak{M}$ is compact.

(iii) There exists a universal vector bundle $\mathcal{V} \to X \times \mathfrak{M}$.

Note that the bundle $\mathcal{V}$ is only well defined up to twisting by a line bundle of the form $\pi_2^* L$, where L is an arbitrary line bundle on $\mathfrak{M}$. However, the vector bundle ad $\mathcal{V}$ is well defined, independent of the twisting. In fact, we can define ad $\mathcal{V}$ even if $\mathcal{V}$ is not defined. To do so, one can show that, following the discussion in Chapter 6, there is always a universal $\mathbb{P}^1$-bundle $\pi \colon P \to X \times \mathfrak{M}$, which equals $\mathbb{P}(\mathcal{V})$ in case there is a universal bundle $\mathcal{V}$. Thus, we can define the rank 3 bundle $\pi_* T_{P/X \times \mathfrak{M}}$, where $T_{P/X \times \mathfrak{M}}$ is the relative tangent sheaf, and this rank 3 bundle equals ad $\mathcal{V}$ in case $\mathcal{V}$ exists. In any case, we have the class $p_1(\text{ad}\,\mathcal{V}) \in H^4(X \times \mathfrak{M})$. Here of course $p_1(\text{ad}\,\mathcal{V}) = -c_2(\text{ad}\,\mathcal{V})$ by definition in case there is a universal bundle $\mathcal{V}$. Now given a class $\xi \in H^4(X \times \mathfrak{M})$, we have the "slant product map" from $H_2(X)$ to $H^2(\mathfrak{M})$, defined roughly as follows: take the class ξ, and look at the part of it which lies in $H^2(X) \otimes H^2(\mathfrak{M})$ under the Künneth decomposition. Neglecting torsion, $H^2(X) \otimes H^2(\mathfrak{M}) \cong \text{Hom}(H_2(X), H^2(\mathfrak{M}))$, and the homomorphism corresponding to ξ is by definition slant product with ξ. In particular, we can take slant product with $-\frac{1}{4}p_1(\text{ad}\,\mathcal{V})$. Here the factor $-\frac{1}{4}$ is chosen since, in case $w = 0$ and a universal bundle exists, one can check that $p_1(\text{ad}\,\mathcal{V})$ is always divisible by 4. Indeed, in this case $p_1(\text{ad}\,\mathcal{V})$ is just $-4c_2(\mathcal{V})$. The end result is the μ-map

$$\mu \colon H_2(X) \to H^2(\mathfrak{M}).$$

Using the μ-map, we can then define the *Donaldson polynomial* $D = D_{w,p}^X$ as follows: it is a multilinear $\mathbb{Q}$-valued form on $H_2(X)$ given by

$$D_{w,p}^X(\alpha) = \int_{\mathfrak{M}} \mu(\alpha)^d,$$

where the integral means evaluation of $\mu(\alpha)^d$ on the fundamental class of $\mathfrak{M}$ (with its natural orientation). Of course, *a priori* the value of $D_{w,p}^X$ also depends on H, the ample divisor which we used to define stability. We will discuss the dependence on H shortly.

For example, if $d = 0$, then under our assumptions on $\mathfrak{M}$, $D_{w,p}^X$ simply counts the number of points in $\mathfrak{M}$. More generally, $D_{w,p}^X(H)$ can be interpreted as the degree of $\mathfrak{M}$ in a suitable projective embedding. Thus, it is an important numerical invariant of the moduli space. These numbers have been calculated for certain classes of surfaces, for example, $K3$ and elliptic surfaces. For $\mathbb{P}^2$, let H be the class of a line. Then there are two sequences of numbers $D_{0,c}^{\mathbb{P}^2}(H)$ and $D_{w,p}^{\mathbb{P}^2}(H)$, where w is the nontrivial element of $H^2(\mathbb{P}^2; \mathbb{Z}/2\mathbb{Z})$. Assuming certain conjectures on the transition functions of Donaldson polynomials, Göttsche has shown that these sequences are defined by a very nonobvious generating function involving modular functions; see [53].

We have already seen that Assumption (iii) above is not necessary in order to define the μ-map, and thus it is not necessary in order to define $D_{w,p}^X$. In fact, it is easy to dispense with (i) as well. For example, if $\mathfrak{M}$ is generically reduced of the expected dimension, or in other words if each component of $\mathfrak{M}$ contains a good bundle, then the above discussion goes over word-for-word to define an element $\mu(\alpha) \in H^2(\mathfrak{M})$. Now every compact projective variety has a fundamental class, and so $D_{w,p}^X$ can be defined as before. It might happen that $\mathfrak{M}$ is not reduced, or worse still, has possibly nonreduced components of the wrong dimension. In these cases, there is still a procedure for defining $D_{w,p}^X$ [40], [16].

The most serious assumption above is (ii), that $\mathfrak{M}$ is compact. Typically $\mathfrak{M}$ is not compact, and it must be compactified by adding in Gieseker semistable torsion free sheaves. In this case, one can prove that the classes $\mu(\alpha)$ extend in a natural way to the compactification [115], [100], [80], and then $D_{w,p}^X$ may be defined as before.

The above construction can be made for a general smooth oriented 4-manifold M, together with a Riemannian metric g. Let P be a principal $SU(2)$-bundle over M, with the unique characteristic class $c_2(P) = c \in H^4(M; \mathbb{Z}) \cong \mathbb{Z}$. More generally, we can consider principal $SO(3)$-bundles P, with characteristic classes $w_2(P) \in H^2(M; \mathbb{Z}/2\mathbb{Z})$ and $p_1(P) \in H^4(M; \mathbb{Z}) \cong \mathbb{Z}$. However, for simplicity we shall just stick to the case of $SU(2)$. Let A be a connection on the principal $SU(2)$-bundle P. Then the curvature F_A is a 2-form with values in the vector bundle $\mathrm{ad}\, P$. Using the metric g, there is an associated Hodge $*$-operator from $A^k(M)$ to $A^{4-k}(M)$, where, in the notation of Chapter 4, $A^k(M)$ is the bundle of C^∞ 2-forms on M. In particular, $*: A^2(M) \to A^2(M)$ satisfies $*^2 = \mathrm{Id}$, and thus as in the last section of Chapter 4 we can take the decomposition of Ω_M^2 into the $+1$ and -1 eigenspaces for $*$:

$$A^2(M) = \Omega_+^2(M) \oplus \Omega_-^2(M).$$

There is an induced decomposition of the space $A^2(M)(\mathrm{ad}\, P)$. We call A anti-self-dual if $F_A \in \Omega_-^2(M)(\mathrm{ad}\, P)$. If $b_2^+(M) > 0$, for a generic metric g, the set of all anti-self-dual connections A on P, modulo the action of the group of C^∞ bundle automorphisms of P, is a finite-dimensional manifold

$\mathcal{M}(P,g) = \mathcal{M}$, and morally speaking there is a universal $SU(2)$-bundle $\mathcal{P}$ over $M \times \mathcal{M}$. (Just as in the holomorphic case, we might have to use an associated $SO(3)$-bundle $\mathrm{ad}\,\mathcal{P}$ instead, which always exists even if the bundle $\mathcal{P}$ does not.) Thus, we can again use slant product with $c_2(\mathcal{P})$, or with $-\frac{1}{4}p_1(\mathrm{ad}\,\mathcal{P})$ if no universal $SU(2)$-bundle exists, to define a μ-map

$$\mu \colon H_2(M) \to H^2(\mathcal{M}).$$

In case M is a complex surface and g is the Hodge metric associated to an ample divisor H, we have identified $\mathcal{M}$ with the moduli space $\mathfrak{M}$ of H-stable rank 2 bundles V with $c_1(V) = 0$ and $c_2(V) = c_2(P)$ in Theorem 26 of Chapter 4. While a Kähler metric need not be generic in the sense that $\mathcal{M}$ need not be a manifold, it turns out that $\mathcal{M}$ always can be locally modeled on a real analytic space and that in this sense the spaces $\mathcal{M}$ and $\mathfrak{M}$ are isomorphic. Moreover, one can identify the corresponding μ-maps.

   Just as with $\mathfrak{M}$, the manifold $\mathcal{M}$ is typically noncompact, even for a generic metric. One can construct a natural compactification $X(P,g)$, the *Uhlenbeck compactification*, for $\mathcal{M}$. The space $X(P,g)$ is not in general a manifold, but it is naturally a stratified space and carries a fundamental class $[X(P,g)]$. Furthermore, the classes $\mu(\alpha), \alpha \in H_2(M)$, extend in a unique way to classes in $H^2(X(P,g))$, also denoted by $\mu(\alpha)$. Thus, we can define the Donaldson polynomials $D_{0,c}^M$ for a general smooth oriented 4-manifold M as before, by evaluating $\mu(\alpha)^d$ over $[X(P,g)]$. (To be completely precise, we also need to choose an orientation for $\mathcal{M}$, but shall not describe the procedure for doing so here.) In order for the above evaluation to be meaningful, d should be one-half the real dimension of $X(P,g)$, or equivalently of the manifold $\mathcal{M}$. It follows from the Atiyah-Singer index theorem that $\dim_{\mathbb{R}} \mathcal{M}$ is even exactly when $b_2^+(M) - b_1(M)$ is odd, and thus Donaldson polynomials can be defined in this case. Of course, in case M is an algebraic surface, then $b_1(M)$ is even and, by the Hodge index theorem, $b_2^+(M)$ is odd. Thus, the Donaldson polynomials are defined in this case, and one natural choice of orientation is to always choose the complex orientation on $\mathcal{M} = \mathfrak{M}$. Similar constructions define the more general Donaldson polynomials $D_{w,p}^M$.

   In case M is an algebraic surface and g is a Hodge metric, there are two potentially different definitions of the Donaldson polynomial, corresponding to the choice of either the Gieseker or the Uhlenbeck compactification. These two compactifications are quite different, but the resulting definitions of the Donaldson polynomial agree, as has been shown by Morgan [100] and Li [80]. Both [100] and [80] investigate the relationship between the two compactifications, and in [80] there is an algebro-geometric definition of the Uhlenbeck compactification, which is a generalized blowdown of the Gieseker compactification.

   To produce actual invariants of the C^∞ structure on the 4-manifold M, we need to analyze the dependence of the moduli space $\mathcal{M}(P,g)$ and its compactification $X(P,g)$ on the metric g. The moduli space $\mathcal{M}(P,g)$

acquires singularities when the bundle P admits reducible anti-self-dual connections. By Hodge theory, this happens when there exists a complex line bundle L on M such that the complex 2-plane bundle V over M associated to the standard representation of $SU(2)$ is isomorphic to $L \oplus L^{-1}$, and such that $c_1(L)$ is orthogonal to every self-dual harmonic 2-form with respect to the metric g. The first condition is simply the condition that $c = c_2(P) = -\zeta^2$, where $\zeta = c_1(L) \in H^2(M; \mathbb{Z})$. The second condition is vacuous if the intersection pairing on $H_2(M)$ is negative definite, and indeed the singularities forced on the moduli space in this case for $c = 1$ are a crucial ingredient in Donaldson's famous theorem that every smooth definite 4-manifold has a diagonalizable intersection form. In general, the condition that a metric g admit reducible anti-self-dual connections has codimension $b_2^+(M)$ in the (infinite-dimensional) space of all metrics on M. Thus, as we noted above, these singularities do not appear in case $b_2^+(M) > 0$ and g is generic.

Since $\mathcal{M}$ is noncompact, we also have to consider what happens to the (singular) space $X(P, g)$. The smaller-dimensional strata of $X(P, g)$ involve moduli spaces $\mathcal{M}(P', g)$, with $c_2(P') < c_2(P)$. In case P' admits reducible anti-self-dual connections, these will contribute "extra" singularities to $X(P, g)$. By the above discussion, such singularities will arise whenever there are classes $\zeta \in H^2(M; \mathbb{Z})$ orthogonal to every self-dual harmonic 2-form and satisfying

$$-c \leq \zeta^2 < 0.$$

By the Hodge theory arguments of Chapter 4, if M is a Kähler surface and ω is the Kähler form of a Kähler metric, an integral class orthogonal to every self-dual harmonic 2-form on M is an integral $(1, 1)$-class orthogonal to ω, and in particular it is the first Chern class of a holomorphic line bundle. Now if V is a strictly ω-semistable rank 2 bundle on M, then there is an exact sequence

$$0 \to \mathcal{O}_M(D) \to V \to \mathcal{O}_M(-D) \otimes I_Z \to 0,$$

with $\omega \cdot D = 0$, and V is S-equivalent to the torsion free sheaf $\mathcal{O}_M(D) \oplus \mathcal{O}_M(-D) \otimes I_Z$. The double dual of this sheaf is a vector bundle of Chern class $-\zeta^2 \leq c$, with a reducible anti-self-dual connection. Thus, singularities of the moduli space are related to strictly semistable bundles.

Returning to the case of a general 4-manifold M, if $b_2^+(M) > 1$, a generic path of metrics linking two generic metrics will not contain any metrics admitting reducible connections, for $\mathcal{M}(P, g)$ or for any of the boundary pieces of $X(P, g)$ which involve moduli spaces $\mathcal{M}(P', g)$ for smaller values of $c_2(P')$. One can then show that the value of the Donaldson polynomial is unchanged along this path, and thus that there is a well-defined Donaldson polynomial $D_{0,c}^M$. Note that this picture is quite different from the one described in Chapter 4 for the behavior of the moduli space under a change of polarization. In that case, a generic path of metrics linking two

Hodge metrics on an algebraic surface X with $b_2^+(X) \geq 3$ will not involve reducible connections, except possibly at the endpoints. However, every path of metrics linking the two Hodge metrics which itself consists entirely of Hodge metrics may well involve metrics which admit reducible connections. In particular, in the algebro-geometric case, there is no distinction between the case $b_2^+(X) = 1$ (i.e., $p_g(X) = 0$) and $b_2^+(X) > 1$ ($p_g(X) \neq 0$). Thus, it is not a priori clear that algebro-geometric definition of the Donaldson invariant sketched above, in the case X is an algebraic surface with $p_g \neq 0$, via moduli spaces of stable bundles, is independent of the ample line bundle used to define stability. In fact, aside from a few special cases such as $K3$ surfaces, it is a challenging open problem to see this via algebraic geometry! However, the equivalent C^∞ version does not depend on the choice of metric provided that $b_2^+(M) > 1$, or equivalently $p_g(M) > 0$.

In case $b_2^+(M) = 1$, a generic path linking two generic metrics will usually contain metrics admitting reducible connections. Thus, the Donaldson polynomial is not independent of the metric, but depends on a "chamber structure" in the positive cone

$$\mathcal{C} = \{\, x \in H^2(M; \mathbb{R}) : x^2 > 0 \,\}.$$

In this case, the induced chamber structure on the ample cone is the one described in Chapter 4. The change in the Donaldson polynomial as we cross a wall of the chamber is surprisingly complicated [30].

The 2-dimensional invariant

Let us use the discussion of vector bundles on the elliptic surface X to calculate its 2-dimensional Donaldson invariant. In this case, the moduli space $\mathfrak{M}$ birationally consists of the choice of one point p lying in a fiber of X. Of course, this last condition is automatically satisfied since every point of X lies in a (unique) fiber, and as we shall see the moduli space is exactly X itself. In particular it is compact. We could also see that the moduli space is compact as follows: as there are no strictly semistable bundles, the moduli space can only be compactified by adding in torsion free, non-locally free sheaves whose double duals are stable. If $\mathfrak{M}$ were not compact, boundary points would correspond to torsion free sheaves V with $p_1(V^{\vee\vee}) \geq p_1(V) + 4$. This would imply that $V^{\vee\vee}$ lives in a moduli space of expected dimension at most -2, contradicting the fact that for the bundles we consider the moduli space is always smooth of the expected dimension.

The recipe for computing the Donaldson invariant in this case is as follows: let $\mathcal{V}$ be a universal sheaf over $X \times X$. Here the first factor X is to be viewed as the original surface and the second is the moduli space, and $\mathcal{V}$ is a vector bundle over $X \times X$. The universal property of $\mathcal{V}$ may be expressed as follows. Let V_0 be the unique stable vector bundle on X for which $p_1(\mathrm{ad}\, V_0) = 3\chi(\mathcal{O}_X)$ and $\det V_0 \cdot f = 1$. Given $p \in X$, let f be the fiber through p and let $i: f \to X$ be the inclusion. Then, if p does not lie on a singular point of a fiber, $\mathcal{V}|X \times \{p\} = V_p$, where by definition there is an exact sequence

$$0 \to V_p \to V_0 \to i_* \mathcal{O}_f(p) \to 0,$$

and the map $V_0 \to i_* \mathcal{O}_f(p)$ is unique mod scalars. If p is the singular point of its fiber, then we replace $\mathcal{O}_f(p)$ by $n_* \mathcal{O}_{\tilde{f}}$, where $n: \tilde{f} \to f$ is the normalization map. We also have $n_* \mathcal{O}_{\tilde{f}} \cong Hom(\mathfrak{m}_p, \mathcal{O}_f)$ (Exercise 7). By Lemma 11 in the singular case, there is again a unique map $V_0 \to n_* \mathcal{O}_{\tilde{f}}$, and it is surjective. Now take $p_1(\operatorname{ad} \mathcal{V}) \in H^4(X \times X)$ (all cohomology with rational coefficients). Using the projection $H^4(X \times X) \to H^2(X) \otimes H^2(X)$, and given $\alpha \in H_2(X)$, we can then define $\mu(\alpha) \in H^2(X)$ by taking slant product with $\frac{1}{4} p_1(\operatorname{ad} \mathcal{V})$. By definition, then,

$$D(\alpha) = \mu(\alpha)^2 \in H^4(X) \cong \mathbb{Q}.$$

To carry out this program we will have to construct the bundle $\mathcal{V}$. First we must fit together the sheaves $i_* \mathcal{O}_f(p)$ as p ranges over X. To do so, let D be the fiber product $X \times_{\mathbb{P}^1} X \subset X \times X$. Then D is smooth except at a point (p, p), where p is a singular point of a singular fiber. Near such a point, D has the local equation $xy = zw$, which is a hypersurface singularity, in fact an ordinary double point in dimension 3. Let $\mathbb{D}$ be the diagonal inside $X \times X$. Then $\mathbb{D} \subset D$, and we consider the sheaf $\mathcal{O}_D(\mathbb{D})$. In the complement of the singular points of D, $\mathbb{D}$ is a Cartier divisor in D and the notation $\mathcal{O}_D(\mathbb{D})$ is meaningful. Moreover, if $i: D \to X \times X$ is the inclusion, we clearly have $i_* \mathcal{O}_D(\mathbb{D})|X \times \{p\} = i_* \mathcal{O}_f(p)$, provided that p is not a singular point of a singular fiber. Thus, $\mathcal{O}_D(\mathbb{D})$ has accomplished the goal of fitting the sheaves $i_* \mathcal{O}_f(p)$.

In case p is the singular point of a singular fiber, we must still assign meaning to $i_* \mathcal{O}_f(p)$. First we should interpret $\mathcal{O}_f(p)$ as the unique torsion free rank 1 sheaf on f of degree 1 which is not locally free at p. Thus, $\mathcal{O}_f(p) = Hom(\mathfrak{m}_p, \mathcal{O}_f) = n_* \mathcal{O}_{\mathbb{P}^1}$. One way to fit this sheaf into the family constructed above is to take the well-defined sheaf $I_{\mathbb{D}}/I_D$ and then set $\mathcal{O}_D(\mathbb{D}) = Hom(I_{\mathbb{D}}/I_D, \mathcal{O}_D)$. Local calculations (Exercise 12) show that $\mathcal{O}_D(\mathbb{D})$ is locally of the form $I_{\mathbb{E}}/I_D$, where $\mathbb{E}$ is locally defined in coordinates by the ideal $(x-w, y-z) \subseteq I_D$. Moreover, again by Exercise 12, if we dualize the inclusion $I_{\mathbb{D}}/I_D \to \mathcal{O}_D$, we obtain (locally) an exact sequence

$$0 \to \mathcal{O}_D \to \mathcal{O}_D(\mathbb{D}) \to \mathcal{O}_{\mathbb{D}} \to 0.$$

In particular $\mathcal{O}_D(\mathbb{D})$ is flat over the second factor X in $X \times X$ since $\mathcal{O}_{\mathbb{D}}$ is flat over X (the projection is an isomorphism) and $\mathcal{O}_D$ is flat over X by the local criterion of flatness, since D is a hypersurface inside the smooth family $X \times X \to X$, and all fibers of the morphism $D \to X$ have dimension 1. For p a point in a smooth fiber f, or a smooth point of a singular fiber f, we clearly have $\mathcal{O}_D(\mathbb{D})|X \times \{p\} = \mathcal{O}_f(p)$, viewed as a sheaf on X. In case p is the singular point of a singular fiber f, $\mathcal{O}_D(\mathbb{D})|X \times \{p\} = Hom(\mathfrak{m}_p, \mathcal{O}_f)$.

Indeed, such a statement is true locally since $I_{\mathbb{E}}/I_D|X \times \{p\}$ has the local form $\mathfrak{m}_p \cong \tilde{O}_f$, so that $\mathcal{O}_D(\mathbb{D})|X \times \{p\}$ is the push-forward of a non-locally free rank 1 sheaf on f. Thus, it is specified by its degree, and it suffices to show that the degree is 1. But by flatness $\deg(\mathcal{O}_D(\mathbb{D})|X \times \{p\})$ is independent of p, and this degree is 1 if p is a smooth point in its fiber. Thus, it is 1 in the case where p is the singular point as well.

Next, let π_1 and π_2 be the projections from $X \times X$ to the first and second factors. We seek a surjective map $\pi_1^* V_0 \to \mathcal{O}_D(\mathbb{D})$. If such a map existed, it would induce on every slice $X \times \{p\}$ a surjection $V_0 \to i_* \mathcal{O}_f(p)$ (with the appropriate modifications if p is a singular point of a singular fiber) and so we could take $\mathcal{V}$ to be the kernel of the surjection. The existence of the bundle $\mathcal{V}$, by the properties of a coarse moduli space, implies that there is a morphism $X \to \mathfrak{M}$, where $\mathfrak{M}$ is the coarse moduli space corresponding to the 2-dimensional problem. By the classification results of the preceding section, the map $X \to \mathfrak{M}$ is a bijection and $\mathfrak{M}$ is smooth. Thus, the map $X \to \mathfrak{M}$ is an isomorphism, identifying X with the moduli space.

Unfortunately, the map $\pi_1^* V_0 \to \mathcal{O}_D(\mathbb{D})$ need not exist! Instead, for every $p \in X$, it follows from Lemma 11 that

$$\dim \operatorname{Hom}(\pi_1^* V_0, \mathcal{O}_D(\mathbb{D})|X \times \{p\}) = \dim H^0((\pi_1^* V_0)^\vee \otimes \mathcal{O}_D(\mathbb{D})|X \times \{p\}) = 1.$$

Thus, by general properties of cohomology and base change, the coherent sheaf

$$R^0 \pi_{2*} \mathcal{H}om(\pi_1^* V_0, \mathcal{O}_D(\mathbb{D})) = \mathcal{L}^{-1}$$

is a line bundle on X. The trivial section of $R^0 \pi_{2*} \mathcal{H}om(\pi_1^* V_0, \mathcal{O}_D(\mathbb{D})) \otimes \mathcal{L}$ then gives a global section of $\mathcal{H}om(\pi_1^* V_0, \mathcal{O}_D(\mathbb{D}) \otimes \pi_2^* \mathcal{L})$, and thus a map

$$\pi_1^* V_0 \to \mathcal{O}_D(\mathbb{D}) \otimes \pi_2^* \mathcal{L}.$$

This map is surjective because its restriction to each fiber is surjective. Thus, we may finally define $\mathcal{V}$ via the kernel of this map:

$$0 \to \mathcal{V} \to \pi_1^* V_0 \to \mathcal{O}_D(\mathbb{D}) \otimes \pi_2^* \mathcal{L} \to 0.$$

To calculate $D(\alpha)$, we must therefore calculate $p_1(\operatorname{ad} \mathcal{V})$. Now $\mathcal{V}$ is given by an elementary modification, and so the general formulas of Chapter 2 tell us in principle how to calculate $p_1(\operatorname{ad} \mathcal{V})$: using the formulas for $c_1(\mathcal{V})$ and $c_2(\mathcal{V})$ in Lemma 16 of Chapter 2, we find that

$$p_1(\operatorname{ad} \mathcal{V}) = p_1(\operatorname{ad} \pi_1^* V_0) + 2c_1(\pi_1^* V_0) \cdot D + D^2 - 4i_* c_1(\pi_2^* \mathcal{L}) - 4\mathbb{D}.$$

Strictly speaking this formula holds away from the singular locus of D, where the sheaf $\mathcal{O}_D(\mathbb{D}) \otimes \pi_2^* \mathcal{L}$ is not necessarily a line bundle. However, the singular locus of D has high codimension in $X \times X$, so that the formula above, which holds a priori on $X \times X - \operatorname{Sing} D$, is actually valid on $X \times X$. So we must calculate these terms. Also, a term of the form $x \otimes 1$ or $1 \otimes x$, for $x \in H^4(X)$, will not affect slant product as far as $H_2(X)$ is concerned (although we would need to keep track of these terms if we were interested in the four-dimensional class) so we will omit such terms as needed. For example, $p_1(\mathrm{ad}\,\pi_1^*V_0) = \pi_1^*p_1(\mathrm{ad}\,V_0)$ is of the form $x \otimes 1$ and will not affect slant product.

Next we have the following lemma:

Lemma 13. *The cohomology class of D is given by*

$$[D] = f \otimes 1 + 1 \otimes f.$$

Proof. It suffices to show that, if $\alpha \in H^2(X)$, then $\pi_1^*\alpha \cup [D] = f \cdot \alpha$, and similarly for $\pi_2^*\alpha$ (note that X is simply connected so that $H^1(X) = 0$). Dually, for a general point $x \in X$, if α is dual to the homology class represented by a smoothly embedded 2-manifold C, we must compute the intersection number $(C \times \{x\}) \cdot D$, which is clearly equal to $C \cdot f$. □

Thus, we see that $D^2 = 2f \otimes f$ and that, since by assumption $c_1(V_0) = \sigma + kf$ for some k,

$$2c_1(\pi_1^*V_0) \cdot D = 2[(\sigma + kf) \otimes 1] \cdot [f \otimes 1 + 1 \otimes f] = 2(\sigma + kf) \otimes f.$$

We come now to the main term of interest,

$$4i_*c_1(\pi_2^*\mathcal{L}) = 4i_*i^*\pi_1^*c_1(\mathcal{L}) = 4[D] \cdot (1 \otimes \lambda)$$
$$= (f \otimes 1 + 1 \otimes f) \cdot (1 \otimes \lambda),$$

where $\lambda = c_1(\mathcal{L}) \in H^2(X)$. Up to a term not affecting slant product this is just $4f \otimes \lambda$. Thus, it suffices to compute λ. Now

$$-\lambda = c_1(\mathcal{L}^{-1}) = c_1(R^0\pi_{2*}Hom(\pi_1^*V_0, \mathcal{O}_D(\mathbb{D}))).$$

Thus, we must compute c_1 of a direct image sheaf. Moreover, in this case it follows from Lemma 11 that $R^1\pi_{2*}Hom(\pi_1^*V_0, \mathcal{O}_D(\mathbb{D})) = 0$. Since π_2 has relative dimension 1, all other $R^i\pi_{2*}$'s are 0 also, and we are set up to use the Grothendieck-Riemann-Roch theorem to find

$$\sum_i (-1)^i c_1(R^i\pi_{2*}Hom(\pi_1^*V_0, \mathcal{O}_D(\mathbb{D}))) = c_1(R^0\pi_{2*}Hom(\pi_1^*V_0, \mathcal{O}_D(\mathbb{D})))$$

$$= -\lambda.$$

Lemma 14. *We have the following formula for $-\lambda$:*

$$-\lambda = -\sigma + \left(\frac{-p_g - k - 1}{2}\right)f.$$

Before we give the proof of Lemma 14, which will be a lengthy if standard calculation, let us complete the calculation of $D(\alpha)$. If we collect all the terms for $-p_1(\mathrm{ad}\, V)$ which will affect slant product, we see that $-4\mu(\alpha)$ is given by slant product with

$$2(\sigma + kf) \otimes f + 2f \otimes f - 4f \otimes \lambda - 4\mathbb{D}$$

$$=2(\sigma + kf) \otimes f + 2f \otimes f - 4f \otimes \sigma - 4\left(\frac{p_g + k + 1}{2}\right)f \otimes f - 4\mathbb{D}$$

$$=2\sigma \otimes f - 4f \otimes \sigma + (2k + 2 - 2p_g - 2k - 2)f \otimes f - 4\mathbb{D}$$

$$=2\sigma \otimes f - 4f \otimes \sigma - 2p_g f \otimes f - 4\mathbb{D}.$$

Taking the slant product with α gives our formula for $\mu(\alpha)$:

Proposition 15. $-4\mu(\alpha) = 2(\sigma \cdot \alpha)f - 4(f \cdot \alpha)\sigma - 2p_g(f \cdot \alpha)f - 4\alpha.$ $\square$

Squaring this term, we find that

$$16\mu(\alpha)^2 = -16(\sigma \cdot \alpha)(f \cdot \alpha) - 16(\sigma \cdot \alpha)(f \cdot \alpha) + 16(f \cdot \alpha)^2(\sigma^2)$$

$$+ 16p_g(f \cdot \alpha)^2 + 32(f \cdot \alpha)(\sigma \cdot \alpha) + 16p_g(f \cdot \alpha)^2 + 16\alpha^2$$

$$= -16(p_g + 1)(f \cdot \alpha)^2 + 32p_g(f \cdot \alpha)^2 + 16\alpha^2$$

$$=16\alpha^2 + 16(p_g - 1)(f \cdot \alpha)^2.$$

Thus, we have the desired formula for the degree 2 Donaldson polynomial:

Theorem 16. $D(\alpha) = \alpha^2 + (p_g - 1)(f \cdot \alpha)^2.$ $\square$

Proof of Lemma 14. Recall the Grothendieck-Riemann-Roch formula, which says that $\mathrm{ch}((\pi_2)_!((\pi_1^*V_0)^\vee \otimes \mathcal{O}_D(\mathbb{D}))) \,\mathrm{Todd}\, X$ is equal to

$$\pi_{2*}\left[\mathrm{ch}(\pi_1^*V_0)^\vee \otimes \mathcal{O}_D(\mathbb{D}))\,\mathrm{Todd}(X \times X)\right].$$

Here ch is the Chern character, so that ch is additive over direct sums and multiplicative over tensor products, provided that one of the factors is a vector bundle. For a line bundle L, $\mathrm{ch}\, L = 1 + c_1(L) + c_1(L)^2/2 + \cdots$. For a rank r bundle V, $\mathrm{ch}\, V = r + c_1(V) + (c_1(V)^2 - 2c_2(V))/2 + \cdots$. Moreover, $\mathrm{Todd}\, V$ is the Todd class of V, which is again multiplicative, and $\mathrm{Todd}\, X$ is by definition the Todd class of the tangent bundle of X. In general $\mathrm{Todd}\, V = 1 + c_1(V)/2 + (c_1(V)^2 + c_2(V))/12 + \cdots$. Finally, $(\pi_2)_!\mathcal{F}$ denotes the formal sum of coherent sheaves $\sum_i(-1)^i R^i\pi_{2*}\mathcal{F}$. In our case the only such term which is nonzero is the $R^0\pi_{2*}$ term, and it is a line bundle. Thus, the formula says that $c_1(R^0\pi_{2*}(\pi_1^*V_0)^\vee \otimes \mathcal{O}_D(\mathbb{D}))$ is equal to the degree 1 term in

$$\pi_{2*}\left[\mathrm{ch}((\pi_1^*V_0)^\vee \otimes \mathcal{O}_D(\mathbb{D})) \cdot \mathrm{Todd}(X \times X)\right](\mathrm{Todd}\, X)^{-1}$$

$$= \pi_{2*}\left[\mathrm{ch}((\pi_1^*V_0)^\vee \otimes \mathcal{O}_D(\mathbb{D})) \cdot \mathrm{Todd}(X \times X) \cdot \pi_2^*(\mathrm{Todd}\, X)^{-1}\right]$$

$$= \pi_{2*}\left[\mathrm{ch}((\pi_1^*V_0)^\vee \otimes \mathcal{O}_D(\mathbb{D})) \cdot \pi_1^*\mathrm{Todd}\, X \cdot \pi_2^*\mathrm{Todd}\, X \cdot \pi_2^*(\mathrm{Todd}\, X)^{-1}\right]$$

$$= \pi_{2*}\left[\mathrm{ch}((\pi_1^*V_0)^\vee \otimes \mathcal{O}_D(\mathbb{D})) \cdot \pi_1^*\mathrm{Todd}\, X\right]$$

$$= \pi_{2*} \left[\text{ch}(\pi_1^* V_0)^\vee \cdot \text{ch}(\mathcal{O}_D(\mathbb{D})) \cdot \pi_1^* \text{Todd} \, X \right]$$
$$= \pi_{2*} \left[\pi_1^* (\text{ch}(V_0)^\vee \cdot \text{Todd} \, X) \cdot \text{ch}(\mathcal{O}_D(\mathbb{D})) \right].$$

We proceed to calculate these terms. First note that we have supposed that $c_1(V_0) = \sigma + kf$, where k is well-defined mod 2. Thus,

$$c_1(V_0)^2 = -(1 + p_g) + 2k,$$

and moreover $p_1(\text{ad} \, V_0) = c_1(V_0)^2 - 4c_2(V_0)$. Since $-p_1(\text{ad} \, V_0) - 3(1 + p_g)$ is the expected dimension of the moduli space for V_0, which is smooth of dimension 0, we have $p_1(\text{ad} \, V_0) = -3(1 + p_g)$. Thus,

$$-(1 + p_g) + 2k - 4c_2(V_0) = -3(1 + p_g),$$

so that solving for $c_2(V_0)$ we find that

$$c_2(V_0) = \frac{p_g + 1 + k}{2}.$$

It follows that

$$\text{ch}(V_0^\vee) = 2 - c_1(V_0) + \frac{c_1(V_0)^2 - 2c_2(V_0)}{2}$$

$$= 2 - (\sigma + kf) + \frac{1}{2}(-(1 + p_g) + 2k - (p_g + 1 + k))\text{pt}$$

$$= 2 - (\sigma + kf) + \left(-(p_g + 1) + \frac{k}{2} \right) \text{pt}.$$

Moreover, using the fact that $c_1(X) = -K_X = -(p_g - 1)f$ and Noether's formula, that $c_1(X)^2 + c_2(X) = 12\chi(\mathcal{O}_X) = 12(p_g + 1)$, we find that

$$\text{Todd} \, X = 1 - \left(\frac{p_g - 1}{2} \right) f + (p_g + 1)\text{pt}.$$

Putting these two calculations together, we see that

$$\pi_1^* (\text{ch}(V_0)^\vee) \cdot \text{Todd} \, X = 2 - (\sigma + (p_g - 1 + k)f) \otimes 1 + N\text{pt} \otimes 1,$$

where $N = (3p_g + 1 + k)/2$.

Next we need a formula for $\text{ch}(\mathcal{O}_D(\mathbb{D}))$:

Lemma 17. Let $i \colon D \to X \times X$ be the inclusion, and similarly for $j \colon \mathbb{D} \to X \times X$. Then up through complex codimension 3,

$$\text{ch}(i_* \mathcal{O}_D(\mathbb{D})) = D - \frac{D^2}{2} + D - \left(\frac{p_g + 3}{2} \right) j_* f + \cdots.$$

Proof. We have an exact sequence

$$0 \to \mathcal{O}_D \to \mathcal{O}_D(\mathbb{D}) \to \mathcal{O}_{\mathbb{D}}(\mathbb{D}) \to 0.$$

Thus, $\mathrm{ch}(i_*\mathcal{O}_D(\mathbb{D})) = \mathrm{ch}(i_*\mathcal{O}_D) + \mathrm{ch}(j_*\mathcal{O}_{\mathbb{D}}(\mathbb{D}))$, where $j \colon \mathbb{D} \to X \times X$ is the inclusion. Now from the exact sequence

$$0 \to \mathcal{O}_{X \times X}(-D) \to \mathcal{O}_{X \times X} \to \mathcal{O}_D \to 0,$$

we see that

$$\mathrm{ch}(i_*\mathcal{O}_D) = 1 - \mathrm{ch}(\mathcal{O}_{X \times X}(-D))$$
$$= 1 - (1 - D + D^2/2) = D - D^2/2,$$

where we have used the fact that $D^3 = 0$. As for the term $\mathrm{ch}(j_*\mathcal{O}_{\mathbb{D}}(\mathbb{D}))$, we ignore the fact that D has singularities since they will occur in high codimension and our arguments will work as if $\mathbb{D}$ were a Cartier divisor on D. Applying Grothendieck-Riemann-Roch to the embedding j, we obtain

$$\mathrm{ch}(j_*\mathcal{O}_{\mathbb{D}}(\mathbb{D})) = j_*(\mathrm{ch}(\mathcal{O}_{\mathbb{D}}(\mathbb{D})) \cdot \mathrm{Todd}(N_{\mathbb{D}/X \times X})^{-1})$$
$$= j_*\left(1 - \frac{p_g - 1}{2}f + (p_g + 1)\mathrm{pt}\right)^{-1} \cdot (1 + \mathbb{D}|_{\mathbb{D}} + \cdots)$$
$$= j_*\left(1 + \frac{p_g - 1}{2}f + \mathbb{D}|_{\mathbb{D}} + \cdots\right)$$
$$= \mathbb{D} + \frac{p_g - 1}{2}j_*f + i_*(\mathbb{D})^2_{\mathbb{D}} + \cdots.$$

Here we have used the fact that the normal bundle of $\mathbb{D}$ in $X \times X$ is the tangent bundle of X to calculate $\mathrm{Todd}(N_{\mathbb{D}/X \times X})$, and the term $(\mathbb{D})^2_{\mathbb{D}}$ refers to the self-intersection of $\mathbb{D}$ on D, which is meaningful in the complement of the finitely many singular points of D. To calculate the term $(\mathbb{D})^2_{\mathbb{D}}$, we must calculate the normal bundle to $\mathbb{D}$ in D, at least at the smooth points. Using the exact sequence

$$0 \to N_{\mathbb{D}/D} \to N_{\mathbb{D}/X \times X} \to N_{D/X \times X}|\mathbb{D} \to 0,$$

we see that

$$(\mathbb{D})^2_{\mathbb{D}} = c_1(N_{\mathbb{D}/D}) = c_1(X) - (f \otimes 1 + 1 \otimes f)|\mathbb{D},$$

again using the fact that $N_{\mathbb{D}/X \times X}$ is the tangent bundle to $X \cong \mathbb{D}$. Thus,

$$(\mathbb{D})^2_D = -(p_g - 1)f - 2f = -(p_g + 1)f.$$

Putting this together we see that

$$\mathrm{ch}(j_*\mathcal{O}_{\mathbb{D}}(\mathbb{D})) = \mathbb{D} + \left[\frac{p_g - 1}{2} - (p_g + 1)\right]j_*f = \mathbb{D} - \left(\frac{p_g + 3}{2}\right)j_*f.$$

Thus,

$$\mathrm{ch}(i_*\mathcal{O}_D(\mathbb{D})) = D - \frac{D^2}{2} + \mathbb{D} - \left(\frac{p_g + 3}{2}\right)j_*f + \cdots. \qquad \Box$$

To complete the proof, we need to apply π_{2*} to the degree 3 term in

$$(2 - (\sigma + (p_g - 1 + k)f) \otimes 1 + N\mathrm{pt} \otimes 1) \cdot \left(D - \frac{D^2}{2} + \mathbb{D} - \left(\frac{p_g + 3}{2}\right)j_*f\right).$$

We have $D = f \otimes 1 + 1 \otimes f$ and $D^2 = 2f \otimes f$. Also, $(\alpha \otimes 1) \cdot \mathbb{D} = j_* \alpha$. Thus, the degree 3 term in the above product is

$$-(p_g + 3)j_* f + \text{pt} \otimes f - j_*(\sigma + (p_g - 1 + k)f) + N\text{pt} \otimes f.$$

Applying π_{2*} gives

$$-(p_g + 3)f + f - \sigma - (p_g - 1 + k)f + Nf = -\sigma + (-2p_g - 1 - k + N)f$$

$$= -\sigma + \left(\frac{-p_g - k - 1}{2}\right)f.$$

Thus, we have established the formula for $-\lambda$.   □

Finally, we describe briefly how to extend these results in case there are multiple fibers. For simplicity we assume that X is an elliptic surface over $\mathbb{P}^1$ with just two multiple fibers, with relatively prime multiplicities m_1 and m_2. We shall consider bundles V over X such that $c_1(V) \cdot f$ is odd, where f is a general fiber of X. Since $f = m_1 m_2 \kappa$, where κ is a primitive cohomology class, we see that $c_1(V) \cdot f$ is odd only if both m_1 and m_2 are odd. Much of the analysis of the first part of this chapter goes through to show that the moduli space is always smooth and irreducible of the expected dimension (which is even), and thus that the 0-dimensional moduli space is a single reduced point. The formula for the 2-dimensional Donaldson invariant is as follows [36]:

Theorem 18. *Let X be a simply connected elliptic surface over $\mathbb{P}^1$ with two multiple fibers with relatively prime multiplicities m_1 and m_2. Let D be the Donaldson invariant corresponding to a 2 dimensional moduli space of vector bundles V on X with $c_1(V) \cdot f$ odd, where f is the class of a general fiber on X. Let κ be the cohomology class such that $f = m_1 m_2 \kappa$. Then*

$$D(\alpha) = \alpha^2 + ((m_1 m_2)^2 (p_g + 1) - m_1^2 m_2^2)(\kappa \cdot \alpha)^2.$$ □

The idea of the proof is to imitate the calculation in the case of a section as far as possible, using an approximate moduli space. This determines $D(\alpha)$ up to correction terms which only depend on an analytic neighborhood of the multiple fibers. Now a simply connected elliptic surface with $p_g = 0$ and just one multiple fiber of multiplicity m is a rational surface, and its Donaldson polynomial is therefore the same as the corresponding polynomial for a rational elliptic surface with a section. But we have computed these polynomials above. Using this information, we can then determine the correction terms and complete the calculation for Theorem 18.

Moduli spaces via extensions

In this section we shall describe a different approach to constructing moduli spaces. While this approach only works in case X has a section, it is still instructive to redo the calculations in this way. We begin with a lemma on various cohomology groups:

Lemma 19. *Let X be an elliptic surface over $\mathbb{P}^1$ with a section σ and let $p_g(X) = p_g$. Then, for all $n \in \mathbb{Z}$,*

$$h^1(X; \mathcal{O}_X(-\sigma + (p_g + 1 - n)f)) = \begin{cases} 0, & \text{if } n > 0, \\ -n + 1, & \text{if } n \leq 0. \end{cases}$$

and

$$h^2(X; \mathcal{O}_X(-\sigma + (p_g + 1 - n)f)) = \begin{cases} n - 1, & \text{if } n \geq 2, \\ 0, & \text{if } n \leq 1. \end{cases}$$

Proof. Clearly, $R^0\pi_*\mathcal{O}_X(-\sigma + (p_g + 1 - n)f) = 0$ for all n and $R^2\pi_*\mathcal{O}_X(-\sigma + (p_g + 1 - n)f) = 0$ for all n since π has relative dimension 1. Thus,

$$H^i(X; \mathcal{O}_X(-\sigma + (p_g + 1 - n)f)) = H^{i-1}(\mathbb{P}^1; R^1\pi_*\mathcal{O}_X(-\sigma + (p_g + 1 - n)f))$$

by the Leray spectral sequence. So we must determine the line bundle

$$R^1\pi_*\mathcal{O}_X(-\sigma + (p_g + 1 - n)f)$$

on $\mathbb{P}^1$. Note that

$$R^1\pi_*\mathcal{O}_X(-\sigma + (p_g + 1 - n)f) = R^1\pi_*\mathcal{O}_X(-\sigma) \otimes \mathcal{O}_{\mathbb{P}^1}(p_g + 1 - n)$$

by the projection formula. Thus, it suffices to determine $R^1\pi_*\mathcal{O}_X(-\sigma)$. Now apply $R^i\pi_*$ to the exact sequence

$$0 \to \mathcal{O}_X(-\sigma) \to \mathcal{O}_X \to \mathcal{O}_\sigma \to 0.$$

Since $R^0\pi_*\mathcal{O}_X \cong R^0\pi_*\mathcal{O}_\sigma = \mathcal{O}_\sigma$ and $R^1\pi_*\mathcal{O}_\sigma = 0$, we see that

$$R^1\pi_*\mathcal{O}_X(-\sigma) \cong R^1\pi_*\mathcal{O}_X = \mathcal{O}_{\mathbb{P}^1}(-(p_g + 1)),$$

by the formulas in the last chapter. Thus, $R^1\pi_*\mathcal{O}_X(-\sigma + (p_g + 1 - n)f) = \mathcal{O}_{\mathbb{P}^1}(-n)$, and the statement of the lemma follows from the formulas for the cohomology of $\mathbb{P}^1$. $\square$

Corollary 20. *With notation as above, there is a unique bundle V_0 on X such that V_0 is a nonsplit extension*

$$0 \to \mathcal{O}_X \to V_0 \to \mathcal{O}_X(\sigma - (p_g + 1)f) \to 0.$$

Moreover, V_0 has stable restriction to every fiber. Finally, $c_1(V_0) = \sigma - (p_g + 1)f$, $c_2(V_0) = 0$, and $-p_1(\operatorname{ad} V_0) = 3(p_g + 1)$.

Proof. The set of all possible extensions as above is classified by

$$H^1(\mathcal{O}_X(\sigma - (p_g + 1)f)^\vee \otimes \mathcal{O}_X) = H^1(\mathcal{O}_X(-\sigma + (p_g + 1)f)).$$

By the above lemma, this group has dimension 1, proving that there is a unique nonsplit extension. To see what happens on a fiber f, the restriction of V_0 to f is an extension of $\mathcal{O}_f(p)$ by $\mathcal{O}_f$, where $p = \sigma \cap f$ is a smooth point of f. Thus, $V_0|f$ is stable provided this extension does not split. To analyze the restriction of the extension, we consider the map

$$H^1(X; \mathcal{O}_X(-\sigma + (p_g + 1)f)) \to H^1(f; \mathcal{O}_f(-p)).$$

It is easy to see that the induced map on cohomology groups corresponds to restricting extensions. Thus, we must show that the image of a nonzero element of $H^1(X; \mathcal{O}_X(-\sigma + (p_g + 1)f))$ is nonzero in $H^1(f; \mathcal{O}_f(-p))$. But the kernel of the restriction map is

$$H^1(X; \mathcal{O}_X(-\sigma + (p_g + 1)f) \otimes \mathcal{O}_X(-f)) = H^1(X; \mathcal{O}_X(-\sigma + p_g f)).$$

Again by the lemma, this last group is zero (it corresponds to $n = 1$).

Clearly, $c_1(V_0)$ and $c_2(V_0)$ are as claimed. Thus,

$$p_1(\operatorname{ad} V_0) = c_1(V_0)^2 = \sigma^2 - 2(p_g + 1) = -3(p_g + 1).$$

This concludes the proof of the corollary. □

The bundle V_0 above is described in [36] as well as by Kametani and Sato [66] (with slightly different normalizations).

Now let us describe the bundles corresponding to the 2-dimensional moduli space. If V is such a bundle, then up to a twist V is obtained from V_0 by an elementary modification along a single fiber. Thus, we may assume that $c_1(V) = \sigma - (p_g + 2)f$ and that $-p_1(\operatorname{ad} V) = 3(p_g + 1) + 2$.

Proposition 21. *Let V be a stable rank 2 bundle on X with $c_1(V) = \sigma - (p_g + 2)f$ and $-p_1(\operatorname{ad} V) = 3(p_g + 1) + 2$. Then either V is given as an extension*

$$0 \to \mathcal{O}_X(-f) \to V \to \mathcal{O}_X(\sigma - (p_g + 1)f) \otimes \mathfrak{m}_x \to 0,$$

where $\mathfrak{m}_x$ is the ideal of a point on X, or V is given as an extension

$$0 \to \mathcal{O}_X \to V \to \mathcal{O}_X(\sigma - (p_g + 2)f) \to 0.$$

Proof. By the proof of Proposition 9, we obtain V from V_0 by making a single allowable elementary modification. Thus, there is an exact sequence

$$0 \to V \to V_0 \to i_* \mathcal{O}_f(x) \to 0,$$

at least if f is a smooth fiber, with a similar result holding if f is singular. In particular, we have the map $\mathcal{O}_X \to V_0$, and either its image is contained in V, or there is at least an induced map $\mathcal{O}_X(-f) \to V$. If the induced map $\mathcal{O}_X(-f) \to V$ vanishes along a curve, it must do so along a curve supported in fibers, for otherwise the map $\mathcal{O}_X \to V_0$ would have to vanish along a curve. Thus, if the map $\mathcal{O}_X(-f) \to V$ vanishes along a curve C, the only possibility is that $C = f$ and the map $\mathcal{O}_X \to V_0$ has image in V with torsion free quotient. Thus, either we can find an exact sequence

$$0 \to \mathcal{O}_X(-f) \to V \to \mathcal{O}_X(\sigma - (p_g + 1)f) \otimes I_Z \to 0$$

for some 0-dimensional subscheme Z, or we can find an exact sequence

$$0 \to \mathcal{O}_X \to V \to \mathcal{O}_X(\sigma - (p_g + 2)f) \otimes I_Z \to 0.$$

Keeping track of the Chern classes shows that Z is a reduced point in the first case and that $Z = \emptyset$ in the second case. $\square$

Next we must determine when extensions as described above are stable:

Proposition 22.

(i) *If V is an extension of the form*

$$0 \to \mathcal{O}_X(-f) \to V \to \mathcal{O}_X(\sigma - (p_g + 1)f) \otimes \mathfrak{m}_x \to 0,$$

then V is locally free if and only if the extension is nonsplit, and there is a unique such V. Moreover, V is stable if and only if $x \notin \sigma$. If V is not stable, then there is an exact sequence

$$0 \to \mathcal{O}_X(\sigma - (p_g + 2)f) \to V \to \mathcal{O}_X \to 0,$$

where the inclusion $\mathcal{O}_X(\sigma - (p_g + 2)f) \to V$ gives the maximal destabilizing sub-line bundle of V.
(ii) *An extension of the form*

$$0 \to \mathcal{O}_X \to V \to \mathcal{O}_X(\sigma - (p_g + 2)f) \to 0$$

is stable if and only if it is nonsplit. Moreover, the set of all such nonsplit extensions is parametrized by a $\mathbb{P}^1$ which we may identify with σ.

Proof. (i) Suppose that V is an extension as in (i) above. We have an exact sequence

$$0 \to H^1(\mathcal{O}_X(-\sigma + p_g f)) \to \operatorname{Ext}^1(\mathcal{O}_X(\sigma - (p_g + 1)f) \otimes \mathfrak{m}_x, \mathcal{O}_X(-f))$$
$$\to H^0(\mathbb{C}_x) \to H^2(\mathcal{O}_X(-\sigma + p_g f)).$$

By Lemma 19, $H^1(\mathcal{O}_X(-\sigma + p_g f)) = H^2(\mathcal{O}_X(-\sigma + p_g f)) = 0$. Thus, $\operatorname{Ext}^1(\mathcal{O}_X(\sigma - (p_g + 1)f) \otimes \mathfrak{m}_x, \mathcal{O}_X(-f))$ maps isomorphically onto $H^0(\mathbb{C}_x)$ and so has dimension 1. Thus, up to isomorphism there is a unique nonsplit extension and it is locally free.

Next suppose that V as above is not stable. Then a destabilizing sub-line bundle of V would restrict to a destabilizing sub-line bundle on the generic fiber. Thus, it must be of the form $\mathcal{O}_X(\sigma + af)$ for some a. Since there is a nonzero map from $\mathcal{O}_X(\sigma + af)$ to V, whose image clearly cannot be contained in the sub-line bundle $\mathcal{O}_X(-f)$, there is an induced nonzero map $\mathcal{O}_X(\sigma + af) \to \mathcal{O}_X(\sigma - (p_g + 1)f) \otimes \mathfrak{m}_x$. This is only possible if there is a section of $\mathcal{O}_X(-(p_g+1+a)f) \otimes \mathfrak{m}_x$, or in other words if $a < -(p_g + 1)$. Now the maximal destabilizing sub-line bundle of V has torsion free quotient, so that there is an exact sequence

$$0 \to \mathcal{O}_X(\sigma + af) \to V \to \mathcal{O}_X(-(p_g + 2 + a)f) \otimes I_Z \to 0$$

for some 0-dimensional subscheme Z. A calculation gives

$$-p_1(\mathrm{ad}\, V) = -(\sigma + (p_g + 2 + 2a)f)^2 + 4\ell(Z) \geq 3(p_g + 1) + 2 + 4\ell(Z),$$

with equality holding if and only if $a = -(p_g + 2)$ and $Z = \emptyset$. But for the V under consideration, we have assumed that $-p_1(\mathrm{ad}\, V) = 3(p_g + 1) + 2$. Thus, we must have $Z = \emptyset$ and the maximal destabilizing sub-line bundle is $\mathcal{O}_X(\sigma - (p_g + 2)f)$. Note that, in this case, there is an exact sequence

$$0 \to \mathcal{O}_X(\sigma - (p_g + 2)f) \to V \to \mathcal{O}_X \to 0,$$

which is the reverse of the extensions in the second possibility (ii) for V above.

By the above argument, an extension of the form

$$0 \to \mathcal{O}_X(-f) \to V \to \mathcal{O}_X(\sigma - (p_g + 1)f) \otimes \mathfrak{m}_x \to 0$$

is unstable if the natural map $\mathcal{O}_X(\sigma - (p_g + 2)f) \to \mathcal{O}_X(\sigma - (p_g + 1)f) \otimes \mathfrak{m}_x$ lifts to a map $\mathcal{O}_X(\sigma - (p_g + 2)f) \to V$. Equivalently, we ask if the nonzero section of $\mathcal{O}_X(f) \otimes \mathfrak{m}_x$ lifts to a section of $V \otimes \mathcal{O}_X(-\sigma + (p_g + 2)f)$. Now $V \otimes \mathcal{O}_X(-\sigma + (p_g + 2)f)$ is an extension of $\mathcal{O}_X(f) \otimes \mathfrak{m}_x$ by $\mathcal{O}_X(-\sigma + (p_g + 1)f)$. Furthermore, $\mathrm{Ext}^1(\mathcal{O}_X(f) \otimes \mathfrak{m}_x, \mathcal{O}_X(-\sigma + (p_g + 1)f))$ is equal to

$$\mathrm{Ext}^1(\mathcal{O}_X(\sigma - (p_g + 1)f) \otimes \mathfrak{m}_x, \mathcal{O}_X(-f)),$$

which as we have seen has dimension 1. The induced map $\mathcal{O}_X \to \mathcal{O}_X(f) \otimes \mathfrak{m}_x$ induces a commutative diagram

$$\mathrm{Hom}(\mathcal{O}_X(f) \otimes \mathfrak{m}_x, \mathcal{O}_X(f) \otimes \mathfrak{m}_x) \longrightarrow \mathrm{Ext}^1(\mathcal{O}_X(f) \otimes \mathfrak{m}_x, \mathcal{O}_X(-\sigma + (p_g + 1)f))$$

$$\downarrow \qquad\qquad\qquad\qquad\qquad\qquad \downarrow$$

$$\mathrm{Hom}(\mathcal{O}_X, \mathcal{O}_X(f) \otimes \mathfrak{m}_x) \qquad \longrightarrow \qquad \mathrm{Ext}^1(\mathcal{O}_X, \mathcal{O}_X(-\sigma + (p_g + 1)f)),$$

and the kernel of the lower map

$$\mathrm{Hom}(\mathcal{O}_X, \mathcal{O}_X(f) \otimes \mathfrak{m}_x) \to \mathrm{Ext}^1(\mathcal{O}_X, \mathcal{O}_X(-\sigma + (p_g + 1)f))$$

is the image of

$$\mathrm{Hom}(\mathcal{O}_X, V \otimes \mathcal{O}_X(-\sigma + (p_g + 2)f)) = \mathrm{Hom}(\mathcal{O}_X(\sigma - (p_g + 2)f), V).$$

By definition $\mathrm{Id} \in \mathrm{Hom}(\mathcal{O}_X(f) \otimes \mathfrak{m}_x, \mathcal{O}_X(f) \otimes \mathfrak{m}_x)$ maps to the extension class ξ corresponding to V. Clearly, Id restricts to a nonzero element of $\mathrm{Hom}(\mathcal{O}_X, \mathcal{O}_X(f) \otimes \mathfrak{m}_x)$, which is mod scalars the same as the unique nonzero section of $\mathcal{O}_X(f) \otimes \mathfrak{m}_x$. Chasing through the diagram we see that the nonzero section of $\mathcal{O}_X(f) \otimes \mathfrak{m}_x$ lifts to give a nonzero map $\mathcal{O}_X(\sigma - (p_g + 2)f) \to V$ if and only if the extension class ξ corresponding to V maps to 0 in $\mathrm{Ext}^1(\mathcal{O}_X, \mathcal{O}_X(-\sigma + (p_g + 1)f))$. Now, in case x is a smooth point of f, the quotient of $\mathcal{O}_X(f) \otimes \mathfrak{m}_x$ by $\mathcal{O}_X$ is easily seen to be $\mathcal{O}_f(-x)$, where f is the unique fiber containing x, as we see by looking at the exact sequence

$$0 \to (\mathcal{O}_X(f) \otimes \mathfrak{m}_x)/\mathcal{O}_X \to \mathcal{O}_X(f)/\mathcal{O}_X \to \mathbb{C}_x \to 0.$$

Similarly, if x is the singular point of a singular fiber, then $(\mathcal{O}_X(f) \otimes \mathfrak{m}_x)/\mathcal{O}_X$ is the ideal sheaf of x in f. Assuming for simplicity that x is a smooth point of f, the restriction map

$$\mathrm{Ext}^1(\mathcal{O}_X(f) \otimes \mathfrak{m}_x, \mathcal{O}_X(-\sigma + (p_g + 1)f)) \to \mathrm{Ext}^1(\mathcal{O}_X, \mathcal{O}_X(-\sigma + (p_g + 1)f))$$

fits into the exact sequence

$$\mathrm{Hom}(\mathcal{O}_X, \mathcal{O}_X(-\sigma + (p_g + 1)f)) \to \mathrm{Ext}^1(\mathcal{O}_f(-x), \mathcal{O}_X(-\sigma + (p_g + 1)f)) \to$$
$$\mathrm{Ext}^1(\mathcal{O}_X(f) \otimes \mathfrak{m}_x, \mathcal{O}_X(-\sigma + (p_g + 1)f)) \to \mathrm{Ext}^1(\mathcal{O}_X, \mathcal{O}_X(-\sigma + (p_g + 1)f)).$$

Furthermore

$$\mathrm{Hom}(\mathcal{O}_X, \mathcal{O}_X(-\sigma + (p_g + 1)f)) = H^0(\mathcal{O}_X(-\sigma + (p_g + 1)f)) = 0.$$

By Chapter 2, Exercise 15, we see that

$$\mathrm{Ext}^1(\mathcal{O}_f(-x), \mathcal{O}_X(-\sigma + (p_g + 1)f)) = H^0(\mathcal{O}_f(x - p)),$$

where $p = \sigma \cap f$, with a similar result holding if x is a singular point of f. Thus, this group has dimension 0 if $x \neq p$ and has dimension 1 if $x = p$. Since $\mathrm{Ext}^1(\mathcal{O}_X(f) \otimes \mathfrak{m}_x, \mathcal{O}_X(-\sigma + (p_g + 1)f))$ also has dimension 1, we see that the map

$$\mathrm{Ext}^1(\mathcal{O}_X(f) \otimes \mathfrak{m}_x, \mathcal{O}_X(-\sigma + (p_g + 1)f)) \to \mathrm{Ext}^1(\mathcal{O}_X, \mathcal{O}_X(-\sigma + (p_g + 1)f))$$

is injective if $x \notin \sigma$, and is the zero map if $x \in \sigma$. Thus, there is no possible destabilizing sub-line bundle if $x \notin \sigma$, whereas V has the destabilizing sub-line bundle $\mathcal{O}_X(\sigma - (p_g + 2)f)$ if $x \in \sigma$.

(ii) In this case $\mathrm{Ext}^1(\mathcal{O}_X(\sigma - (p_g + 2)f), \mathcal{O}_X) = H^1(\mathcal{O}_X(-\sigma + (p_g + 2)f))$. By Lemma 19, the cohomology group has dimension 2, and the set of nonsplit extensions is therefore $\mathbb{P}^1$. Arguing as in the proof of (i) above, a destabilizing sub-line bundle would have to be of the form $\mathcal{O}_X(\sigma + af)$ with $a \leq -(p_g + 2)$, and the case $a = -(p_g + 2)$ would split the exact sequence. Assuming that the destabilizing sub-line bundle has torsion free quotient, we reach a contradiction by working out $p_1(\mathrm{ad}\, V)$ using the destabilizing sub-line bundle along the lines of the proof of the corresponding fact for (i) above; the details are left to the reader. Thus, the bundle V is stable if and only if the extension is not split.

The proof of Lemma 19 identifies $H^1(\mathcal{O}_X(-\sigma + (p_g + 2)f))$ with $H^0(\mathbb{P}^1, \mathcal{O}_{\mathbb{P}^1}(1))$, where we can think of the $\mathbb{P}^1$ as the base of the elliptic surface, identified with σ. A section $s \in H^0(\mathbb{P}^1, \mathcal{O}_{\mathbb{P}^1}(1))$ vanishes at a unique point p of $\mathbb{P}^1$. Running through the identifications, this shows that, if V is the extension of $\mathcal{O}_X(\sigma - (p_g + 2)f)$ by $\mathcal{O}_X$ corresponding to $s \in H^0(\mathbb{P}^1, \mathcal{O}_{\mathbb{P}^1}(1))$, then the restriction of the extension V to a fiber f splits if and only if f lies above p. Note that $V|f$ is given by the extension

$$0 \to \mathcal{O}_f \to V|f \to \mathcal{O}_f(p) \to 0,$$

where we identify $p \in \mathbb{P}^1$ with $p = \sigma \cap f$, the point of σ lying over p. Thus, if s vanishes at p and f is the fiber over p, then $V|f = \mathcal{O}_f \oplus \mathcal{O}_f(p)$, and otherwise $V|f$ is a nonsplit extension. $\square$

We will refer to the stable bundles satisfying the hypotheses of (i) of Proposition 22 as bundles of Type (1), and those satisfying the hypotheses of (ii) as Type (2) bundles. We can interpret Proposition 22 as follows: the 2-dimensional moduli space $\mathfrak{M}$ we are considering breaks up into two pieces: one piece corresponds to the extensions of Type (1) in Proposition 22, where $x \notin \sigma$, and the other piece corresponds to Type (2). Now given $x \in X$, there is a unique Type (1) extension V corresponding to x, so that the first piece looks like $X - \sigma$. As for nonsplit Type (2) extensions, we have seen that they are naturally identified with σ. Our goal now will be to describe how to fit together the two pieces $X - \sigma$ and σ so as to obtain a compact moduli space isomorphic to X.

We begin by constructing a moduli space for all Type (1) extensions. The natural way to fit together extensions of the form

$$0 \to \mathcal{O}_X(-f) \to V \to \mathcal{O}_X(\sigma - (p_g + 1)f) \otimes \mathfrak{m}_x \to 0$$

is to work on $X \times X$, where we view the first factor as the surface and the second as the moduli space. We try to find an extension of the form

$$0 \to \pi_1^* \mathcal{O}_X(-f) \to \mathcal{W} \to \pi_1^* \mathcal{O}_X(\sigma - (p_g + 1)f) \otimes I_{\mathbb{D}} \to 0,$$

where as usual $\mathbb{D}$ is the diagonal. Such an extension would restrict to give a Type (1) extension on every slice $X \times \{x\}$. To carry out the construction, we would need to find an everywhere generating section of

$$\mathcal{E}xt^1(\pi_1^* \mathcal{O}_X(\sigma - (p_g + 1)f) \otimes I_{\mathbb{D}}, \pi_1^* \mathcal{O}_X(-f))$$

and then lift this section to an element of

$$\mathrm{Ext}^1(\pi_1^* \mathcal{O}_X(\sigma - (p_g + 1)f) \otimes I_{\mathbb{D}}, \pi_1^* \mathcal{O}_X(-f)).$$

Now as we have seen in Chapter 2, Exercise 16, $\mathcal{E}xt^1(\pi_1^* \mathcal{O}_X(\sigma - (p_g + 1)f) \otimes I_{\mathbb{D}}, \pi_1^* \mathcal{O}_X(-f))$ is isomorphic to

$$\det N_{\mathbb{D}/X \times X} \otimes (\pi_1^* \mathcal{O}_X(-\sigma + p_g f))|\mathbb{D}.$$

Since $N_{\mathbb{D}/X \times X}$ is just the tangent bundle on X, under the natural identification of $\mathbb{D}$ with X, we see that the Ext sheaf is just the line bundle

$$K_X^{-1} \otimes \mathcal{O}_X(-\sigma + p_g f) = \mathcal{O}_X(-\sigma + f).$$

Thus, we instead try to find $\mathcal{W}$ as an extension

$$0 \to \pi_1^* \mathcal{O}_X(-f) \otimes \pi_2^* \mathcal{O}_X(\sigma - f) \to \mathcal{W} \to \pi_1^* \mathcal{O}_X(\sigma - (p_g + 1)f) \otimes I_{\mathbb{D}} \to 0.$$

The point of adding the extra factor $\pi_2^* \mathcal{O}_X(\sigma - f)$ to the first term in the exact sequence is to cancel out the Ext sheaf, and indeed we now have

$$Ext^1(\pi_1^* \mathcal{O}_X(\sigma - (p_g + 1)f) \otimes I_{\mathbb{D}}, \pi_1^* \mathcal{O}_X(-f) \otimes \pi_2^* \mathcal{O}_X(\sigma - f)) = \mathcal{O}_{\mathbb{D}},$$

which has an everywhere generating section, unique mod scalars. The obstruction to lifting this section back to $Ext^1(\pi_1^* \mathcal{O}_X(\sigma - (p_g + 1)f) \otimes I_{\mathbb{D}}, \pi_1^* \mathcal{O}_X(-f) \otimes \pi_2^* \mathcal{O}_X(\sigma - f))$ lives in

$$H^2(Hom(\pi_1^* \mathcal{O}_X(\sigma - (p_g + 1)f) \otimes I_{\mathbb{D}}, \pi_1^* \mathcal{O}_X(-f) \otimes \pi_2^* \mathcal{O}_X(\sigma - f))$$
$$= H^2(\pi_1^* \mathcal{O}_X(-\sigma + p_g f) \otimes \pi_2^* \mathcal{O}_X(\sigma - f)).$$

An easy application of the Leray spectral sequence (or the Künneth formula) and Lemma 19 shows that this group is zero. Thus, we may construct the extension $\mathcal{W}$.

Unfortunately, $\mathcal{W}$ is not the right universal bundle, since on every slice $X \times \{x\}$ where $x \in \sigma$ the bundle $\mathcal{W}|X \times \{x\}$ is unstable. However, let us record the Chern classes of $\mathcal{W}$ anyway:

$$c_1(\mathcal{W}) = \pi_1^*(\sigma - (p_g + 2)f) + \pi_2^*(\sigma - f),$$
$$c_2(\mathcal{W}) = \pi_1^*(\sigma - (p_g + 1)f) \cdot \pi_2^*(\sigma - f) - \pi_1^* pt + \mathbb{D},$$
$$p_1(ad\,\mathcal{W}) = -2\pi_1^*(\sigma - p_g f) \cdot \pi_2^*(\sigma - f) - 4\mathbb{D} + \cdots,$$

where as usual the omitted terms do not affect slant product.

We now deal with the problem that the restriction of $\mathcal{W}$ to the slice $X \times \{p\}$ is unstable for every $p \in \sigma$. In fact, if W_p is this restriction, then by Proposition 22 there is an exact sequence

$$0 \to \mathcal{O}_X(\sigma - (p_g + 2)f) \to W_p \to \mathcal{O}_X \to 0.$$

Since the destabilizing quotient bundle is unique, we have $Hom(W_p, \mathcal{O}_X) \cong \mathbb{C}$ for all $p \in \sigma$. Thus, if $p_2 \colon X \times \sigma \to \sigma$ is the projection, then

$$R^0 p_{2*} Hom(\mathcal{W}|X \times \sigma, \mathcal{O}_{X \times \sigma}) = \mathcal{L}$$

is a line bundle on σ. We will not have to know what $\mathcal{L}$ is, though! It follows that the induced map

$$\mathcal{W}|X \times \sigma \to p_2^* \mathcal{L}^{-1}$$

is surjective. If $j \colon X \times \sigma \to X \times X$ is the inclusion, we can therefore make the elementary modification defined by

$$0 \to \mathcal{V} \to \mathcal{W} \to j_*(p_2^* \mathcal{L}^{-1}) \to 0.$$

For each $p \in \sigma$, there is thus an exact sequence for V_p, the restriction of $\mathcal{V}$ to the slice $X \times \{p\}$, reversing the destabilizing sequence for W_p:

$$0 \to \mathcal{O}_X \to V_p \to \mathcal{O}_X(\sigma - (p_g + 2)f) \to 0.$$

Thus, if this sequence does not split, then V_p is a Type (2) extension, and we have solved the problem of how to fit together the two different types of extensions. We shall just state the result we need, referring to [36] for the proof:

Proposition 23. *In the above notation, the restriction of $\mathcal{V}$ to every slice $X \times \{x\}$ is stable, and $\mathcal{V}$ identifies the second factor X with the corresponding moduli space of stable bundles.* $\square$

Now let us calculate the μ-map. By general formulas for elementary modifications from Chapter 2, Lemma 16, we find that

$$p_1(\operatorname{ad} \mathcal{V}) = p_1(\operatorname{ad} \mathcal{W}) + 2c_1(\mathcal{W}) \cdot [X \times \sigma] + [X \times \sigma]^2 - 4j_*c_1(p_2^*\mathcal{L}^{-1}).$$

Omitting terms that do not affect slant product, we obtain

$$-2\pi_1^*(\sigma - p_g f) \cdot \pi_2^*(\sigma - f) - 4\mathbb{D} + 2\pi_1^*(\sigma - (p_g + 2)f) \cdot \pi_2^*\sigma + \cdots$$

and so $-4\mu(\alpha)$ is equal to

$$[-2(\sigma \cdot \alpha) + 2p_g(f \cdot \alpha)](\sigma - f) - 4\alpha + 2(\sigma \cdot \alpha)\sigma - 2(p_g + 2)(f \cdot \alpha)\sigma$$
$$= 2(\sigma \cdot \alpha)f - 4(f \cdot \alpha)\sigma - 2p_g(f \cdot \alpha)f - 4\alpha.$$

This agrees with our previous calculation for $-4\mu(\alpha)$ in Proposition 15 of the last section.

Similar arguments will identify a Zariski open subset of the moduli space in general. We have the following straightforward generalization of the arguments of this section:

Theorem 24. *Let Z consist of t points on X lying in distinct fibers of π, such that $Z \cap \sigma = \emptyset$. Then up to isomorphism there is a unique vector bundle V given as an extension*

$$0 \to \mathcal{O}_X(-tf) \to V \to \mathcal{O}_X(\sigma - (p_g + 1)f) \otimes I_Z \to 0.$$

Moreover, V is stable. Finally, $h^0(V \otimes \mathcal{O}_X(tf)) = 1$, so that Z is uniquely determined by V. $\square$

Theorem 24 gives a second identification of a Zariski open subset of the moduli space with a Zariski open subset of $\operatorname{Hilb}^t X$, and it is easy to check that this identification agrees with the one given by Theorem 12. For $t = 2$, the moduli space is in fact isomorphic to $\operatorname{Hilb}^2 X$ [36], but this is no longer the case for $t > 2$.

Vector bundles with trivial determinant

In this section, we begin the description of stable rank 2 vector bundles V over X with $\det V = \mathcal{O}_X$, or more generally such that $\det V$ is pulled back from the base curve. For simplicity, we return to the convention that the base curve of the elliptic surface X is $\mathbb{P}^1$, that all fibers of π are either smooth or nodal, and that X has a section σ. For more general results along these lines, we refer to [42]. In the next section, we will discuss what happens when there are multiple fibers.

If V is stable with respect to a suitable ample divisor, then $V|f$ is semistable for almost all f. After a sequence of allowable elementary modifications, we can assume that $V|f$ is semistable for all f. Note however that $V|f$ is a rank 2 vector bundle of degree 0 on f, and hence is never actually stable. In fact, $V|f$ also fails to be simple, by Theorem 25 below. Thus, the semistability of $V|f$ on one or all fibers does not guarantee the stability or semistability of V. It is also possible that we begin with a stable V and make a sequence of allowable elementary modifications, and the final result is actually unstable. We shall give a sufficient condition for V and all of its elementary modifications to be stable in Theorem 30 below. Finally and most interestingly, in the case of fiber degree 1, the set of stable bundles on a fixed fiber f with a given determinant of degree 1 is a single point, and this led to the bundle V_0 of Proposition 9, which was unique up to twisting by the pullback of a line bundle of the base. In the case of degree 0, the moduli space of semistable bundles is positive-dimensional (in fact it is isomorphic to $\mathbb{P}^1$), and this allows for much more complicated behavior.

Following the above comments, we begin by analyzing the moduli space of semistable bundles of determinant zero on a fiber f. If f is smooth, the answer is given by Theorem 6 of Chapter 2:

Theorem 25. *Let E be a smooth elliptic curve and let V be a semistable rank 2 vector bundle over E. Then either $V = \lambda \oplus \lambda^{-1}$, where λ is a line bundle of degree 0 on E, or V is of the form $\mathcal{E} \otimes \lambda$, where $\mathcal{E}$ is the unique nonsplit extension of $\mathcal{O}_E$ and λ is one of the four line bundles on E with $\lambda^{\otimes 2} = \mathcal{O}_E$. Moreover, as $\mathbb{C}$-algebras,*

$$\operatorname{Hom}(V,V) = \begin{cases} \mathbb{C} \times \mathbb{C}, & \text{if } V \cong \lambda \oplus \lambda^{-1}, \lambda \neq \lambda^{-1}, \\ \mathbb{C}[t]/t^2, & \text{if } V \cong \mathcal{E} \otimes \lambda, \lambda^{\otimes 2} = \mathcal{O}_E, \\ M_2(\mathbb{C}), & \text{if } V \cong \lambda \oplus \lambda, \lambda = \lambda^{-1}. \end{cases}$$

Proof. All but the last statement is contained in the statement of Theorem 6 of Chapter 2 (see also Theorem 9 in Chapter 4). To prove the statement about the homomorphisms of V, note that, if $\lambda \neq \lambda^{-1}$, then as $\deg \lambda = 0$, every homomorphism from λ to λ^{-1} is 0, and the same is true for every homomorphism from λ^{-1} to λ. Thus, (as algebras)

$$\text{Hom}(\lambda \oplus \lambda^{-1}, \lambda \oplus \lambda^{-1}) = \text{Hom}(\lambda, \lambda) \oplus \text{Hom}(\lambda^{-1}, \lambda^{-1}) = \mathbb{C} \oplus \mathbb{C}.$$

For the other statements, we can assume that $\lambda = \mathcal{O}_E$ and must calculate $\text{Hom}(\mathcal{E}, \mathcal{E})$ and $\text{Hom}(\mathcal{O}_E \oplus \mathcal{O}_E, \mathcal{O}_E \oplus \mathcal{O}_E)$. Clearly, the last group is isomorphic to $M_2(\mathbb{C})$, the algebra of 2×2 complex matrices, and it suffices to calculate $\text{Hom}(\mathcal{E}, \mathcal{E})$. There is a rank 1 endomorphism $\varphi \colon \mathcal{E} \to \mathcal{E}$ defined by $\mathcal{E} \to \mathcal{O}_f \to \mathcal{E}$, i.e., take the defining quotient map to $\mathcal{O}_f$, followed by the defining inclusion of $\mathcal{O}_f$ in $\mathcal{E}$. Clearly, φ is a nontrivial element of $\text{Hom}(\mathcal{E}, \mathcal{E})$ of square 0, and it suffices to show that every endomorphism can be uniquely written as $a \,\text{Id} + b\varphi$. Since the coboundary map ∂ arising from the long exact sequence

$$0 \to H^0(\mathcal{O}_f) \to H^0(\mathcal{E}) \to H^0(\mathcal{O}_f) \xrightarrow{\partial} H^1(\mathcal{O}_f)$$

is nonzero (its image is the extension class defining $\mathcal{E}$), $h^0(\mathcal{E}) = 1$. Fix a nonzero section s of $\mathcal{E}$ and let $\psi \in \text{Hom}(\mathcal{E}, \mathcal{E})$. Then $\psi(s) = as$ for some $a \in \mathbb{C}$, and thus $(\psi - a \,\text{Id})(s) = 0$. It follows that $\psi - a \,\text{Id}$ factors through the quotient map and defines a morphism $\mathcal{E}/\mathcal{O}_f \cong \mathcal{O}_f \to \mathcal{E}$. Necessarily the image of this map is contained in the distinguished subbundle $\mathcal{O}_f$ of $\mathcal{E}$, and from this it is clear that $\psi - a \,\text{Id}$ is some multiple $b\varphi$ of φ. Thus, we have shown that $\text{Hom}(\mathcal{E}, \mathcal{E}) \cong \mathbb{C}[t]/(t^2)$, as claimed. $\square$

Note that the bundles $\mathcal{E} \otimes \lambda$ and $\lambda \oplus \lambda$ are S-equivalent in the sense of Chapter 6. Thus, they must be identified as points of the moduli space of semistable bundles on f. After this identification, the moduli space is then $\text{Pic}^0 f/ \pm 1 \cong \mathbb{P}^1$, where $\text{Pic}^0 f$, the set of line bundles on f of degree 0, is identified with f via the choice of an origin p. As we shall see shortly, if we try to make a universal bundle over $f \times \mathbb{P}^1$, we must use the bundles $\mathcal{E} \otimes \lambda$ and not $\lambda \oplus \lambda$. One reason for this is that $\dim \text{Hom}(\mathcal{E} \otimes \lambda, \mathcal{E} \otimes \lambda) = 2$, which is the same as the dimension of the algebra of homomorphisms for the generic bundle. Another and related reason is that, in a certain sense, the bundles $\lambda \oplus \lambda$ are more special than $\mathcal{E} \otimes \lambda$: the bundle $\lambda \oplus \lambda$ has a small deformation which is isomorphic to $\mathcal{E} \otimes \lambda$, but not the other way around.

Before we discuss the problem of trying to find a universal bundle for a smooth elliptic curve f, let us give the analogue of Theorem 25 in the singular case, whose proof is left as an exercise:

Theorem 26. *Let f be a nodal curve of arithmetic genus 1. Then the semistable rank 2 vector bundles V on f with $\det V = \mathcal{O}_f$ are the bundles of the form:*

(i) $\lambda \oplus \lambda^{-1}$, *where λ is a line bundle on f of degree 0;*
(ii) $\mathcal{E} \otimes \lambda$, *where $\mathcal{E}$ is the unique nontrivial extension of $\mathcal{O}_f$ by $\mathcal{O}_f$ and λ is one of the two line bundles on f with $\lambda^{\otimes 2} = \mathcal{O}_f$;*

(iii) *The unique nontrivial locally free extension $\mathcal{G}$ with $\det \mathcal{G} = \mathcal{O}_f$ of the form*

$$0 \to F \to \mathcal{G} \to F \to 0,$$

where F is the unique torsion free rank 1 sheaf of degree 0 which is not locally free.

Moreover, in cases (i) *and* (ii) *the algebra $\mathrm{Hom}(V,V)$ is as described in Theorem 25. In case* (iii), *$\mathrm{Hom}(\mathcal{G},\mathcal{G}) \cong \mathbb{C}[t]/(t^2)$.* $\square$

Note that, in case f is nodal, Theorem 26 says that up to S-equivalence the moduli space of semistable bundles on f with trivial determinant is just $\mathrm{Pic}^0 f/\pm 1$ plus the remaining bundle $\mathcal{G}$. Now $\mathrm{Pic}^0 f \cong \mathbb{C}^*$, and $\mathrm{Pic}^0 f/\pm 1 \cong \mathbb{C}$. Adding in the remaining bundle $\mathcal{G}$ then compactifies $\mathbb{C}$ to a $\mathbb{P}^1$ as in the smooth case.

For both smooth and nodal fibers f, the bundles $\mathcal{E} \otimes \lambda$ are distinguished within a given S-equivalence class by the requirement that $\dim \mathrm{Hom}(V,V)$ is as small as possible. We make this a definition:

Definition 27. Let f be a smooth elliptic curve or a nodal curve of arithmetic genus 1, and let V be a rank 2 semistable vector bundle on f with $\det V = \mathcal{O}_f$. Then V is *regular* if $\dim \mathrm{Hom}(V,V) = 2$.

Next we want to reinterpret the moduli space $\mathbb{P}^1$, in both the smooth and the nodal case, to make clear that it behaves like a coarse moduli space. One way to do so, given the choice of origin $p \in f$, is the following: if λ is a line bundle of degree 0 on f, then we can write $\lambda = \mathcal{O}_f(q - p)$ for a unique $q \in f$, and similarly $\lambda^{-1} = \mathcal{O}_f(r - p)$. The condition that the line bundles $\mathcal{O}_f(q - p)$ and $\mathcal{O}_f(r - p)$ are inverse to each other is exactly the condition that $q + r$ is linearly equivalent to $2p$. Moreover, the map which associates to the vector bundle $\lambda \oplus \lambda^{-1}$ the point $q + r \in |2p| \cong \mathbb{P}^1$ defines an isomorphism from $\mathrm{Pic}^0 f/\pm 1$ to $|2p|$, in case f is smooth. To handle the singular case, as well as to make clear that the map in question really is a morphism, we have the following:

Theorem 28. Let f be a reduced irreducible curve of arithmetic genus 1, and fix a smooth point $p \in f$. Let V be a semistable vector bundle on f of rank 2 with $\det V = \mathcal{O}_f$. Then $h^0(V \otimes \mathcal{O}_f(p)) = 2$, and the natural map

$$\varphi : \mathcal{O}_f^2 = H^0(V \otimes \mathcal{O}_f(p)) \otimes_{\mathbb{C}} \mathcal{O}_f \to V \otimes \mathcal{O}_f(p)$$

is an isomorphism on a Zariski open subset of f. Thus, $\det \varphi$ is a well-defined nonzero section of $\mathcal{O}_f(2p)$, and so V defines an element of $|2p|$ agreeing with the one given above in case $V = \lambda \oplus \lambda^{-1}$. More generally, let S be a reduced scheme and let $\mathcal{V} \to f \times S$ be a vector bundle such that $\mathcal{V}|f \times \{s\}$ is a semistable rank 2 vector bundle over f with trivial *determinant for all $s \in S$. Then there is an induced morphism $\Phi \colon S \to |2p|$ such that $\Phi(s)$ is the point defined by $\mathcal{V}|f \times \{s\}$ for every $s \in S$.*

Proof. By Riemann-Roch, $h^0(V \otimes \mathcal{O}_f(p)) - h^1(V \otimes \mathcal{O}_f(p)) = 2$. On the other hand, using Serre duality, $h^1(V \otimes \mathcal{O}_f(p)) = h^0(V \otimes \mathcal{O}_f(-p)) = 0$, since by semistability there are no nonzero maps $\mathcal{O}_f(p) \to V$. Thus, $h^0(V \otimes \mathcal{O}_f(p)) = 2$. Now suppose that the image of φ is contained in a rank 1 subsheaf I of $V \otimes \mathcal{O}_f(p)$. By semistability, $\deg I \leq 1$, and by construction $h^0(I) = h^0(V \otimes \mathcal{O}_f(p)) = 2$. But by definition $\deg I = h^0(I) - h^1(I) \leq 1$, and using Serre duality $h^1(I) = \dim \operatorname{Hom}(I, \mathcal{O}_f) = 0$, by (ii) of Lemma 3. Hence $h^0(I) \leq 1$. This contradicts the assumption that $h^0(I) = 2$. It follows that the image of φ is a rank 2 subsheaf of V (not necessarily locally free), as claimed, and thus that $\det \varphi$ is a nonzero section of $\det(\mathcal{O}_f^2)^{-1} \otimes \det(V \otimes \mathcal{O}_f(p)) = \mathcal{O}_f(2p)$.

In case $V = \lambda \oplus \lambda^{-1}$, we can write $V \otimes \mathcal{O}_f(p) = \mathcal{O}_f(q) \oplus \mathcal{O}_f(r)$, and the map φ fails to be an isomorphism exactly at q and r. Thus, $\det \varphi$ vanishes at $q + r$, as claimed. (It is straightforward to check that, if $V = \mathcal{E} \otimes \lambda$ with $\lambda^{\otimes 2} = \mathcal{O}_f$ and $\lambda = \mathcal{O}_f(q - p)$, then $\det \varphi$ vanishes at $2q$. If f is nodal and $V = \mathcal{G}$, then $\det \varphi$ is the unique element of $|2p|$ supported at the singular point.)

In the relative case, since $\det \mathcal{V}$ is trivial on every fiber $f \times \{s\}$, $\det \mathcal{V} = \pi_2^* \mathcal{M}$ for some line bundle $\mathcal{M}$ on S. It follows from base change that $\pi_{2*}(\mathcal{V} \otimes \pi_1^* \mathcal{O}_f(p))$ is a rank 2 vector bundle on S, and that the natural map $\pi_2^* \pi_{2*}(\mathcal{V} \otimes \pi_1^* \mathcal{O}_f(p)) \to \mathcal{V} \otimes \pi_1^* \mathcal{O}_f(p)$ restricts on every fiber to give the map φ defined above. Thus, the determinant of the above map is a section of $\pi_2^* \mathcal{L}^{-1} \otimes \pi_2^* \mathcal{M} \otimes \pi_1^* \mathcal{O}_f(2p)$, where $\mathcal{L}$ is the determinant of the rank 2 bundle $\pi_{2*}(\mathcal{V} \otimes \pi_1^* \mathcal{O}_f(p))$ on S. There is then an induced morphism $\Phi \colon S \to |2p|$, which agrees with the above construction on every fiber. $\square$

Suppose now that $\pi \colon X \to \mathbb{P}^1$ is an elliptic surface with a section σ, and such that all fibers are smooth or nodal, and that V is a rank 2 vector bundle over X such that $\det V$ is pulled back from the base $\mathbb{P}^1$. Assume further that the restriction of V to every fiber f of π is semistable. For each f, we have the space $H^0(f; \mathcal{O}_f(2p))$ and the corresponding projective space $|2p|$. Globally over X, there is the rank 2 vector bundle $\pi_* \mathcal{O}_X(\sigma)$ and the associated $\mathbb{P}^1$-bundle $\mathbb{P}(\pi_* \mathcal{O}_X(\sigma))$. In the last chapter, we identified $\pi_* \mathcal{O}_X(\sigma)$ with $\mathcal{O}_{\mathbb{P}^1} \oplus L^{-2}$, where $L = \mathcal{O}_{\mathbb{P}^1}(p_g + 1)$. Thus, setting $k = p_g + 1$, we see that $\mathbb{P}(\pi_* \mathcal{O}_X(\sigma))$ is the rational ruled surface $\mathbb{F}_{2k}$. By restriction, the bundle V defines a point in $\pi^{-1}(x)$ for every $x \in \mathbb{P}^1$. In fact, we have the more precise result:

Theorem 29. *Suppose that $\pi \colon X \to \mathbb{P}^1$ is an elliptic surface as above, and that V is a rank 2 vector bundle over X such that $\det V = \pi^* M$ is pulled back from the base $\mathbb{P}^1$ and such that the restriction of V to every fiber f*

of π is semistable. Then V defines a section $A = A(V)$ of the rational ruled surface $\mathbb{F}_{2k}$.

Proof. Arguing as in the last part of the proof of Theorem 28, the determinant of the natural map $\pi^*\pi_*(V \otimes \mathcal{O}_X(\sigma)) \to V \otimes \mathcal{O}_X(\sigma)$ defines a section of $\det V \otimes \mathcal{O}_X(2\sigma)$ whose restriction to every fiber is nonzero. The identifications

$$H^0(\det V \otimes \mathcal{O}_X(2\sigma)) = H^0(\pi^*M \otimes \mathcal{O}_X(2\sigma)) = H^0(M \otimes \pi_*\mathcal{O}_X(2\sigma))$$

define a nowhere zero section of $M \otimes \pi_*\mathcal{O}_X(2\sigma)$ and thus a section of $\mathbb{P}(\pi_*\mathcal{O}_X(\sigma)) = \mathbb{F}_{2k}$. □

The surface X is a double cover of the rational ruled surface $\mathbb{F}_{2k}$, by Exercise 16 of the last chapter. For example, we can identify $\mathbb{F}_{2k}$ with the quotient X/ι, where ι is the involution corresponding to taking -1 in every fiber. The scheme-theoretic inverse image C of the section $A = A(V)$ in X is a double cover of A, possibly reducible or even non-reduced, and thus it defines a double cover of the base $\mathbb{P}^1$, which we will call the *spectral cover* of $\mathbb{P}^1$. In fact, if V is merely assumed to have semistable restriction to the generic fiber f, we can still define the spectral cover of V as follows: the method of Theorem 29 defines a meromorphic section of $\mathbb{F}_{2k}$, which then completes to an actual section since the base is a curve. Of course, the spectral cover of V, with this definition, is the same as the spectral cover of every elementary modification of V along a fiber.

In general, as we remarked above, it is hard to give a complete characterization of the stability of V, but we do have the following sufficient condition in terms of the spectral cover C:

Theorem 30. *Let V be a rank 2 vector bundle over X such that $\det V = \pi^*M$ is pulled back from the base $\mathbb{P}^1$ and such that the restriction of V to a generic fiber f of π is semistable. Suppose that the spectral cover C of $A(V)$ is reduced and irreducible. If (w, p) are the invariants of V, then V is stable with respect to every (w, p)-suitable ample divisor H.*

Proof. By (iii) of Theorem 5 in Chapter 6, if V is not H-stable, then there exists a divisor D on X and an exact sequence

$$0 \to \mathcal{O}_X(D) \to V \to \mathcal{O}_X(-D + af) \otimes I_Z \to 0,$$

such that $D \cdot f = 0$. The divisor D defines a section Q of X such that D is linearly equivalent to $Q - \sigma$ on the generic fiber, by looking at the support of the natural map $\pi^*\pi_*\mathcal{O}_X(D + \sigma) \to \mathcal{O}_X(D + \sigma)$, and similarly $-D$ defines a section R. (Note that $Q = R$ if and only if D has order 2 on the generic fiber.) By construction $C = Q + R$. Thus, C is either nonreduced or reducible. Conversely, if C is reduced and irreducible, then V must be H-stable. □

We now turn to the problem of constructing bundles over X. As in the above discussion, we begin with the problem of finding a universal bundle $\mathcal{U}$ over $f \times \mathbb{P}^1$. In fact, we will find not one universal bundle but rather an infinite family of such. The reason is the following: suppose that we were given a moduli space $\mathfrak{M}$ of bundles on f, all of which were simple, and two different universal bundles $\mathcal{U}$, $\mathcal{U}'$ over $f \times \mathfrak{M}$, whose restrictions to every slice $f \times \{x\}$ were isomorphic. By simplicity, $R^0\pi_{2*}Hom(\mathcal{U},\mathcal{U}') = L$ is a line bundle on $\mathfrak{M}$, and it is then easy to see that there is a map $\mathcal{U} \otimes \pi_2^* L \to \mathcal{U}'$ which is an isomorphism on every fiber, and thus an isomorphism. Thus, two universal bundles can only differ by twisting by the pullback of a line bundle from the $\mathfrak{M}$ factor. In our situation, however, the bundles V corresponding to points of $\mathbb{P}^1$ are not simple. Thus, given two universal bundles $\mathcal{U}$, $\mathcal{U}'$ as above, the most we can say is that $R^0\pi_{2*}Hom(\mathcal{U},\mathcal{U}')$ is a rank 2 locally free sheaf over $\mathbb{P}^1$. In fact, it is easy to see that this bundle has rank 1 over the sheaf of algebras $\mathcal{A} = R^0\pi_{2*}Hom(\mathcal{U},\mathcal{U})$, which is itself a sheaf of rank 2 commutative algebras over $\mathbb{P}^1$. The sheaf $\mathcal{A}$ defines a double cover of $\mathbb{P}^1$, which is unbranched where the fiber is $\mathbb{C} \oplus \mathbb{C}$ and branched at the four points of $\mathbb{P}^1$ where the fiber is $\mathbb{C}[t]/(t^2)$. Thus, the double cover is a copy of f again. (Special care must be taken when f is a nodal curve.) We will now give two methods for constructing universal bundles, first by pushing forward line bundles on the double cover $f \times f \to f \times \mathbb{P}^1$, and second by taking universal extensions of $\mathcal{O}_f(p)$ by $\mathcal{O}_f(-p)$.

To describe the first method, we assume that f is smooth. Let $\nu: f \to \mathbb{P}^1$ be the quotient of f under the involution $\iota = -\operatorname{Id}$. Consider the line bundle $\mathcal{L} = \mathcal{O}_{f \times f}(\Delta - \{p\} \times f)$. The restriction $\mathcal{L}|f \times \{q\} = \mathcal{O}_f(q - p)$. The pushforward $(\operatorname{Id} \times\nu)_*\mathcal{L}$ is then a rank 2 bundle $\mathcal{U}_0$ on $f \times \mathbb{P}^1$. If $x \in \mathbb{P}^1$ is a point such that $\nu^{-1}(x) = \{q,r\}$ consists of two distinct points, then $\nu_*\mathcal{L}|f \times \{x\} \cong \mathcal{O}_f(q - p) \oplus \mathcal{O}_f(r - p)$. Thus, away from the branch points of ν, $\mathcal{U}_0$ fits together the line bundles $\lambda \oplus \lambda^{-1}$. Of course, we will have to analyze what happens at the branch points. More generally, let $\mathbf{d}$ be a divisor on f. Then we can consider $\mathcal{U}_\mathbf{d} = (\operatorname{Id} \times\nu)_*(\mathcal{L} \otimes \pi_2^*\mathcal{O}_f(\mathbf{d}))$. It is clear that $\mathcal{U}_0$ and $\mathcal{U}_\mathbf{d}$ have isomorphic restrictions to every slice $f \times \{x\}$ of $f \times \mathbb{P}^1$.

To describe what happens at the branch points of ν, we have the following result, for whose proof we refer to [42]:

Proposition 31. *For all $x \in \mathbb{P}^1$, the restriction $\mathcal{U}_\mathbf{d}|f \times \{x\}$ is a regular semistable bundle. It is isomorphic to $\mathcal{O}_f(q-p) \oplus \mathcal{O}_f(r-p)$ if $\nu^{-1}(x) = \{q,r\}$ has two distinct points, and isomorphic to $\mathcal{E} \otimes \mathcal{O}_f(q - p)$ if x is a branch point of ν with preimage q.* $\square$

Although $\mathcal{U}_0$ and $\mathcal{U}_\mathbf{d}$ have isomorphic restrictions to every slice, they have different Chern classes in general:

Lemma 32. *With notation as above, suppose that* $\deg \mathbf{d} = d$. *Then*

$$\det \mathcal{U}_\mathbf{d} = \pi_2^* \mathcal{O}_{\mathbb{P}^1}(d-1),$$
$$c_2(\mathcal{U}_\mathbf{d}) = 1.$$

Proof. Fixing $\mathbf{d}$, let D be the divisor $\Delta - (\{p\} \times f) + \pi_2^* \mathbf{d}$. Then by Proposition 27 of Chapter 2, $c_1(\mathcal{U}_\mathbf{d})$ is represented by the divisor $(\nu \times \mathrm{Id})_* D - 2f \times \{x\}$, where x is an arbitrary point of $\mathbb{P}^1$. (Here the branch divisor of $\nu \times \mathrm{Id}$ is linearly equivalent to $4f \times \{x\}$.) Now $(\nu \times \mathrm{Id})_* \Delta = \Gamma_\nu$ is the graph of $\nu \colon f \to \mathbb{P}^1$. Since $\mathrm{Pic}(f \times \mathbb{P}^1) \cong \pi_1^* \mathrm{Pic}\, f \oplus \mathbb{Z}$, Γ_ν is linearly equivalent to $a(f \times \{x\}) + \pi_1^* \mathbf{e}$ for some divisor $\mathbf{e}$ on f. Clearly, $\Gamma_\nu \cap (f \times \{x\}) = \nu^{-1}(x) \times \{x\}$ and $\Gamma_\nu \cap (p \times \mathbb{P}^1) = (p, \nu(p))$ (transverse intersections if x is not a branch point of ν). Thus, $\Gamma_\nu = (f \times \{x\}) + 2(\{p\} \times \mathbb{P}^1)$. Moreover, $(\nu \times \mathrm{Id})_*(\{p\} \times f) = 2(\{p\} \times \mathbb{P}^1)$, since ν has degree 2, and $(\nu \times \mathrm{Id})_*(\pi_2^* \mathbf{d}) = d(f \times \{x\})$. Combining, we see that $(\nu \times \mathrm{Id})_* D - 2(f \times \{x\})$ is equal to

$$(f \times \{x\}) + 2(\{p\} \times \mathbb{P}^1) - 2(\{p\} \times \mathbb{P}^1) + d(f \times \{x\}) - 2(f \times \{x\})$$
$$= (d-1)(f \times \{x\}),$$

as claimed. As for $c_2(\mathcal{U}_\mathbf{d})$, by Proposition 28 of Chapter 2 and the fact that $(\nu_* D)^2 = \nu_* D \cdot B = 0$, where B is the branch locus of $\nu \times \mathrm{Id}$, we see that

$$c_2(\mathcal{U}_\mathbf{d}) = -\tfrac{1}{2}(\nu \times \mathrm{Id})_*(D^2).$$

Since $\Delta^2 = (\{p\} \times f)^2 = (\pi_2^* \mathbf{d})^2 = 0$,

$$D^2 = -2 - 2d + 2d = -2.$$

Thus, $c_2(\mathcal{U}_\mathbf{d}) = 1$. $\square$

The Chern class $c_2(\mathcal{U}_\mathbf{d})$ is defined more generally inside the Chow group $A^2(f \times \mathbb{P}^1)$. Since rational equivalence is essentially trivial on the fibers $\mathbb{P}^1$, it is elementary to show that $A^2(f \times \mathbb{P}^1) \cong f \times \mathbb{Z}$. Working inside $A^2(f \times \mathbb{P}^1) \otimes \mathbb{Z}[\tfrac{1}{2}]$, the proof of Lemma 32 shows that $c_2(\mathcal{U}_\mathbf{d})$ is represented by the 0-cycle $-\pi_1^*(\mathbf{d} - (d+1)p) \cdot \pi_2^*(x)$. In fact, this equality already holds in $A^2(f \times \mathbb{P}^1)$. The main interest in this more refined calculation is that the universal bundle $\mathcal{U}_\mathbf{d}$ detects not only the integer d but also the full linear equivalence class of the divisor $\mathbf{d}$. In this sense, twisting by line bundles on the spectral cover has led to different universal bundles.

Given V, we have an associated section $A = A(V) \subset \mathbb{F}_{2k}$, or equivalently the spectral cover $C \subset X$. Conversely, a section A of $\mathbb{F}_{2k}$ leads to a bisection C of X, not necessarily reduced or irreducible, and we can use C to construct vector bundles on X whose restriction to every fiber is regular and semistable. To do so, we assume for simplicity that C is smooth and does not pass through the singular points of any singular fiber (although neither of these assumptions is necessary). Let $T = C \times_{\mathbb{P}^1} X$ be the pulled back elliptic surface and let $\rho\colon T \to C$ be the natural projection. There is a degree 2 morphism $\nu\colon T \to X$. The diagonal $\mathbb{D} \subset X \times_{\mathbb{P}^1} X$ then pulls back to a Cartier divisor on T, which we denote by Σ. Clearly, Σ is the section of $T \to C$ corresponding to the embedding $C \to C \times_{\mathbb{P}^1} X$ given by taking the identity on the first factor and the inclusion of C in X in the second factor. Note also that the preimage of $C \subset X$ under the degree 2 morphism $\nu\colon T \to X$ splits into Σ and another component Σ', and Σ and Σ' meet transversally along the intersection of the branch locus of ν with Σ. The section σ of π pulls back to a section Σ_0 of T. The analogue of the construction of the universal bundle $\mathcal{U}_0$ on a single fiber is then the following: let μ be a line bundle on C, and define

$$V(A, \mu) = \nu_* \left(\mathcal{O}_T(\Sigma - \Sigma_0) \otimes \rho^* \mu \right).$$

The proof of the following may then be found in [42]:

Lemma 33. *The restriction of the bundle $V(A, \mu)$ to every fiber f is regular and semistable with trivial determinant. Moreover, if V is a rank 2 vector bundle on X such that the restriction of V to every fiber f is regular and semistable with trivial determinant, and $A = A(V)$, then there exists a unique line bundle μ on C such that $V = V(A, \mu)$.* $\square$

In the situation of Lemma 33, suppose that V is a rank 2 vector bundle on X such that the restriction of V to every fiber f is semistable with trivial determinant. We would like to find a criterion for when the restriction to each fiber is actually regular. The following result is proved in [35] and [40]:

Theorem 34. *Let V be a rank 2 vector bundle on X such that the restriction of V to every fiber f is semistable with trivial determinant. Let $A = A(V)$ be the corresponding section of $\mathbb{F}_{2k}$. If A meets the branch locus B of the double cover $\iota\colon X \to \mathbb{F}_{2k}$ transversally, then the restriction of V to every fiber f is regular, and hence $V = V(A, \mu)$ for a unique line bundle μ on C.* $\square$

Our next goal is to compute the Chern classes of $V(A, \mu)$. Before we do so, however, let us record some basic facts about the geometry of $\mathbb{F}_{2k}$ and the double cover map $\eta\colon X \to \mathbb{F}_{2k}$:

Lemma 35. *Let σ_0 be the negative section of $\mathbb{F}_{2k}$, let ℓ be a fiber of the ruled surface $\mathbb{F}_{2k}$, and let $A \in |\sigma_0 + (2k + r)\ell|$ be a section with $A \neq \sigma_0$, i.e., $r \geq 0$.*

(i) *$A^2 = 2k + 2r$ and $\dim |A| = 2k + 2r + 1$;*
(ii) *If B is the branch locus of the double cover map $\eta\colon X \to \mathbb{F}_{2k}$, then B is linearly equivalent to $4\sigma_0 + 6k\ell$ and $A \cdot B = 6k + 4r$;*
(iii) *If $C = \eta^* A$ is the preimage of a generic section A, so that C is smooth, then $g(C) = 3k + 2r - 1$.*

Proof. Parts (i) and (ii) are routine calculations, and (iii) follows from the Riemann-Hurwitz formula. $\square$

Theorem 36. With $V(A, \mu)$ defined as above, let $m = \deg \mu$. Then we have the following formulas for the Chern classes of $V(A, \mu)$:

(i) $\det V(A, \mu) = \mathcal{O}_X((-k - r + m)f)$;
(ii) $c_2(V(A, \mu)) = 2k + r$.

Proof. The map $C \to \mathbb{P}^1 \cong A$ is ramified at the points $B \cap A$ by construction. The branch locus in X of the double cover map $\rho\colon T \to X$ is linearly equivalent to $(6k + 4r)f = (A \cdot B)f$, by (ii) of Lemma 35 above. Thus, using Proposition 27 in Chapter 2, $\det V(A, \mu)$ corresponds to the divisor

$$\nu_*(\Sigma) - \nu_*(\Sigma_0) + mf - (3k + 2r)f.$$

Clearly, $\nu_*\Sigma = C$ and $\nu_*\Sigma_0 = \nu_*\nu^*\sigma = 2\sigma$. To determine the divisor class of C, note that (as divisor classes)

$$C = \eta^* A = \eta^*\sigma_0 + (2k + r)\eta^*\ell = 2\sigma + (2k + r)f.$$

Combining, we see that

$$\nu_*(\Sigma) - \nu_*(\Sigma_0) + mf - (3k+2r)f = (2k+r+m-3k-2r)f = (-k-r+m)f,$$

as claimed in (i). In particular, $\nu_*(\Sigma - \Sigma_0) + mf$ is linearly equivalent to a multiple of f.

To see (ii), we use Proposition 28 in Chapter 2: if $D = \Sigma - \Sigma_0 + \rho^*\mathbf{m}$, where $\mathbf{m}$ is a divisor on C corresponding to μ, then

$$c_2(V(A, \mu)) = \tfrac{1}{2}\left((\nu_*D)^2 - \nu_*(D^2) - (\nu_*D) \cdot (3k + 2r)f\right).$$

Since ν_*D is a multiple of f, $(\nu_*D)^2 = (\nu_*D) \cdot f = 0$, and so $c_2(V(A, \mu)) = -\tfrac{1}{2}\nu_*(D^2)$. Since Σ and Σ_0 are both sections of T, $f \cdot (\Sigma - \Sigma_0) = 0$ (here we also use f to denote a fiber of ρ), and since $f^2 = 0$ as well, we see that

$$D^2 = (\Sigma - \Sigma_0)^2 = \Sigma^2 + \Sigma_0^2 - 2(\Sigma \cdot \Sigma_0).$$

As both Σ and Σ_0 are sections of T, $\Sigma^2 = \Sigma_0^2 = (\nu^*\sigma)^2 = -2k$, and thus $\Sigma^2 + \Sigma_0^2 = -4k$. Moreover,

$$\begin{aligned}
2(\Sigma \cdot \Sigma_0) &= 2(\Sigma \cdot \nu^*\sigma) = 2(\nu_*\Sigma \cdot \sigma) \\
&= 2(C \cdot \sigma) = (C \cdot 2\sigma) = (\eta^* A \cdot \eta^*\sigma_0) \\
&= 2(\sigma_0 + (2k + r)\ell) \cdot \sigma_0 = 2r.
\end{aligned}$$

Thus,

$$c_2(V(A, \mu)) = -\tfrac{1}{2}(-4k - 2r) = 2k + r,$$

as claimed. $\square$

We can now describe the birational structure of the moduli space of stable bundles on X whose determinant is pulled back from the base. The proof is given in [35]:

Theorem 37. *Let $\Delta = af$ be a multiple of the fiber on X and let $c = 2k + r$, where $k = p_g(X) + 1$ and $r \geq 0$. Let (w, p) be the corresponding invariants. Then, for every (w, p)-suitable ample divisor H, the moduli space of H-stable rank 2 vector bundles V with invariants (w, p) contains a Zariski open and dense subset M which fibers over a nonempty Zariski open subset U of $|\sigma_0 + (2k + r)\ell| \cong \mathbb{P}^{2k+2r+1}$. The fiber over a point $A \in U$ is isomorphic to the Jacobian $J(C)$, where C is the double cover $\eta^{-1}(A)$.* $\square$

Thus, the elliptic fibration $X \to \mathbb{P}^1$ has become a fibration over a projective space, whose fibers are the Jacobians of the hyperelliptic curves C.

The second method for constructing bundles is via extensions. As before, we begin with the case of a single curve f, possibly singular. We shall give a construction of bundles corresponding to points of $\mathbb{P}^1$, more in keeping with the interpretation of the moduli space $\mathbb{P}^1$ as the linear system $|2p|$.

Lemma 38. *Let V be an extension*

$$0 \to \mathcal{O}_f(-p) \to V \to \mathcal{O}_f(p) \to 0.$$

(i) *V is semistable if and only if the extension is nonsplit.*
(ii) *If V is an extension $\lambda \oplus \lambda^{-1}$, then $\lambda \neq \lambda^{-1}$.*
(iii) *Conversely, the bundles $\lambda \oplus \lambda^{-1}$, $\lambda \neq \lambda^{-1}$, $\mathcal{E} \otimes \lambda$, $\lambda^{\otimes 2} = \mathcal{O}_f$, and (in case f is singular) the bundle $\mathcal{G}$ of Theorem 26 are all extensions of $\mathcal{O}_f(p)$ by $\mathcal{O}_f(-p)$.*

Proof. (i) If V is unstable, then the only possible destabilizing line bundle must be $\mathcal{O}_f(p)$, but in this case the extension would split.

(ii) It suffices to show that, for all λ of degree 0, $h^0(V \otimes \lambda) \leq 1$. But this is clear from the defining exact sequence for V, since $h^0(\mathcal{O}_f(-p) \otimes \lambda) = 0$ and $h^0(\mathcal{O}_f(p) \otimes \lambda) = 1$.

(iii) It suffices to show that, with V any of the bundles enumerated in (iii), there exists a subbundle of V isomorphic to $\mathcal{O}_f(-p)$, since then the quotient $V/\mathcal{O}_f(-p)$ is automatically isomorphic to $\mathcal{O}_f(p)$. For example, if $V = \lambda \oplus \lambda^{-1}$, then

$$\text{Hom}(\mathcal{O}_f(-p), V) \cong H^0(\mathcal{O}_f(p) \otimes \lambda) \oplus H^0(\mathcal{O}_f(p) \otimes \lambda^{-1}).$$

Now $\mathcal{O}_f(p) \otimes \lambda \cong \mathcal{O}_f(q)$ for a unique point $q \in f$, and likewise $\mathcal{O}_f(p) \otimes \lambda^{-1} \cong \mathcal{O}_f(r)$. The hypothesis that $\lambda \neq \lambda^{-1}$ means exactly that $q \neq r$. Every nonzero section of $\mathcal{O}_f(q)$ vanishes just at q, and likewise every nonzero section of $\mathcal{O}_f(r)$ vanishes just at r. Thus, a general element of $\mathrm{Hom}(\mathcal{O}_f(-p), V)$ does not vanish anywhere and so its image is a subbundle of V.

Next consider the case $V = \mathcal{E} \otimes \lambda$. We shall just write out the case where λ is trivial. Using the exact sequence

$$0 \to \mathcal{O}_f(p) \to \mathcal{E} \otimes \mathcal{O}_f(p) \to \mathcal{O}_f(p) \to 0,$$

and the fact that $h^1(\mathcal{O}_f(p)) = 0$, we see that there exists a section s of $\mathcal{E} \otimes \mathcal{O}_f(p)$ which projects onto a nonzero section of $\mathcal{O}_f(p)$. Clearly, s can only vanish at p, and if s is nonvanishing it defines a subbundle of $\mathcal{E} \otimes \mathcal{O}_f(p)$ isomorphic to $\mathcal{O}_f$ and thus a subbundle of $\mathcal{E}$ isomorphic to $\mathcal{O}_f(-p)$. In this case, the quotient is necessarily $\mathcal{O}_f(p)$. So it suffices to show that s does not vanish at p. But if s vanishes at p, the inclusion $\mathcal{O}_f \to \mathcal{E} \otimes \mathcal{O}_f(p)$ factors through a map $\mathcal{O}_f(p) \to \mathcal{E} \otimes \mathcal{O}_f(p)$, such that the composition

$$\mathcal{O}_f(p) \to \mathcal{E} \otimes \mathcal{O}_f(p) \to \mathcal{O}_f(p)$$

is nonzero and thus is an isomorphism. It follows that there is a nonzero map $\mathcal{O}_f \to \mathcal{E}$ splitting the extension, contrary to hypothesis.

The case of $\mathcal{G}$ is similar and will be left to the reader. $\square$

Suppose that V corresponds to a nonzero extension class

$$\xi \in H^1(f; \mathcal{O}_f(-2p)).$$

How do we decide what bundle V corresponds to in terms of the classification of Theorem 25? The answer is given by the following lemma, where for simplicity we will not work out the case of $\mathcal{G}$ for a singular f:

Lemma 39. *Let V correspond to a nonzero extension class $\xi \in H^1(f; \mathcal{O}_f(-2p))$. Then V is of the form $\lambda \oplus \lambda^{-1}$ or $\mathcal{E} \otimes \lambda$, where $\lambda = \mathcal{O}_f(q - p)$, if and only if $\xi = \nu(q)$ under the natural morphism $\nu \colon f \to \mathbb{P}(H^0(\mathcal{O}_f(2p)^\vee)) = \mathbb{P}(H^1(\mathcal{O}_f(-2p)))$.*

Proof. V is of the form $\lambda \oplus \lambda^{-1}$ or $\mathcal{E} \otimes \lambda$, where $\lambda = \mathcal{O}_f(q - p)$, if and only if $h^0(V \otimes \lambda) = 1$, if and only if a nonzero section of $\mathcal{O}_f(q)$ lifts to a section of $V \otimes \lambda$, if and only if the image of $H^0(\mathcal{O}_f(q))$ in $H^1(\mathcal{O}_f(q - 2p))$ is zero under the coboundary map coming from the exact sequence

$$0 \to \mathcal{O}_f(q - 2p) \to V \otimes \lambda \to \mathcal{O}_f(q) \to 0.$$

Now the commutative diagram of exact sequences

$$
\begin{array}{ccccccccc}
0 & \longrightarrow & \mathcal{O}_f(-2p) & \longrightarrow & V \otimes \mathcal{O}_f(-p) & \longrightarrow & \mathcal{O}_f & \longrightarrow & 0 \\
 & & \downarrow & & \downarrow & & \downarrow & & \\
0 & \longrightarrow & \mathcal{O}_f(q - 2p) & \longrightarrow & V \otimes \lambda & \longrightarrow & \mathcal{O}_f(q) & \longrightarrow & 0
\end{array}
$$

gives rise to a commutative diagram

$$
\begin{array}{ccc}
H^0(\mathcal{O}_f) & \xrightarrow{\ \partial\ } & H^1(\mathcal{O}_f(-2p)) \\
\downarrow & & \downarrow \\
H^0(\mathcal{O}_f(q)) & \xrightarrow{\ \partial\ } & H^1(\mathcal{O}_f(q-2p)),
\end{array}
$$

where the vertical maps are the natural inclusions, and the image of $1 \in H^0(\mathcal{O}_f)$ is the extension class ξ. Since the left-hand vertical map is an isomorphism, we see that $\partial(H^0(\mathcal{O}_f(q))) = 0$ if and only if the image of ξ under the natural map $H^1(\mathcal{O}_f(-2p)) \to H^1(\mathcal{O}_f(q-2p))$ is 0, if and only if ξ is dual to the hyperplane which is the image of $H^0(\mathcal{O}_f(2p-q))$ in $H^0(\mathcal{O}_f(2p))$, if and only if $\xi = \nu(q)$ under the natural morphism $\nu \colon f \to \mathbb{P}(H^0(\mathcal{O}_f(2p)^\vee))$. $\square$

Corollary 40. *If $\xi_1, \xi_2 \in H^1(\mathcal{O}_f(-2p))$ define isomorphic vector bundles, then ξ_2 and ξ_2 differ by a nonzero scalar.* $\square$

We can then make a universal extension of $\mathcal{O}_f(p)$ by $\mathcal{O}_f(-p)$ which fits into an exact sequence

$$
0 \to \pi_1^* \mathcal{O}_f(-p) \otimes \pi_2^* \mathcal{O}_{\mathbb{P}^1}(1) \to \mathcal{U} \to \pi_1^* \mathcal{O}_f(p) \to 0.
$$

Note that such extensions are classified by

$$
H^1(\pi_1^* \mathcal{O}_f(-p) \otimes \pi_1^* \mathcal{O}_f(-p) \otimes \pi_2^* \mathcal{O}_{\mathbb{P}^1}(1)) = H^1(\pi_1^* \mathcal{O}_f(-2p) \otimes \pi_2^* \mathcal{O}_{\mathbb{P}^1}(1))
$$
$$
\cong H^1(f; \mathcal{O}_f(-2p)) \otimes H^0(\mathbb{P}^1; \mathcal{O}_{\mathbb{P}^1}(1)),
$$

by the Künneth formula. Now canonically the $\mathbb{P}^1$ is $\mathbb{P}H^1(f; \mathcal{O}_f(-2p))$, and thus $H^0(\mathbb{P}^1; \mathcal{O}_{\mathbb{P}^1}(1)) = H^1(f; \mathcal{O}_f(-2p))^\vee = X$, say, where X is a 2-dimensional vector space. We can then take the element in

$$
X \otimes X^\vee \cong \operatorname{Hom}(X, X)
$$

corresponding to the identity, to define an extension $\mathcal{U}$ which restricts on every slice $f \times \xi$ to the extension defined by ξ. (Note that if we had not twisted the subbundle by $\pi_2^* \mathcal{O}_{\mathbb{P}^1}(1)$, the possible extensions of $\pi_1^* \mathcal{O}_f(p)$ by $\pi_1^* \mathcal{O}_f(-p)$ would then all be pulled back from a single fixed extension of $\mathcal{O}_f(p)$ by $\mathcal{O}_f(-p)$.)

Computing Chern classes of $\mathcal{U}$, we see that $\det \mathcal{U} = \pi_2^* \mathcal{O}_{\mathbb{P}^1}(1)$ and that, in $A^2(f \times \mathbb{P}^1)$, $c_2(\mathcal{U}) = \pi_1^*(p) \cdot \pi_2^*(x)$. From this, one can show that $\mathcal{U} = \mathcal{U}_{2p}$, in the notation of Lemma 32.

Similar but more involved constructions yield a universal relative extension over X. However, without using line bundles on the spectral cover, we cannot construct a Zariski open subset of the moduli space in this way, but only a section of the fibration described in Theorem 37.

Even fiber degree and multiple fibers

In this section, we want to generalize the results of the last section to the case where X does not necessarily have a section. For simplicity, we shall just limit ourselves to the case where X is simply connected. Thus, the base curve of X is $\mathbb{P}^1$ and X has at most two multiple fibers, of relatively prime multiplicities m_1 and m_2. We shall also assume that the multiple fibers have smooth reduction and that all other singular fibers are irreducible nodal curves. As usual, stability is assumed to be with respect to a suitable polarization. We would like to outline the description of stable vector bundles V over X with $c_1(V) \cdot f \equiv 0 \pmod 2$ for the general fiber f. More detailed arguments can be found in [35], [40], and [36].

As in the last section, we shall just discuss V under the assumption that $V|f$ is semistable for all fibers f. Working on a fixed smooth fiber f, if E is a semistable rank 2 bundle on f with $\det E = \delta$, where $\deg \delta = 2d$, then either

$$E \cong \lambda \oplus (\delta \otimes \lambda^{-1})$$

for a line bundle λ with $\deg \lambda = \deg \delta / 2$, or E is given as a nonsplit extension

$$0 \to \lambda \to E \to \lambda \to 0,$$

where λ is one of the four line bundles on f such that $\lambda^{\otimes 2} = \delta$. Moreover, such E are identified in the moduli space with the direct sum $\lambda \oplus \lambda$. Thus, the moduli space of semistable bundles is isomorphic to $J^d(f)/\iota$, where $J^d(f)$ is the group of line bundles of degree d on f, which is a principal homogeneous space over $J^0(f)$, and ι is the involution of $J^d(f)$ defined by $\lambda \mapsto \delta \otimes \lambda^{-1}$. The involution ι has four fixed points, and the quotient $J^d(f)/\iota$ is isomorphic to $\mathbb{P}^1$.

Now we can fit this picture together over X: define the surface $J^d(X)$ to be the elliptic surface over $\mathbb{P}^1$ whose fiber over t is naturally $J^d(\pi^{-1}(t))$. In terms of the classification scheme of the last chapter, if X corresponds to $\xi \in H^1(G, B)$, then $J^d(X)$ corresponds to the class $d\xi \in H^1(G, B)$. For example, if $d = 0$, then $J^0(X) = J(X)$ is the Jacobian surface of X. Thus, if X has a multiple fiber of multiplicity m at t, corresponding to the image of ξ in $H^1(G_t, B_t) \cong (\mathbb{Q}/\mathbb{Z})^2$ having order m, then the image of $d\xi$ has order $m/\gcd(d, m)$ in $(\mathbb{Q}/\mathbb{Z})^2$. In our case, we assume that there exists a divisor Δ on X with $\Delta \cdot f = 2d$ (here Δ will be the determinant of our rank 2 bundles V). Thus, $m_1 m_2$ divides $2d$, since f is divisible by $m_1 m_2$ in $\operatorname{Num} X$, and furthermore m_1 and m_2 are relatively prime. We see that there are two cases: either $m_1 m_2$ divides d, in which case $J^d(X)$ has no multiple fibers, or $m_1 m_2$ does not divide d. In the second case, one of m_1, m_2 is odd, since they are relatively prime. Say m_2 is odd. Since $m_1 m_2$ divides $2d$ but $m_1 m_2$ does not divide d, m_1 is even, say $m_1 = 2m_1'$, and $2m_1'$ divides $2d$ so that m_1' divides d and d/m_1' is odd. Clearly, the gcd of

$m_1 = 2m_1'$ and d is m_1', so that $m_1/\gcd(d, m_1) = 2$. It follows that in this case $J^d(X)$ has a multiple fiber of multiplicity 2 at the point corresponding to m_1 and no other multiple fibers. Finally, if κ is the primitive class such that $f = m_1 m_2 \kappa$, then as $\Delta \cdot f = 2d$,

$$\Delta \cdot \kappa = \frac{2d}{m_1 m_2}.$$

In particular, $J^d(X)$ has no multiple fibers if $\Delta \cdot \kappa$ is even and has a single multiple fiber of multiplicity 2 if $\Delta \cdot \kappa$ is odd.

Now a stable rank 2 bundle V gives a bisection C of $J^d(X)$ (possibly reducible or of multiplicity 2), defined as follows: for each fiber f over $t \in \mathbb{P}^1$, we associate to t the pair of line bundles $\{\lambda, \delta \otimes \lambda^{-1}\}$ of degree d such that $V|f \cong \lambda \oplus (\delta \otimes \lambda^{-1})$, in case $V|f$ is a direct sum (and $\delta = \Delta|f$), and if $V|f$ is a nonsplit extension of λ by λ we assign to t the line bundle λ with multiplicity 2. Clearly, C is invariant under the natural involution of $J^d(X)$ corresponding to the involution ι on each fiber. Thus, C is the pullback of a curve A on the quotient $J^d(X)/\iota$. Note that $J^d(X)/\iota$ is birational to a ruled surface $\mathbb{F}$ over $\mathbb{P}^1$ and A is a section of the ruling. In general, as we vary the bundles V, the curves A will move in a linear system on some surface birational to a ruled surface, and we will denote this linear system by $|A|$. Conversely, a section A defines a bisection C of $J^d(X)$ via pullback, where C may be reducible or multiple with multiplicity 2. Note however that if $J^d(X)$ has a multiple fiber of multiplicity 2, then for every bisection C of $J^d(X)$, the induced map $C \to \mathbb{P}^1$ must be branched at the point of C meeting the multiple fiber, whereas this phenomenon does not happen if $J^d(X)$ has no multiple fibers.

Given the bisection C, how close is it to determining V? Assume that C is smooth and irreducible, and let $Y = C \times_{\mathbb{P}^1} X \to C$ be the induced elliptic surface. We assume that $C \to \mathbb{P}^1$ is generic (not branched over the points corresponding to singular or multiple fibers of X) so that Y is a smooth elliptic surface. On X, since C is irreducible, there is no way to distinguish λ from $(\delta \otimes \lambda^{-1})$. But if $\nu: Y \to X$ is the double cover map, then one shows that there is a divisor D on Y such that, for a given fiber f on X, if $V|f = \lambda \oplus (\delta \otimes \lambda^{-1})$, where $\delta = \Delta|f$, if $f' \cong f$ is a fiber above f on Y, then $\mathcal{O}_Y(D)|f'$ is either λ or $(\delta \otimes \lambda^{-1})$ under the natural identification. Thus, on the double cover we can coherently fit together the line bundles λ to obtain a well-defined line bundle $\mathcal{O}_Y(D)$. It then follows easily that there is an exact sequence

$$0 \to \mathcal{O}_Y(D) \to \nu^* V \to \mathcal{O}_Y(\nu^* \Delta - D) \to 0.$$

(Here we use the assumption that $V|f$ is semistable for all f to conclude that the quotient $\nu^* V/\mathcal{O}_Y(D)$ is actually a line bundle.) The most difficult part of the analysis is to show the following [40], [35]:

Theorem 41. *Suppose that $V|f$ is semistable for all f and that the corresponding bisection C is generic. Let F be the branch locus of the map $Y \to X$. Then $V = \nu_* \mathcal{O}_Y(D + F)$. Conversely, starting with a generic bisection C of $J^d(X)$ and letting $g \colon Y \to X$ be the induced double cover, if D is a divisor on Y such that $\det \nu_* \mathcal{O}_Y(D+F) = \Delta$, then the rank 2 vector bundle $V = \nu_* \mathcal{O}_Y(D + F)$ is a stable rank 2 bundle on X with $c_1(V) = \Delta$.*

Thus, we must analyze the condition that $\det \nu_* \mathcal{O}_Y(D + F) = \Delta$. Let G be the divisor class on X such that $2G$ is linearly equivalent to the branch locus on X. In other words, $\nu^* G = F$. Using the formulas of Chapter 2, Proposition 27, we see that $\det \nu_* \mathcal{O}_Y(D + F) = \nu_* D + \nu_* F - G$, where $\nu^* G = F$ and thus $\nu_* F = \nu_* \nu^* G = 2G$. It follows that

$$\det \nu_* \mathcal{O}_Y(D + F) = \nu_* D + G = \Delta.$$

Now the set of all divisor classes D such that $\nu_* D = \Delta - G$ is a principal homogeneous space over the set of all divisor classes E on Y such that $\nu_* E = 0$. A straightforward argument identifies this group with an extension of $\mathrm{Pic}^0(C)$ by the torsion in $H^2(Y; \mathbb{Z})$.

To describe the torsion in $H^2(Y; \mathbb{Z})$, note that Y has two multiple fibers F_1' and F_1'' of multiplicity m_1 and two multiple fibers F_2' and F_2'' of multiplicity m_2, if C is not branched over either multiple fiber. In this case, $F_1' - F_1''$ is a torsion element of order m_1, and similarly $F_2' - F_2''$ is a torsion element of order m_2. In fact, it is easy to show, using Proposition 27 of the last chapter, that these two elements generate the torsion subgroup of $H^2(Y; \mathbb{Z})$, which is isomorphic to $\mathbb{Z}/m_1\mathbb{Z} \times \mathbb{Z}/m_2\mathbb{Z} = \mathbb{Z}/m_1 m_2 \mathbb{Z}$. In this case, we arrive at the following picture for a Zariski open subset of the moduli space (at least for large enough c_2): it fibers over a Zariski open subset of the projective space $|A|$, and the fibers are $m_1 m_2$ copies of a complex torus $\mathrm{Pic}^0(C)$ of dimension $g = g(C)$. Once again, we see the ghost of the elliptic structure on X in the birational structure of the moduli space.

In case $m_1 = 2m_1'$ and $\Delta \cdot \kappa$ is odd, the picture is slightly different: the map $C \to \mathbb{P}^1$ is always branched at the point corresponding to the multiple fiber F_1. Thus, Y has a single multiple fiber of multiplicity m_1' and two multiple fibers of multiplicity m_2 lying over F_2. The torsion subgroup of $H^2(Y; \mathbb{Z})$ is then just isomorphic to $\mathbb{Z}/m_2\mathbb{Z}$. Otherwise, the picture is the same: the moduli space fibers birationally over a projective space, and the fibers are m_2 copies of a complex torus.

Using the above and an analysis of the case of odd fiber degree, one can determine the "leading coefficient" a_n of the Donaldson polynomial $D_{w,p}^X$ of a simply connected elliptic surface. Here it follows from very general arguments [40] that we can write

$$D_{w,p}^X = \sum_{i \leq n} a_i q_X^i \kappa^{d-2i},$$

where q_X is the intersection form on X and the above expression lives in the symmetric algebra on $H^2(X)$. We assume that n is chosen so that $a_n \neq 0$, and refer to a_n as the *leading coefficient* of $D^X_{w,p}$. The final answer is the following:

Theorem 42. *Let X be a simply connected elliptic surface with $p_g(X) = p_g > 0$ and multiple fibers of multiplicities m_1 and m_2. Let a_n be the leading coefficient of the Donaldson polynomial corresponding to invariants w, p. Up to universal combinatorial factors,*

$$a_n = \begin{cases} (m_1 m_2)^{p_g}, & \text{if } w \cdot \kappa \equiv 0 \mod 2, \\ (m_1 m_2)^{p_g} m_2, & \text{if } w \cdot \kappa \equiv 1 \mod 2 \text{ and } m_1 = 2m_1' \text{ is even}, \\ 1, & \text{if } w \cdot \kappa \equiv 1 \mod 2 \text{ and } m_1 \text{ and } m_2 \text{ are odd}. \end{cases}$$

Roughly speaking, the idea of the calculation is as follows. We have defined the map $\mu \colon H_2(X) \to H^2(\mathfrak{M})$, where $\mathfrak{M}$ is the appropriate moduli space. On the Zariski open subset of $\mathfrak{M}$ that we have described above, one shows that $\mu(f)$ is represented by the pullback of a hyperplane in the projective space $|A|$. Thus, if $\dim |A| = a$, $\mu(f)^a$ is represented by a number of copies of the Jacobian $J(C)$ of the corresponding bisection C, and $\mu(f)^b = 0$ if $b > a$. In fact, this statement holds over the full moduli space and its Uhlenbeck compactification, not just on the dense open subset described above. Now if X has a section σ, then $\mu(f)^a$ is represented by a single copy of $J(C)$, and one shows that $\mu(\sigma)$ restricts on $J(C)$ to the theta divisor Θ of $J(C)$. In this way, we can calculate $\mu(f)^b \cdot \mu(\sigma)^{d-b}$ for all $b \geq a$: it is 0 for $b > a$ and $(d-a)!$ for $b = a$. If X has multiple fibers, there are no sections, but one can show that, for every multisection τ, if $t = \tau \cdot f$, then $\mu(\tau)$ restricts to $t \cdot \Theta$ on each copy of $J(C)$, and so we can make the calculation as before.

Combining Theorem 42 and Theorem 18, it is an algebraic exercise to show that we can recover the integers m_1 and m_2 from the Donaldson polynomials of X in case $p_g > 0$. A more detailed analysis, which involves working out the four-dimensional invariant in the case where m_1 and m_2 are odd, also handles the case where $p_g = 0$.

Beyond the leading coefficient, the second coefficient of the Donaldson polynomial has been determined in the case $p_g = 1$ by Bauer [8] and Morgan and O'Grady [102], and this information has been used to determine the second coefficient in the case of trivial determinant for all $p_g \geq 1$ [101], [83], [142]. Independently, Fintushel and Stern have worked out the entire Donaldson series for all $p_g \geq 1$, without using algebraic geometry [33].

Exercises

1. If $0 \to L' \to L \to \tau \to 0$ is an exact sequence of sheaves on C, where L' and L are torsion free rank 1 sheaves on C and τ is supported on a

finite number of points, then $\deg L' \leq \deg L$, with equality holding if and only if $L' = L$.

2. If L' and L are torsion free rank 1 sheaves on C and $\operatorname{Hom}(L', L) \neq 0$, then $\deg L' \leq \deg L$ with equality only if $L = L'$.

3. If V is a rank 2 vector bundle and

$$0 \to L' \to V \to L \to 0$$

is exact, where L' and L are torsion free rank 1 sheaves, then $\deg L' + \deg L = \deg V$.

4. If $n \colon \tilde{C} \to C$ is the normalization and $\tilde{L}$ is a line bundle on $\tilde{C}$, show that $\deg n_* \tilde{L} = \deg \tilde{L} + \delta$.

5. Suppose that λ is a line bundle on C. Show that $\deg(L \otimes \lambda) = \deg L + \deg \lambda$.

6. Let C be a curve with only one singular point, which is an ordinary double point, and let $n \colon \tilde{C} \to C$ be the normalization. Show that every torsion free rank 1 sheaf on C is either a line bundle or of the form $n_* \tilde{L}$, where L is a line bundle on $\tilde{C}$. (Let $R = \mathbb{C}\{x, y\}$ and let $S = R/(xy)R$. If L is a torsion free rank 1 S-module, then let $M = xL \oplus yL$ and let $N = L/xL \oplus L/yL$. The hypothesis that L is torsion free implies that there are inclusions $M \subseteq L \subseteq N$. Both M and N are $\tilde{S}$-modules, where $\tilde{S} = \mathbb{C}\{x\} \oplus \mathbb{C}\{y\}$, and since they are torsion free we may assume that $M \cong \tilde{S}$ and $M = \mathfrak{m} \cdot \tilde{S} = \mathfrak{m}$, where $\mathfrak{m}$ is the maximal ideal of S and also of $\tilde{S}$. Thus, either $L \cong \tilde{S}$ or $L \cong \mathfrak{m} \cong \tilde{S}$ (locally), or $\tilde{S}/L$ has length 1. Show that we can then find an isomorphism $L \cong S$.)

7. Let $S = R/(xy)R$ as in the previous exercise, and let $\tilde{S} = R/xR \oplus R/yR$. Show that

$$\cdots \to S \oplus S \to S \oplus S \to \tilde{S} \to 0$$

is a resolution of the S-module $\tilde{S}$, where the maps alternate between $(a, b) \mapsto (xa, yb)$ and $(a, b) \mapsto (ya, xb)$. Conclude that $\operatorname{Ext}^i_S(\tilde{S}, S) = 0$ for all $i > 0$. Verify that $\operatorname{Hom}(\tilde{S}, S) = \mathfrak{m}$ and that $\operatorname{Hom}(\mathfrak{m}, S) = \tilde{S}$ (canonically). Finally, show that $\operatorname{Ext}^1(\tilde{S}, \tilde{S}) \cong \tilde{S}/S$ as S- and $\tilde{S}$-modules. Show that, in $\mathbb{P} \operatorname{Ext}^1(\tilde{S}, \tilde{S}) \cong \mathbb{P}^1$, there are two points corresponding to non-free S-modules, both isomorphic to $\tilde{S} \oplus S$, that $\tilde{S}$ acts transitively on the remaining $\mathbb{C}^*$, and that all of the points of this $\mathbb{C}^*$ correspond to free S-modules $\cong S \oplus S$.

8. Let C be irreducible with only nodes as singularities. Let L' and L be two torsion free rank 1 sheaves on C. Show that $Hom(L', L)$ is again a torsion free rank 1 sheaf on C, and that $\deg L - \deg L' \leq \deg Hom(L', L) \leq \deg L - \deg L' + \delta$. When does equality at either end hold? When is $Hom(L', L)$ locally free?

9. Let C be a reduced irreducible curve with normalization $\tilde{C}$ and let A, B be rank 1 torsion free sheaves on C of degrees a and b, respectively. If $\operatorname{Hom}(A, B) \neq 0$, then $a \leq b$, and in this case $\dim \operatorname{Hom}(A, B) \leq b - a + 1$. (If $\operatorname{Hom}(A, B) \neq 0$, then there is a nonzero map, necessarily

an injection, from A to B and thus an exact sequence

$$0 \to A \to B \to Q \to 0.$$

Thus, $b = a + \ell(Q) \geq a$. To prove the second statement, argue by induction on $b-a$. In case $b-a = 0$, either $\dim \operatorname{Hom}(A, B) = 0$ or $B \cong A$, the claim is that $\dim \operatorname{Hom}(A, A) = 1$. In this case, $\mathcal{H}om(A, A)$ is a coherent sheaf of commutative torsion free rank 1 $\mathcal{O}_C$-algebras, which agrees with $\mathcal{O}_C$ at the smooth points of C, and thus is a subalgebra of $\mathcal{O}_{\tilde{C}}$. Thus, $\dim \operatorname{Hom}(A, A) = h^0(C; \mathcal{H}om(A, A)) \leq H^0(\mathcal{O}_{\tilde{C}}) = 1$. For the inductive step, assume that the result has been shown for all B' such that $\deg B' - a < n$. Given B such that $\deg B - a = b - a = n$, let x be a smooth point of C and define B' by the exact sequence

$$0 \to B' \to B \to \mathcal{O}_x \to 0.$$

Thus, $\deg B' = b - 1$ and there is an exact sequence

$$0 \to \operatorname{Hom}(A, B') \to \operatorname{Hom}(A, B) \to \operatorname{Hom}(A, \mathcal{O}_x).$$

By assumption $\dim \operatorname{Hom}(A, B') \leq b - 1 - a + 1 = b - a$. Since x is a smooth point of C, A is a line bundle in a neighborhood of x and so $\dim \operatorname{Hom}(A, \mathcal{O}_x) = 1$. It follows that $\dim \operatorname{Hom}(A, B) \leq b - a + 1$.)

10. Let C be an irreducible curve with only ordinary double points such that $p_a(C) = 1$. Let L be a torsion free rank 1 sheaf on C. Show that if $\deg L > 0$ or if $\deg L = 0$ and L is not equal to $\mathcal{O}_C$, then $h^0(L) = \deg L$ and $h^1(L) = 0$. What happens if $L = \mathcal{O}_C$ or if $\deg L < 0$?

11. Show that, if V is an unstable rank 2 vector bundle on the irreducible curve C, then there is a unique maximal destabilizing torsion free rank 1 subsheaf L in the sense of Proposition 20 of Chapter 4.

12. Let $S = \mathbb{C}\{x, y, z, w\}$ be the ring of convergent power series in four variables and let $R = S/(xy - zw)$. Thus, R is the analytic local ring of an ordinary double point of dimension 3. Let $I = I_{\mathbb{D}}R = (x - z, y - w)R$. We seek a free resolution of the R-module I. Let $A = \begin{pmatrix} y & w \\ z & x \end{pmatrix}$, where we view the entries as lying in R, and let $B = \begin{pmatrix} x & -w \\ -z & y \end{pmatrix}$. Note that $\det A = \det B = 0$ in R and that $A \cdot B = B \cdot A = 0$ as well. Show that

$$\cdots \xrightarrow{B} R^2 \xrightarrow{A} R^2 \xrightarrow{B} R^2 \xrightarrow{A} R^2 \xrightarrow{f} I \to 0$$

is exact, where $f(a, b) = a(x - z) + b(y - w)$. Dualizing the above, we find that

$$0 \to \operatorname{Hom}_R(I, R) \to R^2 \xrightarrow{{}^tA} R^2 \xrightarrow{{}^tB} R^2 \to \cdots$$

is exact. Conclude that $\operatorname{Ext}^i_R(I, R) = 0, i > 0$, and that $\operatorname{Hom}_R(I, R) \cong I_{\mathbb{E}}R = (x - w, y - z)R$. Arguing similarly, show that $\operatorname{Ext}^i(R/I, R) = 0, i \neq 1$, and that $\operatorname{Ext}^1(R/I, R) \cong R/I$. Show also that I is reflexive.

Is it true that the R-module $I = I_{\mathbb{D}}R(x - z, y - w)R$ is isomorphic to the R-module $I_{\mathbb{E}}R(x - w, y - z)R$?

13. Let C be an irreducible curve with only ordinary double points such that $p_a(C) = 1$. Let L_1 and L_2 be torsion free, non-locally free rank 1 sheaves on C, with $\deg L_i = d_i$. Show that, if $d_1 \geq d_2$, then there exist locally free extensions

$$0 \to F_1 \to V \to F_2 \to 0,$$

with $\det V$ any line bundle of degree $d_1 + d_2$. (First show that one such V exists, using Exercise 7, and then tensor with an appropriate line bundle of degree 0.)

9

Bogomolov's Inequality and Applications

Statement of the theorem

Our main goal in this chapter will be to give a proof of the following:

Theorem 1 (Bogomolov). *Let X be an algebraic surface and let H be an ample divisor on X. Suppose that V is an H-stable rank 2 vector bundle on X. Then $c_1^2(V) \leq 4c_2(V)$. Equivalently, $p_1(\mathrm{ad}\,V) \leq 0$.*

The argument given here can be adapted, with a few technical modifications, to prove the general theorem: if, in the above, V is assumed to have rank r, then

$$(r-1)c_1^2(V) \leq 2rc_2(V).$$

There is a further generalization to higher-dimensional varieties: if X is a smooth projective variety of dimension n and V is an H-stable vector bundle of rank r on X, then

$$(r-1)c_1^2(V) \cdot H^{n-2} \leq 2rc_2(V) \cdot H^{n-2}.$$

Bogomolov's original proof of Theorem 1 appears in [10]; see also [129] for an account of the proof in the case of rank 2. Gieseker [46] gave a proof which involves studying the restriction of V to a general hyperplane section and reducing mod p. The proof we give here is a combination of the two proofs and is essentially the same as Miyaoka's proof [97]. A deep approach to the theorem, as discussed in Chapter 4, is via the Donaldson-Uhlenbeck-Yau theorem that a stable V carries a Hermitian-Einstein connection, in which case the theorem becomes an exercise in differential geometry. Moreover, this proof shows that, in case equality holds and $c_1(V) = 0$, then V is a unitary flat vector bundle, in other words a bundle associated to a unitary representation of $\pi_1(X)$. Finally, Fernandez del Busto [32] has recently given a proof in the rank 2 case based on the Kawamata-Viehweg vanishing theorem.

Next we give some easy applications in the rank 2 case.

Corollary 2. *Suppose that X is an algebraic surface and that V is a rank 2 vector bundle such that $c_1^2(V) > 4c_2(V)$. Then there is a unique sub-line bundle $\mathcal{O}_X(D)$ of V and an exact sequence*

$$0 \to \mathcal{O}_X(D) \to V \to \mathcal{O}_X(D') \otimes I_Z \to 0,$$

where Z is a codimension 2 local complete intersection subscheme of X, such that $(D - D')^2 > 0$ and $(D - D') \cdot H > 0$ for every ample divisor H.

Proof. Fix an ample divisor H_0. Theorem 1 implies that V is not H_0-stable, and in fact V cannot be H_0-semistable, since a strictly H_0-semistable vector bundle V satisfies Bogomolov's inequality (Exercise 1 of Chapter 6). Thus, V is H_0-unstable, and there exists an exact sequence

$$0 \to \mathcal{O}_X(D) \to V \to \mathcal{O}_X(D') \otimes I_Z \to 0,$$

where $(D - D') \cdot H_0 > 0$. Note moreover that

$$c_1^2(V) - 4c_2(V) = (D - D')^2 - 4\ell(Z) > 0,$$

and thus $(D - D')^2 > 0$. But now Lemma 19 of Chapter 1 implies that $(D - D') \cdot H > 0$ for every ample divisor H. The uniqueness is clear from the uniqueness of the destabilizing sub-line bundle for H. $\square$

The above says that $D - D'$ is in the cone dual to the ample cone, which is exactly the cone spanned by the effective divisors. In fact, by Lemma 12 in Chapter 1, some multiple of $D - D'$ is effective.

The next corollary, due to Bogomolov, is a strong version of a restriction theorem for rank 2 vector bundles due to Mumford and Mehta and Ramanathan, which says that for large k and a *generic* curve $C \in |kH|$, the restriction $V|C$ is semistable. In fact, in the course of the proof of Bogomolov's inequality, we shall prove a weak version of the theorem of Mumford and Mehta and Ramanathan. The final inequality then allows us to prove the following much more precise restriction theorem:

Corollary 3. *Let X be an algebraic surface, let H be an ample divisor on X, and let V be an H-stable rank 2 vector bundle on X with $p_1(\operatorname{ad} V) = p$. Suppose that $k \geq -p$. Then for every smooth curve $C \in |kH|$, the vector bundle $V|C$ is stable.*

Proof. Suppose instead that $V|C$ is not stable. Then there exists a quotient line bundle L of $V|C$ such that $2 \deg L < \deg(V|C) = kc_1(V) \cdot H$. If $j: C \to X$ is the inclusion, there is a surjection $V \to j_*L$. Let V' be the kernel of this surjection, so that there is an exact sequence

$$0 \to V' \to V \to j_*L \to 0,$$

and V' is an elementary modification of V. By Lemma 16 in Chapter 2, $c_1(V') = c_1(V) - kH$ and $c_2(V') = c_2(V) - c_1(V) \cdot (kH) + \deg L$. A calculation gives (cf. also Lemma 9 in Chapter 6):

$$p_1(\text{ad } V') = p_1(\text{ad } V) + 2c_1(V) \cdot (kH) + (kH)^2 - 4 \deg L$$
$$> p + k^2(H^2) \geq p + p^2 \geq 0.$$

Thus, V' does not satisfy Bogomolov's inequality and is therefore unstable. Let $\mathcal{O}_X(D) \to V'$ be the maximal destabilizing sub-line bundle, so that there is an exact sequence

$$0 \to \mathcal{O}_X(D) \to V' \to \mathcal{O}_X(D') \otimes I_Z \to 0,$$

with $H \cdot (D - D') > 0$. In particular

$$0 \leq p + k^2(H^2) < p_1(\text{ad } V') = (D - D')^2 - 4\ell(Z) \leq (D - D')^2.$$

Now V' is a subsheaf of V, and so there is a nonzero map from $\mathcal{O}_X(D)$ to V. Since V is H-stable, $2D \cdot H < c_1(V) \cdot H$. Thus, if $m = -(2D - c_1(V)) \cdot H$, then $m > 0$. Moreover, the Hodge index theorem (Exercise 10 of Chapter 1) implies that $(2D - c_1(V))^2 H^2 \leq m^2$.

Since $D + D' = c_1(V') = c_1(V) - kH$, we have $D - D' = 2D - c_1(V) + kH$. Using $H \cdot (D - D') > 0$, we have $m = -(2D - c_1(V)) \cdot H \leq kH^2$. Thus,

$$p + k^2(H^2) < (D - D')^2 = (2D - c_1(V))^2 + 2k(2D - c_1(V)) \cdot H + k^2 H^2$$
$$\leq \frac{m^2}{H^2} - 2km + k^2 H^2,$$

so that $2km < m^2/H^2 - p$. Using $0 < m < kH^2$ and $-p \leq k$, we finally have

$$2k < \frac{m}{H^2} - \frac{p}{m} \leq k + \frac{k}{m} \leq 2k,$$

a contradiction. Thus, $V|C$ is stable.   □

The proof of Corollary 3 goes over unchanged in case C is just assumed to be reduced and irreducible.

We remark that, if we take $k > -p$, the argument above shows that the bundles $V|C$ become progressively "more stable," in the sense that for every quotient line bundle L of $V|C$, $\mu(L) - \mu(V|C)$ is bounded away from 0 by a fixed positive rational number (depending on the choice of k).

The last easy corollary of Bogomolov's inequality is the following elegant proof (due to Mumford) of Mumford's generalization of the Kodaira vanishing theorem on a surface (Theorem 26 in Chapter 1):

Corollary 4. Let X be a smooth surface and let L be a nef and big line bundle on X. Then $H^1(X; L^{-1}) = 0$.

Proof. Suppose instead that $H^1(X; L^{-1}) \neq 0$. Then there is a rank 2 vector bundle V which is a nonsplit extension of L by $\mathcal{O}_X$; in other words, there is an exact sequence

$$0 \to \mathcal{O}_X \to V \to L \to 0.$$

By the Whitney product formula, $c_1(V) = L$ and $c_2(V) = 0$. As L is big, $c_1(V)^2 = L^2 > c_2(V) = 0$. Thus, V is H-unstable for every ample H. Let $\mathcal{O}_X(D)$ be the maximal destabilizing sub-line bundle. As usual we have an exact sequence

$$0 \to \mathcal{O}_X(D) \to V \to \mathcal{O}_X(-D) \otimes L \otimes I_Z \to 0.$$

The induced map $\mathcal{O}_X(D) \to V \to L$ must be nonzero, for otherwise the map $\mathcal{O}_X(D) \to V$ would factor through $\mathcal{O}_X$, so that $D = -E$ where E is effective. In this case $H \cdot D = -(H \cdot E) \leq 0$, so that $\mathcal{O}_X(D)$ is not destabilizing.

Thus, $\mathcal{O}_X(D) = L \otimes \mathcal{O}_X(-E)$ for some effective divisor E. Since L is nef, $L \cdot E \geq 0$. We can write V as an extension

$$0 \to L \otimes \mathcal{O}_X(-E) \to V \to \mathcal{O}_X(E) \otimes I_Z \to 0,$$

and thus

$$0 = c_2(V) = (L \cdot E) - E^2 + \ell(Z).$$

Rewrite this as $E^2 = (L \cdot E) + \ell(Z) \geq L \cdot E \geq 0$. Now $2H \cdot D > H \cdot L$ for every ample H. Since L is nef, it is a limit of ample divisors, and thus $2L \cdot D = 2L^2 - 2(L \cdot E) \geq L^2$. So $L^2 \geq 2(L \cdot E)$. Thus, $(L^2)(E^2) \geq 2(L \cdot E)(E^2) \geq 2(L \cdot E)^2$. On the other hand, by the Hodge index theorem we have $(L^2)(E^2) \leq (L \cdot E)^2$. So $2(L \cdot E)^2 \leq (L \cdot E)^2$, which is only possible if $L \cdot E = 0$. By the Hodge index theorem, we would have $E^2 \leq 0$, and since we have seen that $E^2 \geq 0$, $E^2 = 0$, and E is numerically equivalent to 0. As E is effective, $E = 0$ and since there is a nontrivial map $L \to L \otimes I_Z$, $Z = \emptyset$. But now there is a map $L \to V$ projecting to a nonzero map $L \to L$, which after adjusting by a scalar we can assume to be the identity. In particular the defining exact sequence for V is split, contrary to assumption. It follows that $H^1(L^{-1}) = 0$. $\square$

In the next section we shall generalize the methods of the above proof to give Reider's analysis of linear systems of the form $|D + K_X|$, where D is nef and big.

The theorems of Bombieri and Reider

Bombieri's famous theorem on the multiple of the canonical bundle needed to define a birational morphism for a surface of general type is the following:

Theorem 5. *Let X be an algebraic surface, and suppose that K_X is nef and big. Let $\varphi_n \colon X \dashrightarrow \mathbb{P}^N$ be the rational map corresponding to the linear system $|nK_X|$, if this is meaningful. Then:*

(i) *For all $n \geq 5$, φ_n is a birational morphism to its image, which is the normal surface $\tilde{X}$ obtained by contracting the smooth rational curves C on X with $C^2 = -2$.*

(ii) *φ_4 is always a morphism (i.e., $|4K_X|$ has no base points), and φ_4 is birational provided that $K_X^2 \geq 2$.*

(iii) *φ_3 is a morphism if $K_X^2 \geq 2$ and is birational provided that $K_X^2 \geq 3$.*

(iv) *φ_2 is a morphism if $K_X^2 \geq 5$ and is birational provided that $K_X^2 \geq 10$, unless there is a pencil of curves on X of genus 2.*

(Prior to Bombieri's theorem, Moishezon had shown in [132] that φ_9 is birational to its image, and Kodaira [73] established a similar result for φ_6.)

We will deduce Bombieri's theorem from the following theorem:

Theorem 6 (Reider). *Let X be an algebraic surface, and let D be a nef and big divisor on X. Then:*

(i) *If $D^2 \geq 5$ and x is a base point of $|K_X + D|$, then there exists a curve E on X with $x \in \operatorname{Supp} E$ such that either $D \cdot E = 0$ and $E^2 = -1$ or $D \cdot E = 1$ and $E^2 = 0$.*

(ii) *If $D^2 \geq 10$ and $x, y \in X$ are two points, possibly infinitely near, such that $|K_X + D|$ does not separate x and y, then there exists a curve E on X such that $x, y \in \operatorname{Supp} E$ such that either $D \cdot E = 0$ and $E^2 = -1$ or -2 or $D \cdot E = 1$ and $E^2 = 0$ or 1 or $D \cdot E = 2$ and $E^2 = 0$.*

Proof that Reider's theorem implies Bombieri's theorem. For $n \geq 2$, the divisor $D = (n-1)K_X$ is nef and big. Note that if E is a curve with $E \cdot D = 0$, then E^2 is even, by the Wu formula, and, if $n > 2$, then $D \cdot E = (n-1)(K_X \cdot E)$ which can never be 1. If $n = 2$ and $D \cdot E = K_X \cdot E = 1$, then again by the Wu formula E^2 is odd, and so we can never have $E^2 = 0$. Thus, we can apply (i) of Reider's theorem as soon as $(n-1)^2 K_X^2 \geq 5$. This inequality is automatic if $n \geq 4$, and it is satisfied for $n = 3$ as soon as $K_X^2 \geq 2$, and for $n = 2$ as soon as $K_X^2 \geq 5$.

Next we consider when φ_n is birational. First note that $(n-1)^2 K_X^2 \geq 10$ as long as $n \geq 5$, or $n = 4$ and $K_X^2 \geq 2$, or $n = 3$ and $K_X^2 \geq 3$, or $n = 2$ and $K_X^2 \geq 10$. Leaving aside the question of the existence of smooth rational curves of self-intersection -2 on X for the moment, we see that the case $D \cdot E = 0, E^2 = -1$ is impossible, as well as the case $D \cdot E = 1, E^2 = 0$. Moreover, if $n \geq 3$, we cannot have $D \cdot E = 1$, since $D \cdot E$ is divisible by $n - 1$. Likewise, if $n \geq 3$, we cannot have $D \cdot E = 2$ and $E^2 = 0$: either $n \geq 4$, in which case $(n-1)|2$, which is impossible, or $n = 3$, $D \cdot E = 2$, and so $K_X \cdot E = 1$. But then by the Wu formula E^2 is odd. Thus, K_X fails to separate p and q if and only if there exists an effective divisor E containing p and q such that $E^2 = -2$, $E \cdot K_X = 0$, or $n = 2$, $K_X \cdot E = 1$, and $E^2 = 1$, or $n = 2$, $K_X \cdot E = 2$, and $E^2 = 0$. There are only finitely many curves E such that $K_X \cdot E = 0$, and so φ_n is always birational away from these. In the remaining exceptional cases, $n = 2$ and either $K_X \cdot E = 1$ and $E^2 = 1$, or $K_X \cdot E = 2$ and $E^2 = 0$. In both cases $p_a(E) = 2$. Either there are only finitely many such curves E, or X has a (possibly irrational) pencil of curves of genus 2.

The remaining point to check is that, if $n \geq 5$, then $\varphi_n(X)$ really is the normal surface $\bar{X}$ obtained by contracting the smooth rational curves of self-intersection -2 on X. This would lead us too far afield into the geometry of rational double points, and so for the proof (as well as for the proof that $\varphi_n(X)$ is projectively normal if $n \geq 5$), we refer to [11] and [7]. $\square$

Proof of Reider's theorem. First suppose that D is nef and big, with $D^2 \geq 5$, and that x is a base point for $|K_X + D|$. We seek a rank 2 vector bundle V which is given as an extension

$$0 \to \mathcal{O}_X \to V \to \mathcal{O}_X(D) \otimes \mathfrak{m}_x \to 0,$$

where $\mathfrak{m}_x$ is the ideal sheaf of the reduced point x. By Theorem 12 in Chapter 2, a locally free extension V exists if and only if every section of $K_X + D$ vanishes at x, i.e., x is a base point for $|K_X + D|$. By assumption, then, we can find such a V, and clearly $c_1(V) = D$ and $c_2(V) = 1$. If $D^2 \geq 5$, Bogomolov's inequality is violated and so V is unstable. Thus, there exists a sub-line bundle $\mathcal{O}_X(F)$ of V such that $V/\mathcal{O}_X(F)$ is torsion free and $2H \cdot F > H \cdot D > 0$ for every ample divisor H. Since D is a limit of ample divisors, $2D \cdot F \geq D^2$. Now the induced map $\mathcal{O}_X(F) \to \mathcal{O}_X(D) \otimes \mathfrak{m}_x$ is nonzero, for otherwise $F = -E$, where E is effective, and thus $2H \cdot F \leq 0$ for every ample H. Thus, $F = D - E$, where E is effective and $x \in \operatorname{Supp} E$. In particular E is not numerically trivial. The inequality $2D \cdot F \geq D^2$ becomes $D^2 \geq 2D \cdot E$. Since there is an exact sequence

$$0 \to \mathcal{O}_X(D - E) \to V \to \mathcal{O}_X(E) \otimes I_Z \to 0$$

for some 0-dimensional subscheme Z, $1 = c_2(V) = D \cdot E - E^2 + \ell(Z)$, and thus $D \cdot E - E^2 \leq 1$. Thus, we have:

$$2D \cdot E \leq D^2, \qquad D \cdot E - E^2 \leq 1,$$
$$D \cdot E \geq 0, \qquad D^2 E^2 \leq (D \cdot E)^2,$$

where the third inequality holds since D is nef and E is effective, and the fourth follows from the Hodge index theorem (Exercise 10 in Chapter 1). Putting this together,

$$D \cdot E \leq 1 + E^2 \leq 1 + \frac{(D \cdot E)^2}{D^2} \leq 1 + \frac{D \cdot E}{2},$$

and so $D \cdot E \leq 2$, with equality if and only if D and E are proportional. In this case $E^2 \geq 1$ and so $D^2 \leq 4$, a case excluded by the hypotheses of Reider's theorem. Thus, either $D \cdot E = 0$ or $D \cdot E = 1$. If $D \cdot E = 0$, then $E^2 \leq 0$ by the Hodge index theorem and $E^2 = 0$ if and only if E is numerically trivial, which does not hold since E is a nonzero effective divisor. Since $-E^2 \leq 1$, in fact $E^2 = -1$.

The remaining possibility is that $D \cdot E = 1$. In this case, $E^2 \leq 1/D^2 \leq 1/5$, so that $E^2 \leq 0$. But as $E^2 \geq 1 - (D \cdot E) = 0$, $E^2 = 0$ in this case.

Now assume that $D^2 \geq 10$ and that x, y are distinct points of X such that $|K_X + D|$ does not separate x and y, in other words that every section of $\mathcal{O}_X(K_X + D)$ which vanishes at x also vanishes at y. Again by Theorem 12 of Chapter 2, there exists a vector bundle V and an exact sequence

$$0 \to \mathcal{O}_X \to V \to \mathcal{O}_X(D) \otimes I_{\{x,y\}} \to 0,$$

where $I_{\{x,y\}}$ is the ideal sheaf of the reduced scheme supported on $\{x, y\}$. In this case, $c_1(V) = D$ and $c_2(V) = 2$, and since $D^2 \geq 10$, we have $c_1(V)^2 \geq 4c_2(V) = 8$. Arguing as in the previous case, there is an effective divisor E with $x, y \in \operatorname{Supp} E$ such that

$$2D \cdot E \leq D^2, \qquad D \cdot E - E^2 \leq 2,$$
$$D \cdot E \geq 0, \qquad D^2 E^2 \leq (D \cdot E)^2.$$

Thus,

$$D \cdot E \leq 2 + E^2 \leq 2 + \frac{(D \cdot E)^2}{D^2} \leq 2 + \frac{D \cdot E}{2},$$

and so $D \cdot E \leq 4$, with equality if and only if D and E are proportional. If $D \cdot E = 4$ and $D = \lambda E$ for some positive rational number λ, then $10 \leq D^2 = 16/E^2$, so that $E^2 = 1$, contradicting $D \cdot E - E^2 \leq 2$. If $D \cdot E = 3$, then $E^2 \geq 1$, and so $D^2 \leq 9/E^2 \leq 9$, contradicting $D^2 \geq 10$. Thus, $D \cdot E$ is either $0, 1$, or 2.

If $D \cdot E = 0$, then $E^2 < 0$ by the Hodge index theorem (note that $E^2 = 0$ is impossible since E is effective and nonzero). On the other hand, since $D \cdot E - E^2 \leq 2$, $E^2 = -2$ or -1. If $D \cdot E = 1$, then $E^2 \geq -1$ and $E^2 \leq 1/D^2 \leq 1/10$, so that $E^2 \leq 0$. The only possibilities are $E^2 = -1, 0$. Lastly, if $D \cdot E = 2$, then $E^2 \geq 0$ and $E^2 \leq 4/10$, so that $E^2 = 0$.

Finally, we consider the case where x is a point of X and y is an infinitely near point, in other words a tangent direction at x. In this case there is a 0-dimensional local complete intersection subscheme Z supported at x corresponding to the tangent direction y, with $\ell(Z) = 2$. To say that the sections of $K_X + D$ do not separate the tangent directions at x is to say that every section of $K_X + D$ vanishing at x also vanishes on the direction y. A local argument shows that there is an everywhere generating section of $Ext^1(\mathcal{O}_X(D) \otimes I_Z, \mathcal{O}_X)$ which lifts to an element of $Ext^1(\mathcal{O}_X(D) \otimes I_Z, \mathcal{O}_X)$. Thus, we may construct a locally free extension V as before, and reach a similar conclusion. □

The proof of Bogomolov's theorem

Our goal in this section is to prove the following theorem:

Theorem 7. *Let V be a rank 2 vector bundle on X. Suppose that there exist two smooth curves C_1 and C_2 on X with $C_1 \cdot C_2 > 0$ and $C_2^2 \geq 0$ such that $V|C_i$ is semistable for $i = 1, 2$. Then $c_1(V)^2 \leq 4c_2(V)$.*

In the theorem, C_1 is allowed to equal C_2, in which case the statement simply reads $C_1^2 > 0$. We will first prove the theorem, and then show in the next section that if V is stable it fulfills the hypotheses of the theorem. We will also postpone the proof of the following basic result to the next section:

Theorem 8. *Let C be a smooth curve and let W be a semistable vector bundle of rank 2 on C such that $\deg W = 0$. If L is a line subbundle of $\operatorname{Sym}^n W$, then $\deg L \leq 0$. If W is a semistable rank 2 vector bundle with $\deg W$ arbitrary, and L is a line subbundle of $\operatorname{Sym}^{2n} W \otimes (\det W)^{-n}$, then $\deg L \leq 0$.*

We note that the theorem is an easier special case of the following fact: if W is semistable, then so is $\operatorname{Sym}^n W$. We shall discuss this more fully in the next section.

Next we have the following easy lemma:

Lemma 9. *Let C be a smooth curve, and let W be a semistable vector bundle of rank 2 on C such that $\deg W = 0$. Then $h^0(\operatorname{Sym}^n W) \leq n + 1$. If W is a semistable rank 2 vector bundle with $\deg W$ arbitrary, then $h^0(\operatorname{Sym}^{2n} W \otimes (\det W)^{-n}) \leq 2n + 1$.*

Proof. We shall just give the proof for the case where $\deg \det W = 0$; the other case is similar. We shall prove the following: if S is an r-dimensional subspace of $H^0(\operatorname{Sym}^n W)$, then the corresponding map $\mathcal{O}_C^r \to \operatorname{Sym}^n W$ is injective. In particular, taking $r = n + 1$, if there is a subspace of $H^0(\operatorname{Sym}^n W)$ of dimension $n + 1$, then the induced map $\mathcal{O}_C^{n+1} \to \operatorname{Sym}^n W$ is injective, and its determinant, which is a section of $\det(\operatorname{Sym}^n W)$, is nonzero. Now it follows from the next lemma (and is easy to check directly) that $\deg \det(\operatorname{Sym}^n W) = 0$. Thus, if there is a nonzero section of $\det(\operatorname{Sym}^n W)$, it is nowhere vanishing, and the corresponding homomorphism $\mathcal{O}_C^{n+1} \to \operatorname{Sym}^n W$ is an isomorphism. In particular $h^0(\operatorname{Sym}^n W) = n + 1$. Hence, if $h^0(\operatorname{Sym}^n W) \geq n + 1$, then $h^0(\operatorname{Sym}^n W) = n + 1$, and in all cases $h^0(\operatorname{Sym}^n W) \leq n + 1$.

To see the claim on r-dimensional subspaces of $H^0(\operatorname{Sym}^n W)$, we argue by induction on r. The case $r = 1$ is clear. Now suppose that S is an $(r + 1)$-dimensional subspace of $h^0(\operatorname{Sym}^n W)$ such that the induced map

$\mathcal{O}_C^{r+1} \to \operatorname{Sym}^n W$ is not injective. Denote its image by U, so that U is a torsion free subsheaf of $\operatorname{Sym}^n W$ and is therefore a vector bundle of rank at most r. Choosing an r-dimensional subspace of S and applying induction, there is an injection $\mathcal{O}_C^r \to \operatorname{Sym}^n W$ whose image is contained in U. Thus, U has rank r and properly contains the image of $\mathcal{O}_C^r$. There exists a point $t \in C$ such that the induced map on fibers $\mathbb{C}^r = (\mathcal{O}_C^r)_t \to U_t$ fails to be injective. If v is in the kernel of this map, there is a global section s of $\mathcal{O}_C^r$ such that $s(t) = v$, and so the induced map $\mathcal{O}_C \to \operatorname{Sym}^n W$ defined by s vanishes at t. It follows that there is an effective nonzero divisor $\mathbf{d}$ on C such that the map $\mathcal{O}_C \to \operatorname{Sym}^n W$ factors through the inclusion $\mathcal{O}_C \to \mathcal{O}_C(\mathbf{d})$. But then $\operatorname{Sym}^n W$ has the line subbundle $\mathcal{O}_C(\mathbf{d})$, which has positive degree, contradicting the previous theorem. $\square$

More generally, if W is a semistable bundle of rank r and degree 0 on a smooth curve C, then arguments very similar to those given above show that $h^0(W) \le r+1$, and the lemma would follow from this if we had proved directly that, starting out with a semistable rank 2 vector bundle W, then $\operatorname{Sym}^n W$ is also semistable.

We turn now to the proof of Theorem 7. The basic idea of the proof is as follows. Fix a vector bundle V and two smooth curves C_1 and C_2 with $C_1 \cdot C_2 > 0$. Suppose that $V|C_1$ is semistable but that $c_1(V)^2 > 4c_2(V)$. First suppose that $\det V = \mathcal{O}_X$, and that $0 = c_1(V)^2 > 4c_2(V)$, in other words that $c_2(V) < 0$. We shall show that, for all $n \ge 1$, $h^0(\operatorname{Sym}^n V) \ge n^3/12 + O(n^2)$. Thus, for $n \gg 0$, $h^0(\operatorname{Sym}^n V) > n + 1$. While $\operatorname{Sym}^n V$ has many sections, it follows from Lemma 9 that $h^0(\operatorname{Sym}^n V|C_1) \le n+1$. Thus, we can eventually find a nonzero section of $\operatorname{Sym}^n V$ vanishing on C_1, and then it is easy to see that $V|C_2$ cannot be semistable. There is a standard device for handling the case where $\det V$ is not necessarily trivial: we consider instead the vector bundle $\operatorname{Sym}^{2n} V \otimes (\det V)^{-n}$, which has rank $2n + 1$ and (as we shall see) trivial determinant. Note that, if we replace V by a twist $V \otimes L$, where L is a line bundle, then $\operatorname{Sym}^{2n} V \otimes (\det V)^{-n}$ is unchanged, and in particular it is unchanged if we replace V by $V^{\vee}$. In this case, we shall show that $h^0(\operatorname{Sym}^{2n} V \otimes (\det V)^{-n}) \ge n^3/6 + O(n^2)$. Thus, for $n \gg 0$, $h^0(\operatorname{Sym}^{2n} V \otimes (\det V)^{-n}) > 2n + 1$, and we can reach a contradiction as before.

To prove these statements, we begin with the following Chern class calculations:

Lemma 10. *Let V be a rank 2 vector bundle on X.*

(i) *If $\det V$ is trivial and $c_2(V) = c$, then for all $n \ge 1$,*

$$c_1(\operatorname{Sym}^n V) = 0,$$

$$c_2(\mathrm{Sym}^n V) = \left(\sum_{\substack{0 \le i \le n \\ i \equiv n \ (\mathrm{mod}\ 2)}} i^2 \right) c.$$

(ii) *In general, if $p_1(\mathrm{ad}\ V) = p$, then for all $n \ge 1$,*

$$c_1(\mathrm{Sym}^{2n} V \otimes (\det V)^{-n}) = 0,$$

$$c_2(\mathrm{Sym}^{2n} V \otimes (\det V)^{-n}) = \left(\sum_{i=0}^{n} i^2 \right)(-p).$$

Proof. (i) By the splitting principle, we may assume that V is a direct sum of line bundles L_1 and L_2 with $c_1(L_1) = \alpha$ and $c_1(L_2) = -\alpha$. Thus, $c_2(V) = -\alpha^2$. In this case, $\mathrm{Sym}^n V$ is a direct sum of the line bundles $L_1^i \otimes L_2^{n-i}$, with $c_1(L_1^i \otimes L_2^{n-i}) = i\alpha + (n-i)(-\alpha) = (2i-n)\alpha$. Thus,

$$c_1(\mathrm{Sym}^n V) = \sum_{i=0}^{n}(2i-n)\alpha = \left(\sum_{\substack{-n \le i \le n \\ i \equiv n \ (\mathrm{mod}\ 2)}} i \right)\alpha = 0.$$

Likewise,

$$c_2(\mathrm{Sym}^n V) = \left(\sum_{0 \le i < j \le n} (2i-n)(2j-n) \right)\alpha^2.$$

We see that it suffices to prove the identity

$$\sum_{0 \le i < j \le n} (2i-n)(2j-n) = - \sum_{\substack{0 \le i \le n \\ i \equiv n \ (\mathrm{mod}\ 2)}} i^2.$$

In the sum on the left, consider those i, j with $i < j$ and $i + j = n$. Such terms contribute

$$(2i-n)(2(n-i)-n) = -(n-2i)^2$$

to the sum. Clearly, such terms exist for $i = 0, \ldots, [n/2]$ and their sum is exactly the right-hand side of the claimed equality above. Thus, it suffices to show that the sum of the remaining terms is 0. Given $i < j$ with $i+j \ne n$, we have $n - j \ne i$. If $n - j = j$, then $2j - n = 0$ and the corresponding term in the sum is 0. Otherwise, either $i < n-j$ or $n-j < i$, but in any case the corresponding term in the sum is $(2i-n)(2(n-j)-n) = -(2i-n)(2j-n)$ which cancels the term $(2i-n)(2j-n)$. Thus, the nonzero terms of this form cancel each other off in pairs, and so the sum of all the terms with $i + j \ne n$ is indeed 0.

The proof of (ii) is similar and will be left to the reader. $\square$

Corollary 11. *Let V be a rank 2 vector bundle on X.*

(i) *If $c_1(V) = 0$ and $c_2(V) = c$, then $\chi(\mathrm{Sym}^n V) = -cn^3/6 + O(n^2)$.*

(ii) *In general, if $p_1(\text{ad } V) = p$, then $\chi(\text{Sym}^{2n} V \otimes (\det V)^{-n}) = pn^3/3 + O(n^2)$.*

Proof. To see (i), note by the Riemann-Roch formula that

$$\chi(\text{Sym}^n V) = -c_2(\text{Sym}^n V) + (n+1)\chi(\mathcal{O}_X).$$

Thus, by (i) of Lemma 10, it suffices to see that

$$\left(\sum_{\substack{0 \le i \le n \\ i \equiv n \,(\text{mod } 2)}} i^2 \right) = n^3/6 + O(n^2),$$

which is an elementary exercise left to the reader. The argument for (ii) is similar. $\square$

Lemma 12. *Let V be a rank 2 vector bundle on X.*

(i) *If $c_1(V) = 0$ and $c_2(V) = c$, then $h^0(\text{Sym}^n V) \ge -cn^3/12 + O(n^2)$.*
(ii) *In general, if $p_1(\text{ad } V) = p$, then $h^0(\text{Sym}^{2n} V \otimes (\det V)^{-n}) \ge pn^3/6 + O(n^2)$.*

Proof. As usual we shall only write down the case (i). Note that $(\text{Sym}^n V)^\vee = \text{Sym}^n(V^\vee) \cong \text{Sym}^n V$. Using Serre duality and Corollary 11,

$$h^0(\text{Sym}^n V) + h^0(\text{Sym}^n V \otimes K_X) \ge -cn^3/6 + O(n^2).$$

Thus, it suffices to show that $h^0(\text{Sym}^n V) = h^0(\text{Sym}^n V \otimes K_X) + O(n^2)$. More generally, let L be an arbitrary line bundle on X. We claim that $h^0(\text{Sym}^n V) = h^0(\text{Sym}^n V \otimes L) + O(n^2)$. Write $L = \mathcal{O}_X(D_1 - D_2)$, where D_1 and D_2 are smooth curves on X. We shall show that $h^0(\text{Sym}^n V \otimes L) \le h^0(\text{Sym}^n V) + O(n^2)$; the proof of the other inequality is similar. From the exact sequence

$$0 \to \text{Sym}^n V \otimes \mathcal{O}_X(-D_2) \to \text{Sym}^n V \otimes \mathcal{O}_X(D_1 - D_2)$$
$$\to \text{Sym}^n V|D_1 \otimes \mathcal{O}_{D_1}(D_1 - D_2) \to 0,$$

we see that

$$h^0(\text{Sym}^n V \otimes L) \le h^0(\text{Sym}^n V \otimes \mathcal{O}_X(-D_2)) + h^0(\text{Sym}^n V|D_1 \otimes \mathcal{O}_{D_1}(D_1 - D_2)),$$

and using the inclusion $\text{Sym}^n V \otimes \mathcal{O}_X(-D_2) \subseteq \text{Sym}^n V$, we see that $h^0(\text{Sym}^n V \otimes \mathcal{O}_X(-D_2)) \le h^0(\text{Sym}^n V)$. Thus, it suffices to show that

$$h^0(\text{Sym}^n V|D_1 \otimes \mathcal{O}_{D_1}(D_1 - D_2)) = O(n^2).$$

Now on D_1 we can write V as an extension of two line bundles:

$$0 \to L_1 \to V|D_1 \to L_2 \to 0,$$

with deg $L_1 = \ell = -\deg L_2$, say. It follows that $\operatorname{Sym}^n V | D_1 \otimes \mathcal{O}_{D_1}(D_1 - D_2)$ has a filtration whose successive quotients are $L_1^{n-i} \otimes L_2^i \otimes \mathcal{O}_{D_1}(D_1 - D_2)$. Thus,

$$h^0(\operatorname{Sym}^n V | D_1 \otimes \mathcal{O}_{D_1}(D_1 - D_2)) \le \sum_{i=0}^n h^0(L_1^{n-i} \otimes L_2^i \otimes \mathcal{O}_{D_1}(D_1 - D_2)).$$

If $\deg \mathcal{O}_{D_1}(D_1 - D_2) = d$, then $\deg L_1^{n-i} \otimes L_2^i \otimes \mathcal{O}_{D_1}(D_1 - D_2) = (n - 2i)\ell + d$. If $L_1^{n-i} \otimes L_2^i \otimes \mathcal{O}_{D_1}(D_1 - D_2)$ is a nonspecial line bundle (i.e., if $h^1(L_1^{n-i} \otimes L_2^i \otimes \mathcal{O}_{D_1}(D_1 - D_2)) = 0$), then by Riemann-Roch on the smooth curve D_1,

$$h^0(L_1^{n-i} \otimes L_2^i \otimes \mathcal{O}_{D_1}(D_1 - D_2)) = (n - 2i)\ell + d + (1 - g),$$

where $g = g(D_1)$. On the other hand, if $L_1^{n-i} \otimes L_2^i \otimes \mathcal{O}_{D_1}(D_1 - D_2)$ is effective and special, then by Clifford's theorem

$$h^0(L_1^{n-i} \otimes L_2^i \otimes \mathcal{O}_{D_1}(D_1 - D_2)) \le \tfrac{1}{2}((n - 2i)\ell + d) + 1.$$

In both cases we have a bound which is linear in $n - 2i$, with the implied constants independent of n. Thus, summing over i, we see that $h^0(\operatorname{Sym}^n V | D_1 \otimes \mathcal{O}_{D_1}(D_1 - D_2) \le O(n^2)$. □

Now we can complete the proof of Theorem 7. For simplicity we shall just write down the case where $\det V = 0$. Suppose that $c_2(V) = c < 0$, and so $h^0(\operatorname{Sym}^n V) = -cn^3/12 + O(n^2)$. Choose n so that $h^0(\operatorname{Sym}^n V) > n + 1$. By Lemma 9, since $V | C_1$ is a semistable vector bundle of rank 2 and degree 0,

$$h^0(\operatorname{Sym}^n V | C_1) \le n + 1.$$

It follows that there exists a nonzero section of $\operatorname{Sym}^n V$ vanishing on C_1, in other words that $h^0(\operatorname{Sym}^n V \otimes \mathcal{O}_X(-C_1)) \ne 0$. This is equivalent to the existence of a nonzero map

$$\mathcal{O}_X(C_1) \to \operatorname{Sym}^n V.$$

There is an effective curve D such that the map $\mathcal{O}_X(C_1) \to \operatorname{Sym}^n V$ vanishes along D, and such that the induced map $\mathcal{O}_X(C_1 + D) \to \operatorname{Sym}^n V$ vanishes at only finitely many points. In particular, the restricted map

$$\mathcal{O}_X(C_1 + D) | C_2 \to \operatorname{Sym}^n V | C_2$$

is injective. Note that C_2 is nef, since $C_2^2 \ge 0$, and hence $C_2 \cdot D \ge 0$. Thus, $\operatorname{Sym}^n V | C_2$ has the sub-line bundle $\mathcal{O}_X(C_1 + D) | C_2$, which has degree $(C_1 \cdot C_2) + (D \cdot C_2) > 0$. By Theorem 8, this contradicts the semistability of $V | C_2$. □

Symmetric powers of vector bundles on curves

In this section, we study $\operatorname{Sym}^n W$, where W is a vector bundle on the smooth curve C. We begin with the following lemmas, due to Gieseker [46]:

Lemma 13. Let $p\colon \tilde{C} \to C$ be a finite morphism between two smooth curves $\tilde{C}$ and C, and let W be a rank 2 vector bundle on C. Then W is semistable if and only if p^*W is semistable.

Proof. First assume that W is unstable, and let L' be a destabilizing line subbundle. Thus, $\mu(L') > \mu(W)$. Now $\mu(p^*W) = (\deg p)\mu(W)$ and likewise $\mu(p^*L') = (\deg p)\mu(L')$. Thus, p^*L' is a destabilizing line subbundle of p^*W, which is therefore unstable.

Conversely, assume that p^*W is unstable. After passing to a finite cover, we may assume that $p\colon \tilde{C} \to C$ is a Galois cover, with Galois group G. The first part of the proof shows that, after passing to the Galois cover, the pulled back bundle (which we continue to denote by p^*W) is still unstable. Let L be the maximal destabilizing line subbundle. Then for all $\sigma \in G$, σ^*L is the maximal destabilizing subsheaf of $\sigma^*p^*W = W$. It follows that $\sigma^*L = L$ for all $\sigma \in G$. By Theorem 28 (to be proved in the Appendix), there exists a line subbundle L' of W such that $p^*L' = L$. Again using $\mu(p^*W) = (\deg p)(\mu(W)$ and $\mu(L) = \mu(p^*L') = (\deg p)(\mu(L'))$, we see that L' is destabilizing. Thus, W is unstable as well. $\square$

In the above situation, it is possible for W to be stable and p^*W to be strictly semistable. For a typical example, suppose that $p\colon \tilde{C} \to C$ is an étale double cover with covering involution ι, and let L be a line bundle of degree 0 on $\tilde{C}$ such that $\iota^*L \neq L$. If W is the rank 2 vector bundle on C defined by p_*L, then we leave it as an exercise to show that $W = p_*L$ is stable but that $p^*W = L \oplus \iota^*L$, which is strictly semistable.

We can interpret the example above in terms of unitary representations of $\pi_1(C)$ (compare the discussion of the theorem of Narasimhan and Seshadri in the last section of Chapter 4). If W is stable, corresponding to a unitary representation ρ from $\pi_1(C)$ to $U(2)$, then the homomorphism $\pi_1(\tilde{C}) \to \pi_1(C)$ induces a unitary representation $\tilde{\rho}\colon \pi_1(\tilde{C}) \to U(2)$, whose associated vector bundle is just p^*W. However, even if ρ is irreducible, it does not necessarily follow that $\tilde{\rho}$ is irreducible. Thus, p^*W may split as a direct sum of line bundles.

We now relate the stability of a rank 2 vector bundle W over the curve C to the geometry of the corresponding ruled surface $\mathbb{P}(W)$.

Lemma 14. Let W be a semistable rank 2 vector bundle over the smooth curve C, and let $\pi\colon X = \mathbb{P}(W) \to C$ be the associated ruled surface. Let $\mathcal{O}_X(1)$ be the tautological quotient line bundle of $\pi^*W^\vee$ on X. If $\deg W =$

0, then $\mathcal{O}_X(1)$ is nef and $c_1(\mathcal{O}_X(1))^2 = 0$. More generally, if $\deg W$ is arbitrary, let $L = K_{X/C}^{-1} = \mathcal{O}_X(2) \otimes \pi^* \det W$. Then L is nef and $L^2 = 0$.

Proof. We shall just consider the general case. Let D be an irreducible curve on X. If $D = f$ is a fiber of π, then $L \cdot f = 2$. If $D \neq f$, then the restriction of p to D defines a finite morphism $p \colon \tilde{D} \to C$, where $\tilde{D}$ is the normalization of D. Let $\nu \colon \tilde{D} \to X$ be the induced map, so that $p = \pi \circ \nu$. By the previous lemma, $p^* W$ is semistable on $\tilde{D}$, and thus $p^* W^\vee$ is also semistable. Since $\mathcal{O}_X(1)$ is a quotient of $\pi^* W^\vee$ on X, $\nu^* \mathcal{O}_X(1)$ is a quotient of $p^* W^\vee$. Moreover, $\mu(\nu^* \mathcal{O}_X(1)) = \deg \mathcal{O}_X(1)|D = L \cdot D \geq \mu(p^* W^\vee)$. This says that

$$2 \deg(\mathcal{O}_X(1)|D) \geq \deg(p^* W^\vee) = -\deg(\pi^* \det W|D),$$

and so $\deg(L|D) = 2 \deg(\mathcal{O}_X(1)|D) + \deg(\pi^* \det W|D) \geq 0$.

To prove the statements about $c_1(\mathcal{O}_X(1))^2$ and L^2, note that

$$L^2 = 4c_1(\mathcal{O}_X(1))^2 + 4 \deg W.$$

Thus, it suffices to show that, for an arbitrary rank 2 bundle W on C, we have $c_1(\mathcal{O}_X(1))^2 = -\deg W$. This is a very special case of a general formula about the tautological bundle on the projectivization of a vector bundle [55], [45]. In this special case, we can also use Exercise 4 of Chapter 5. □

Note that the line bundle $L = K_{X/C}^{-1}$ defined above is unchanged if we replace W by a twist, and in particular if we replace W by $W^\vee$. The converse to Lemma 14 is an easy exercise: let W be a rank 2 vector bundle on C and let L be defined as in Lemma 14. If L is nef, then W is semistable.

Now we recall the correspondence between curves D on X, not containing any fibers in their support, and with $D \cdot f = n > 0$ and line subbundles of $\mathrm{Sym}^n W^\vee$. Suppose that D is an irreducible curve on X with $D \cdot f = n > 0$. Then we can write $\mathcal{O}_X(D) = \mathcal{O}_X(n) \otimes \pi^* \lambda$ for a uniquely determined line bundle λ on C. The exact sequence

$$0 \to \mathcal{O}_X \to \mathcal{O}_X(D) \to \mathcal{O}_D(D) \to 0$$

together with the isomorphisms $R^0 \pi_* \mathcal{O}_X = \mathcal{O}_C$, $R^1 \pi_* \mathcal{O}_X = 0$, and $R^0 \pi_* \mathcal{O}_X(D) = (R^0 \pi_* \mathcal{O}_X(n)) \otimes \lambda = \mathrm{Sym}^n W^\vee \otimes \lambda$, shows that there is an exact sequence

$$0 \to \mathcal{O}_C \to \mathrm{Sym}^n W^\vee \otimes \lambda \to \pi_* \mathcal{O}_D(D) \to 0.$$

Thus, D gives a line subbundle λ^{-1} of $\mathrm{Sym}^n W^\vee$. Conversely, a line subbundle of $\mathrm{Sym}^n W^\vee$, which we can write as λ^{-1} for some line bundle λ, gives a nowhere vanishing section of $\mathrm{Sym}^n W^\vee \otimes \lambda$ and thus a section of $\mathcal{O}_X(n) \otimes \pi^* \lambda$, which, as is easy to check, does not vanish along any fibers (but may be reducible).

Theorem 15. *If W is semistable and $\deg W = 0$, then $\operatorname{Sym}^n W^\vee$ has no line subbundles of positive degree. If $\deg W$ is arbitrary, then $\operatorname{Sym}^{2n} W \otimes (\det W)^{-n}$ has no line subbundles of positive degree.*

Proof. First consider the case where $\deg W = 0$. Note that

$$\mu(\operatorname{Sym}^n W^\vee) = \frac{1}{n+1} \deg(\operatorname{Sym}^n W^\vee) = -\frac{1}{n+1} \frac{n(n+1)}{2} \deg W = 0,$$

by the splitting principle (as applied in Lemma 10). Now suppose given a line subbundle of $\operatorname{Sym}^n W^\vee$, which we write as λ^{-1}. We claim that

$$\deg(\lambda^{-1}) = -\deg \lambda \le 0,$$

or in other words that $\deg \lambda \ge 0$. Let s be the section of $\mathcal{O}_X(n) \otimes \pi^* \lambda$ corresponding to the inclusion $\lambda^{-1} \to \operatorname{Sym}^n W^\vee$, and suppose that s vanishes along the effective curve D. Then $\mathcal{O}_X(D) = \mathcal{O}_X(n) \otimes \pi^* \lambda$, and since $\mathcal{O}_X(1)$ is nef,

$$0 \le c_1(\mathcal{O}_X(1)) \cdot D = n c_1(\mathcal{O}_X(1))^2 + \deg \lambda = 0 + \deg \lambda,$$

so that $\deg \lambda \ge 0$.

The case where $\deg W$ is arbitrary is similar, noting that a line subbundle λ^{-1} of $\operatorname{Sym}^{2n} W \otimes (\det W)^{-n}$ corresponds to a section of $\operatorname{Sym}^{2n} W \otimes (\det W)^{-n} \otimes \lambda$. Moreover,

$$\operatorname{Sym}^{2n} W \otimes (\det W)^{-n} = \operatorname{Sym}^{2n}(W^\vee \otimes \det W) \otimes (\det W)^{-n}$$

$$= \operatorname{Sym}^{2n}(W^\vee) \otimes (\det W)^n.$$

Thus, the line subbundle λ^{-1} gives a section of $\operatorname{Sym}^{2n}(W^\vee) \otimes (\det W)^n \otimes \lambda$, and thus a section of $L^n \otimes \lambda$ vanishing along a curve D. In this case we have

$$0 \le L \cdot D = nL^2 + 2 \deg \lambda = 2 \deg \lambda,$$

and so again $\deg \lambda \ge 0$. $\square$

A much deeper fact is that, if W is semistable, then $\operatorname{Sym}^n W$ is semistable for all n, and indeed all tensor powers of W are semistable. More generally still, if W_1 and W_2 are two semistable bundles on C, then $W_1 \otimes W_2$ is again semistable. Note that it suffices to prove this statement when W_1 and W_2 are actually stable. In case $\deg W_i = 0, i = 1, 2$, this follows easily from the fact that W_i is associated to an irreducible unitary representation ρ_i of $\pi_1(C)$, and thus $W_1 \otimes W_2$ is associated to the (not necessarily irreducible) unitary representation $\rho_1 \otimes \rho_2$. A purely algebraic proof (in characteristic 0) is given by Gieseker in [46], based on Hartshorne's theory of ample vector bundles [60].

Restriction theorems

Our goal in this section is to prove the following weak restriction theorem:

Theorem 16. *Let H be a very ample divisor on the smooth surface X and let V be an H-stable rank 2 vector bundle with $p_1(\operatorname{ad} V) = p$. Then either $p \leq 0$ or there exists a smooth curve $C \in |H|$ such that $V|C$ is semistable.*

As a corollary, we see that Bogomolov's theorem holds for V, either automatically because $p \leq 0$ or by Theorem 7 (taking $C_1 = C_2 = C$).

Proof of Theorem 16. Choose a pencil contained in $|H|$ whose general member is smooth. Blowing up the base points, we obtain a smooth surface $\tilde{X}$ which is a blowup of X and a morphism $\pi\colon \tilde{X} \to \mathbb{P}^1$ whose general fiber is a smooth curve C with $C \in |H|$. Let $\rho\colon \tilde{X} \to X$ be the birational morphism from $\tilde{X}$ to X. To prove Theorem 16, we argue as follows: first we shall show that $\rho^* V$ is stable for some ample divisor H_0 on $\tilde{X}$. On the other hand, we have defined suitable ample divisors in Chapter 6 (with respect to the morphism $\pi\colon \tilde{X} \to \mathbb{P}^1$). Let H_1 be a (w, p)-suitable ample divisor, where $w = w_2(\rho^* V)$ and $p = p_1(\rho^* \operatorname{ad} V)$. Either $\rho^* V$ is also stable with respect to H_1, or by Proposition 22 of Chapter 4 $\rho^* V$ is strictly semistable with respect to some ample divisor H which is a convex rational combination of H_0 and H_1. But a strictly semistable vector bundle is easily seen to satisfy Bogomolov's inequality, by Exercise 1 in Chapter 6, and so in this case $p \leq 0$. Otherwise, we can assume that $\rho^* V$ is in fact stable with respect to a suitable ample divisor. In this case, by (i) of Theorem 5 in Chapter 6 (for which we shall give a complete proof below), the restriction of $\rho^* V$ to almost all fibers of π is semistable. In particular, the restriction of V to a general smooth fiber C of π is semistable, which concludes the proof of Theorem 16. $\square$

The two points which remain to be proved are first, that $\rho^* V$ is stable with respect to some ample divisor on $\tilde{X}$, and second, the proof of (i) of Theorem 5 in Chapter 6. To deal with the first point, since ρ is a sequence of blowups, it will suffice to handle the case of one blowup at a time. In fact, we have defined stability with respect to nef and big divisors in Chapter 4, and the proof of Proposition 22 in Chapter 4 works equally well in case H_0 is simply assumed to be nef and big. Take $H_0 = \rho^* H$. We must show that $\rho^* V$ is $\rho^* H$-stable. But $\mu_{\rho^* H}(\rho^* V) = \mu_H(V)$. If $\mathcal{O}_{\tilde{X}}(\tilde{D})$ is a sub-line bundle of $\rho^* V$, then writing $\tilde{D} = \rho^* D + aE$, we see that $\mathcal{O}_X(D) = \rho_* \mathcal{O}_{\tilde{X}}(\tilde{D})^{\vee\vee}$ is a sub-line bundle of $\rho_* \rho^* V = V$. Thus,

$$\rho^* H \cdot \tilde{D} = H \cdot D < \mu_H(V) = \mu_{\rho^* H}(\rho^* V),$$

and so ρ^*V is ρ^*H-stable. In fact, there is the following more precise result which says that ρ^*V is actually stable with respect to an appropriate ample divisor on $\tilde{X}$:

Theorem 17. *Let Y be a smooth surface and let H be an ample divisor on Y. Suppose that $\rho: \tilde{Y} \to Y$ is the blowup of Y at a point x, and let E be the exceptional divisor on $\tilde{Y}$. Let V be an H-stable rank 2 vector bundle on Y with $\det V = \Delta$ and $p_1(\operatorname{ad} V) = p$. Then for all N such that $N\rho^*H - E$ is ample on $\tilde{Y}$ and $N > \sqrt{|p|/2}$, the vector bundle ρ^*V is stable with respect to $N\rho^*H - E$.*

Proof. Suppose that $\mathcal{O}_{\tilde{Y}}(D)$ is a sub-line bundle of ρ^*V. We must show that $(N\rho^*H - E) \cdot D < 0$. We may assume that $\rho^*V/\mathcal{O}_{\tilde{Y}}(D)$ is torsion free, so that there is an exact sequence

$$0 \to \mathcal{O}_{\tilde{Y}}(D) \to \rho^*V \to \mathcal{O}_{\tilde{Y}}(\rho^*\Delta - D) \otimes I_Z \to 0.$$

Let $D = \rho^*D' + aE$ for a uniquely specified divisor class D' on Y. There is an inclusion $\rho_*\mathcal{O}_{\tilde{Y}}(D) \subset \rho_*\rho^*V$, and by the projection formula $\rho_*\rho^*V = V \otimes \rho_*\mathcal{O}_{\tilde{Y}} = V$. Moreover, $\rho_*\mathcal{O}_{\tilde{Y}}(D)$ is either of the form $\mathcal{O}_Y(D')$ or $\mathcal{O}_Y(D') \otimes \mathfrak{m}_x^n$ for some positive integer n, and so its double dual, which is just $\mathcal{O}_Y(D')$, includes into V. The quotient $V/\mathcal{O}_Y(D')$ can have torsion only at x, since it is isomorphic to $\rho^*V/\mathcal{O}_{\tilde{Y}}(D)$ away from x, and so it is torsion free by Proposition 5 of Chapter 2. Thus, there is an exact sequence

$$0 \to \mathcal{O}_Y(D') \to V \to \mathcal{O}_Y(\Delta - D') \otimes I_W \to 0,$$

where W is a 0-dimensional subscheme. By the stability of V, $H \cdot (2D' - \Delta) < 0$. Since $p_1(\operatorname{ad} V) = p_1(\operatorname{ad} \rho^*V) = p$, we have:

$$p = (2D - \Delta)^2 - 4\ell(Z) \leq (2D - \Delta)^2 = (2D' - \Delta)^2 - 4a^2 \leq (2D' - \Delta)^2.$$

Moreover, $(N\rho^*H - E) \cdot D = N(H \cdot (2D' - \Delta)) + a$. Thus, if $a \leq 0$, then automatically

$$(N\rho^*H - E) \cdot D = N(H \cdot (2D' - \Delta)) + a \leq N(H \cdot (2D - \Delta)) < 0.$$

So we may assume that $a > 0$. If $(2D' - \Delta)^2 \leq 0$, then

$$4a^2 \leq (2D' - \Delta)^2 - p \leq -p \leq |p|,$$

and so $a \leq \sqrt{|p|}/2$. Thus,

$$(N\rho^*H - E) \cdot D = N(H \cdot (2D' - \Delta)) + a \leq a - N < \sqrt{|p|}/2 - \sqrt{|p|}/2 \leq 0.$$

The remaining possibility is $(2D' - \Delta)^2 > 0$. Since, as we have just seen above, $(N\rho^*H - E) \cdot D \leq a - N$, and $N > \sqrt{|p|/2}$ by hypothesis, we may also assume that $a \geq \sqrt{|p|/2}$. Then $2a^2 \geq |p|$, and so

$$(2D' - \Delta)^2 \geq p + 4a^2 \geq 4a^2 - |p| \geq 2a^2.$$

Now by the Hodge index theorem (Exercise 10 in Chapter 1),

$$H^2(2D' - \Delta)^2 \le (H \cdot (2D' - \Delta))^2,$$

and so, since $H^2 \ge 1$ and $H \cdot (2D' - \Delta) < 0$,

$$H \cdot (2D' - \Delta) \le - \left[(2D' - \Delta)^2\right]^{1/2}.$$

Thus, since $N \ge 1$, we have

$$(N\rho^*H - E) \cdot D = N(H \cdot (2D' - \Delta)) + a$$
$$\le - \left[(2D' - \Delta)^2\right]^{1/2} + a$$
$$\le -\sqrt{2a^2} + a < 0.$$

Thus, in all cases $(N\rho^*H - E) \cdot D < 0$ and so ρ^*V is stable with respect to $N\rho^*H - E$. □

Theorem 17 says that ρ^*V is stable for every ample divisor in every chamber of type (w, p) on $\tilde{Y}$ containing ρ^*H in its closure.

The final point is the following (Theorem 5 in Chapter 6):

Theorem 18. *Let $\pi: X \to C$ be a morphism from the smooth surface X to a smooth curve C, let V be a rank 2 vector bundle on X with invariants (w, p), and let H be a (w, p)-suitable ample divisor on X. Suppose that V is H-semistable. Then for all but finitely many fibers f of π, $V|f$ is semistable.*

Proof. Suppose instead that $V|f$ is unstable for infinitely many fibers f. We have seen in the discussion of the proof of Theorem 5 in Chapter 6 that it is enough to show that there exists a sub-line bundle L of V such that the sub-line bundle $L|f$ of $V|f$ is destabilizing for a single smooth fiber f, for then V is H-unstable. To find L, we shall first find a finite Galois cover $q: \tilde{C} \to C$ and a sub-line bundle L' of $q^*X = \tilde{X}$ with the property that, if $p: \tilde{X} \to X$ is the induced Galois morphism, then $L'|f$ destabilizes $p^*V|f$ for infinitely many fibers of $q^*X \to \tilde{C}$. We can then argue as in Lemma 13 that L' descends to a line bundle L on X with the same property.

Let $d = \deg(V|f) = \det V \cdot f$. If f is a smooth fiber such that $V|f$ is unstable, then there exists a sub-line bundle of $V|f$ with degree greater than $d/2$. On the other hand, there is an upper bound independent of the choice of f for the degree of a sub-line bundle of $V|f$. In fact, we can always write V as an extension

$$0 \to \mathcal{O}_X(D) \to V \to \mathcal{O}_X(\Delta - D) \otimes I_Z \to 0$$

for some divisor D and 0-dimensional subscheme Z (as we have seen in Chapter 2). Let $e_1 = D \cdot f$ and $e_2 = (\Delta - D) \cdot f$. For all but the finitely many fibers f for which $f \cap \operatorname{Supp} Z \neq \emptyset$, there is an exact sequence

$$0 \to L_1 \to V|f \to L_2 \to 0,$$

where $\deg L_i = e_i$. Thus, any sub-line bundle of $V|f$ has degree at most $e = \max\{e_1, e_2\}$. The cases where $f \cap \operatorname{Supp} Z \neq \emptyset$ can then be handled directly (or we can ignore these finitely many fibers in what follows).

Let T be any Zariski open subset of C such that π is smooth over T. We will for the moment replace X by $\pi^{-1}(T)$, so that we can assume that $\pi\colon X \to T$ is smooth. We would like to take the Picard varieties $\operatorname{Pic}^n f$, for each smooth fiber f of π, and fit them together into a variety $\operatorname{Pic}^n(X/T)$ mapping to T. The main property that we need is the existence of a Poincaré line bundle $\mathcal{P}$ on $X \times_T \operatorname{Pic}^n(X/T)$, with the property that, if $\xi \in \operatorname{Pic}^n(X/T)$ corresponds to a line bundle L on the fiber f lying over the point of T corresponding to ξ, then $\mathcal{P}|f \times_T \{\xi\}$ is naturally just L. Unfortunately, such a construction is not always possible. However, if $\pi\colon X \to T$ has a section, then the scheme $\operatorname{Pic}^n(X/T)$ and the line bundle $\mathcal{P}$ exist (see, for example, [58] and [91]). Moreover, the fibers of the morphism $\operatorname{Pic}^n(X/T) \to T$ are the Picard varieties of the fibers, and the morphism $\operatorname{Pic}^n(X/T) \to T$ is smooth and proper. Note that, if T is a smooth curve, then there is always a finite morphism $\tilde{T} \to T$, where $\tilde{T}$ is again a smooth curve, such that, if $\tilde{X} = X \times_T \tilde{T}$, then $\tilde{X}$ is smooth (since π is smooth) and $\tilde{\pi}\colon \tilde{X} \to \tilde{T}$ has a section. In this case, the fibers of $\tilde{\pi}$ may be identified with the corresponding fibers of π. Let $p\colon \tilde{X} \to X$ be the induced finite morphism.

For a smooth fiber f of π or $\tilde{\pi}$, $V|f$ is unstable if and only if $V|f$ has a sub-line bundle L of degree greater than $d/2$, where $d = \deg(V|f)$, and every sub-line bundle L of $V|f$ has degree at most e for some fixed integer e. Let n be an integer with $d/2 < n \leq e$. Over $\tilde{X} \times_{\tilde{T}} \operatorname{Pic}^n(\tilde{X}/\tilde{T})$, we have the Poincaré line bundle $\mathcal{P}$. Let $\pi_1\colon \tilde{X} \times_{\tilde{T}} \operatorname{Pic}^n(\tilde{X}/\tilde{T}) \to \tilde{X}$ and $\pi_2\colon \tilde{X} \times_{\tilde{T}} \operatorname{Pic}^n(\tilde{X}/\tilde{T}) \to \operatorname{Pic}^n(\tilde{X}/\tilde{T})$ be the projections to the first and second factors. Note that $\pi_2\colon \tilde{X} \times_{\tilde{T}} \operatorname{Pic}^n(\tilde{X}/\tilde{T}) \to \operatorname{Pic}^n(\tilde{X}/\tilde{T})$ is smooth and proper, and therefore flat. For a fiber f of $\tilde{\pi}$, $p^*V|f$ is unstable if and only if there exists an n with $d/2 < n \leq e$ and a $\xi \in \operatorname{Pic}^n(\tilde{X}/\tilde{T})$ lying over the same point as f such that, if L is the line bundle corresponding to ξ, then $(V|f) \otimes L^{-1}$ has a section, in other words $h^0(\pi_1^* p^* V \otimes \mathcal{P}^{-1}|\pi_2^{-1}(\xi)) \geq 1$. Let $Z_n \subseteq \operatorname{Pic}^n(\tilde{X}/\tilde{T})$ be defined by

$$Z_n = \{\xi \in \operatorname{Pic}^n(\tilde{X}/\tilde{T}) : h^0(\pi_1^* p^* V \otimes \mathcal{P}^{-1}|\pi_2^{-1}(\xi)) \geq 1\}.$$

By the semicontinuity theorem, Z_n is a closed subvariety of $\operatorname{Pic}^n(\tilde{X}/\tilde{T})$, and since $\operatorname{Pic}^n(\tilde{X}/\tilde{T}) \to \tilde{T}$ is proper, the image Y_n of Z_n in $\tilde{T}$ is also closed. Thus, since $\tilde{T}$ is a curve, either Y_n is finite or $Y_n = \tilde{T}$. Clearly, $\bigcup_{d/2 < n \leq e} Y_n$ is the set of $t \in \tilde{T}$ such that $p^*V|\tilde{\pi}^{-1}(t)$ is unstable.

Suppose that Y_n is finite for every n such that $d/2 < n \leq e$. Since there are only finitely many such n, there are only finitely many $t \in \tilde{T}$ such that $p^*V|\tilde{\pi}^{-1}(t)$ is unstable, and hence only finitely many fibers f of π such that $V|f$ is unstable. Conversely, if $V|f$ is unstable for infinitely many fibers f of π, then $Y_n = \tilde{T}$ for some n with $d/2 < n \leq e$. Choose such an n. Since $Z_n \to \tilde{T}$ is surjective, there exists a smooth curve $\hat{T}$ in Z_n mapping onto $\tilde{T}$. Since we are free to replace $\tilde{T}$ by the finite cover $\hat{T}$, we may assume that the morphism $Z_n \to \tilde{T}$ has a section σ. Using the section σ to identify $\tilde{T}$ with a subvariety of $\mathrm{Pic}^n(\tilde{X}/\tilde{T})$, we may then identify $\tilde{X}$ with a subvariety of $\tilde{X} \times_{\tilde{T}} \mathrm{Pic}^n(\tilde{X}/\tilde{T})$. The line bundle $\mathcal{P}$ then restricts to a line bundle over $\tilde{X}$. After passing to a finite cover we can assume that $\tilde{T} \to T$ is Galois. By assumption $R^0 \tilde{\pi}_*(p^*V \otimes \mathcal{P}^{-1})$ is a nonzero torsion free sheaf $\mathcal{F}$ on $\tilde{T}$. If $\mathcal{L}$ is a sub-line bundle of $\mathcal{F}$, then there is a section of $\mathcal{F} \otimes (\mathcal{L})^{-1}$ and thus of $p^*V \otimes (\mathcal{P} \otimes \tilde{\pi}^*\mathcal{L})^{-1}$. Since we are free to replace $\mathcal{P}$ by $\mathcal{P} \otimes \tilde{\pi}^*\mathcal{L}$, we may assume that there is a nonzero map $\mathcal{P} \to p^*V$ with $\mathcal{P}|f$ a destabilizing sub-line bundle of $V|f$ for every fiber f for which the induced map is nonzero. After replacing $\mathcal{P}$ by $\mathcal{P} \otimes \mathcal{O}_{\tilde{X}}(E)$ for some effective divisor E, we may further assume that the quotient $p^*V/\mathcal{P}$ is torsion free. Note that this will not affect the condition that $\mathcal{P}|f$ is destabilizing, since the only problem is when E contains some multiple of a fiber f as a component and $\deg \mathcal{O}_{\tilde{X}}(f)|f = 0$.

Let us recapitulate the situation. Starting with the smooth morphism $\pi\colon X \to T$, where T is a smooth curve, we find a Galois cover $\tilde{T} \to T$, where $\tilde{T}$ is again a smooth curve, with the following property: Making the base change $p\colon \tilde{X} = X \times_T \tilde{T} \to X$, there exists a line bundle $\mathcal{P}$ on $\tilde{X}$ such that $\mathcal{P}$ is a sub-line bundle of p^*V and $p^*V/\mathcal{P}$ is torsion free. It follows that, for a generic fiber f, $(V|f)/(\mathcal{P}|f)$ is torsion free. Thus, for a generic fiber f, $\mathcal{P}|f$ is the maximal destabilizing sub-line bundle of $V|f$.

Let G be the Galois group of $\tilde{T}$ over T, let η be the generic point of T, so that $\eta = \mathrm{Spec}\, k$, where k is the function field of T, and let $\tilde{\eta} = \mathrm{Spec}\, \tilde{k}$ be the generic point of $\tilde{T}$. The surface $X \to T$ restricts to a curve X_η over η, in other words a curve defined over the field k, and likewise $\tilde{X}$ restricts to a curve $X_{\tilde{\eta}}$, the pullback of X_η to $\tilde{\eta}$. Likewise, we can restrict V to a vector bundle V_η over X_η and p^*V to a bundle $V_{\tilde{\eta}}$. The sub-line bundle $\mathcal{P}$ restricts to a sub-line bundle $\mathcal{L}'$ of $V_{\tilde{\eta}}$, such that $V_{\tilde{\eta}}/\mathcal{L}'$ is torsion free, and $\mathcal{L}'$ is the maximal destabilizing subbundle of $V_{\tilde{\eta}}$. It follows that, for all $\sigma \in G$, $\sigma^*\mathcal{L}' = \mathcal{L}'$ as a subsheaf of $V_{\tilde{\eta}} = \sigma^*V_{\tilde{\eta}}$. By Proposition 23 of the Appendix, there is a line bundle $\mathcal{L}$ defined over η and an inclusion $\mathcal{L} \subseteq V_\eta$, such that $\mathcal{L}$ pulls back to $\mathcal{L}'$. In particular $\mathcal{L}$ is destabilizing, in the sense that $\deg \mathcal{L} > d/2$.

We now consider the global case, where T is an open subset of C and X is defined over all of C, not just over T. The coherent subsheaf $\mathcal{L}$ defined on X_η extends to a coherent subsheaf L_U of $V|U$ defined on some open subscheme U of X. For example, taking an affine open subset of T of the form $\mathrm{Spec}\, R$, where k is the quotient field of R, it is easy to see that the sheaf $\mathcal{L}$ involves only finitely many denominators in the ring R, and thus there is a single $f \in R$ such that $\mathcal{L}$ can be defined over $\pi^{-1}(\mathrm{Spec}\, R_f)$. The coherent subsheaf L_U of $V|U$ then extends to a coherent subsheaf L of V defined on all of X (see for example [61, p. 126, ex. 5.15(c)]). Here L is a torsion free rank 1 sheaf, and after replacing it by its double dual we can assume that it is a line bundle. By construction $L \cdot f = \deg \mathcal{L} > d/2$, so that L is the desired destabilizing line bundle. This then concludes the proof of Theorem 18. □

Appendix: Galois descent theory

We begin by considering the following situation: k is a field, and K is a finite Galois extension of k with Galois group G. Let V be a k-vector space, not necessarily finite-dimensional over k, and suppose that $W = V \otimes_k K$. In this case, the k-linear action of $\sigma \in G$ on K extends to a k-linear automorphism $\varphi_\sigma: W \to W$, such that $\varphi_\sigma(v \otimes \alpha) = v \otimes \sigma(\alpha)$. In particular, $\varphi_{\mathrm{Id}} = \mathrm{Id}$, $\varphi_\sigma \circ \varphi_\tau = \varphi_{\sigma\tau}$, and, for all $\alpha \in K$ and $w \in W$, $\varphi_\sigma(\alpha \cdot w) = \sigma(\alpha)\varphi_\sigma(w)$. More generally, we make the following definition:

Definition 19. Let W be a K-vector space. A *twisted G-representation* is a homomorphism from G to $\mathrm{Aut}_k(W)$, the group of k-linear automorphisms of W, such that, if φ_σ is the image of σ, then for all $\alpha \in K$ and $w \in W$, $\varphi_\sigma(\alpha \cdot w) = \sigma(\alpha)\varphi_\sigma(w)$. In case $W = V \otimes_k K$ for some k-vector space V, and φ_σ is as defined above, then we will call φ the *standard twisted G-representation associated to V.*

For a twisted G-representation W, we let

$$W^G = \{w \in W : \varphi_\sigma(w) = w \text{ for all } \sigma \in G\}.$$

Thus, W^G is a k-vector subspace of W.

We define homomorphisms and isomorphisms of twisted G-representations in the obvious way: Given W_1 and W_2 two twisted G-representations, a *morphism* of twisted G-representations from W_1 to W_2 is a K-linear map F such that $F \circ \varphi_\sigma = \varphi_\sigma \circ F$ for all $\sigma \in G$. Equivalently $F \in \mathrm{Hom}_K(W_1, W_2)^G$, where $F \mapsto \varphi_\sigma \circ F \circ \varphi_\sigma^{-1}$ defines a twisted G-representation on $\mathrm{Hom}_K(W_1, W_2)$.

Example 1. Suppose W is a 1-dimensional twisted G-representation, with basis element w. Thus, (since $\varphi_\sigma(w) \neq 0$) $\varphi_\sigma(w) = a(\sigma)w$ for a unique $a(\sigma) \in K^*$. Now

$$a(\sigma\tau)w = \varphi_{\sigma\tau}(w) = \varphi_\sigma \circ \varphi_\tau(w) = \varphi_\sigma(a(\tau)w) = \sigma(a(\tau))a(\sigma)w.$$

Hence $a(\sigma\tau) = \sigma(a(\tau))a(\sigma)$, so that $a(\sigma)$ is a 1-cocycle for G. The element w is well-defined up to the choice of $\lambda \in K^*$. Replacing w by λw gives

$$\varphi_\sigma(\lambda w) = \sigma(\lambda)a(\sigma)w = \sigma(\lambda)\lambda^{-1}a(\sigma)$$

so that the 1-cocycle $a(\sigma)$ is multiplied by the coboundary $\sigma(\lambda)\lambda^{-1}$. Thus, we may identify isomorphism classes of 1-dimensional twisted G-representations with the Galois cohomology group $H^1(G, K^*)$. If $W =$

$V \otimes_k K$, then we can choose $w = v \otimes 1$ and $\varphi_\sigma(w) = w$ for all σ, so that $a(\sigma) = 1$ for all σ.

Example 2. Let W be an arbitrary K-vector space, and, for $\sigma \in G$, let W^σ be the K-vector space whose underlying abelian group is W, but with scalar multiplication given by the formula

$$\alpha \cdot_\sigma w = \sigma(\alpha)w.$$

Note that $(W^\sigma)^\tau = W^{\sigma\tau}$. The identity map $W \to W^\sigma$ is k-linear. There is an induced K-linear homomorphism $\Phi \colon W \otimes_k K \to \bigoplus_{\sigma \in G} W^\sigma$ defined as follows: thinking of the elements of $\bigoplus_{\sigma \in G} W^\sigma$ as functions $\mathbf{w} \colon G \to W$, define $\Phi(w \otimes \alpha)(\sigma) = \sigma(\alpha)w$. It is easy to check that Φ is well-defined and K-linear. In fact, we claim that Φ is an isomorphism, which follows by reducing to the case where $\dim_K W = 1$ and using the Galois theory isomorphism $K \otimes_k K \cong \bigoplus_{\sigma \in G} K^\sigma$. Note that Φ carries the standard twisted G-representation of G on $W \otimes_k K$ to the twisted G-representation on $\bigoplus_{\sigma \in G} W^\sigma$ defined as follows: thinking of the elements of $\bigoplus_{\sigma \in G} W^\sigma$ as functions $\mathbf{w} \colon G \to W$, define $\varphi_\tau(\mathbf{w}(\sigma)) = \mathbf{w}(\sigma\tau)$. Since by definition $\alpha \cdot \mathbf{w}(\sigma) = \sigma(\alpha)\mathbf{w}(\sigma)$, $\varphi_\tau(\alpha \cdot \mathbf{w}) = \tau(\alpha) \cdot \mathbf{w}$ (so that we have indeed defined a twisted G-representation on $\bigoplus_{\sigma \in G} W^\sigma$) and that $\Phi(w \otimes \tau(\alpha)) = \varphi_\tau \cdot \Phi(w \otimes \alpha)$.

We remark that a k-linear isomorphism $\varphi_\sigma \colon W \to W$ such that $\varphi_\sigma(\alpha w) = \sigma(\alpha)w$ is the same thing as a K-linear isomorphism $W \to W^\sigma$. For all $\tau \in G$, φ_σ also induces a K-linear isomorphism $W^\tau \to W^{\sigma\tau}$, and the condition that φ defines a twisted G-representation is just the condition that, for all $\sigma, \tau \in G$, $\varphi_\sigma \circ \varphi_\tau = \varphi_{\sigma\tau}$ as K-linear isomorphisms $W \to W^{\sigma\tau}$

Lemma 20. *Let V be a k-vector space. For the standard twisted G-representation on $V \otimes_k K$, we have*

$$(V \otimes_k K)^G = V.$$

Let V_1 and V_2 be k-vector spaces and let $W_i = V_i \otimes_k K$, viewed as a standard twisted G-representation. If $F \colon W_1 \to W_2$ is a morphism of twisted G-representations, then there is a unique k-linear map $f \colon V_1 \to V_2$ such that $F = f \otimes \mathrm{Id}$.

Proof. To see the first statement, let $v \in V$. Then $\varphi_\sigma((k \cdot v) \otimes_k K) = (k \cdot v) \otimes_k K$ for all $\sigma \in G$. Thus, if we choose a basis for V, say $V = \bigoplus_i k \cdot v_i$, then $V \otimes_k K = \bigoplus_i K \cdot v_i$ and φ_σ preserves the direct sum. So it is enough to consider the case where V is 1-dimensional. In this case $V \cong k$ and $V \otimes_k K \cong K$, with the natural G-action. So we are reduced to the statement that $K^G = k$, which is clear by Galois theory.

As for the second assertion, let $f = F|(W_1)^G = F|V_1$. Clearly, $F(W_1^G) \subseteq (W_2)^G = V_2$, and we let $f \colon V_1 \to V_2$ be the induced map. Since $W_1 =$

$V_1 \otimes_k K$, it is easy to see that $f \otimes \mathrm{Id} = F$ and that f is the unique map with this property. $\square$

The main result of this Appendix is a converse to Lemma 20:

Theorem 21. *Let W be a twisted G-representation, where W is a K-vector space. Then the natural map from $W^G \otimes_k K$ to W is an isomorphism of twisted G-representations. In particular, every twisted G-representation is isomorphic to a standard one.*

We note that Theorem 21 is a generalization of the fact that the Galois cohomology group $H^1(G, K^*) = 0$.

Proof of Theorem 21. There is a K-linear map from $W^G \otimes_k K$ to W, defined by $w \otimes \alpha \mapsto \alpha w$, and it is a homomorphism of twisted G-representations, since for $\sigma \in G$ and $w \in W^G$,

$$w \otimes \sigma(\alpha) \mapsto \sigma(\alpha)w = \sigma(\alpha)\varphi_\sigma(w) = \varphi_\sigma(\alpha w).$$

Next, considering W as a k-vector space, note that the twisted G-representation on W induces a K-linear representation of G on the K-vector space $W \otimes_k K$, by letting G act on the first factor and K on the second.

Claim. *The bilinear map $(w, \alpha) \mapsto \sum_{\sigma \in G} \alpha \varphi_{\sigma^{-1}}(w) \cdot \sigma$ induces a K-linear isomorphism*

$$\rho \colon W \otimes_k K \to W \otimes_K K[G] = \bigoplus_{\sigma \in G} W \cdot \sigma$$

which is equivariant with respect to the K-linear action of G.

Proof of the claim. By definition, ρ is K-linear. To check that it is G-equivariant, let $\tau \in G$. It is enough to check that, on generators $w \otimes \alpha$, we have $\rho(\varphi_\tau(w) \otimes \alpha) = \tau(\rho(w \otimes \alpha))$. Now

$$\rho(\varphi_\tau(w) \otimes \alpha) = \sum_{\sigma \in G} \alpha \varphi_{\sigma^{-1}}(\varphi_\tau(w)) \cdot \sigma = \sum_{\sigma \in G} \alpha \varphi_{\sigma^{-1}\tau}(w) \cdot \sigma.$$

Make the change of variable by replacing $\sigma \in G$ by $\tau\sigma$:

$$\sum_{\sigma \in G} \alpha \varphi_{\sigma^{-1}\tau}(w) \cdot \sigma = \sum_{\sigma \in G} \alpha \varphi_{(\tau\sigma)^{-1}\tau}(w) \cdot \tau\sigma$$

$$= \sum_{\sigma \in G} \alpha \varphi_{\sigma^{-1}}(w) \cdot \tau\sigma = \tau\left(\sum_{\sigma \in G} \alpha \varphi_{\sigma^{-1}}(w) \cdot \sigma\right).$$

Thus, $\rho(\varphi_\tau(w) \otimes \alpha) = \tau(\rho(w \otimes \alpha))$.

To see that ρ is an isomorphism, we must show that, given an element $\sum_\sigma w_\sigma \cdot \sigma \in \bigoplus_\sigma W \cdot \sigma$, then there exists a unique element of $W \otimes_k K$, necessarily of the form $\sum_{i=1}^{n} w_i \otimes \alpha_i$, such that $\rho(\sum_{i=1}^{n} w_i \otimes \alpha_i) = \sum_{\sigma} w_\sigma \cdot \sigma$. In other words, given a collection of elements $w_\sigma \in W$ indexed by $\sigma \in G$, we must find $\alpha_i \in K$, $w_i \in W$, for $i = 1, \ldots, n$ such that $w_\sigma = \sum_i \alpha_i \varphi_{\sigma^{-1}}(w_i)$ for every σ. Applying φ_σ to both sides, we see that, given the collection w_σ, we must find w_i and α_i such that

$$\varphi_\sigma(w_\sigma) = \sum_i \varphi_\sigma(\alpha_i \varphi_{\sigma^{-1}}(w_i)) = \sum_i \sigma(\alpha_i) w_i,$$

and such that the element $\sum_i \alpha_i \otimes w_i$ is uniquely defined in $W \otimes_k K$. It suffices to show that the map $\pi \colon W \otimes_k K \to \bigoplus_{\sigma \in G} W \cdot \sigma$ defined by $w \otimes \alpha \mapsto \sum_\sigma \sigma(\alpha) w \cdot \sigma$ is an isomorphism, for then the element $\sum_\sigma \varphi_\sigma(w_\sigma) \cdot \sigma$ is the image of $\sum_i w_i \otimes \alpha_i$ for a unique element $\sum_i w_i \otimes \alpha_i \in W \otimes_k K$. On the other hand, the map $\pi \colon W \otimes_k K \to \bigoplus_{\sigma \in G} W \cdot \sigma$ given by $w \otimes \alpha \mapsto \sum_\sigma \sigma(\alpha) w \cdot \sigma$ can be defined for every K-vector space W, regardless of whether W has a twisted G-representation, and π clearly commutes with taking (possibly infinite) direct sums. So it is enough to prove that π is an isomorphism in case W is 1-dimensional over K, say $W = K$. In this case we are reduced to considering the map $K \otimes_k K \to \bigoplus_{\sigma \in G} K \cdot \sigma$ defined by $\beta \otimes \alpha \mapsto \sum_\sigma \sigma(\alpha) \beta \cdot \sigma$. By standard Galois theory, this map is an isomorphism. Thus, ρ is an isomorphism.   □

Returning to the proof of Theorem 21, the claim shows that there is an isomorphism

$$(W \otimes_k K)^G \cong \left(\bigoplus_{\sigma \in G} W \cdot \sigma \right)^G.$$

Clearly, $\left(\bigoplus_{\sigma \in G} W \cdot \sigma \right)^G = W$, by taking the diagonal embedding of W in $\bigoplus_{\sigma \in G} W \cdot \sigma$. On the other hand, we claim that the natural map $W^G \otimes_k K \to (W \otimes_k K)^G$ is an isomorphism: consider the exact sequence

$$0 \to W^G \to W \to \prod_{\sigma \in G} W,$$

where the first map is the inclusion and the second is $w \mapsto (\varphi_\sigma(w) - w)_{\sigma \in G}$. This exact sequence of k-vector spaces remains exact when we tensor with K, so that

$$0 \to W^G \otimes_k K \to W \otimes_k K \to \prod_{\sigma \in G} W \otimes_k K$$

is exact. Thus, $W^G \otimes_k K$ is identified with the kernel of $W \otimes_k K \to \prod_{\sigma \in G} W \otimes_k K$, namely $(W \otimes_k K)^G$. Putting this together, we have showed that the map ρ identifies $W^G \otimes_k K$ with the diagonal embedding of W in $\bigoplus_{\sigma \in G} W \cdot \sigma$. For $w \otimes \alpha \in W^G \otimes_k K$, $\rho(w \otimes \alpha) = \sum_\sigma (\alpha w) \cdot \sigma$. Thus, the map $W^G \otimes_k K \to W$ defined by $w \otimes \alpha \mapsto \alpha w$ is an isomorphism, which concludes the proof of Theorem 21.   □

Corollary 22. *Let K be a finite Galois extension of k and let U be a k-vector space. Suppose that W is a K-vector subspace of $U \otimes_k K$ such that, for all $\sigma \in G$, $\varphi_\sigma(W) = W$, where φ_σ is the standard twisted G-representation on $U \otimes_k K$. Then W^G is a k-vector subspace of U and the map $W^G \otimes_k K \to W$ is an isomorphism.*

Proof. By hypothesis the standard twisted G-representation on $U \otimes_k K$ restricts to a twisted G-representation on W, for which $W^G \subseteq (U \otimes_k K)^G = U$. Thus, the corollary is immediate. $\square$

Next we want to find circumstances where we can apply Theorem 21 to sheaves over schemes defined over a subfield of an algebraically closed field. Suppose that X is a (separated) scheme over $\operatorname{Spec} k$, and that K is a finite Galois extension of k. Let $X' = X \times_{\operatorname{Spec} k} \operatorname{Spec} K$, and let $p \colon X' \to X$ be the natural morphism. Locally $X = \operatorname{Spec} R$, where R is a k-algebra, and thus locally $X' = \operatorname{Spec} R \otimes_k K$, with the morphism p corresponding to the inclusion $R \to R \otimes_k K$. An element $\sigma \in G$ defines a morphism $\bar\sigma \colon X' \to X'$ with $\overline{\sigma\tau} = \bar\tau\bar\sigma$. Let $\mathcal{F}$ be a coherent sheaf on X'. Locally $\mathcal{F}$ corresponds to an $(R \otimes_k K)$-module M. We can form the pulled back sheaf $\bar\sigma^* \mathcal{F}$, and $\bar\sigma^* \bar\tau^* = (\bar\tau\bar\sigma)^* = (\overline{\sigma\tau})^*$. If $\mathcal{F}$ locally corresponds to the $(R \otimes_k K)$-module M, then it is eaasy to check that $\bar\sigma^* \mathcal{F}$ corresponds to the $(R \otimes_k K)$-module $M^{\sigma^{-1}}$. Note that the analogue of $W \otimes_k K \cong \bigoplus_\sigma W^\sigma$ is the isomorphism $p^* p_* \mathcal{F} \cong \bigoplus_\sigma \bar\sigma^* \mathcal{F}$.

Suppose that we are given a coherent sheaf $\mathcal{G}$ on X, locally corresponding to the R-module N. Then we have the sheaf $p^* \mathcal{G}$ on X', and it corresponds locally to $N \otimes_R (R \otimes_k K) = N \otimes_k K$. We define a *twisted G-representation on $\mathcal{F}$* to be a collection of isomorphisms $\varphi_\sigma \colon \bar\sigma^* \mathcal{F} \to \mathcal{F}$, such that $\varphi_{\mathrm{Id}} = \mathrm{Id}$, and such that, for all $\sigma, \tau \in G$, if we also denote by φ_τ the induced isomorphism

$$(\overline{\sigma\tau})^* \mathcal{F} = \bar\sigma^* \bar\tau^* \mathcal{F} \to \bar\sigma^* \mathcal{F},$$

then $\varphi_\sigma \circ \varphi_\tau = \varphi_{\sigma\tau}$. Equivalently we seek a homomorphism $\varphi \colon G \to \operatorname{Aut} p_* \mathcal{F}$ such that, for every local section $\alpha \in p_* \mathcal{O}_{X'}$ and $s \in p_* \mathcal{F}$, $\varphi_\sigma(\alpha s) = \sigma(\alpha) s$ under the natural action of the sheaf of algebras $p_* \mathcal{O}_{X'}$ on $p_* \mathcal{F}$. In the local case $X = \operatorname{Spec} R$ and $X' = \operatorname{Spec}(R \otimes_k K)$, if $\mathcal{F}$ corresponds to the $(R \otimes_k K)$-module M, then φ_σ is the same thing as an R-module homomorphism $M^{\sigma^{-1}} \to M$, or equivalently $M \to M^\sigma$, and φ is equivalent to a twisted G-representation on M commuting with the R-module structure.

Proposition 23. *Suppose that X is a scheme over $\operatorname{Spec} k$ and that $X' = X \times_{\operatorname{Spec} k} \operatorname{Spec} K$.*

(i) *Let $\mathcal{F}'$ be a sheaf on X' with a twisted G-representation on $\mathcal{F}'$. Then there is a sheaf $\mathcal{F}$ on X such that $p^* \mathcal{F}$ is isomorphic as a coherent sheaf with a twisted G-representation to $\mathcal{F}'$.*

(ii) *Suppose that $\mathcal{F}$ is a coherent sheaf on X and that $\mathcal{G}'$ is a coherent subsheaf of $p^*\mathcal{F}$ such that $\sigma^*\mathcal{G}' = \mathcal{G}'$, as a subsheaf of $p^*\mathcal{F}$, for every $\sigma \in G$. Then there exists a coherent subsheaf $\mathcal{G}$ of $\mathcal{F}$ such that $\mathcal{G}'$ is the subsheaf $p^*\mathcal{G}$ of $p^*\mathcal{F}$.*

Proof. We shall just prove (i); the proof of (ii) is similar. First suppose that $X = \operatorname{Spec} R$ and that $X' = \operatorname{Spec}(R \otimes_k K)$. Then M has a twisted G-representation commuting with the R-module structure, and thus $N = M^G$ is an R-module satisfying $N \otimes_R (R \otimes_k K) \cong N \otimes_k K \cong M$ as twisted G-representations. In this case, we can take $\mathcal{G}$ to be the $\mathcal{O}_X$-module corresponding to N. We leave it as an exercise to show that N is finitely generated.

In general, take an affine open cover $\{U_i\}$ of X, where $U_i = \operatorname{Spec} R_i$, $U_{ij} = U_i \cap U_j = \operatorname{Spec} R_{ij}$, and $U_{ijk} = U_i \cap U_j \cap U_k = \operatorname{Spec} R_{ijk}$. Over each U_i, we have found a coherent sheaf $\mathcal{G}_i$ and an isomorphism $\eta_i \colon p^*\mathcal{G}_i \cong \mathcal{F}|U_i$ as sheaves with twisted G-representations. Thus, over U_{ij}, we have isomorphisms

$$\tilde{\psi}_{ij} = (\eta_j|U_{ij})^{-1} \circ (\eta_i|U_{ij}) \colon p^*\mathcal{G}_i|U_{ij} \cong \mathcal{F}|U_{ij} \cong p^*\mathcal{G}_j|U_{ij}.$$

By Lemma 20, there exist isomorphisms $\psi_{ij} \colon \mathcal{G}_i|U_{ij} \to \mathcal{G}_i|U_{ij}$ inducing $\tilde{\psi}_{ij}$. Moreover, over U_{ijk},

$$\tilde{\psi}_{jk} \circ \tilde{\psi}_{ij} = \eta_k^{-1} \circ \eta_j \circ \eta_j^{-1} \circ \eta_i = \eta_k^{-1} \circ \eta_i = \tilde{\psi}_{ik}.$$

Thus, by the uniqueness part of Lemma 20, $\psi_{jk} \circ \psi_{ij} = \psi_{ik}$ on U_{ijk}, and so the sheaves $\mathcal{G}_i$ glue together to give a sheaf $\mathcal{G}$, such that $p^*\mathcal{G} = \mathcal{F}$. $\square$

We next make some remarks which we will not need in the applications. In practice, except under the circumstances of (ii) of Proposition 23 above, it is hard to find an explicit twisted G-representation on a sheaf $\mathcal{F}$. Instead we consider the following situation. Suppose that X and X' are as above and that $\mathcal{F}$ is a coherent sheaf on X' with $\operatorname{Aut} \mathcal{F} \cong K^*$. For example, if X is proper over k and geometrically integral, then every line bundle on X' satsifies this condition. Suppose further that $\bar{\sigma}^*\mathcal{F} \cong \mathcal{F}$ for every $\sigma \in G$, and let $\psi_\sigma \colon \bar{\sigma}^*\mathcal{F} \to \mathcal{F}$ be such an isomorphism. By hypothesis ψ_σ is unique up to multiplication by an element of K^*. In particular, $\psi_\sigma \circ \psi_\tau \circ \psi_{\sigma\tau}^{-1} \colon \mathcal{F} \to \mathcal{F}$ is multiplication by an element $c(\sigma, \tau) \in K^*$.

Lemma 24. *The element $c(\sigma, \tau)$ defines a 2-cocycle for G with values in K^*, and the associated cohomology class in $H^2(G, K^*)$ is independent of the choice of the ψ_σ. Moreover, the cohomology class is trivial if and only if there is a choice of ψ_σ such that $\psi_\sigma \circ \psi_\tau = \psi_{\sigma\tau}$, in other words for which ψ_σ defines a twisted G-representation.*

Proof. We compute: $c(\sigma, \tau)$ is a 2-cocycle if and only if, for all $\sigma, \tau, \rho \in G$, we have

$$c(\sigma, \tau) = c(\sigma, \tau\rho)\sigma(c(\tau, \rho))c(\sigma\tau, \rho)^{-1},$$

which we can write as $c(\sigma, \tau)^{-1}c(\sigma, \tau\rho)c(\sigma\tau, \rho)^{-1}\sigma(c(\tau, \rho)) = 1$. Multiplication on $\mathcal{F}$ by the term $c(\sigma, \tau)^{-1}c(\sigma, \tau\rho)c(\sigma\tau, \rho)^{-1}$ is the same as the isomorphism $\mathcal{F} \to \mathcal{F}$ given by

$$\psi_{\sigma\tau} \circ \psi_\tau^{-1} \circ \psi_\sigma^{-1} \circ \psi_\sigma \circ \psi_{\tau\rho} \circ \psi_{\sigma\tau\rho}^{-1} \circ \psi_{\sigma\tau\rho} \circ \psi_\rho^{-1} \circ \psi_{\sigma\tau}^{-1}$$

$$= \psi_{\sigma\tau} \circ \psi_\tau^{-1} \circ \psi_{\tau\rho} \circ \psi_\rho^{-1} \circ \psi_{\sigma\tau}^{-1}.$$

Now multiplication by $\sigma(c(\tau, \rho))$ on $\bar{\sigma}^*\mathcal{F}$ is the isomorphism $\bar{\sigma}^*\mathcal{F} \to \bar{\sigma}^*\mathcal{F}$ defined by $\psi_\tau \circ \psi_\rho \circ \psi_{\tau\rho}^{-1}$. We leave to the reader the verification that $c(\sigma, \tau)^{-1}c(\sigma, \tau\rho)c(\sigma\tau, \rho)^{-1}\sigma(c(\tau, \rho))$ corresponds to the automorphism of $\mathcal{F}$ defined by

$$\psi_{\sigma\tau} \circ \psi_\tau^{-1} \circ \psi_\tau \circ \psi_\rho \circ \psi_{\tau\rho}^{-1} \circ \psi_{\tau\rho} \circ \psi_\rho^{-1} \circ \psi_{\sigma\tau}^{-1} = \mathrm{Id}.$$

Thus, $c(\sigma, \tau)$ is a 2-cocycle. If we replace ψ_σ by $\lambda_\sigma\psi_\sigma$ for some $\lambda \in K^*$, then we multiply $c(\sigma, \tau)$ by the coboundary $\lambda_{\sigma\tau}^{-1}\sigma(\lambda_\tau)\lambda_\sigma$. Thus, the cohomology class is independent of the choice of ψ_σ, and the class is trivial if and only if there is a choice of ψ_σ for which $c(\sigma, \tau) = 1$, in other words $\psi_\sigma \circ \psi_\tau = \psi_{\sigma\tau}$. $\square$

Corollary 25. *Suppose that $H^2(G, K^*) = 0$ and that $\mathcal{F}$ is a coherent sheaf on X' with $\mathrm{Aut}\,\mathcal{F} \cong K^*$ and such that $\bar{\sigma}^*\mathcal{F} \cong \mathcal{F}$ for all $\sigma \in G$. Then there exists a coherent sheaf $\mathcal{G}$ on X such that $\mathcal{F} = p^*\mathcal{G}$. In particular, if X is a geometrically integral scheme proper over k and L' is a line bundle on X' such that $\bar{\sigma}^*L' \cong L'$ for all $\sigma \in G$, then $L' = p^*L$ for a line bundle L on X.* $\square$

We turn now to finding circumstances under which $H^2(G, K^*) = 0$. In general the groups $H^2(G, K^*)$ are extremely complicated. For example, if k is a local field or a number field, then the determination of $H^2(G, K^*)$ is the basic information in local or global class field theory. However, in certain cases we can say that $H^2(G, K^*) = 0$:

Definition 26. An *algebraic function field* k in one variable is the field of meromorphic functions on an algebraic curve over an algebraically closed field k_0. Equivalently, k is a finite separable extension of $k_0(x)$, the field of rational functions in one variable over the algebraically closed field k_0. If k is an algebraic function field in one variable over k_0 and L is a subfield of k containing k_0 such that $[k : L]$ is finite, then L is also an algebraic function field in one variable over k_0, and likewise every finite extension of an algebraic function field in one variable over k_0 is again an algebraic function field in one variable over k_0.

Theorem 27. *If k is a finite field or an algebraic function field in one variable, and K is a finite Galois extension of k with Galois group G, then $H^2(G, K^*) = 0$.*

Proof. By standard Galois theory, $H^1(G, K^*) = 0$. (This also follows from Theorem 21 and the discussion in Example 1 following Definition 19.) In case k is a finite field or an algebraic function field in one variable, the norm map $N: K^* \rightarrow k^*$ is surjective; we shall outline a proof of this fact in Exercise 11. The same also holds when we replace G by a subgroup H of G and k by the fixed field of H, or G by G/H, where H is a normal subgroup of G, and K by the fixed field of H. It is then a standard argument in the cohomology theory of finite groups that $H^2(G, K^*) = 0$. Indeed, if G is cyclic $H^2(G, K^*)$ is isomorphic to $k^*/N(K^*)$, by [137, p. 141], and thus is trivial. If G is solvable, then using the result for cyclic groups and induction, it follows from the inflation-restriction sequence and the fact that $H^1(G, K^*) = 0$ (see [137, p. 126, Prop. 5]) that $H^2(G, K^*) = 0$. In particular, if G is a p-group, then $H^2(G, K^*) = 0$. Now, for a general Galois group G and a prime p, let G_p be the p-Sylow subgroup of G. Then by [137, Cor. on p. 148], since $H^2(G_p, K^*) = 0$ for every p, $H^2(G, K^*) = 0$ as well. □

Lastly we give a corollary of Theorem 21 which was used in the proof of Lemma 13. Let $p: X \rightarrow Y$ be a finite morphism of schemes. Suppose that G is a finite group of automorphisms of Y with $p \circ \sigma = p$ for all $\sigma \in G$ and such that $(p_* \mathcal{O}_X)^G = \mathcal{O}_Y$. In this case we shall refer to p as a *Galois morphism* of schemes with Galois group G. Locally, if $U = \operatorname{Spec} R$ is an affine open subset of Y, then $p^{-1}(U) = \operatorname{Spec} R'$ where R' is a finite R-algebra with an action of G such that $(R')^G = R$.

For example, if $p: \tilde{C} \rightarrow C$ is a morphism between two smooth projective curves such that the induced map on function fields is a Galois extension, then p is a Galois morphism.

Theorem 28. *Let $p: X \rightarrow Y$ be a finite flat Galois morphism with Galois group G, and suppose that X and Y are both integral. Let $\mathcal{F}$ be a locally free sheaf on Y and suppose that $\mathcal{G}'$ is a coherent subsheaf of $p^* \mathcal{F}$ such that:*

(i) For all $\sigma \in G$, $\sigma^* \mathcal{G}' = \mathcal{G}'$ as a subsheaf of $p^* \mathcal{F}$;
(ii) $p^* \mathcal{F}/\mathcal{G}'$ is torsion free.

Then there exists a coherent subsheaf $\mathcal{G}$ of $\mathcal{F}$ such that $\mathcal{G}' = p^* \mathcal{G}$ as a subsheaf of $p^* \mathcal{F}$.

Proof. Suppose first that $Y = \operatorname{Spec} R$, $X = \operatorname{Spec} R'$, $\mathcal{F}$ corresponds to the R-module M, and $\mathcal{G}'$ corresponds to an R'-submodule N' of $M \otimes_R R'$ which is closed under the action of G. We may assume that $Y = \operatorname{Spec} R$

is chosen small enough so that M is free. Let $N = (N')^G$. Then N is an R-module, and it is a submodule of $(M \otimes_R R')^G$. Since $M \cong R^n$ for some n, $(M \otimes_R R')^G \cong ((R')^n)^G = R^n = M$, and so N is an R-submodule of M. We shall show that the natural map $N \otimes_R R' \to N'$ is an isomorphism.

Since R' is flat over R, the induced map $N \otimes_R R' \to M$ is injective, and thus so is the map $N \otimes_R R' \to N'$. We claim first that the cokernel of the map $N \otimes_R R' \to N'$ is a torsion module. Let k be the quotient field of R and let K be the quotient field of R'. Then G acts on K and $K^G = k$, and so K is a finite Galois extension of k. Indeed we have the following lemma, whose proof is left as an exercise (compare also [5, p. 68, Ex. 12]):

Lemma 29. *Let R' be a commutative ring, let G be a finite group of automorphisms of R', and let $R = (R')^G$. Suppose that S' is a multiplicative subset of R' such that $\sigma(S') = S'$ for all $\sigma \in G$, and set $S = (S')^G$. Let N' be an R'-module with an action of G such that $\sigma(rn) = \sigma(r)\sigma(n)$ for all $r \in R'$ and $n \in N'$. Then there is a natural isomorphism*

$$\left[(S')^{-1} N'\right]^G \cong S^{-1} \left[(N')^G\right].$$

In particular, if R' is an integral domain with fraction field K, taking $S' = R' - \{0\}$ and $S = R - \{0\}$, then K^G is the quotient field k of R. □

Let $V = M \otimes_R k$, so that $(M \otimes_R R') \otimes_{R'} K = V \otimes_k K$. Let $W' = (N') \otimes_{R'} K \subseteq V \otimes_k K$. Thus, W' is a K-vector subspace of $V \otimes_k K$ such that $\sigma(W') = W'$ for all $\sigma \in G$. By Corollary 22, $W = (W')^G$ is a k-vector subspace of V such that $W \otimes_k K = W'$. By Lemma 29, since W' is the localization of N' with respect to the multiplicative subset $R' - \{0\}$, W is the localization of N with respect to $R - \{0\}$, in other words $W = N \otimes_R k$. Thus, if we tensor the inclusion $N \otimes_R R' \to N'$ with K, it becomes the isomorphism $W \otimes_k K = W'$. So the cokernel T of the map $N \otimes_R R' \to N'$ is a torsion R'-module. Now by assumption $Q' = (M \otimes_R R')/N'$ is torsion free. Applying the functor $(\cdot)^G$ to the exact sequence

$$0 \to N' \to M \otimes_R R' \to Q' \to 0,$$

we obtain an exact sequence

$$0 \to N \to M \to (Q')^G.$$

Since Q' is torsion free, $(Q')^G$ is a torsion free R-module. Thus, so is $Q = M/N$. Since R' is flat over R, there is an exact sequence

$$0 \to N \otimes_R R' \to M \otimes_R R' \to Q \otimes_R R' \to 0.$$

So $Q \otimes_R R' \cong (M \otimes_R R')/(N \otimes_R R')$. In particular, $Q \otimes_R R'$ has a submodule $N'/(N \otimes_R R') = T$, which as we have seen is a torsion module. On the other hand, we claim that $Q \otimes_R R'$ is torsion free: since Q is torsion free, it is a submodule of a free R-module, by Proposition 20 of Chapter 2. Thus, $Q \otimes_R R'$ is a submodule of a free R'-module, and so it is torsion free.

Hence $N'/(N \otimes_R R')$ is a torsion submodule of a torsion free module, and is therefore 0. Thus, $N' = N \otimes_R R'$.

This concludes the proof in the affine case (where $\mathcal{F}$ is trivialized), and in the general case it is easy to check that the isomorphisms over an affine cover which trivializes $\mathcal{F}$ piece together as in the proof of Theorem 23. $\square$

Exercises

1. Using Reider's theorem, show that, if X is a $K3$ surface and D is a nef and big divisor on X, then $|2D|$ has no base points and $|3D|$ defines a birational morphism from X to $\mathbb{P}^N$ for some N.

2. Let X be an algebraic surface and let H be an ample divisor on X. Show that, for $n \geq 4$, $nH + K_X$ is very ample on X.

3. Let X be an algebraic surface and let V be a vector bundle of rank r on X, or more generally a torsion free sheaf of rank r. Define the *Bogomolov number* $B(V) = 2rc_2(V) - (r-1)c_1(V)^2$. Thus, if V is stable with respect to some ample divisor, then Bogomolov's inequality is the statement $B(V) \geq 0$. Show that $B(V) = B(V \otimes L)$ for every line bundle L, $B(V) = B(V^\vee)$ if V is locally free, and $B(V) \geq B(V^\vee) = B(V^{\vee\vee})$ in general. Now suppose that there is an exact sequence of torsion free sheaves

$$0 \to W_1 \to W \to W_2 \to 0,$$

with $c_1(W_i) = \Delta_i$. Show that

$$B(W) = \frac{r}{r_1} B(W_1) + \frac{r}{r_2} B(W_2) - \frac{1}{r_1 r_2}(r_2 \Delta_1 - r_1 \Delta_2)^2.$$

4. Let X be an algebraic surface and let V be a vector bundle of rank r on X. Suppose that Bogomolov's inequality holds for every stable bundle on X of rank less than r, and that V is strictly semistable with respect to an ample line bundle H. Using the previous exercise, show that V also satisfies Bogomolov's inequality (compare Exercise 1 in Chapter 6). Working a little harder, show that, if V is H_0-stable and H_1-unstable for two ample divisors H_0, H_1, then H_0 and H_1 are separated by a wall W^ζ with

$$-\frac{r^2}{4}B(V) \leq \zeta^2 < 0,$$

where $B(V)$ is the Bogomolov number of V. (Using the first part and the previous exercise, it is enough to show that V is strictly semistable with respect to some convex combination $H_t = (1-t)H_0 + tH_1$. Using Exercise 14 in Chapter 4, show that the set $\{t \in [0,1] : V$ is H_t-stable$\}$ is open, and likewise $\{t \in [0,1] : V$ is H_t-unstable$\}$ is open. Thus, there is a point $t_0 \in [0,1]$ for which V is H_{t_0}-strictly semistable.)

5. Let $p: \tilde{C} \to C$ be an étale double cover with covering involution ι, and let L be a line bundle of degree 0 on $\tilde{C}$ such that $\iota^* L \neq L$. If

$W = p_*L$, show that $W = p_*L$ is stable but that $p^*W = L \oplus \iota^*L$ is strictly semistable.

6. Let C be a smooth curve and let W be a rank 2 vector bundle over C. Let $X = \mathbb{P}(W)$ and $L = \mathcal{O}_X(2) \otimes \pi^* \det W$ as in Lemma 14. If L is nef, show that W is semistable.

7. Let C be a smooth curve and let $\mathcal{V}$ a rank 2 vector bundle over $C \times T$, where T is a scheme. For each $t \in T$, let V_t be the vector bundle on C corresponding to $\mathcal{V}|C \times \{t\}$. Arguing as in the proof of Theorem 18, show that the set of $t \in T$ such that V_t is semistable (or stable) is an open subset of T. (Note that instead of $\mathrm{Pic}^n(C \times T/T)$ we can work directly with $\pi_1^* \mathrm{Pic}^n C$ in this case.)

8. Let R be a commutative ring and R' an R-algebra. We say that R' is *faithfully flat* over R if it is flat over R and if, for every R-module M, $M \otimes_R R' = 0$ if and only if $M = 0$.
   (a) Show that, if k is a field, K is an extension field of k, and R is a k-algebra, then $R' = R \otimes_k K$ is a faithfully flat R-algebra.
   (b) Let R' be a faithfully flat R-algebra and M an R-module. Show that M is a finitely generated R-module if and only if $M \otimes_R R'$ is a finitely generated R'-module.

9. Prove Lemma 29. (Suppose that $n/s \in (S')^{-1}N'$. Then

$$n/s = \frac{n \prod_{\sigma \neq 1} \sigma(s)}{\prod_{\sigma \in G} \sigma(s)}.$$

So every element of $(S')^{-1}N'$ can be written as n/s with $\sigma(s) = s$ for all $\sigma \in G$.)

10. Let $p: \tilde{C} \to C$ be a degree 2 morphism between two smooth curves, and let x be a branch point of p. Suppose that ι is the involution corresponding to p. Show that $\iota^*\mathcal{O}_{\tilde{C}}(-x)$ is a subsheaf of $\mathcal{O}_{\tilde{C}} = p^*\mathcal{O}_C$, invariant under the Galois group, but that there is no subsheaf $\mathcal{G}$ of $\mathcal{O}_C$ with $p^*\mathcal{G} = \mathcal{O}_{\tilde{C}}(-x)$. Thus, Hypothesis (ii) in Theorem 28 is necessary.

11. Let k be a finite field or an algebraic function field in one variable, and let K be a finite Galois extension of k with Galois group G. Then the norm map $N: K^* \to k^*$ is surjective, where $N(\alpha) = \prod_{\sigma \in G} \sigma(\alpha)$. More generally, let F be a homogeneous polynomial of degree d in $k[x_1, \ldots, x_n]$. If $n > d$, then F has a nontrivial zero in k^n.
    (To begin, the second statement implies the first, by taking $d = [K : k]$ and $F(x_1, \ldots, x_{d+1}) = N(\sum_{i=1}^d x_i\alpha_i) - \lambda x_{d+1}^d$, where $\alpha_1, \ldots, \alpha_d$ is a k-basis for K, noting that $N(\alpha)$ is nonzero if $\alpha \neq 0$. The second statement is due to Chevalley in case k is finite and may be found as an exercise in Lang [78, p. 213]. Of course, it is easy to check directly that the norm map is surjective for finite fields. For a function field k in one variable, the second statement is Tsen's theorem. Prove Tsen's theorem first in case k is the function field of $\mathbb{P}^1$, in other words, $k = k_0(t)$ where k_0 is algebraically closed. In this case $F \in k_0(t)[x_1, \ldots, x_n]$, and

after clearing denominators we can assume that $F \in k_0[t][x_1, \ldots, x_n]$.
Let N be the largest degree of a coefficient of F. For a given natural
number m, consider $F(p_1, \ldots, p_n)$, where each p_i is a polynomial of
degree $\leq m$. The space of all such vectors of polynomials is $\mathbb{A}^{n(m+1)}$,
and since F is homogeneous $F = 0$ is well-defined in $\mathbb{P}^{n(m+1)-1}$. Now
$F(p_1, \ldots, p_n)$ is a polynomial of degree $\leq dm + N$ whose coefficients
are homogeneous of degree d in the coefficients of the p_i. Thus, to say
$F = 0$ is $dm + N$ equations in $\mathbb{P}^{n(m+1)-1}$, which will have a solution
provided $dm + N \leq nm + n - 1$, or in other words if we choose

$$m \geq \frac{N - n + 1}{n - d}.$$

More generally, if $F_1, \ldots, F_r \in k_0(t)[x_1, \ldots, x_n]$ are homogeneous of
degree d_i, then there is a common zero of the F_i provided that $n > \sum_{i=1}^{r} d_i$.

In case k is a general function field, then we can write k as a finite extension of $k_0(t)$ for k_0 algebraically closed. Let $r = [k : k_0(t)]$. Choosing
a basis $\alpha_1, \ldots, \alpha_r$ for k over $k_0(t)$, we may write $F = \sum_i F_i \alpha_i$, where
each F_i is a function on $k^n = (k_0(t))^{rn}$, which is homogeneous of de-
gree d. Since $rd > rn$, we may apply the above to find a common zero
of the F_i.)

Classification of Algebraic Surfaces and of Stable Bundles

In this chapter, we outline the major results in the classification theory of surfaces, and then proceed to fill in the details of the proofs. While the proofs given here do not rely on Mori theory, we give a brief description of the corresponding results for threefolds, whose proofs rely heavily on Mori theory. In the last section, we survey some of the known results on the structure of the moduli space of stable bundles over a surface, and try to relate these results to the geometry of the original surface.

Outline of the classification of surfaces

In this section we shall list the various results that go under the general heading of classification of algebraic surfaces. The proofs will be given in the following sections. We begin with the following definition:

Definition 1. Let X be an algebraic surface. Define the *Kodaira dimension* $\kappa(X)$ of X as follows:

$$\kappa(X) = \min\{k \in \mathbb{Z} : P_n(X)/n^k \text{ is a bounded function of } n \geq 1\}.$$

For example, it follows formally that if $P_n(X) = 0$ for all n, then $\kappa(X) = -\infty$. If $P_n(X) \neq 0$ for some n, then since $P_{nm}(X) \geq P_n(X)$, it follows that $\kappa(X) \geq 0$. It is easy to see that $\kappa(X) \leq 2$ (and we shall prove much more precise statements later). Thus, the possible values for $\kappa(X)$ are $-\infty, 0, 1, 2$. By Corollary 5 of Chapter 3, $\kappa(X)$ is a birational invariant.

We can similarly define the Kodaira dimension $\kappa(X)$ of any smooth variety X of dimension n, and show that it is either $-\infty$ or an integer between 0 and n. For example, if C is a smooth curve, then $\kappa(C) = -\infty$ if $g(C) = 0$, $\kappa(C) = 0$ if $g(C) = 1$, and $\kappa(C) = 1$ if $g(C) \geq 2$. A variety X is *of general type* if $\kappa(X) = \dim X$.

The idea that the asymptotic behavior of the plurigenera has a deep influence on the structure of the surface X was already known to the Italians (see, for example, Enriques' book [31]), and was used systematically by Kodaira in the "Kodaira classification" of surfaces. However, the notation $\kappa(X)$ was first introduced in the Shafarevich seminar [132].

Let us now list some of the major results in the classification theory. One of the first major results is the intrinsic characterization of those surfaces which are rational, in other words birational to $\mathbb{P}^2$. If C is a smooth curve, then C is rational, in other words its function field is $\mathbb{C}(t)$ for some transcendental element t, if and only if its genus $g(C)$ is 0. For surfaces X, the analogous result is:

Theorem 2 (Castelnuovo). *Let X be an algebraic surface with $P_2(X) = q(X) = 0$. Then X is rational.*

We note that it does not suffice to assume that $p_g(X) = q(X) = 0$. There are examples of algebraic surfaces with torsion in H^2 for which $p_g = q = 0$. The first such example was constructed by Enriques: there exist $K3$ surfaces Y with a fixed point free holomorphic automorphism ι of order 2, and the quotient $X = Y/\iota$ is an algebraic surface with $q = p_g = 0$ and $H_1(X; \mathbb{Z}) = \mathbb{Z}/2\mathbb{Z}$. Such a surface is called an *Enriques surface*. Later Godeaux found certain quintic surfaces in $\mathbb{P}^3$ which have a fixed point free automorphism of order 5. The quotients are then surfaces X with $q = p_g = 0$ and $H_1(X; \mathbb{Z}) = \mathbb{Z}/5\mathbb{Z}$. Thus, none of these surfaces can be rational. Dolgachev [22] showed that the logarithmic transform of a rational elliptic surface at two fibers with relatively prime multiplicities is a nonrational algebraic surface X with $q = p_g = 0$ and $H_1(X; \mathbb{Z}) = 0$. By the canonical bundle formula, it follows that X is not rational (for example, $P_n(X) \neq 0$ for some n). These examples are in fact simply connected, and thus homeomorphic to rational surfaces, by Freedman's classification of topological 4-manifolds [34]. By Donaldson theory, it can be shown that they are not diffeomorphic to rational surfaces [24], [38], [119]. Barlow [6] constructed a simply connected surface of general type B with $q = p_g = 0$. Again using Donaldson theory, Kotschick [76] and also Okonek and Van de Ven [120] showed that B is not diffeomorphic to a rational surface. After a considerable amount of work in this direction [122], [127], [128], [124], the author and Qin [43], as well as Pidstrigach [123], showed that an algebraic surface diffeomorphic to a rational surface is necessarily rational. These arguments were greatly simplified by the advent of Seiberg-Witten theory, and could then be used to show that the plurigenera are smooth invariants [41], [16] (see also [118]).

The arguments used to prove Castelnuovo's theorem also determine the minimal models of a rational surface, a result due to Vaccaro and Andreotti:

Theorem 3. *A minimal rational surface is either $\mathbb{P}^2$ or $\mathbb{F}_n$ for some $n \neq 1$. Thus, every rational surface X is the blowup either of $\mathbb{P}^2$ or of $\mathbb{F}_n$ for some $n \neq 1$.*

A problem related to the characterization of rational surfaces is the characterization of ruled surfaces, in other words those surfaces birational to $C \times \mathbb{P}^1$ for some curve C. One basic result in this direction is:

Theorem 4. *Let X be a minimal surface such that K_X is not nef. Then X is rational or ruled, i.e., X is either $\mathbb{P}^2$ or a geometrically ruled surface.*

Mori has introduced a series of new ideas in the classification theory of surfaces, threefolds, and higher-dimensional algebraic varieties. From the viewpoint of Mori's theory, Theorem 4 says the following: let X be an algebraic surface, and suppose that K_X is not nef. Then either there exists an exceptional curve on X (X is not minimal), or there exists a morphism from X to a smooth curve C with all fibers $\mathbb{P}^1$ (X is geometrically ruled), or $X = \mathbb{P}^2$. We shall discuss the analogous classification results for threefolds later in this chapter.

As an immediate corollary to Theorem 4, a minimal surface X such that $K_X^2 < 0$ is a ruled surface X over a curve C of genus at least 2.

The following theorem and its higher-dimensional analogues go by the name of the *abundance theorem*:

Theorem 5. *Let X be a minimal algebraic surface such that K_X is nef. Then $K_X^2 \geq 0$. Moreover:*

(i) *$\kappa(X) = 0$ if and only if K_X is numerically equivalent to 0.*
(ii) *$\kappa(X) = 1$ if and only if $K_X^2 = 0$ but K_X is not numerically equivalent to 0.*
(iii) *$\kappa(X) = 2$, i.e., X is of general type, if and only if $K_X^2 > 0$.*

In all cases, either $P_4(X)$ or $P_6(X) \neq 0$.

Theorem 5 is really quite a surprising statement. For example, it says that, if K_X is numerically trivial, then it has finite order in $\mathrm{Pic}\, X$. Likewise, if $K_X^2 = 0$ but K_X is not numerically trivial, then some multiple of K_X has at least two sections. Thus, for example K_X cannot be linearly equivalent to an irreducible curve D with $D^2 = 0$ but such that the normal bundle of D has infinite order in $\mathrm{Pic}^0 D$. (The remaining statement, that if K_X is nef and big then X is of general type, is much easier.)

Let X be a surface, not necessarily regular. Then X is the blowup of a minimal surface $\bar{X}$. Now either $K_{\bar{X}}$ is nef, in which case $P_4(\bar{X})$ or $P_6(\bar{X}) \neq 0$, or $\bar{X}$ is rational or geometrically ruled. Since the plurigenera are invariant under blowup, and a blowup of a geometrically ruled surface is a ruled surface, we have the following numerical characterization of rational or ruled, which is akin to Castelnuovo's theorem:

Corollary 6 (Enriques). *Let X be an algebraic surface with either $P_4(X) = 0$ or $P_6(X) = 0$. Then X is rational or ruled. In particular, X is rational or ruled if and only if $\kappa(X) = -\infty$.*

Since the least common multiple of 4 and 6 is 12, we could make the slightly stronger assumption that $P_{12}(X) = 0$ in the above corollary.

Using the above corollary, we have the following characterization of algebraic surfaces X with $\kappa(X) \geq 0$:

Proposition 7. *Let X be an algebraic surface. Then the following are equivalent:*

(i) $\kappa(X) \geq 0$.
(ii) X has a unique minimal model.
(iii) There exists a surface X' birational to X such that $K_{X'}$ is nef.
(iv) There exists a minimal model X' of X for which $K_{X'}$ is nef.
(v) For every minimal model X' of X, $K_{X'}$ is nef.
(vi) X is not rational or ruled.

Proof. The implication (i) $\iff$ (vi) is Corollary 6. We have seen that (i) $\implies$ (ii) by Theorem 19 of Chapter 3. Conversely, if $\kappa(X) = -\infty$, then X is either rational or ruled and so does not have a unique minimal model. Thus, (ii) $\implies$ (i).

Next we show that (i) $\implies$ (v). Suppose that $\kappa(X) \geq 0$. Thus, there exists an $n \geq 1$ and a section $\sum_i a_i C_i \in |nK_X|$, where the C_i are distinct irreducible curves and the a_i are positive integers. Clearly, $K_X \cdot C \geq 0$ for every irreducible curve C on X which is not one of the C_i. Thus, if $K_X \cdot C < 0$, then $C = C_i$ for some i. Now

$$K_X \cdot C_i = \frac{1}{n}\left(\sum_{j \neq i} a_j(C_j \cdot C_i) + a_i(C_i)^2\right).$$

Since $a_j(C_j \cdot C_i) \geq 0$ for all $j \neq i$, and $a_i > 0$, if $K_X \cdot C_i < 0$, then $(C_i)^2 < 0$. In this case as $K_X \cdot C_i < 0$ and $(C_i)^2 < 0$, it follows by Lemma 11 of Chapter 3 that C_i is an exceptional curve, so that X is not minimal. Conversely, if X is minimal, then K_X is nef. Thus, (i) $\implies$ (v). The implications (v) $\implies$ (iv) and (iv) $\implies$ (iii) are trivial.

Finally, we show that (iii) $\implies$ (i). If (i) does not hold, then X is rational or ruled, and the discussion of the minimal models of such surfaces shows that there exists a smooth rational curve C on X with $C^2 \geq 0$ and thus $K_X \cdot C = -2 - C^2 \leq -2$. Thus, K_X is not nef. $\square$

For minimal surfaces with $\kappa = -\infty$, the classification scheme is the same as that for rank 2 vector bundles over a curve C, up to twisting by a line bundle. The unstable bundles have a very explicit description in terms of extensions, and the semistable bundles at least fit together into a coarse moduli space. There is also a fine classification in the cases $\kappa = 0, 1$. To deal with the case $\kappa = 0$, recall that an Enriques surface is the quotient of a $K3$ surface by an fixed point free involution of order 2. An Enriques surface is also elliptic. In fact every Enriques surface is an elliptic surface over $\mathbb{P}^1$ with invariant $d = \deg L = 1$ and two multiple fibers, both of multiplicity 2, and conversely. There are also quotients of abelian surfaces by fixed point free automorphisms, called *hyperelliptic surfaces*. All such surfaces are elliptic, in two different ways, and have been described in the exercises to Chapter 7.

Theorem 8. *Let X be a minimal algebraic surface with $\kappa(X) = 0$. Then X is a $K3$ surface, an abelian surface, an Enriques surface, or a hyperelliptic surface.*

In particular, every minimal algebraic surface with $\kappa = 0$ has a finite covering space with trivial canonical bundle.

Theorem 9. *Let X be a minimal algebraic surface with $\kappa(X) = 1$. Then X is an elliptic surface.*

For every minimal elliptic surface X, the canonical bundle K_X is numerically equivalent to rf as a divisor with rational coefficients, where $r \in \mathbb{Q}$. Clearly, K_X is nef if and only if $r \geq 0$, and K_X is numerically equivalent to 0 if and only if $r = 0$. Thus, the elliptic surfaces with $\kappa = -\infty$ correspond to the case $r < 0$ and those with $\kappa = 0$ correspond to $r = 0$. The possibilities for such surfaces have been listed in Exercise 7 of Chapter 7. It follows from this list that the elliptic surfaces with $\kappa = -\infty$ are either rational surfaces or else are ruled surfaces over an elliptic base (although not every geometrically ruled surface over an elliptic curve is actually an elliptic surface). The elliptic surfaces with $\kappa = 0$ are the Enriques surfaces, the hyperelliptic surfaces, and certain $K3$ and abelian surfaces. Thus, with few exceptions, an elliptic surface has Kodaira dimension 1. Sometimes the surfaces with Kodaira dimension 1 are thus called the *properly elliptic surfaces*. In any case, all elliptic surfaces can be fairly explicitly described by the classification scheme of Chapter 7.

No such fine classification theory exists for surfaces of general type. There are restrictions on the Chern numbers of minimal surfaces of general type. Classically, there is Noether's inequality

$$K_X^2 \geq 2p_g(X) - 4.$$

The cases of equality have been analyzed by Horikawa [64] and surfaces where equality is attained are called *Horikawa surfaces*.

A much deeper result is the Bogomolov-Miyaoka-Yau inequality

$$K_X^2 \leq 3c_2(X) = 3\chi(X),$$

with equality holding if and only if X is the quotient of the unit ball in $\mathbb{C}^2$ by a discrete group of holomorphic automorphisms acting freely and with compact quotient [95], [96], [147].

Beyond these numerical restrictions, the study of surfaces of general type largely consists in studying examples and goes under the name "geography" (for deciding which Chern numbers or other topological invariants arise as the invariants of a minimal surface of general type), and "botany" (for describing all of the deformation types of surfaces within a fixed topological type). For a survey of some of these results, see [121].

Let us record the following useful result on surfaces of general type:

Proposition 10. *Let X be an algebraic surface such that K_X is nef and big. Then X is a minimal surface of general type, and the plurigenera P_n of X, for $n \geq 2$, are given by the formula*

$$P_n(X) = \frac{n(n-1)}{2}(K_X)^2 + \chi(\mathcal{O}_X).$$

Conversely, if X is a surface of general type, the following are equivalent:

(i) *X is minimal.*

(ii) *K_X is nef.*

(iii) *For all $n \geq 2$, $P_n(X) = \dfrac{n(n-1)}{2}(K_X)^2 + \chi(\mathcal{O}_X)$.*

(iv) *There exists an $n \geq 2$ such that $P_n(X) = \dfrac{n(n-1)}{2}(K_X)^2 + \chi(\mathcal{O}_X)$.*

(v) *For all $n \geq 2$, $H^1(X; nK_X) = 0$.*

(vi) *There exists an $n \geq 2$ such that $H^1(X; nK_X) = 0$.*

Proof. Let X be an algebraic surface such that K_X is nef and big. Since K_X is nef, X is minimal. For $n \geq 2$, $H^1(X; nK_X)$ is Serre dual to $H^1(X; (1-n)K_X)$. Since $(1-n)K_X$ is the negative of the nef and big divisor $(n-1)K_X$, the Mumford vanishing theorem implies that $H^1(X; (1-n)K_X) = 0$. Moreover, $H^2(X; nK_X)$ is Serre dual to $H^0(X; (1-n)K_X)$. As K_X is big, $K_X \cdot ((1-n)K_X) < 0$. Since K_X is nef, $(1-n)K_X$ cannot be effective, and so $H^0(X; (1-n)K_X) = 0$. It follows that, for $n \geq 2$,

$$P_n(X) = h^0(nK_X) = \chi(\mathcal{O}_X(nK_X)).$$

Applying Riemann-Roch, we have $\chi(\mathcal{O}_X(nK_X)) = [n(n-1)/2](K_X)^2 + \chi(\mathcal{O}_X)$. Thus, we have established the formula for P_n, and clearly $\kappa(X) = 2$ since P_n is a quadratic polynomial in n. Thus, X is of general type. (Another way to see that X is of general type is to note that K_X induces an ample divisor on the surface X obtained by contracting the -2-curves orthogonal to K_X.)

To see the equivalences in the remaining statement of the theorem, note that K_X is nef implies that X is minimal, so that (ii) $\Longrightarrow$ (i). Conversely, for a minimal surface with $\kappa \geq 0$, we know by Proposition 7 that K_X is nef. Thus, (i) $\Longrightarrow$ (ii). Moreover, if X is minimal, then the formula in (iii) holds by the first statement of Proposition 10, so (i) $\Longrightarrow$ (iii). We have also seen that (i) $\Longrightarrow$ (v) as well. The implications (iii) $\Longrightarrow$ (iv) and (v) $\Longrightarrow$ (vi) are trivial. Next, for all n, the dimension of $H^0(X; (1-n)K_X)$ is a birational invariant, and the argument above shows that this dimension is 0 if X is minimal and $n \geq 2$. Thus, $H^0(X; (1-n)K_X)$ is always 0 if $n \geq 2$. Let X_0 be the minimal model of X. Thus, $P_n(X) = P_n(X_0)$ and $K_X^2 \leq K_{X_0}^2$, with equality holding if and only if $X = X_0$. Let us show that (iv) $\Longrightarrow$ (i). Assuming (iv), we have

$$P_n(X) = \frac{n(n-1)}{2}(K_X)^2 + \chi(\mathcal{O}_X)$$

$$= \frac{n(n-1)}{2}(K_{X_0})^2 + \chi(\mathcal{O}_{X_0}),$$

since $P_n(X)$ is a birational invariant. As $\chi(\mathcal{O}_X) = \chi(\mathcal{O}_{X_0})$ and $n \geq 2$, it follows that $(K_X)^2 = (K_{X_0})^2$ and hence that $X = X_0$ is minimal. Thus, (iv) $\Longrightarrow$ (i). Finally, the implication (vi) $\Longrightarrow$ (iv) is an easy consequence of the Riemann-Roch formula and the fact that $H^0(X; (1-n)K_X) = 0$ and so $H^2(X; nK_X) = 0$. This concludes the proof. $\square$

One final result concerning surfaces with $\kappa \geq 0$ goes under the general name of the Castelnuovo-deFranchis theorem:

Theorem 11. *Let X be an algebraic surface with $\kappa(X) \geq 0$. Then $c_2(X) \geq 0$, with equality holding if and only if X is a minimal elliptic surface whose only singular fibers are multiple fibers with smooth reduction or X is an abelian surface.*

Corollary 12. *Let X be an algebraic surface with $\kappa(X) \geq 0$. Then $\chi(\mathcal{O}_X) \geq 0$, with equality holding if and only if X is an elliptic surface whose only singular fibers are multiple fibers with smooth reduction or X is an abelian surface.*

Proof of the corollary. It suffices to prove the statement under the assumption that X is minimal, since $\chi(\mathcal{O}_X)$ is a birational invariant. In this case, by Noether's formula, $12\chi(\mathcal{O}_X) = c_1(X)^2 + c_2(X)$ with $c_1(X)^2 \geq 0$ since K_X is nef and $c_2(X) \geq 0$ by the above theorem. Thus, $\chi(\mathcal{O}_X) \geq 0$, and if equality holds $c_1(X)^2 = c_2(X) = 0$. Hence X is a minimal elliptic surface whose only singular fibers are multiple fibers with smooth reduction, or X is an abelian surface. Conversely, if X is a minimal elliptic surface whose only singular fibers are multiple fibers with smooth reduction or X is an abelian surface, then $c_1(X)^2 = c_2(X) = 0$. $\square$

Proof of Castelnuovo's theorem

Let X be a surface with $q(X) = P_2(X) = 0$. Note that $p_g(X) = 0$ as well, so that $H^1(\mathcal{O}_X) = H^2(\mathcal{O}_X) = 0$. From the exponential sheaf sequence, $\operatorname{Pic} X = H^2(X; \mathbb{Z})$. In this case, we shall prove:

Theorem 13. *Let X be a surface with $q(X) = P_2(X) = 0$. Then there exists a smooth rational curve C on X with $C^2 \geq 0$.*

It is not even *a priori* obvious that a rational surface has this property! In fact, if $X \dashrightarrow \mathbb{P}^2$ is a birational isomorphism, then the strict transform of a line in $\mathbb{P}^2$ on X may well be singular. Thus, we cannot just use the strict transform of a line to find the curve C.

Proof that Theorem 13 implies Castelnuovo's theorem. Let X be a surface with $q(X) = P_2(X) = 0$ and let C be a smooth rational curve on X with $C^2 \geq 0$. From the exact sequence

$$0 \to \mathcal{O}_X \to \mathcal{O}_X(C) \to \mathcal{O}_C(C) \to 0,$$

we see that $\dim |C| \geq 1$ and that $|C|$ has no fixed components. Choose a pencil inside $|C|$. After blowing up the base locus, there is a morphism $\tilde{X} \to \mathbb{P}^1$ whose general fiber is a smooth rational curve. Thus, there is by Lemma 8 of Chapter 5 a blowdown $\tilde{X} \to Y$, where $Y \to \mathbb{P}^1$ is a geometrically ruled surface over $\mathbb{P}^1$. Hence Y is birational to $\mathbb{P}^1 \times \mathbb{P}^1$ and thus to $\mathbb{P}^2$. (Of course, by Theorem 9 of Chapter 5 and the classification of vector bundles over $\mathbb{P}^1$, $Y = \mathbb{F}_n$ for some n.) Thus, Y and therefore X are rational. $\square$

Proof of Theorem 13. We begin the proof with the famous "termination of adjunction" lemma:

Lemma 14. *Let X be a minimal surface such that $q(X) = P_2(X) = 0$. Then for every divisor D on X, and for all $n \gg 0$, $|D + nK_X| = \emptyset$.*

Proof. First suppose that $K_X^2 \geq 0$. By the Riemann-Roch theorem,

$$\chi(\mathcal{O}_X(-K_X)) = \tfrac{1}{2}(-K_X)(-2K_X) + 1 \geq 1,$$

and thus either $h^0(-K_X)$ or $h^2(-K_X)$ is nonzero. But $h^2(-K_X) = h^0(2K_X) = P_2(X) = 0$, so that $h^0(-K_X) \neq 0$. Since $h^0(K_X) = 0$, $-K_X$ is not the trivial divisor, and thus $-K_X = C$ for some nonzero effective curve C. Thus, for every divisor D, if H is an ample divisor and $n \gg 0$,

$H \cdot (D + nK_X) = (H \cdot D) - n(H \cdot C) < 0$. It follows that $D + nK_X$ is not effective if $n \gg 0$.

If $K_X^2 < 0$ and D is a fixed divisor, suppose that $D + mK_X$ is effective. Note that $(D + nK_X) \cdot K_X < 0$ for all $n \gg 0$. In particular $D + nK_X$ is not the trivial divisor for $n \gg 0$. If m is sufficiently large and $D + mK_X$ is linearly equivalent to the effective nonzero divisor $\sum_i a_i C_i$, where the C_i are irreducible curves and $a_i > 0$, then as $(D + mK_X) \cdot K_X < 0$ we must have $C_i \cdot K_X < 0$ for some i. As X is minimal, it follows that $C_i^2 \geq 0$ and thus that C_i is nef. In particular, $C_i \cdot (D + nK_X) \geq 0$ for all n such that $D + nK_X$ is effective and $(D + nK_X) \cdot K_X < 0$. But since $C_i \cdot K_X < 0$, $C_i \cdot (D + nK_X) < 0$ for all $n \gg 0$, contradicting $C_i \cdot (D + nK_X) \geq 0$. Thus, for $n \gg 0$ the divisor $D + nK_X$ is not linearly equivalent to an effective divisor.   □

Returning to the proof of Castelnuovo's theorem, we claim that there exists a very ample divisor H such that H is not an integer multiple of K_X. In fact, if every very ample divisor is an integer multiple of K_X, then as every divisor is linearly equivalent to a difference of very ample divisors it would follow that $\mathrm{Pic}\, X = \mathbb{Z}[K_X]$. But we have seen that $\mathrm{Pic}\, X \cong H^2(X; \mathbb{Z})$, so that $H^2(X; \mathbb{Z}) \cong \mathbb{Z}[K_X]$ as well. By Poincaré duality, $K_X^2 = 1$. Moreover, $b_2(X) = 1$ and $b_1(X) = b_3(X) = 0$, so that $c_2(X) = 3$. On the other hand, from Noether's formula we have

$$c_1^2(X) + c_2(X) = 1 + 3 = 12\chi(\mathcal{O}_X) = 12,$$

which is absurd.

Thus, we may choose a very ample H which is not an integer multiple of K_X, so that $h^0(\mathcal{O}_X(H)) \neq 0$ or equivalently $|H| \neq \emptyset$. Using Lemma 14, there exists an $n \geq 0$ such that $|H + nK_X| \neq \emptyset$ but $|H + (n+1)K_X| = \emptyset$. Since H is not an integer multiple of K_X, $H + nK_X$ is not the trivial divisor. Thus, there exists $\sum_i a_i C_i \in |H + nK_X|$, where the C_i are irreducible curves and $a_i > 0$. We shall show that all of the C_i are smooth rational curves and that $C_i^2 \geq 0$ for some i.

Since $|\sum_i a_i C_i + K_X| = \emptyset$ and $|\sum_i a_i C_i + K_X|$ contains $|C_i + K_X|$ for every i, we must have $|C_i + K_X| = \emptyset$ for every i. On the other hand, by applying adjunction to C_i, we have the exact sequence

$$0 \to \mathcal{O}_X(K_X) \to \mathcal{O}_X(C_i + K_X) \to \omega_{C_i} \to 0.$$

As $h^1(K_X) = h^1(\mathcal{O}_X) = q(X) = 0$, the map $H^0(\mathcal{O}_X(C_i + K_X)) \to H^0(\omega_{C_i})$ is surjective. But since $H^0(\mathcal{O}_X(C_i + K_X)) = 0$, $H^0(\omega_{C_i}) = 0$ as well, in other words $p_a(C_i) = 0$. It follows that C_i is a smooth rational curve for every i.

Finally, we must show that $C_i^2 \geq 0$ for some i. If $C_i^2 < 0$, then, since X is minimal, $K_X \cdot C_i \geq 0$. Thus, if $C_i^2 < 0$ for all i, then $K_X \cdot (\sum_i a_i C_i) \geq 0$.

Let $D = \sum_i a_i C_i$. Then $D = H + nK_X$, $K_X \cdot D \geq 0$, and

$$K_X \cdot D + D^2 = K_X \cdot D + H \cdot D + n(K_X \cdot D) > 0,$$

since H is ample and $D \neq 0$. By assumption $h^0(K_X + D) = 0$, and thus $h^2(-D) = 0$. Since D is effective and nonzero, $h^0(-D) = 0$. Thus, $\chi(\mathcal{O}_X(-D)) \leq 0$. On the other hand, applying the Riemann-Roch theorem, we find that

$$\chi(\mathcal{O}_X(-D)) = \tfrac{1}{2}\left((-D)^2 + (K_X \cdot D)\right) + 1 = \tfrac{1}{2}\left(K_X \cdot D + D^2\right) + 1 > 0,$$

a contradiction. Thus, for some i the curve C_i is a smooth rational curve and $C_i^2 \geq 0$. □

The above argument also shows the following:

Corollary 15. *The minimal models of $\mathbb{P}^2$ are exactly the surfaces $\mathbb{P}^2$ and $\mathbb{F}_n, n \neq 1$.*

Proof. Clearly, the surfaces $\mathbb{P}^2$ and $\mathbb{F}_n, n \neq 1$, are in fact minimal. Conversely, suppose that X is a minimal rational surface. By Castelnuovo's theorem, X contains a smooth rational curve C with $C^2 \geq 0$. Considering the exact sequence

$$0 \to \mathcal{O}_X \to \mathcal{O}_X(C) \to \mathcal{O}_C(C) \to 0$$

and using $h^1(\mathcal{O}_X) = 0$, we see that $h^0(\mathcal{O}_X(C)) \geq 2$. Choose $C' \in |C|, C' \neq C$. Then C and C' span a pencil inside $|C|$ with no fixed curves. First suppose that every element in the pencil is a smooth rational curve. Then there is a blowup of X, say $\tilde{X}$, and a morphism $f \colon \tilde{X} \to \mathbb{P}^1$ such that all of the fibers of f are smooth. In fact, we can take the $\mathbb{P}^1$ to be the parameter space of the pencil and take $\tilde{X}$ to be the incidence correspondence

$$\{(x, t) \in X \times \mathbb{P}^1 : x \in C_t\},$$

where C_t is the curve in $|C|$ corresponding to $t \in \mathbb{P}^1$. The projections of $\tilde{X}$ to the first and second factors of $X \times \mathbb{P}^1$ induce a birational morphism $\tilde{X} \to X$ and a morphism $\pi \colon \tilde{X} \to \mathbb{P}^1$ such that $\pi^{-1}(t) = C_t$. Since C_t is smooth for every t, a local calculation shows that $\tilde{X}$ is smooth, and so $\tilde{X}$ is a blowup of X by the factorization of birational morphisms. Since all fibers of π are smooth rational curves, $\tilde{X} \to \mathbb{P}^1$ is a geometrically ruled surface, and thus is $\mathbb{F}_n$ for some n. If $\tilde{X} = X$, then we are done: We must have $X \cong \mathbb{F}_n, n \neq 1$, since X is minimal. If $X \neq \tilde{X}$, then, as $\tilde{X}$ is a blowup of X, it contains an exceptional curve E. Since the only curve of negative self-intersection on $\mathbb{F}_n$ is the negative section σ, with $\sigma^2 = -n$, necessarily $n = 1$, and there is an induced birational morphism from the contraction of E, namely $\mathbb{P}^2$, to X. As $\mathbb{P}^2$ is already minimal, $X = \mathbb{P}^2$.

Now suppose that every pencil in $|C|$ contains a reducible element $\sum_i a_i C_i$. From the exact sequence

$$0 \to \mathcal{O}_X(K_X) \to \mathcal{O}_X(K_X + C) \to \omega_C \to 0,$$

we see that $|K_X + C| = \emptyset$, and thus, as in the proof of Theorem 13, it follows that $|K_X + C_i|$, which is a subseries of $|K_X + \sum_i a_i C_i| = |K_X + C|$, is empty for every i. As in the proof of Theorem 13, C_i must be a smooth rational curve for every i. We claim that $C_i^2 \geq 0$ for some i. The argument parallels that in Theorem 13: if $C_i^2 < 0$, then since X is minimal, $C_i \cdot K_X \geq 0$. Thus, if $C_i^2 < 0$ for every i, then $0 \leq K_X \cdot (\sum_i a_i C_i) = K_X \cdot C$. But $K_X \cdot C = -2 - C^2 \leq -2$, which is a contradiction. Thus, $C_i^2 \geq 0$ for some i. Now $C^2 = \sum_j a_j(C \cdot C_j) \geq a_i(C \cdot C_i)$, since C is nef, and

$$a_i(C \cdot C_i) = a_i^2 C_i^2 + a_i \sum_{j \neq i} a_j(C_i \cdot C_j) \geq a_i^2 C_i^2.$$

Thus, $C_i^2 \leq C^2$, with equality if and only if C_i meets no other C_j and $a_i = 1$. By the connectedness theorem, C_i meets no other C_j if and only if $C = a_i C_i$, and, if $a_i = 1$, then $C_i = C$ is smooth, which we have ruled out. Thus, $0 \leq C_i^2 < C^2$. If there exists a pencil in $|C_i|$ such that every member is smooth, we are done. Otherwise, we may continue this procedure, noting that at each stage the self-intersection is nonnegative and strictly decreases. Thus, the procedure must eventually terminate. By the first part of the argument, we then have $X = \mathbb{P}^2$ or $\mathbb{F}_n, n \neq 1$. $\square$

The following corollary to Castelnuovo's theorem is the affirmative answer to the Lüroth problem for an algebraically closed field of characteristic 0. (It fails over algebraically closed fields of positive characteristic as well as over fields of characteristic 0 which are not algebraically closed.)

Corollary 16. *Let X be a rational surface, and let $f: X \to Y$ be a generically finite morphism to the smooth surface Y. Then Y is rational. In terms of function fields, let x_1 and x_2 be algebraically independent over $\mathbb{C}$. Let k be a subfield of $\mathbb{C}(x_1, x_2)$ containing $\mathbb{C}$ and such that*

$$[\mathbb{C}(x_1, x_2) : k] < \infty.$$

Then k is a pure transcendental extension of $\mathbb{C}$ of transcendence degree 2.

Proof. Since f is generically finite, its differential is an isomorphism on a Zariski open subset of Y. Thus, $f^*: H^0(Y; \Omega_Y^1) \to H^0(X; \Omega_X^1)$ is injective. It follows that f^* defines inclusions $H^1(Y; \mathcal{O}_Y) \subseteq H^1(X; \mathcal{O}_X)$ and $H^0(Y; \mathcal{O}_Y(2K_Y)) \subseteq H^0(X; \mathcal{O}_X(2K_X))$. Thus, if $q(X) = P_2(X) = 0$, then $q(Y) = P_2(Y) = 0$ as well, and we can apply Castelnuovo's theorem to Y. $\square$

The Albanese map

The main tool which we need to handle the classification for irregular surfaces is the Albanese variety of X.

Definition 17. Let X be an algebraic surface (or more generally a compact Kähler manifold). Define the *Albanese variety* $\text{Alb}\,X$ to be the compact complex torus $H^0(X;\Omega^1_X)^*/H_1(X;\mathbb{Z})$. More precisely, there is a natural map

$$H_1(X;\mathbb{Z}) \to H^0(X;\Omega^1_X)^*$$

defined by integration: if $\gamma \in H_1(X;\mathbb{Z})$, then, since holomorphic forms are closed on a compact Kähler manifold, for $\varphi \in H^0(X;\Omega^1_X)$,

$$\varphi \mapsto \int_\gamma \varphi$$

is a well-defined linear function on $H^0(X;\Omega^1_X)$. By Hodge theory, the kernel of the induced map $H_1(X;\mathbb{Z}) \to H^0(X;\Omega^1_X)^*$ is the torsion subgroup of $H_1(X;\mathbb{Z})$, and its image is a discrete subgroup of $H^0(X;\Omega^1_X)^*$ with compact quotient. By definition this quotient is then $\text{Alb}\,X$. One can show that it is in fact an abelian variety (in other words, a complex torus which is also an algebraic variety).

Proposition 18. *For every algebraic surface or compact Kähler manifold X and for every choice of a base point $p \in X$, there is a morphism $\alpha\colon X \to \text{Alb}\,X$ with the following properties:*

(i) *$\alpha^*\colon H^0(\text{Alb}\,X;\Omega^1_{\text{Alb}\,X}) \to H^0(X;\Omega^1_X)$ is an isomorphism.*
(ii) *For every complex torus T, if $f\colon X \to T$ is a morphism such that $f(p) = 0 \in T$, then there is a unique morphism of complex tori $g\colon \text{Alb}\,X \to T$ such that $f = g \circ \alpha$.*

Proof. Define the morphism α via integration: given $x \in X$, choose a piecewise C^∞ path $\sigma\colon [0,1] \to X$ such that $\sigma(0) = p$ and $\sigma(1) = x$, and set

$$\alpha(x) = \left(\varphi \mapsto \int_\sigma \varphi\right),$$

well defined up to the choice of the path σ. Since two choices of σ differ by a piecewise C^∞ closed curve γ, $\alpha(x)$ is well defined up to the image of $H_1(X;\mathbb{Z})$, and thus as an element of $\text{Alb}\,X$. By the fundamental theorem of calculus, α is holomorphic and its complex derivative at x is the natural map $T_{X,x} \to T_{\text{Alb}\,X,\alpha(x)} = H^0(\Omega^1_X)^*$ defined by

$$\xi \in T_{X,x} \mapsto \left(\varphi \in H^0(\Omega^1_X) \mapsto \varphi_x(\xi)\right).$$

Thus, α^* is given by the natural map

$$\Omega^1_{\mathrm{Alb}\,X,\alpha(x)} = H^0(\mathrm{Alb}\,X;\Omega^1_{\mathrm{Alb}\,X}) = H^0(X;\Omega^1_X) \to \Omega^1_{X,x}.$$

In particular, $\alpha^* \colon H^0(\mathrm{Alb}\,X;\Omega^1_{\mathrm{Alb}\,X}) \to H^0(X;\Omega^1_X)$ is an isomorphism.

Now suppose that T is a complex torus and $f\colon X \to T$ is a holomorphic map. Then f^* induces a complex linear map $H^0(T;\Omega^1_T) \to H^0(X;\Omega^1_X)$, and thus a linear map $f_* \colon (H^0(X;\Omega^1_X))^* \to (H^0(T;\Omega^1_T))^*$. Moreover, $f_*(H_1(X;\mathbb{Z})) \subseteq H_1(T;\mathbb{Z})$ and so there is an induced morphism $g\colon \mathrm{Alb}\,X \to (H^0(T;\Omega^1_T))^*/H_1(T;\mathbb{Z}) = T$. By construction $f(p) = 0 = g \circ \alpha(p)$ and $f^* = (g \circ \alpha)^*$. Thus, the difference map $f - (g \circ \alpha)$ has differential equal to 0, and so it is constant with image 0. It follows that $f = g \circ \alpha$. We leave the uniqueness of g as an exercise. $\square$

The morphism $\alpha\colon X \to \mathrm{Alb}\,X$ defined above, which is unique up to the choice of a base point, will be called the *Albanese map*.

Let X be a surface. Two basic properties of the Albanese map $\alpha\colon X \to \mathrm{Alb}\,X$ which we shall need are:

Proposition 19. *Let X be a surface and let $\alpha\colon X \to \mathrm{Alb}\,X$ be the Albanese map.*

(i) *If $\alpha(X)$ has dimension 2, then $p_g(X) \neq 0$.*
(ii) *If $\alpha(X)$ is a curve C, then C is smooth and $\alpha\colon X \to C$ has connected fibers. In this case $\mathrm{Alb}(X) = J(C)$, the Jacobian of the curve C.*

Proof. If $\alpha(X)$ has dimension 2, there exists a point x such that α^* has rank 2. In particular, there exists an $\omega_{\alpha(x)} \in \Omega^2_{\mathrm{Alb}\,X,\alpha(x)}$ such that $\alpha^*(\omega_{\alpha(x)}) \neq 0$. Since $\Omega^1_{\mathrm{Alb}\,X}$ is trivial, $\Omega^2_{\mathrm{Alb}\,X}$ is trivial as well, and so we can find a global holomorphic 2-form ω on $\mathrm{Alb}\,X$ such that ω induces $\omega_{\alpha(x)}$. Thus, $\alpha^*\omega$ is a nonzero holomorphic 2-form on X so that $p_g(X) \neq 0$.

If $\alpha(X)$ is a curve C, let $\tilde{C}$ be the normalization of C. Note that $g(\tilde{C}) \geq 1$. Since X is normal, the map $X \to C$ factors through a morphism $X \to \tilde{C}$, and we have the composite map $X \to \tilde{C} \to J(\tilde{C})$, where $J(\tilde{C})$ is the Jacobian of $\tilde{C}$. By the universal property of $\mathrm{Alb}\,X$, there is an induced map $\mathrm{Alb}\,X \to J(\tilde{C})$. The map $\tilde{C} \to C \to \mathrm{Alb}\,X \to J(\tilde{C})$ then agrees with the map $\tilde{C} \to J(\tilde{C})$, and so the image of $\tilde{C}$ in $J(\tilde{C})$ is C. Since the map from a curve of genus at least 1 to its Jacobian is an embedding, $C = \tilde{C}$, and in particular C is smooth. There is the map $g\colon \mathrm{Alb}\,X \to J(C)$, and the morphism $C \to \mathrm{Alb}\,X$ induces a morphism $f\colon J(C) \to \mathrm{Alb}\,X$ (essentially because $J(C)$ is the Albanese variety of C and by an argument similar to the proof of Proposition 18). The composite map $C \to J(C) \xrightarrow{f} \mathrm{Alb}\,X \xrightarrow{g} J(C)$ is by construction the natural map $C \to J(C)$, and thus by the universal property of $J(C)$ we must have $f \circ g = \mathrm{Id}$. Likewise, the map $X \to \mathrm{Alb}\,X \xrightarrow{g} J(C) \xrightarrow{f} \mathrm{Alb}\,X$ is the map $\alpha\colon X \to \mathrm{Alb}\,X$, and so $g \circ f = \mathrm{Id}$ as well. Thus,

$\mathrm{Alb}\, X \cong J(C)$. Finally, let us show that the fibers of α are connected. Let $X \to D \to C$ be the Stein factorization of $X \to C$. Thus, the fibers of $X \to D$ are connected and $D \to C$ is finite; moreover D is normal since C is normal. Applying the above argument with C replaced by D, we see that the map $D \to \mathrm{Alb}\, X$ induces an isomorphism $J(D) \to \mathrm{Alb}\, X$, and thus that the map $J(D) \to J(C)$ is an isomorphism. Thus, $D = C$ and the fibers of α are connected.   □

Proofs of the classification theorems for surfaces

Throughout the rest of this chapter we shall always take X to be a *minimal* algebraic surface unless otherwise noted. The first main result of this section shows that, for a minimal surface X, K_X is nef if and only if X is not rational or ruled.

Theorem 20. *If X is a minimal algebraic surface and K_X is not nef, then X is rational or ruled.*

Proof. First we claim that, if X is minimal and K_X is not nef, then $P_n(X) = 0$ for all $n \geq 1$. This follows from the proof of the implication (i) $\Longrightarrow$ (v) of Proposition 7 (the proof did not require any of the classification results).

Next suppose that $q(X) = 0$. Since $P_2(X) = 0$, X is rational by Castelnuovo's theorem. Thus, we may assume that $q(X) > 0$. Let $\alpha \colon X \to \mathrm{Alb}\, X$ be the Albanese map. If $\alpha(X)$ is a surface, then $p_g(X) \neq 0$, by Proposition 19. Thus, $\alpha(X) = C$ is a smooth curve and the Albanese map is a fibration $\pi \colon X \to C$. Moreover, $\mathrm{Alb}\, X = J(C)$ and thus $g(C) = q(X)$. We also may assume that $K_X^2 \leq 0$: Suppose that $K_X^2 > 0$. By Noether's formula, we have $K_X^2 + (2 - 4q + b_2(X)) = 12(1 - q)$. Thus,

$$8q + b_2(X) = 10 - K_X^2 \leq 9.$$

It follows that (since we have assumed that $q \geq 1$) $q = 1$ and $b_2(X) = 1$. But X maps onto its Albanese, which is a curve, and, for a general fiber f, $f^2 = 0$. Since $K_X^2 > 0$, K_X is not a multiple of f, and so the classes $[f]$ and $[K_X]$ span a rank 2 subgroup of $H^2(X; \mathbb{Z})$. But then $b_2(X) \geq 1$, a contradiction. Hence we may assume that $K_X^2 \leq 0$. Our goal now will be to show that the fibers of the Albanese map are smooth rational curves, so that X is ruled.

Claim 1. *Let $\pi \colon X \to C$ be the Albanese fibration of X, and let D be an irreducible curve on X such that $K_X \cdot D < 0$ and $|K_X + D| = \emptyset$. Then either X is ruled or D is smooth with $g(D) = q$, $\pi|D \colon D \to C$ is unramified, and $\pi|D$ is an isomorphism if $q > 1$.*

Proof of Claim 1. From the exact sequence

$$0 \to \mathcal{O}_X(K_X) \to \mathcal{O}_X(K_X + D) \to \omega_D \to 0,$$

and the fact that $H^0(\mathcal{O}_X(K_X + D)) = 0$ by assumption, it follows that the map $H^0(\omega_D) \to H^1(K_X)$ is injective. However, $h^1(K_X) = q$ and $h^0(\omega_D) = p_a(D)$, so that $p_a(D) \le q = g(C)$. Thus, if $\tilde{D}$ is the normalization of D, then as $g(\tilde{D}) \le p_a(D)$, $g(\tilde{D}) \le g(C)$ as well. On the other hand, there is a morphism $\tilde{D} \to C$. If this morphism were constant, then D would be contained in a fiber of π and so $D^2 \le 0$. But since X is minimal and $K_X \cdot D < 0$, $D^2 \ge 0$, so that $D^2 = 0$ and nD is a fiber of π for some n. In this case $K_X \cdot f < 0$ for a general fiber f of π, and thus $K_X \cdot f + f^2 < 0$. It follows that f is a smooth rational curve, and so X is ruled.

Thus, if X is not ruled, the morphism $\tilde{D} \to C$ is onto. But then $g(\tilde{D}) \ge g(C)$, by the Riemann-Hurwitz formula, since $g(C) \ge 1$, and so $g(\tilde{D}) = p_a(D) = g(C)$. Moreover, since $g(C) \ne 0$, again by the Riemann-Hurwitz formula either the morphism $D \to C$ has degree 1 or $g(C) = 1$ and $D \to C$ is unramified.

Claim 2. *If X is not ruled, there exists an irreducible curve D on X such that $K_X \cdot D < -1$ and $|K_X + D| = \emptyset$.*

Proof of Claim 2. Since K_X is not nef, there exists an irreducible curve C on X with $K_X \cdot C \le -1$. As X is minimal, $C^2 \ge 0$ and thus C is nef. By assumption $K_X^2 \le 0$. Thus, for all $n \ge 0$, $(2C + nK_X) \cdot K_X \le -2$. Now $|2C| \ne \emptyset$, but for $n \gg 0$, $|2C + nK_X| = \emptyset$ since $(2C + nK_X) \cdot C < 0$ for $n \gg 0$ and C is nef. Thus, there exists an $n \ge 0$ such that $|2C + nK_X| \ne \emptyset$ and $|2C + (n+1)K_X| = \emptyset$. Let $D = \sum_i n_i C_i \in |2C + nK_X|$. Then D is effective, $K_X \cdot D < -1$, and $|D + K_X| = \emptyset$. If $0 \le D' \le D$, then $|D' + K_X| = \emptyset$ as well. Thus, we may assume that $C_i \cdot K_X < 0$ for all i. By Claim 1, C_i is smooth and $g(C_i) = q$ for all i. We shall show in this case that $D = C_i$ for some i, and thus that D is irreducible.

First we claim that $n_i = 1$ for all i. Indeed, if $n_i \ge 2$ for some i, then $|2C_i + K_X| = \emptyset$, and thus $h^2(\mathcal{O}_X(-2C_i)) = 0$. From the exact sequence

$$0 \to \mathcal{O}_X(-2C_i) \to \mathcal{O}_X \to \mathcal{O}_{2C_i} \to 0,$$

it follows that the map from $H^1(\mathcal{O}_X)$ to $H^1(\mathcal{O}_{2C_i})$ is surjective, and thus that $h^1(\mathcal{O}_{2C_i}) \le q$. On the other hand, using the exact sequence

$$0 \to \mathcal{O}_{C_i}(-C_i) \to \mathcal{O}_{2C_i} \to \mathcal{O}_{C_i} \to 0,$$

and the fact that $H^0(\mathcal{O}_{2C_i}) \to H^0(\mathcal{O}_{C_i}) = \mathbb{C}$ is surjective, it follows that

$$h^1(\mathcal{O}_{2C_i}) = h^1(\mathcal{O}_{C_i}) + h^1(\mathcal{O}_{C_i}(-C_i)) = q + h^1(\mathcal{O}_{C_i}(-C_i)).$$

Thus, $h^1(\mathcal{O}_{C_i}(-C_i)) = 0$. But $p_a(C_i) \ge 1$, and so $K_X \cdot C_i + C_i^2 \ge 0$. As $K_X \cdot C_i < 0$, $C_i^2 > 0$. So $\mathcal{O}_{C_i}(-C_i)$ is a line bundle on C_i of strictly negative degree, say $\deg \mathcal{O}_{C_i}(-C_i) = -d$ with $d > 0$. By Riemann-Roch on C_i,

$$h^1(\mathcal{O}_{C_i}(-C_i)) = h^0(\mathcal{O}_{C_i}(-C_i)) + d + g(C_i) - 1 \geq d > 0,$$

contradicting the fact that $h^1(\mathcal{O}_{C_i}(-C_i)) = 0$. It follows that $n_i = 1$ for all i.

Finally, we show that there is just one curve C_i. If, say, D contains the curve $C_1 + C_2$, then again $|C_1 + C_2 + K_X| = \emptyset$. Arguing as above with the exact sequence

$$0 \to \mathcal{O}_X(-C_1 - C_2) \to \mathcal{O}_X \to \mathcal{O}_{C_1+C_2} \to 0,$$

we find that $h^1(\mathcal{O}_{C_1+C_2}) \leq q = g(C_i), i = 1, 2$. From the exact sequence

$$0 \to \mathcal{O}_{C_1}(-C_2) \to \mathcal{O}_{C_1+C_2} \to \mathcal{O}_{C_2} \to 0,$$

and the fact that $H^0(\mathcal{O}_{C_1+C_2}) \to H^0(\mathcal{O}_{C_2})$ is surjective we have $h^1(\mathcal{O}_{C_1}(-C_2)) = 0$. But $\deg \mathcal{O}_{C_1}(-C_2) = -d \leq 0$, with equality holding only if $\mathcal{O}_{C_1}(-C_2)$ is the trivial line bundle and in this case $h^0(\mathcal{O}_{C_1}(-C_2)) = 1$. Thus, by Riemann-Roch again

$$h^1(\mathcal{O}_{C_1}(-C_2)) = h^0(\mathcal{O}_{C_1}(-C_2)) + d + g(C_1) - 1 \geq 1.$$

We again contradict $h^1(\mathcal{O}_{C_1}(-C_2)) = 0$. It follows that there is just one curve C_i, and that $D = C_i$ is an irreducible curve with $K_X \cdot D < -1$ and $|K_X + D| = \emptyset$. $\square$

Completion of the proof of Theorem 20. Choose a curve D as in Claim 2. By Claim 1, if X is not ruled, either D is a section of the fibration π or $q = 1$ and $D \to C$ is unramified. First suppose that D is a section of π, so that $D \cdot f = 1$ for a general fiber f of π. In all cases, by the Riemann-Roch theorem,

$$h^0(D) + h^2(D) \geq 1 - q + \tfrac{1}{2}\left(D^2 - D \cdot K_X\right)$$
$$= 1 - q + \tfrac{1}{2}\left(D^2 + D \cdot K_X\right) - (D \cdot K_X)$$
$$= -q + p_a(D) - (D \cdot K_X) = -q + q - (D \cdot K_X) \geq 2.$$

Since $h^2(D) = h^0(K_X - D)$ and $h^0(K_X) = 0$, $h^2(D) = 0$. Thus, $|D|$ has dimension at least 1, and has no fixed curves since it contains the reduced irreducible element D. Thus, the base locus of $|D|$ is finite. On the other hand, choose a general fiber f of π such that $f \cap D$ is not a base point for $|D|$. Then the image of $|D|$ defines a morphism from f to $\mathbb{P}^1$ of degree 1, so that f is rational and X is ruled.

In the remaining case, $q = 1$, $p_g = 0$, and so $\chi(\mathcal{O}_X) = 0$. Take $\tilde{X}$ to be the fiber product $X \times_C D$. Then $f \colon \tilde{X} \to X$ is an unramified covering of X of degree d, say, with a section $\tilde{D}$ corresponding to the morphism $D \to X$. Since $f^* K_X = K_{\tilde{X}}$,

$$K_{\tilde{X}} \cdot \tilde{D} = f^* K_X \cdot \tilde{D} = K_X \cdot f_* \tilde{D} = K_X \cdot D < -1,$$

and thus $K_{\tilde{X}}$ is not nef. Moreover, $\tilde{X}$ is still minimal: an exceptional curve E on $\tilde{X}$ must lie in a fiber of $\tilde{X} \to D$, since it is rational, and thus $f|E\colon E \to f(E)$ is an isomorphism. Moreover, $f(E)^2 = E^2 = -1$. It follows that $f(E)$ is a smooth rational curve on X, necessarily exceptional, contradicting the minimality of X. Since $\tilde{X}$ is a minimal surface such that $K_{\tilde{X}}$ is not nef, we have seen that $p_g(\tilde{X}) = 0$ and so $h^2(\tilde{D}) = h^0(K_{\tilde{X}} - \tilde{D}) = 0$. On the other hand, it is easily checked from Noether's formula that $\chi(\mathcal{O}_{\tilde{X}}) = d\chi(\mathcal{O}_X) = 0$ and the Riemann-Roch argument given above in the case where D is a section implies that $h^0(\mathcal{O}_X(\tilde{D})) \geq 2$. It then follows by the above arguments that the general fiber of $\tilde{X} \to \tilde{D}$ is a smooth rational curve, and thus the same is true for the fiber of $X \to C$. Hence X is ruled. $\square$

We have now dealt with the case where K_X is not nef. We may thus assume that K_X is nef. In particular it follows that $K_X^2 \geq 0$. Let us restate the abundance theorem in a more precise form:

Theorem 21. *Let X be a minimal algebraic surface with K_X nef. Then $K_X^2 \geq 0$. Moreover:*

(i) *$\kappa(X) = 0$ if and only if K_X is numerically equivalent to 0. In this case X is a K3 surface, an abelian surface, an Enriques surface, or a hyperelliptic surface.*

(ii) *$\kappa(X) = 1$ if and only if $K_X^2 = 0$ but K_X is not numerically equivalent to 0. In this case X is an elliptic surface.*

(iii) *$\kappa(X) = 2$, i.e., X is of general type, if and only if $K_X^2 > 0$.*

In all cases, either $P_4(X)$ or $P_6(X) \neq 0$.

Proof. We consider the various possibilities, starting with (iii):

Case I: K_X is nef and $K_X^2 > 0$.

In this case, X is of general type by Proposition 10. We show that $P_4(X)$ or $P_6(X)$ is nonzero. If $p_g(X) \neq 0$, we are done. If $p_g(X) = 0$, we claim that in fact $P_2(X) \neq 0$. First we use the following slightly more general lemma:

Lemma 22. *Let X be a surface such that K_X is nef and $p_g(X) = 0$. Then $q(X) \leq 1$. If K_X is nef and $p_g(X) = 1$, then $q(X) \leq 2$.*

Proof. By Noether's formula,

$$c_1^2 + c_2 = K_X^2 + 2 - 4q + 2p_g + h^{1,1} = 12(1 - q + p_g)$$

and so $8q + h^{1,1} + K_X^2 = 10 + 10p_g$. As K_X is nef, $K_X^2 \geq 0$. Thus, if $p_g = 0$, then $q \leq 1$, and if $p_g = 1$, then $q \leq 2$. $\square$

Now by Proposition 10, if K_X is nef and big, then $P_2(X) = K_X^2 + \chi(\mathcal{O}_X)$. If moreover $p_g(X) = 0$, then $q \leq 1$, $\chi(\mathcal{O}_X) \geq 0$, and thus $P_2 \geq K_X^2 \geq 1$. In particular, $P_2 \neq 0$ in this case.

Case II: K_X is numerically equivalent to 0 and $q(X) = 0$.

By Castelnuovo's theorem, $P_2(X) \neq 0$. Thus, either K_X is trivial or K_X is not trivial but $2K_X$ is trivial. In the first case X is a $K3$ surface. In the second case there is an unramified double cover $\tilde{X}$ for which $K_{\tilde{X}}$ is trivial. Moreover, $c_2(X) = 12$. We shall shortly show that $\tilde{X}$ is a $K3$ surface, so that by definition X is an Enriques surface.

Case III: K_X is numerically equivalent to 0 and $q(X) > 0$.

In this case, consider the Albanese map $\alpha \colon X \to \mathrm{Alb}\, X$. If the image is not a curve, then $p_g(X) \neq 0$. It follows that K_X is trivial and that $p_g = 1$. By Lemma 22, $q(X) \leq 2$, and hence $q = 2$ since $\dim \mathrm{Alb}\, X \geq 2$. Thus, $\mathrm{Alb}\, X$ is a complex torus of dimension 2. Since the pullback of a nonzero 2-form on $\mathrm{Alb}\, X$ is nonzero, α is unramified. It follows that X is a complex torus and that α is an isomorphism.

Otherwise, the Albanese map is a fibration $X \to C$. If f is a general fiber, then $2g(f) - 2 = f^2 + K_X \cdot f = 0$. Thus, f is an elliptic curve and X is an elliptic surface. The elliptic surfaces X for which K_X is numerically trivial have been classified in Exercises 7 and 8 of Chapter 7. In all cases either $4K_X$ or $6K_X$ is trivial, and if $q(X) > 0$, then X is a hyperelliptic surface or a complex torus.

Finally, we have seen that if K_X is trivial, then X is a $K3$ surface or a complex torus. If X is a complex torus, then $b_2(X) = 6$. If Y is a quotient of X by a finite group, with $\alpha \colon X \to Y$ the quotient map, then $b_2(Y) \leq b_2(X)$, because the map $\alpha_* \alpha^*$ is multiplication by the degree of the cover and thus the kernel of α^* is torsion. Hence the case $q = p_g = 0$, $P_2 \neq 0$ above corresponds to Enriques surfaces (since in this case $b_2(Y) = 12 > 6$).

Case IV: K_X is not numerically equivalent to 0, $K_X^2 = 0$ and $q(X) = 0$.

In this case we shall prove that X is elliptic. We begin with two lemmas:

Lemma 23. *Let X be a surface such that K_X is nef and K_X is not numerically equivalent to 0. Then for all $n \geq 1$, $h^0(-nK_X) = 0$ and for all $n \geq 2$, $h^2(nK_X) = 0$. More generally, suppose that D is a divisor on X numerically equivalent to $nK_X + D'$, where D' is effective. If $n \geq 2$ or if $n = 1$ and D' is not 0, then $h^2(D) = 0$.*

Proof. Since K_X is not numerically trivial, there must exist an ample divisor H such that $K_X \cdot H \neq 0$, since every divisor is a difference of ample divisors. Moreover, since K_X is nef, we must have $K_X \cdot H > 0$. Thus, $H \cdot (-nK_X) < 0$ for every $n \geq 1$ and so $-nK_X$ is not effective. If $n \geq 2$, $h^2(nK_X) = h^0((1 - n)K_X)$ and $(1 - n)K_X$ is a negative multiple of K_X.

Thus, $(1 - n)K_X$ cannot be effective for $n > 1$ and so $h^2(nK_X) = 0$. The case of D numerically equivalent to $nK_X + D'$ is similar. $\square$

Lemma 24. *Let X be a surface such that K_X is nef, K_X is not numerically equivalent to 0, and $K_X^2 = 0$. If $P_n(X) \geq 2$ for some n, then X is elliptic.*

Proof. Choose an n such that $\dim |nK_X| \geq 1$. Quite generally, if $\sum_i a_i C_i \in |nK_X|$, then $K_X \cdot C_i \geq 0$ since K_X is nef and $K_X \cdot (\sum_i a_i C_i) = nK_X^2 = 0$. Thus, $K_X \cdot C_i = 0$ for all i. Since K_X is not numerically equivalent to 0, it follows from the Hodge index theorem that $C_i^2 \leq 0$ for all i, with $C_i^2 = 0$ if and only if C_i is numerically equivalent to a rational multiple of K_X.

Let E be the fixed component of $|nK_X|$. Thus, we can write $nK_X = D + E$, where $|D|$ has no fixed components and D is effective. It follows that $D^2 \geq 0$. On the other hand, since $K_X \cdot D = 0$, $D^2 \leq 0$, so that $D^2 = 0$ and D is numerically a rational multiple of K_X. In particular $|D|$ has no base points. Let $\varphi: X \to \mathbb{P}^N$ be the morphism defined by $|D|$. Since $D^2 = 0$, the image of φ is a curve. Taking the Stein factorization, there is a morphism $\pi: X \to C$ such that all fibers f of π satisfy: $f \cdot K_X = 0$. It follows that a smooth fiber f is elliptic, and thus that X is elliptic. $\square$

Returning to the study of Case IV, first suppose that $p_g(X) \neq 0$. Then for all $n \geq 0$, $h^0(nK_X) + h^2(nK_X) \geq 1 + p_g \geq 2$. On the other hand, for $n \geq 2$, $h^2(nK_X) = 0$ by Lemma 23. It follows that $h^0(nK_X) \geq 2$, and so, by Lemma 24, X is elliptic.

Thus, we can assume that $p_g(X) = 0$ and either that X is elliptic or that $P_n(X) \leq 1$ for every $n \geq 2$. Since $h^2(nK_X) = 0$ for $n \geq 2$, $h^0(nK_X) \geq 1$, and thus in fact $P_n(X) = 1$ for all $n \geq 2$. Let $\sum_i a_i C_i \in |2K_X|$ and let $\sum_i b_i C_i \in |3K_X|$, where $a_i, b_i \geq 0$. Then $3 \sum_i a_i C_i = 2 \sum_i b_i C_i$ in $|6K_X|$, and so there exists $c_i \geq 0$ with $a_i = 2c_i$ and $b_i = 3c_i$. Hence $\sum_i b_i C_i - \sum_i a_i C_i = \sum_i c_i C_i$ is an effective curve in K_X, contradicting the fact that $h^0(K_X) = 0$. So X is elliptic, and moreover we have shown directly (without appealing to Castelnuovo's theorem) that $P_2(X) = 1$, and in particular that $P_2(X) \neq 0$.

Case V: K_X is not numerically equivalent to 0, $K_X^2 = 0$ and $q(X) > 0$.

We shall show in this case that X is elliptic as well. First, if $P_n \geq 2$ for some n, then X is elliptic by Lemma 24. Thus, we may as well assume that $P_n \leq 1$ for all n. In particular $p_g \leq 1$. Moreover, we have $h^2(nK_X) = 0$ for all $n \geq 2$ and thus $h^0(nK_X) \geq \chi(\mathcal{O}_X)$. Thus, we can assume that $\chi(\mathcal{O}_X) \leq 1$. Thus, if $p_g = 1$, then $q \geq 2$, and in fact, by Lemma, 22 $q = 2$. In this case we have:

Lemma 25. *If X is a surface with K_X nef, $p_g(X) = 1$, $q(X) = 2$, and $K_X^2 = 0$ but K_X is not numerically equivalent to 0, then X is elliptic.*

Proof. As in the proof of Lemma 24, if $K_X = \sum_i a_i C_i$, then $K_X \cdot C_i = 0$ and the span of the C_i is negative semidefinite with a radical spanned over $\mathbb{Q}$ by K_X. In particular, $C_i^2 \leq 0$, and either $C_i^2 = 0$ and $C_i \cdot C_j = 0$ for $i \neq j$ or $C_i^2 < 0$, C_i is smooth rational, and all of the curves C_j which lie on the same connected component of $\sum_i a_i C_i$ as C_i are smooth rational curves which span a negative semidefinite lattice whose radical is a rational multiple of K_X.

Consider the Albanese map $\alpha \colon X \to \text{Alb}\, X$. If one of the curves C_i is smooth rational, then $\alpha(C_i) = 0$ and in fact the entire component through C_i must be mapped to a point. By the Hodge index theorem, the image of α is a curve. Hence the Albanese map is a fibration $\pi \colon X \to C$ (over a curve of genus 2) and, if f is a general fiber, then $K_X \cdot f = 0$. Thus, f is an elliptic curve and X is elliptic.

We may therefore assume that all of the curves C_i are numerically equivalent to a rational multiple of K_X and that $C_i^2 = 0$. If $\alpha(C_i)$ is a point for some i, for example if some C_i is singular, then the above argument shows that the Albanese map is an elliptic fibration. Thus, all of the C_i are elliptic curves and $\alpha(C_i)$ is a curve for every i. Since every morphism from one complex torus to another is a homomorphism followed by a translation, we can assume that $\alpha(C_i) = E$ is a smooth elliptic curve passing through the origin of $\text{Alb}\, X$. There is thus an induced morphism $X \to \text{Alb}\, X/E$ for which the image of C_i is a point. Let $\pi \colon X \to C$ be the Stein factorization. Then π is a fibration, and there exists a fiber containing C_i. Since $C_i^2 = 0$, the complete fiber through C_i is of the form aC_i for some positive integer a. It follows that, if f is a general fiber of π, then $K_X \cdot f = 0$ and so f is elliptic. Once again, we see that X is elliptic. $\square$

The remaining case is where K_X is not numerically equivalent to 0, $K_X^2 = 0$, $q(X) > 0$, and $p_g(X) = 0$. If $p_g(X) = 0$, then we have $q \leq 1$ by Lemma 22, and so (since we are assuming $q > 0$) $q = 1$. In this case, the most difficult case in the classification, we have the following:

Theorem 26. *If X is a minimal surface with $p_g(X) = 0$, $q(X) = 1$, and $K_X^2 = 0$, then X is either ruled over an elliptic curve or X is elliptic.*

Proof. Since $c_1^2 = \chi(\mathcal{O}_X) = 0$, $c_2(X) = 0$ as well, and thus, as $b_1(X) = 2$, $b_2(X) = 2$. We may assume that K_X is not numerically equivalent to 0, since hyperelliptic surfaces were shown to be elliptic previously. Let $\pi \colon X \to E$ be the Albanese fibration, where E is a smooth elliptic curve. Since $b_2(X) = 2$, all fibers of π are irreducible. Using the last statement of Exercise 5 in Chapter 7, if π is the Albanese map, and if f is a general fiber of π with $g(f) = g$, then $c_2(X) \geq (2g-2)\chi(E) = 0$, with equality only if all fibers have smooth reduction of genus g. We may assume that $g > 1$, for otherwise X is either ruled over an elliptic base or is elliptic. If nF is a multiple fiber, then $2g - 2 = K_X \cdot f = n(K_X \cdot F) = n(2g-2)$. Thus, $n = 1$

and π is a smooth morphism. The main problem now is that, unlike the previous cases, the elliptic fibration is not in general given by the Albanese fibration, and we shall have to construct it by other means. There are two approaches to finding the desired elliptic fibration, an analytic approach and an algebraic approach. We shall sketch both.

Analytic Proof. We have found a smooth morphism $\pi\colon X \to E$, whose fibers are curves of genus $g > 1$. There is an associated period map: Let $\mathfrak{H}_g$ be the Siegel upper half space of symmetric complex $g \times g$ matrices $Z = X + iY$ such that Y is positive definite. It is known that $\mathfrak{H}_g$ is biholomorphic to a bounded domain in $\mathbb{C}^{g(g+1)/2}$. Let Γ_g be the group $Sp(2g, \mathbb{Z})$. Then Γ_g acts properly discontinuously on $\mathfrak{H}_g$. To a smooth curve C of genus g, its period matrix is well defined mod Γ_g, and thus defines a point of $\Gamma_g \backslash \mathfrak{H}_g$. Given a smooth morphism $X \to E$, all of whose fibers have genus g, there is an associated holomorphic map $p\colon E \to \Gamma_g \backslash \mathfrak{H}_g$, the *period map*. Moreover, p is locally liftable in the following sense: if $\psi\colon \mathbb{C} \to E$ is the universal cover, with covering group $\Lambda \cong \mathbb{Z} \oplus \mathbb{Z}$, then there is a holomorphic map $\tilde{p}\colon \mathbb{C} \to \mathfrak{H}_g$ and a homomorphism $\rho\colon \Lambda \to \Gamma_g$, such that $\tilde{p}(z + \lambda) = \rho(\lambda) \cdot \tilde{p}(z)$. Thus, there is an induced map $\mathbb{C}/\Lambda = E \to \Gamma_g \backslash \mathfrak{H}_g$, which is just p. Moreover, there is a rank g vector bundle V over $\mathfrak{H}_g$, which is holomorphically trivial, together with a natural Γ_g-action. Thus, Λ acts on $\tilde{p}^* V$, and the induced bundle $\tilde{p}^* V / \Lambda$ is equal to the rank g vector bundle $R^0 \pi_* \Omega^1_{X/E}$.

The special feature of our situation is the following: the universal cover of E is $\mathbb{C}$, and there are no nonconstant holomorphic functions from $\mathbb{C}$ into a bounded domain in $\mathbb{C}^N$. Thus, $\tilde{p}$ is constant. If $\operatorname{Im} \tilde{p} = \{x\}$, it follows that $\operatorname{Im} \rho$ is contained in the isotropy subgroup of x in Γ_g. But as Γ_g acts properly discontinuously, the isotropy subgroup of a point is finite. Thus, there is a subgroup Λ' of Λ of finite index such that $\rho(\Lambda') = \{\operatorname{Id}\}$. Let $E' \to E$ be the induced unramified cover and let $\pi'\colon X' \to E'$ be the pulled-back family. It follows that $R^0 (\pi')_* \Omega^1_{X'/E'}$ is the trivial rank g vector bundle.

At this point, there are several ways to proceed. One way is to check that X' is a surface such that $K_{X'}$ is nef, $K_{X'}^2 = 0$, and $K_{X'}$ is not numerically trivial, since $K_{X'}$ is pulled back from X by the unramified covering map. But $p_g(X') \neq 0$, since $K_{X'} = \omega_{X'/E'} \otimes (\pi')^* K_{E'} = \Omega^1_{X'/E'}$, since E' is an elliptic curve and so $K_{E'}$ is trivial. So we can apply the previous classification results to conclude that X' is elliptic, with $\kappa(X') = 1$, and that X is the quotient of X' by a group of holomorphic covering transformations. It is easy to check that the elliptic fibration defined by $|nK_{X'}|$ for n large is preserved by the covering transformations, and thus induces an elliptic structure on X. Another way to proceed is to use the fact that $R^0 (\pi')_* \Omega^1_{X'/E'}$ is a trivial rank g vector bundle, so that $H^0(\Omega^1_{X'/E'}) = \mathbb{C}^g$. There is thus a morphism $X' \to \mathbb{P}^{g-1}$ and its image is a curve C of genus g, since $c_1(\Omega^1_{X'/E'})^2 = c_1(\Omega^1_{X'/E'} \otimes (\pi')^* K_{E'})^2 = K_{X'}^2 = 0$. It follows that there is a morphism $X' \to E' \times C$ which is in fact an isomorphism, and X is a quotient of $E' \times C$. Keeping track of the possible automorphisms shows that X is an elliptic surface over C/G for some subgroup G of the group of automorphisms of C. Thus, X is again elliptic.

Algebraic Proof. We begin with a series of lemmas:

Lemma 27. *Let D be a divisor numerically equivalent to $nK_X + F$, where F is effective and $F^2 = K_X \cdot F = 0$. Suppose that either $n \geq 2$ or $n = 1$ and F is not the 0 divisor. Then $h^0(D) = h^1(D)$.*

Proof. By Lemma 23, we know that $h^2(D) = 0$. By assumption, $\chi(\mathcal{O}_X) = 0$ and $D^2 = D \cdot K_X = 0$ as well. Thus, by Riemann-Roch, $h^0(D) - h^1(D) = 0$. $\square$

Lemma 28. *For all $n \geq 2$, there exists a divisor D numerically equivalent to nK_X such that $h^0(D) \neq 0$.*

Proof. We shall show that there exists a line bundle λ of degree 0 on E such that $h^0(\mathcal{O}_X(nK_X) \otimes \pi^* \lambda) \neq 0$. By Lemma 27, it suffices to find a λ such that $h^1(\mathcal{O}_X(nK_X) \otimes \pi^* \lambda) \neq 0$. To do so, let us fit together all of the line bundles $\mathcal{O}_X(nK_X) \otimes \pi^* \lambda$: choose a point $t \in E$, and let $\mathcal{P}$ be the line bundle $\mathcal{O}_{E \times E}(\Delta - (\{t\} \times E))$ on $E \times E$. Thus, on each slice $E \times \{s\}$, $\mathcal{P}|E \times \{s\}$ is essentially just $\mathcal{O}_E(s - t)$, and as s ranges over E these line bundles range over all of the line bundles on E of degree 0. Let $\pi_1 \colon X \times E \to X$ and $\pi_2 \colon X \times E \to E$ be the first and second projections. There is also the induced map $p \colon X \times E \to E \times E$ defined by $(\pi \circ \pi_1, \pi_2)$. Consider the line bundle $\mathcal{L} = \pi_1^* \mathcal{O}_X(nK_X) \otimes p^* \mathcal{P}$ on $X \times E$. For each fixed $s \in E$, the restriction of $\mathcal{L}$ to $X \times \{s\}$ can be identified with $\mathcal{O}_X(nK_X) \otimes \pi^* \mathcal{O}_E(s-t)$, and as s runs over E the line bundles $\mathcal{O}_E(s - t)$ run over all of the line bundles of degree 0 on E.

Given $n \geq 2$, we may assume that $P_n(X) = 0$, for otherwise we can take λ to be the trivial line bundle. In particular, by the semicontinuity theorem $R^0 \pi_{2*} \mathcal{L}$ is a torsion free sheaf on E which is zero in a neighborhood of t, and thus $R^0 \pi_{2*} \mathcal{L} = 0$. By Lemma 23, $H^2(\mathcal{O}_X(nK_X) \otimes \pi^* \lambda) = 0$ for all λ of degree 0 and hence $R^2 \pi_{2*} \mathcal{L} = 0$. Finally, if λ is trivial, then $H^1(nK_X) = 0$ by Lemma 28, since $h^1(nK_X) = P_n(X) = 0$, and so $R^1 \pi_{2*} \mathcal{L}$ is a torsion sheaf. Moreover, if $H^1(\mathcal{O}_X(nK_X) \otimes \pi^* \lambda) = 0$ for all λ of degree 0, then $R^1 \pi_{2*} \mathcal{L} = 0$. Thus, it suffices to show that $R^1 \pi_{2*} \mathcal{L} \neq 0$.

By applying the Leray spectral sequence to π_2 and $\mathcal{L}$, we see that

$$\chi(\mathcal{L}) = \sum_{p,q} (-1)^{p+q} h^p(E; R^q \pi_{2*} \mathcal{L}) = -h^0(E; R^1 \pi_{2*} \mathcal{L}),$$

since $R^q \pi_{2*} \mathcal{L} = 0$, $q \neq 1$, and $R^1 \pi_{2*} \mathcal{L}$ is a skyscraper sheaf. Thus, it suffices to show that $\chi(\mathcal{L}) \neq 0$. But we can calculate $\chi(\mathcal{L})$ by using the

Riemann-Roch theorem on the threefold $X \times E$:

$$\chi(\mathcal{L}) = \int_{X \times E} \text{ch}(\mathcal{L}) \cdot \text{Todd}(X \times E)$$

$$= \int_{X \times E} \left(1 + \mathcal{L} + \frac{\mathcal{L}^2}{2} + \frac{\mathcal{L}^3}{6}\right) \cdot \pi_1^* \text{Todd}(X) \pi_2^* \text{Todd}(E)$$

$$= \int_{X \times E} \left(1 + \mathcal{L} + \frac{\mathcal{L}^2}{2} + \frac{\mathcal{L}^3}{6}\right) \cdot \left(1 - \frac{\pi_1^* K_X}{2}\right) = \frac{\mathcal{L}^3}{6} - \frac{\mathcal{L}^2 \cdot \pi_1^* K_X}{4},$$

where we have used the multiplicativity of the Todd genus to conclude that

$$\text{Todd}(X \times E) = \pi_1^* \text{Todd}(X) \pi_2^* \text{Todd}(E).$$

Since T_E is trivial, $\text{Todd}(E) = 1$, and since $c_2(X) = 0$, $\text{Todd}(X) = 1 - K_X/2$. Now $\mathcal{L} = \pi_1^*(nK_X) \otimes p^* \mathcal{O}_{E \times E}(\Delta - (\{t\} \times E))$. Moreover, $K_X^2 = 0$ and $(\Delta - (\{t\} \times E))^2 = \Delta^2 - 2\Delta \cdot (\{t\} \times E) = -2$ because $\Delta^2 = 0$ (apply adjunction to $\Delta \subset E \times E$). Thus,

$$\mathcal{L}^3 = 3\pi_1^*(nK_X) \cdot p^*(\Delta - \{t\} \times E)^2 = -6(nK_X \cdot \pi^*(t))$$
$$= -6(nK_X \cdot f) = -6n(2g - 2) = -12n(g - 1).$$

Likewise,

$$\mathcal{L}^2 \cdot \pi_1^* K_X = -2(K_X \cdot \pi^*(t)) = -2(K_X \cdot f) = -4(g - 1),$$

and putting this together we find that

$$\chi(\mathcal{L}) = -2n(g - 1) + (g - 1) = -(2n - 1)(g - 1).$$

It follows that the length of $R^1\pi_{2*}\mathcal{L}$ is $(2n-1)(g-1) \neq 0$, and so there exists a λ on E of degree 0 such that $h^0(\mathcal{O}_X(nK_X) \otimes \pi^*\lambda) \neq 0$, as desired. $\square$

By Lemma 28, there exists an effective divisor D numerically equivalent to nK_X. Suppose that $D = \sum_i a_i C_i$ where the C_i are irreducible curves and $a_i > 0$. As in the proof of Lemma 25, since K_X is nef, the connected components of D are all numerically equivalent up to a positive rational number, and the irreducible curves in a connected component span a negative semidefinite lattice. Thus, either $C_i^2 < 0$ or C_i is a connected component of the support of D and $C_i^2 = 0$. If $C_i^2 < 0$, then C_i is smooth rational and thus lies in a fiber of the morphism $\pi \colon X \to E$. Using the first part of the proof, all of the fibers of π are irreducible curves of arithmetic genus g, so this case is impossible. Thus, C_i is a connected component of the fiber for every i and $C_i^2 = 0$. Hence C_i is an irreducible curve of arithmetic genus 1, numerically equivalent to a positive rational multiple of K_X. (In fact, C_i must be smooth, since otherwise its normalization would be rational, and thus C_i would be contained in a fiber of the Albanese map, which as we have seen is impossible.)

Consider the set of all irreducible curves D on X such that D has arithmetic genus 1 and $D^2 = 0$, or equivalently such that $D \cdot K_X = 0$.

Such a curve must be numerically equivalent to a positive rational multiple of K_X. Moreover, by adjunction $\mathcal{O}_X(K_X + D)|D = \mathcal{O}_D$. Hence $\mathcal{O}_X(nK_X + nD)|D = \mathcal{O}_D$ for all positive integers n. Suppose that D_1 and D_2 are two curves on X of arithmetic genus 1 with $D_i^2 = 0$, $i = 1, 2$. Then both D_1 and D_2 are numerically equivalent to a rational multiple of K_X and in particular $D_1 \cdot D_2 = 0$. Thus, either D_1 and D_2 are disjoint or $D_1 = D_2$. Suppose that we can find $D_1 \neq D_2$. Consider the exact sequence

$$0 \to \mathcal{O}_X(2K_X + D_1 + D_2) \to \mathcal{O}_X(2K_X + 2D_1 + 2D_2) \to \mathcal{O}_{D_1} \oplus \mathcal{O}_{D_2} \to 0.$$

Since $H^2(2K_X + D_1 + D_2) = 0$, it follows that the map $H^1(\mathcal{O}_X(2K_X + 2D_1 + 2D_2)) \to H^1(\mathcal{O}_{D_1}) \oplus H^1(\mathcal{O}_{D_2}) = \mathbb{C} \oplus \mathbb{C}$ is surjective. Moreover, $h^1(\mathcal{O}_X(2K_X + 2D_1 + 2D_2)) = h^0(\mathcal{O}_X(2K_X + 2D_1 + 2D_2))$, and so $|2K_X + 2D_1 + 2D_2|$ contains a pencil. Arguing as in the proof of Lemma 24, it follows that the moving part of the pencil defines a morphism $X \to B$ whose general fiber is an elliptic curve, and thus that X is elliptic.

Thus, we are reduced to showing that there exist two disjoint curves D_1, D_2 on X with $p_a(D_i) = 1$ and $D_i^2 = 0$, $i = 1, 2$. We know that one such curve D_1 exists. For $n \geq 2$, consider the linear system $|nK_X + nD_1|$. By adjunction $K_X \otimes \mathcal{O}_X(D_1)|D_1$ is the trivial line bundle, and thus the same is true for $\mathcal{O}_X(nK_X + nD_1)|D_1$. Consider the exact sequence

$$0 \to \mathcal{O}_X(nK_X + (n-1)D_1) \to \mathcal{O}_X(nK_X + nD_1) \to \mathcal{O}_{D_1} \to 0.$$

Since $H^2(nK_X + (n-1)D_1) = 0$ for all $n \geq 2$ by Lemma 23 and $H^1(\mathcal{O}_{D_1}) = \mathbb{C}$, $h^1(nK_X + nD_1) \geq 1$. By Lemma 27, $|nK_X + nD_1| \neq \emptyset$. We have seen that, if C is an irreducible component of an element of $|nK_X + nD_1|$, then C is a curve of arithmetic genus 1 with $C^2 = C \cdot K_X = C \cdot D_1 = 0$. If there is no such component C which is disjoint from D_1, then, for every n, $nK_X + nD_1 = mD_1$ for some positive integer m. Applying the same argument to $n + 1$, $(n + 1)K_X + (n + 1)D_1$ is linearly equivalent to $m'D_1$ for some integer m', and thus K_X is linearly equivalent to $(m' - m - 1)D_1$. Since $K_X \cdot H > 0$ for every ample H, we must have $m' - m - 1 > 0$. In this case K_X is effective, and so $p_g(X) \neq 0$. But we have already dealt with the case $p_g(X) > 0$ just after the proof of Lemma 24. Thus, in case $p_g(X) = 0$, we have found two disjoint curves as desired.   □

We have showed that, if K_X is nef, $K_X^2 = 0$, $p_g(X) = 0$, and $q(X) = 1$, X is elliptic and then all fibers of the elliptic fibration are smooth or multiple curves with smooth reduction. Such surfaces do exist and are given as follows: by the canonical bundle formula,

$$K_X = \pi^*(L \otimes K_C) \otimes \mathcal{O}_X\left(\sum_i (m_i - 1)F_i\right),$$

where $\deg L = 0$ and the m_i are the multiple fibers of multiplicity m_i. It follows that $p_g(X) = h^0(C; L \otimes K_C)$. Thus, if $g(C) \geq 1$, then $p_g(X) \geq 1$. If $g(C) = 1$, then as L has order 4 or 6 we see that $P_4(X)$ or $P_6(X)$ is nonzero. However, if X has a multiple fiber, then $P_2(X) \neq 0$ in this case. If X does not have a multiple fiber, then K_X is numerically equivalent to 0. If $g(C) = 0$, then L is necessarily trivial. To find an example of such surfaces, let C be a curve of genus $g \geq 2$ such that there exists an automorphism φ of order d with $C/\langle\varphi\rangle = \mathbb{P}^1$ (for example, if $d = 2$, this just says that C is a hyperelliptic curve). Let E be an elliptic curve and ξ a point of order d on E. Then $\mathbb{Z}/d\mathbb{Z}$ acts diagonally on $C \times E$, where the action on the first factor is via φ and on the second by addition by ξ. The quotient $(C \times E)/(\mathbb{Z}/d\mathbb{Z}) = X$ fibers over $C/\langle\varphi\rangle = \mathbb{P}^1$, and the general fiber is a smooth elliptic curve. However, there are multiple fibers at the fixed points of φ, with multiplicity equal to the ramification index. The quotient surface X also fibers over $E/\langle\xi\rangle = E'$, where E' is an elliptic curve, and we leave it as an exercise to show that the morphism $X \to E'$ is the Albanese fibration, with all fibers isomorphic to C.

To conclude the proof of Theorem 21, we must check that either $P_4(X)$ or $P_6(X)$ is nonzero in case $p_g(X) = 0$ and $q(X) = 1$. By the above remarks, we can assume that X is an elliptic surface over $\mathbb{P}^1$ whose associated line bundle L is trivial. Thus, K_X is numerically equivalent to rf, where

$$r = -2 + \sum_i \left(\frac{m_i - 1}{m_i}\right).$$

By assumption $r \geq 0$. The case $r = 0$ has been discussed in Exercise 7 in Chapter 7, and corresponds to multiplicities

$$(2, 2, 2, 2), (2, 4, 4), (2, 3, 6), (3, 3, 3).$$

In all cases $P_4(X)$ or $P_6(X)$ is 1. Again by Exercise 7 in Chapter 7, there are at least three multiple fibers, and the following cases are excluded: $(2, 2, m), m \geq 2$; $(2, 3, 4)$, $(2, 3, 5)$. Now suppose that there are r multiple fibers F_i, with multiplicity $m_i \geq 2$. Thus, $K_X = (r - 2)f - \sum_i F_i$, with $r \geq 3$. If $r \geq 4$, then $2K_X = 2(r - 2)f - 2\sum_i F_i$. Since $r \geq 4$, $2(r - 2) \geq r$, and so $2(r - 2) = r + s$ with $s \geq 0$. Thus,

$$2K_X = sf + \sum_i (f - 2F_i) = sf + \sum_i (m_i - 2)F_i,$$

which is effective as $m_i \geq 2$. If $r = 3$, the possibilities are $3 \leq m_1 \leq m_2 \leq m_3$, $m_1 = 2$, $4 \leq m_2 \leq m_3$ or $m_1 = 2$, $m_2 = 3$, $m_3 \geq 6$. In the first case

$$3K_X = 3f - 3F_1 - 3F_2 - 3F_3 = (m_1 - 3)F_1 + (m_2 - 3)F_2 + (m_3 - 3)F_3,$$

which is effective since $m_i \geq 3$. In the second case,

$$4K_X = 4f - 4F_1 - 4F_2 - 4F_3 = 2f - 2f + (m_2 - 4)F_2 + (m_3 - 4)F_3$$
$$= (m_2 - 4)F_2 + (m_3 - 4)F_3,$$

which is effective since $m_2 \geq 4$ and $m_3 \geq 4$. In the last case,

$$6K_X = 6f - 6F_1 - 6F_2 - 6F_3 = 6f - 3f - 2f - 6F_3 = (m_3 - 6)F_3,$$

which is again effective as $m_3 \geq 6$. Thus, in all cases, either $P_4(X)$ or $P_6(X)$ is nonzero. $\square$

The Castelnuovo-deFranchis theorem

We recall the statement of the Castelnuovo-deFranchis theorem:

Theorem 29. *Let X be an algebraic surface with $\kappa(X) \geq 0$. Then $c_2(X) \geq 0$, with equality holding if and only if X is a minimal elliptic surface whose only singular fibers are multiple fibers with smooth reduction or X is an abelian surface.*

Proof. We may clearly assume that X is minimal, since c_2 can only increase under blowing up. The theorem is clear if $q(X) = 0$, since then $c_2(X) = 2 + b_2(X) > 0$. Next suppose that the Albanese morphism π has image a curve C. Then $g(C) = q \geq 1$ and the general fiber of π has genus $g \geq 1$ since X is not ruled. By Exercise 5 in Chapter 7,

$$c_2(X) = \chi(X) \geq (2 - 2g)(2 - 2q) = (2g - 2)(2q - 2) \geq 0.$$

Moreover, equality can hold only if either $g = 1$ and all fibers of π are multiple fibers with smooth reduction (by Corollary 17 in Chapter 7) or $q = 1$ and $c_2(X) = 0$. In this case $4 = 4q = 2 + 2p_g + h^{1,1}$. Since $h^{1,1} \neq 0$, $p_g(X) = 0$ and $\chi(\mathcal{O}_X) = 0$. Thus, by Noether's formula $K_X^2 = 0$. Since X is minimal, X is elliptic by Theorem 26. Again by Corollary 17 in Chapter 7, the only singular fibers in the elliptic fibration are multiple fibers with smooth reduction. Thus, we have proved the theorem in case the Albanese image of X is a curve, and the above proof also works in case there is a morphism from X to a curve C of genus at least 2.

Next suppose that there exist two linearly independent holomorphic 1-forms η_1, η_2 on X such that $\eta_1 \wedge \eta_2 = 0$. We claim that in this case there is a morphism from X to a smooth curve C of genus at least 2, and hence that the theorem is true in this case as well. More precisely, we have:

Lemma 30. *Suppose that X is a smooth surface and that η_1, η_2 are two linearly independent holomorphic 1-forms on X such that $\eta_1 \wedge \eta_2 = 0$. Then there exists a fibration $\pi \colon X \to C$, where C is a smooth curve of genus $g \geq 2$, and two holomorphic 1-forms ψ_1, ψ_2 on C such that $\eta_i = \pi^* \psi_i$, $i = 1, 2$.*

Proof. Let $\mathbb{C}(X)$ be the function field of X, so that $\mathbb{C}(X)$ is a finitely generated field extension on $\mathbb{C}$ of transcendence degree 2. Let $z_1, z_2 \in \mathbb{C}(X)$ be two algebraically independent meromorphic functions on X. Then the module of Kähler differentials $\Omega^1_{\mathbb{C}(X)/\mathbb{C}}$ is a vector space of dimension 2 over $\mathbb{C}(X)$, with basis dz_1, dz_2. Moreover, two elements $\alpha_1, \alpha_2 \in \Omega^1_{\mathbb{C}(X)/\mathbb{C}}$

are linearly independent over $\mathbb{C}(X)$ if and only if $\alpha_1 \wedge \alpha_2 \neq 0$ in $\Omega^2_{\mathbb{C}(X)/\mathbb{C}}$. Now the 1-forms η_i induce elements of $\Omega^1_{\mathbb{C}(X)/\mathbb{C}}$, by restriction to the generic point of X, which we will continue to denote by η_i, and clearly $\eta_1 \wedge \eta_2 = 0$ in $\Omega^2_{\mathbb{C}(X)/\mathbb{C}}$. Thus, there exists $f \in \mathbb{C}(X), f \neq 0$ such that $\eta_1 = f\eta_2$. A holomorphic 1-form on X is closed, and so

$$0 = d\eta_1 = df \wedge \eta_2.$$

Thus, $\eta_2 = g\, df$ for some $g \in \mathbb{C}(X)$. Since η_2 is closed as well,

$$0 = d\eta_2 = dg \wedge df,$$

and thus df and dg are linearly dependent in $\Omega^1_{\mathbb{C}(X)/\mathbb{C}}$. It follows that $\Omega^1_{\mathbb{C}(f,g)/\mathbb{C}}$ has dimension 1 over $\mathbb{C}(f,g)$, and thus that the transcendence degree of $\mathbb{C}(f,g)$ over $\mathbb{C}$ is 1. The inclusion $\mathbb{C}(f,g) \subset \mathbb{C}(X)$ defines a rational map $\varphi \colon X \dashrightarrow D$, where D is the smooth curve corresponding to the function field $\mathbb{C}(f,g)$. Let $\tilde{X} \to D$ be a blowup for which φ becomes a morphism, and let $\tilde{\pi} \colon \tilde{X} \to C$ be the Stein factorization of $\tilde{X} \to D$. The 1-forms $fg\, df$ and $g\, df$ are meromorphic 1-forms on D which pull back to meromorphic 1-forms ψ_1, ψ_2 on C, and then further under $\tilde{\pi}^*$ to the forms η_1, η_2. Since η_1 and η_2 are holomorphic and linearly independent, it is easy to see that ψ_1 and ψ_2 are two linearly independent holomorphic 1-forms on C. But then necessarily $g(C) \geq 2$. In particular, every exceptional curve on $\tilde{X}$ from the resolution of indeterminacy must map to a point of C, and so $\tilde{\pi}$ factors through a morphism $\pi \colon X \to C$. By construction $\eta_i = \pi^*\psi_i$ on a dense open subset of X, and hence everywhere, as claimed. $\square$

Thus, we may assume that, whenever η_1 and η_2 are two linearly independent holomorphic 1-forms on X, $\eta_1 \wedge \eta_2 \neq 0$. In this case, we claim that $p_g \geq 2q - 3$. To see this, let $V = H^0(\Omega^1_X)$, with $\dim V = q$ and let $W = H^0(\Omega^2_X)$ with $\dim W = p_g$. Inside $\bigwedge^2 V$, there is the cone $\mathcal{C}$ of elements of the form $v_1 \wedge v_2$, where v_1 and v_2 are two linearly independent vectors inside V, and clearly $\mathcal{C}$ is the affine cone over the Grassmannian $G(2,q)$ of 2-dimensional subspaces of V. Thus, $\mathcal{C}$ has dimension $\dim G(2,q) + 1 = 2(q-2) + 1 = 2q - 3$. There is the natural map $\bigwedge^2 V \to W$. Let K be its kernel and let I be the image. By hypothesis, $K \cap \mathcal{C} = \{0\}$, so that $\dim K + \dim \mathcal{C} \leq \dim \bigwedge^2 V$. On the other hand, $\dim K + \dim I = \dim \bigwedge^2 V$. Hence

$$\dim K + \dim \mathcal{C} \leq \dim \bigwedge^2 V = \dim K + \dim I \leq \dim K + p_g.$$

It follows that $\dim \mathcal{C} = 2q - 3 \leq p_g$.

Thus, if X is any surface for which $c_2(X) \leq 0$,

$$c_2(X) = 2 - 4q + 2p_g + h^{1,1} \geq 2 - 4q + 2(2q-3) + h^{1,1} = h^{1,1} - 4 \geq -3.$$

On the other hand, suppose that X is a surface such that $c_2(X) < 0$. Then in particular $q(X) > 0$, so that $H_1(X; \mathbb{Z})$ has rank at least 1. It follows that, for every $n > 0$, there exists a surjective homomorphism $\pi_1(X) \to \mathbb{Z}/n\mathbb{Z}$, and thus an unramified cover $\tilde{X}$ of X of degree n. An elementary Euler characteristic argument shows that $c_2(\tilde{X}) = nc_2(X) < -n$. Applying the above argument to $\tilde{X}$, we see that $-n > c_2(\tilde{X}) \geq -3$, which is absurd as soon as $n \geq 4$. Hence, we must have $c_2(X) \geq 0$.

Finally, we consider the case where $c_2(X) = 0$. If $K_X^2 = 0$ as well, then $\kappa(X) = 0$ or 1, and by the classification of such surfaces either X is an abelian surface or X is an elliptic surface with all fibers smooth or multiple with smooth reduction. Otherwise, $K_X^2 > 0$, and so by Noether's formula $1 + p_g > q$, or in other words $p_g \geq q$. Next, we claim the following:

Lemma 31. *Suppose that X is a smooth surface and that there does not exist a fibration $\pi \colon X \to C$, where C is a smooth curve of genus $g \geq 2$. Then $h^{1,1}(X) \geq 2q - 1$.*

Proof that Lemma 31 implies the Castelnuovo-deFranchis theorem. If there exists a fibration $\pi \colon X \to C$, where C is a smooth curve of genus $g \geq 2$, then either X is elliptic or the fibers f of π have genus at least 2. We have already handled the elliptic case above, and the other case follows from the inequality $c_2 \geq (2g(f) - 2)(2g(C) - 2)$. So we can assume that $p_g \geq 2q - 3$ and that $h^{1,1} \geq 2q - 1$. It follows that

$$c_2(X) = 2 - 4q + 2p_g + h^{1,1} \geq 2q - 5.$$

If $q = 0$ or $q \geq 3$, then automatically $c_2(X) > 0$. If $q = 1$, then $p_g \geq 1$ and $h^{1,1} \geq 1$. Thus, $c_2(X) \geq 2 - 4 + 2 + 1 \geq 1$. If $q = 2$, then $p_g \geq 2$ and $h^{1,1} \geq 3$, so that $c_2(X) \geq 2 - 8 + 4 + 3 \geq 1$. In all cases, $c_2(X) > 0$. $\square$

Proof of Lemma 31. Let V_1 be the complex vector space $H^0(\Omega_X^1) = H^{1,0}(X)$ and let V_2 be the conjugate vector space $H^{0,1}(X)$. Wedge product $(\eta_1, \eta_2) \mapsto \eta_1 \wedge \eta_2$ induces a complex linear map $V_1 \otimes V_2 \to W = H^{1,1}(X)$. Suppose that there exist $\eta_1 \in V_1, \eta_2 \in V_2$ such that $\eta_i \neq 0$, $i = 1, 2$, but that $\eta_1 \wedge \eta_2 = 0$ as an element of $H^{1,1}(X) \subseteq H^2(X; \mathbb{C})$, in other words $\eta_1 \wedge \eta_2$ is exact. It follows that

$$\eta_1 \wedge \eta_2 \wedge \overline{\eta_1 \wedge \eta_2} = \eta_1 \wedge \eta_2 \wedge \bar{\eta}_2 \wedge \bar{\eta}_1$$

is also exact. Set $\eta_1 \wedge \bar{\eta}_2 = \omega$, so that

$$\eta_1 \wedge \eta_2 \wedge \bar{\eta}_2 \wedge \bar{\eta}_1 = \omega \wedge \bar{\omega}.$$

Since ω is a holomorphic 2-form, $\omega \wedge \bar{\omega} = 0$ only if $\omega = 0$. In this case, by Lemma 30, either there is a fibration $\pi \colon X \to C$, where C is a smooth curve of genus $g \geq 2$, or $\bar{\eta}_2$ is a multiple of η_1. In this case $\eta_1 \wedge \bar{\eta}_1$ is exact. But since η_1 is closed, locally we can write $\eta_1 = df$ for some holomorphic function f, and after replacing f by $f + c$ for some constant c we can assume that there is a nonempty open set where $f = z_1$ is part of a coordinate system on X. Thus, $\eta_1 = dz_1$, and clearly $\eta_1 \wedge \bar{\eta}_1 = dz_1 \wedge d\bar{z}_1 \neq 0$.

Thus, we can assume that, for all $v_1 \neq 0$, $v_1 \in V_1$, the induced map $V_2 \to W$ defined by $v_2 \mapsto$ the image of $v_1 \otimes v_2$ in W is injective, and symmetrically in V_2. It is now an easy exercise in algebraic topology, left to the reader, that in this case $\dim V_1 + \dim V_2 - 1 \leq \dim W$. In our case, since $\dim V_1 = \dim V_2 = q$ and $\dim W = h^{1,1}$, this gives $h^{1,1} \geq 2q - 1$. $\square$

Classification of threefolds

The classification of threefolds and higher-dimensional algebraic varieties has been revolutionized by the deep ideas of Mori. We shall just outline some of the known results concerning threefolds here, referring to [18] and [75] for more details and references.

Following the order of the theorems in the first section of this chapter, let us first consider threefolds X with $\kappa(X) = -\infty$. First, we need a preliminary definition:

Definition 32. Let X be a variety of dimension n. We say that X is *unirational* if there exists a dominant morphism $X' \to X$, where X' is a rational variety. Equivalently, X is unirational if there exists a dominant rational map $\mathbb{P}^n \dashrightarrow X$. We say that X is *uniruled* if there exists a dominant rational map $\mathbb{P}^1 \times Y \dashrightarrow X$, where $\dim Y = n - 1$.

Unlike the case of curves or surfaces, a unirational threefold need not be rational. It is easy to see that, if X is uniruled, then $\kappa(X) = -\infty$. Miyaoka has proved the converse to this statement in case $\dim X = 3$:

Theorem 33. *If X is a smooth threefold with $\kappa(X) = -\infty$, then X is uniruled.*

In a curious twist, the proof uses the strong form of Bogomolov's inequality proven by Donaldson (Theorem 27 in Chapter 4): if V is a stable bundle on a surface Y with $c_1(V) = c_2(V) = 0$, then there is a flat connection on V and so V is associated to a unitary representation of the fundamental group $\pi_1(Y, *)$.

We turn now to the analogue of Theorem 4, in other words to the question of what keeps the canonical bundle of a threefold from being nef. The answer is a theorem of Mori, based on his theory of extremal rays:

Theorem 34. *Suppose that X is a smooth threefold such that K_X is not nef. Then one of the following holds:*

(i) *$-K_X$ is ample and $\operatorname{Pic} X$ has rank 1;*
(ii) *There exists a morphism $\pi: X \to Y$, where Y is a surface and the generic fiber of π is a $\mathbb{P}^1$;*

(iii) *There exists a morphism* $\pi: X \to Y$, *where* Y *is a curve and the generic fiber of* π *is a del Pezzo surface;*

(iv) *There exists a birational morphism* $\pi: X \to Y$, *where* Y *is a normal projective variety, whose exceptional set is one of the following five types:*

  (a) Y *is smooth and* π *is the blowup of a smooth curve on* Y;

  (b) Y *is smooth and* π *is the blowup of a smooth point on* Y;

  (c) Y *is singular and* π *contracts a smooth divisor on* X *isomorphic to* $\mathbb{P}^1 \times \mathbb{P}^1$, *with normal bundle* $\pi_1^* \mathcal{O}_{\mathbb{P}^1}(-1) \otimes \pi_2^* \mathcal{O}_{\mathbb{P}^1}(-1)$;

  (d) Y *is singular and* π *contracts a singular divisor* D *on* X *isomorphic to a singular quadric in* $\mathbb{P}^3$, *with normal bundle the restriction to* D *of* $\mathcal{O}_{\mathbb{P}^3}(-1)$;

  (e) Y *is singular and* π *contracts a smooth divisor on* X *isomorphic to* $\mathbb{P}^2$, *with normal bundle* $\mathcal{O}_{\mathbb{P}^1}(-2)$.

In this theorem, Cases (i), (ii), (iii) correspond to the rational and ruled surfaces in Theorem 4. In Case (iv), (a) and (b) are manageable since the threefold Y is again smooth. In the remaining cases, however, Y is singular, and we cannot just keep going until K_X becomes nef. However, a difficult theorem of Mori shows that there is a sequence of contractions and somewhat more general birational operations beginning with X as above such that, at the end, the result is a possibly singular threefold X' with "mild" singularities such that either X' satifies an analogue of Cases (i), (ii), (iii) above (and in fact X' is uniruled) or $K_{X'}$ is nef. Here some care must be taken in defining nef, since X' may be singular and $K_{X'}$ is not always a Cartier divisor. However, some multiple of the Weil divisor $K_{X'}$ will be Cartier, and then nef can be defined as in the smooth case. Such an X' will be called *minimal* or a *minimal model* of X. The precise definition of minimal involves understanding the allowable singularities on X', and will not be given here. Unlike the surface case, there is not always a morphism from a smooth X to a minimal model, which need not be unique.

Finally, we have the generalization of the abundance theorem to dimension 3, proved by Kawamata:

Theorem 35. *Suppose that* X *is a minimal* (*not necessarily smooth*) *threefold such that* K_X *is nef. Then one of the following holds:*

(i) $\kappa(X) = 0$ *and* K_X *is numerically equivalent to 0.*

(ii) $\kappa(X) = 1$, K_X *is not numerically equivalent to 0, and* K_X^2 *is numerically equivalent to 0. In this case, some multiple of* K_X *defines a morphism to a curve whose general fiber is a surface with* $\kappa = 0$.

(iii) $\kappa(X) = 2$, K_X^2 *is not numerically equivalent to 0, and* K_X^3 *is numerically equivalent to 0. In this case, some multiple of* K_X *defines a morphism to a surface whose general fiber is an elliptic curve.*

(iv) $\kappa(X) = 3$. *In this case,* X *is of general type and* $K_X^3 > 0$.

Classification of vector bundles

In this section, our goal is to survey some results concerning the general structure of the moduli space of vector bundles on surfaces. Due to the efforts of many mathematicians, substantial progress has been made. On the other hand, the theory outlined here is not in any sense complete.

Throughout this section, X will denote a smooth algebraic surface and H an ample divisor on X. With X and H understood, given $w \in H^2(X; \mathbb{Z}/2\mathbb{Z})$ and $p \in H^4(X; \mathbb{Z})$, we have defined the moduli space $\mathfrak{M}(w, p)$. We will discuss the overall dependence on H later, and will write $\mathfrak{M}_H(w, p)$ when we wish to make the dependence on H explicit. The first result is an existence result:

Theorem 36. *If* $p \ll 0$, $\mathfrak{M}(w, p) \neq \emptyset$.

General existence results of this type, for surfaces and higher dimensional varieties, go back to Maruyama [85]. In the context of gauge theory, and for general 4-manifolds and principal bundles with arbitrary (compact) structure groups, results of this type were established by Taubes [145]. In case $w = 0$, there are sharper results. For example, one has the following:

Theorem 37. *For all* $c \geq 2p_g(X) + 2$, *there exists an* H-*stable rank* 2 *vector bundle* V *with* $\det V = \mathcal{O}_X$ *and* $c_2(V) = c$.

Theorem 37 was proved by Gieseker [49], under the slightly weaker assumption that $c \geq 4\left([p_g(X)/2] + 1\right)$, and in the form stated above in [37]. All known approaches are deformation-theoretic in nature. Taubes uses the gluing theory of anti-self-dual connections (which is a C^∞ version of deformation theory). Gieseker studies the deformation theory of bundles on the singular scheme which is the union of X blown up at a certain number of points together with the corresponding number of copies of $\mathbb{P}^2$, where each copy of $\mathbb{P}^2$ is glued to the blowup of X by identifying a line in the $\mathbb{P}^2$ to an exceptional divisor in the blowup of X. The approach in [37] is to consider the problem of smoothing certain torsion free sheaves on X to vector bundles.

All of the above constructions actually show that there exist good vector bundles V with the given Chern numbers. If we relax the condition that the vector bundles are good, can we significantly improve the bounds in the theorem? Are the bounds of the theorem optimal for good bundles? Qi Xia, in his Columbia thesis, showed that for simply connected elliptic surfaces X with no multiple fibers, there are no good vector bundles V with $\det V = \mathcal{O}_X$ and $c_2(V) = c$ for $c < 2p_g(X)$ if $p_g(X)$ is even and for $c \leq 2p_g(X)$ if $p_g(X)$ is odd. So the bounds given above are close to the best possible in this case.

There are less explicit existence results in case $\det V$ is not trivial, as well as in the case of higher rank [62], [82].

Once existence results have been established, we can go on to ask about the overall structure of $\mathfrak{M}(w, p)$, in both very general terms and in a refined sense. For example, deformation theory tells us that, once $\mathfrak{M}(w, p)$ is nonempty, it has dimension at least $-p - 3\chi(\mathcal{O}_X)$ at every point. One of the most surprising results is that of Donaldson on the dimension of $\mathfrak{M}(w, p)$:

Theorem 38. *There exists a constant C depending only on X, H, w such that, if $p \leq C$, then all components of $\mathfrak{M}(w, p)$ are good. In particular, $\dim \mathfrak{M}(w, p) = -p - 3\chi(\mathcal{O}_X)$ at all points.*

Donaldson's theorem is given in [26]. For a related approach to the proof, see [37]. The general idea is the following: by deformation theory, we know that $\mathfrak{M}(w, p)$ has dimension at least $-p - 3\chi(\mathcal{O}_X)$ at every point. Donaldson's method is to estimate the dimension of the space of vector bundles which are not good, or in other words for which $h^2(\operatorname{ad} V) \neq 0$. The general goal is to show that the set of such bundles has dimension at most $-\frac{3}{4}p + O(\sqrt{|p|})$. By Serre duality, it is enough to estimate the dimension of the space of V such that $h^2(\operatorname{ad} V \otimes K_X) \neq 0$. In other words, we must consider the V such that there is a homomorphism $\varphi \colon V \to V \otimes K_X$ whose trace is 0 (as a section of K_X). There are two cases to consider. In the first and elementary case, $\det \varphi = 0$ as well, and thus φ has rank 1. Setting $\operatorname{Ker} \varphi = \mathcal{O}_X(D)$, there is an exact sequence

$$0 \to \mathcal{O}_X(D) \to V \to \mathcal{O}_X(\Delta - D) \otimes I_Z \to 0,$$

and an inclusion $\mathcal{O}_X(\Delta - D) \otimes I_Z \subset V \otimes K_X$. Moreover, both V and $V \otimes K_X$ are stable, so that $\mu(D) < \mu(V)$ and $\mu(D) > \mu(\Delta) - \mu(K_X) - \mu(V)$. The argument now involves some straightforward manipulation with the Hodge index theorem.

In case $\det \varphi$ is not 0, the argument is much more subtle, and involves looking at the double cover $f \colon Y \to X$ defined by the eigenvalues of φ. In this case, we can compare V with the pushforward $W = f_*L$ of a certain line bundle L on Y. It turns out that W is a rank 2 subsheaf of V with torsion free quotient, and thus can be viewed as a generalized elementary modification of V. The theory of such modifications can then be used to bound the number of moduli of V for which $h^2(\operatorname{ad} V \otimes K_X) \neq 0$.

The dimension of the moduli space is asymptotically much larger than the dimension of its singular locus, and deformation theory shows that, once the moduli space has the correct dimension it is a local complete intersection. Thus, Donaldson's theorem has the following corollary, observed by Jun Li [81] (and generalized by him to the Gieseker compactification):

Corollary 39. *There exists a constant C depending only on X, H, w such that, if $p \leq C$, then all components of $\mathfrak{M}(w, p)$ are reduced, normal local complete intersections of dimension $-p - 3\chi(\mathcal{O}_X)$ at all points.*

It is not known if there are further restrictions on the singularities of the moduli space $\mathfrak{M}(w, p)$ for $p \ll 0$. It seems reasonable to expect that, in general, the singularities can be quite complicated in high codimension.

In the above discussion, the canonical bundle plays no special role. Thus, similar methods can prove the following generalization:

Theorem 40. *Let M be a fixed line bundle on X. There exists a constant C_M depending only on X, H, w such that, if $p \leq C_M$, then for every component Y of $\mathfrak{M}(w, p)$, there exists a vector bundle $V \in Y$ such that $h^0(\operatorname{ad} V \otimes M) = 0$.*

Let C be a smooth curve on X. If we apply the above to $M = \mathcal{O}_X(C) \otimes K_X$, it says that, as long as p is sufficiently negative, where the implied constant only depends on X, H, C, the general bundle V with $w_2(V) = w$ and $p_1(\operatorname{ad} V) = p$ satisfies $h^0(\operatorname{ad} V \otimes \mathcal{O}_X(C) \otimes K_X) = 0$. By Serre duality, $h^2(\operatorname{ad} V \otimes \mathcal{O}_X(-C)) = 0$. Now the group $H^2(\operatorname{ad} V \otimes \mathcal{O}_X(-C))$ is the cokernel of the natural map

$$H^1(\operatorname{ad} V) \to H^1(\operatorname{ad} V | C)$$

which measures the differential of the restriction map from the deformation theory of V on X to the deformation theory of $V|C$ on C. Thus, if $H^2(\operatorname{ad} V \otimes \mathcal{O}_X(-C)) = 0$, then $H^2(\operatorname{ad} V) = 0$, and the morphism of moduli functors is a submersion in an appropriate sense. Roughly speaking, this says that the deformations of V give a general deformation of $V|C$. Now if $g(C) \geq 2$, the general deformation of a vector bundle on C is stable (with similar results if $g(C) = 0, 1$). Thus, we have:

Theorem 41. *Let C be a smooth curve on X of genus at least 2. Then for all $p \ll 0$ and for every component Y of $\mathfrak{M}(w, p)$, the general bundle V in Y satisfies: $V|C$ is stable.*

We see that the discussion of vector bundles over ruled or elliptic surfaces is in a sense a special case of this result. In a sense, Theorem 41 is dual to the restriction theorems of Mehta and Ramanathan and Bogomolov (Corollary 3 in Chapter 9), which says that, fixing the stable bundle V, the restriction of V to a smooth curve $C \in |kH|$ is stable as soon as k is sufficiently large. In Theorem 41, we fix instead the curve C and conclude that the restriction of the generic bundle $V \in \mathfrak{M}(w, p)$ to C is stable as soon as $|p|$ is sufficiently large.

Donaldson's proof of Theorem 38 does not give any effective information on the size of the constant C. Recently, new ideas have been brought to this subject by Gieseker and Li on one hand and by O'Grady on the other. Using these new ideas, Gieseker and Li were able to show [50]:

Theorem 42. *For $p \ll 0$, $\mathfrak{M}(w, p)$ is irreducible.*

A very closely related result was obtained by O'Grady. Both Gieseker and Li and O'Grady have showed that $\mathfrak{M}(w, p)$ is generically good in the case of higher rank bundles V, and O'Grady has gone on in [116] to show that $\mathfrak{M}(w, p)$ is eventually irreducible in the higher rank case as well, as well as to make many of the constants that appear in his argument effective.

Finally, the estimates involved for the easy case of Donaldson's theorem show the following [126]:

Theorem 43. *Let X be a surface and let $\mathcal{K} \subset \mathcal{A}(X)$ be a compact subset of the ample cone. Then there exists a constant C depending only on X and $\mathcal{K}$ such that, for all ample divisors H and H' whose numerical equivalence classes lie in $\mathcal{K}$, and for all $p \leq C$, the moduli spaces $\mathfrak{M}_H(w, p)$ and $\mathfrak{M}_{H'}(w, p)$ are birational.*

Beyond these very general qualitative results which apply for all surfaces, we can ask about specific classes of surfaces, and in particular how the geometry of the surface is reflected in the geometry of the moduli space. The first general results along these lines are due to Maruyama, Barth, and Hulek, for the case $X = \mathbb{P}^2$; their results have been briefly discussed in Chapter 6, and we recall them again:

Theorem 44. *For $X = \mathbb{P}^2$, the moduli space $\mathfrak{M}(w, p)$ is irreducible, smooth, and unirational of the expected dimension $-p - 3$. In fact, it is rational if $w \neq 0$.*

Vector bundles over ruled surfaces have been described in detail in Chapter 6, although much of this study dates back to the work of Hoppe and Spindler [63] and Brosius [14]. Thus, we have analyzed all minimal surfaces with Kodaira dimension $-\infty$.

In case X is a minimal surface of Kodaira dimension 0, it is either elliptic or a $K3$ or abelian surface. In the $K3$ and abelian case, Mukai showed [103], [104], [105]:

Theorem 45. *Let X be a $K3$ surface. Then $\mathfrak{M}(w, p)$ is a smooth variety of the expected dimension $-p - 6$ and there is a holomorphic nowhere vanishing 2-form Ω on $\mathfrak{M}(w, p)$.*

In fact, Mukai's theorem applies to the Gieseker compactification of $\mathfrak{M}(w, p)$, provided there are no strictly Gieseker semistable torsion free sheaves with the appropriate Chern classes, and shows that the 2-form

Ω extends holomorphically over the compactification. Thus, the essential properties of $K3$ and abelian surfaces (that the canonical bundle is trivial) carry over to the moduli spaces, which are then examples of "symplectic manifolds." We should also add that Gomez and O'Grady have (independently) recently shown the following:

Theorem 46. *Let X be a $K3$ surface. If $w \neq 0$, then for all values of p, the moduli space $\mathfrak{M}(w,p)$ is irreducible.*

O'Grady has also extended his proof to the case of higher rank.

We have discussed the case of elliptic surfaces in great detail, and shall not say much more about them here, except to emphasize that the elliptic fibration on X is reflected (birationally) in a similar fibration on $\mathfrak{M}(w,p)$.

Thus, we have dealt with all possible values of $\kappa(X)$ except $\kappa(X) = 2$, in other words the case of general type. In this direction, Jun Li has shown the following [81] (see also [65, p. 232]):

Theorem 47. *Suppose that X is a minimal surface of general type and that there exists a $C \in |K_X|$ which is reduced (in particular, $p_g(X) \neq 0$). Then for $p \ll 0$ and such that $-p - 3\chi(\mathcal{O}_X)$ is even, $\mathfrak{M}(w,p)$ is of general type. In other words, the Kodaira dimension of $\mathfrak{M}(w,p)$ is equal to its dimension.*

Similar but much easier arguments show that, if X is a minimal surface with $\kappa(X) = 0$, then for $p \ll 0$, $\mathfrak{M}(w,p)$ has Kodaira dimension at most 0, and has Kodaira dimension equal to 0 if X is a $K3$ or abelian surface ([65, p. 234]).

Many questions remain open in the above theorems. For example, there seems to be more room for a detailed analysis of the case of strictly semistable sheaves. Moreover, what if X is of general type and $p_g(X) = 0$? Even the case of elliptic surfaces with $p_g = 0$ is not completely understood: What is the Kodaira dimension of $\mathfrak{M}(w,p)$ in this case? In case X is a properly elliptic surface with $p_g(X) = 0$, f denotes the class of a fiber on X, and Δ is a divisor on X with $\Delta \cdot f = 2e + 1$, then the moduli space $\mathfrak{M}(\Delta, c)$ is birational to $\mathrm{Sym}^t J^{e+1}(X)$ for an appropriate integer t [36]. Now $J^{e+1}(X)$ is an elliptic surface with multiple fibers of the same multplicities as X, since $e+1$ is relatively prime to the multiplicities m_i. (Necessarily $m_i | 2e+1$, and since $1 = 2(e+1) - (e+1)$, the gcd of $e+1$ and $2e+1$ is 1.) Thus, the Kodaira dimension of $J^{e+1}(X)$ is the same as that of X, and it then follows that the Kodaira dimension of $\mathfrak{M}(w,p)$ is equal to t, namely one-half the dimension of $\mathfrak{M}(w,p)$ (This is the maximum possible Kodaira dimension, since $\mathfrak{M}(w,p)$. birationally fibers over $\mathrm{Sym}^t \mathbb{P}^1$ and the fibers are complex tori, which therefore have Kodaira dimension 0.) However, the case where $w \cdot f = 0$ has not been worked out yet, nor are there any clues yet as to what happens for general type surfaces with $p_g = 0$.

Some of the above questions seem to be deeply connected to the relationship between stable bundles and 0-cycles. Let $A^2(X)$ denote the Chow group of 0-cycles on X modulo rational equivalence, and let $Z(X)$ be the subgroup of 0-cycles of degree 0. There is a natural homomorphism from $Z(X)$ to $\mathrm{Alb}(X)$. Mumford has shown that, if $p_g(X) > 0$, then $Z(X)$ is quite large [108]; in fact, it behaves somewhat like an infinite-dimensional algebraic variety. On the other hand, Bloch has conjectured that, if $p_g(X) = 0$, then the homomorphism from $Z(X)$ to $\mathrm{Alb}(X)$ is an isomorphism, and this conjecture has been verified in many cases. Fixing a base point $p_0 \in X$, there is a function $\mathfrak{M}(w,p) \to Z(X)$ defined by

$$V \mapsto c_2(V) - c \cdot p_0,$$

where $c_2(V)$ is computed in the group $A^2(X)$ and $c = \deg c_2(V)$. The induced function $\mathfrak{M}(w,p) \to \mathrm{Alb}(X)$ is in fact a morphism. Fitting together the above functions, there is an induced function

$$\coprod_{p \leq 0} \mathfrak{M}(w,p) \to Z(X),$$

which can be shown to be surjective. What can we say about the fibers of this map? To some extent, we ask if, at least for $p \ll 0$, the moduli space $\mathfrak{M}(w,p)$ behaves more like X or more like the space of 0-cycles on X modulo rational equivalence.

In all of the above discussion, we have tacitly assumed that the surface X is minimal. Let us now discuss the effect of blowing up: suppose that $\rho: \tilde{X} \to X$ is the blowup of X at a point p, and that E is the exceptional divisor. For an ample divisor H on X, we will choose the ample divisor $NH - E$ on $\tilde{X}$, where the exact value of N will depend on p. Let $\mathfrak{M}(w,p)$ be the moduli space of H-stable bundles V on X and let $\widetilde{\mathfrak{M}}(w,p)$ be the corresponding moduli space for $\tilde{X}$, where we identify $w \in H^2(X; \mathbb{Z}/2\mathbb{Z})$ with $\rho^*w \in H^2(\tilde{X}; \mathbb{Z}/2\mathbb{Z})$. In this case, the map $V \mapsto \rho^*V$ defines a morphism $\mathfrak{M}(w,p) \to \widetilde{\mathfrak{M}}(w,p)$. There is the following theorem, which can be proved by using the fact that for $p \ll 0$ the generic $\tilde{V} \in \widetilde{\mathfrak{M}}(w,p)$ satisfies $\tilde{V}|E \cong \mathcal{O}_E \oplus \mathcal{O}_E$:

Theorem 48. *The map ρ^* embeds $\mathfrak{M}(w,p)$ as a Zariski open subset of $\widetilde{\mathfrak{M}}(w,p)$. For $p \ll 0$, this subset is also dense, and thus $\mathfrak{M}(w,p)$ and $\widetilde{\mathfrak{M}}(w,p)$ are birational.*

Instead of considering vector bundles $\tilde{V}$ on $\tilde{X}$ such that $\det \tilde{V}$ is the pullback of a line bundle on X, we can look at general vector bundles $\tilde{V}$. After twisting by $\mathcal{O}_{\tilde{X}}(aE)$ for the appropriate $a \in \mathbb{Z}$, we can assume that $\det \tilde{V} = \mathcal{O}_{\tilde{X}}(\rho^*\Delta)$ or that $\det \tilde{V} = \mathcal{O}_{\tilde{X}}(\rho^*\Delta - E)$. We have already dealt with the first case. In the second case $\det \tilde{V} \cdot E = 1$ and, for $p \ll 0$, the restriction of $\tilde{V}$ to E is $\mathcal{O}_E \oplus \mathcal{O}_E(1)$. In this case, Brussee [15] has shown that $\tilde{V}$ is obtained from a semistable vector bundle V on X by an elementary modification of the form

$$0 \to \tilde{V} \to \rho^* V \to j_* \mathcal{O}_E \to 0,$$

where $j \colon E \to \tilde{X}$ is the inclusion. (If $\det V = \Delta$ and $c_2(V) = p$, then by the formulas for elementary modifications $\det \tilde{V} = \Delta - E$ and $c_2(\tilde{V}) = c_2(V)$.) Since $\mathrm{Hom}(\rho^* V, j_* \mathcal{O}_E) = \mathrm{Hom}(\mathcal{O}_E \oplus \mathcal{O}_E, \mathcal{O}_E)$ has dimension 2, there is a $\mathbb{P}^1$ of vector bundles $\tilde{V}$ corresponding to V. Thus, we have the following:

Theorem 49. *Let $w' = \underline{\rho}^* w + e$, where e is the mod 2 class of $[E]$. For all $p \ll 0$, the moduli space $\mathfrak{M}(w', p)$ is birationally a $\mathbb{P}^1$-bundle over $\mathfrak{M}(w, p)$.*

For more details on the relationship between vector bundles on X and on $\tilde{X}$, see Chapter 9 and [38] and [15]. It would be nice to have more precise results; for example, can we obtain moduli spaces on $\tilde{X}$ from those on X by a series of "flips," or in other words a reasonably explicit sequence of blowups and blowdowns (at least when the moduli spaces for X are all good)?

Beyond these results, are there more concrete instances where the geometry of X is reflected in the geometry of the moduli space $\mathfrak{M}(w, p)$? This question has been studied for genus 2 fibrations (in other words, surfaces X such that there exists a fibration $\pi \colon X \to C$ whose general fiber has genus 2) by D. Gomprecht [52]. By using the description of the moduli space of stable bundles over a genus 2 curve C given by Narasimhan and Ramanan [111] and a really intricate analysis of the restriction picture discussed in Chapters 6 and 8, he showed the following:

Theorem 50. *Let $\pi \colon X \to \mathbb{P}^1$ be a genus 2 fibration such that all singular fibers of π are irreducible nodal curves. For an integer c, let H be a c-suitable polarization on X. Then for $c \gg 0$, there is a Zariski open and dense subset of the moduli space $\mathfrak{M}(0, c)$ of H-stable rank 2 vector bundles V with $\det V = \mathcal{O}_X$ and $c_2(V) = c$ which is a finite cover of a Zariski open subset of a $\mathbb{P}^{c+1}$-bundle over $\mathbb{P}^{3c - \chi(\mathcal{O}_X) - 1}$, and the degree of the cover is $2^{2c - \chi(\mathcal{O}_X) - 2}$.*

A genus 2 fibration X is a double cover of a rational ruled surface, in other words there is a degree 2 morphism from X to a $\mathbb{P}^1$-bundle over $\mathbb{P}^1$. Thus, the moduli space carries a vestige of the structure of the genus 2 fibration.

Exercises

1. Let V_1 and V_2 be two complex vector spaces, and let $F \colon V_1 \otimes V_2 \to W$ be a complex linear map such that, for all nonzero $v_1 \in V_1$, the map $v_2 \mapsto F(v_1 \otimes v_2)$ is injective, and symmetrically for $v_2 \in V_2$. Show that $\dim W \geq \dim V_1 + \dim V_2 - 1$. (There is an induced continuous map

$F: \mathbb{P}(V_1) \times \mathbb{P}(V_2) \to \mathbb{P}(W)$. Calculate F^* in cohomology.) Is the same result true for real vector spaces?

2. Suppose that X is a surface such that K_X is nef and big. Let $\bar{X}$ be the surface obtained by contracting the finitely many -2-curves orthogonal to K_X. Show that K_X is the pullback of an ample divisor on $\bar{X}$, and use this to give another proof of the first part of Proposition 10 (for n large).

3. Let $\pi: X \to C$ be an elliptic surface over the base curve C, with multiple fibers of multiplicities $m_1, \ldots, m_r$. Let L be the dual of the line bundle $R^1\pi_*\mathcal{O}_X$, with $\deg L = d$, and let $g(C) = g$. For $n \geq 2$ and $g \geq 2$, show that

$$P_n(X) = (2g - 2 + d)n + \sum_i \left[\frac{n(m_i - 1)}{m_i} \right] + 1 - g.$$

What are the formulas in the remaining cases? In particular, P_n is not a linear polynomial in n.

4. Let E be an elliptic curve and let G be a finite subgroup of the translation group of E. Let C be a curve on which G acts faithfully, such that $C/G \cong \mathbb{P}^1$. Let G act diagonally on $C \times E$, and let X be the quotient $(C \times E)/G$. Show that there is a morphism $p_1: X \to \mathbb{P}^1$ whose general fiber is E, and which has multiple fibers where G does not act freely. Moreover, there is a smooth morphism $p_2: X \to E/G$ whose fibers are all isomorphic to C. Finally, show that $p_2: X \to E/G$ is the Albanese morphism. (Using the fact that X is an elliptic surface over $\mathbb{P}^1$, show first that $q(X) = 1$. Thus, there is a finite morphism $\text{Alb}\, X \to E/G$. Now use the fact that the fibers of $X \to \text{Alb}\, X$ are connected.)

References

1. E. Arbarello, M. Cornalba, P.A. Griffiths and J. Harris, *Geometry of Algebraic Curves,* vol. I, Springer-Verlag, New York, 1985.

2. M. Artin, *Some numerical criteria for contractibility of curves on algebraic surfaces,* Amer. J. Math. **84** (1962), 485–496.

3. _____, *On isolated rational singularities of surfaces,* Amer. J. Math. **88** (1966), 129–136.

4. M. Atiyah, *Vector bundles over an elliptic curve,* Proc. London Math. Soc. **7** (1957), 414–452.

5. M. Atiyah and I. Macdonald, *Commutative Algebra,* Addison-Wesley, Reading, MA, 1969.

6. R. Barlow, *A simply connected surface of general type with $p_g = 0$,* Invent. Math. **79** (1985), 293–301.

7. W. Barth, C. Peters, and A. Van de Ven, *Compact Complex Surfaces,* Ergebnisse der Mathematik und ihrer Grenzgebiete 3. Folge, vol. 4, Springer-Verlag, Berlin, 1984.

8. S. Bauer, *Diffeomorphism classification of elliptic surfaces with $p_g = 1$,* J. Reine Angew. Math. **451** (1994), 89–148.

9. A. Beauville, *Surfaces Algébriques Complexes,* Astérisque, vol. 54, Société Mathématique de France, Paris, 1978.

10. F. Bogomolov, *Holomorphic tensors and vector bundles on projective varieties,* Math. USSR Izv. **13** (1979), 499–555.

11. E. Bombieri, *Canonical models of surfaces of general type,* Publ. Math. Inst. Hautes Études Sci. **42** (1973), 171–219.

12. A. Borel and J.-P. Serre, *Le théorème de Riemann-Roch,* Bull. Soc. Math. France **86** (1958), 97–136.

13. N. Bourbaki, *Groupes et Algèbres de Lie,* Chaps. 4, 5, and 6, Masson, Paris, 1981.

14. J.E. Brosius, *Rank-2 vector bundles on a ruled surface* I, Math. Annalen **265** (1983), 155–168; II, Math. Ann. **266** (1983), 199–214.

15. R. Brussee, *Stable bundles on blown up surfaces,* Math. Z. **205** (1990), 551–565.

16. _____, *Some C^∞ properties of Kähler surfaces*, Algebraic Geometry e-prints 9503004.

17. F. Catanese, *Pluricanonical-Gorenstein-curves*, Enumerative Geometry and Classical Algebraic Geometry (Nice 1981) (P. LeBarz and Y. Hervier, eds.), Progress in Mathematics, vol. 24, Birkhäuser, Boston, MA, 1982, pp. 51–95.

18. H. Clemens, J. Kollár, and S. Mori, *Higher Dimensional Complex Geometry*, Astérisque, vol. 166, Société Math. de France, Paris, 1988.

19. F. R. Cossec and I. Dolgachev, *Enriques surfaces* I, Progress in Mathematics, vol. 76, Birkhäuser Boston, Boston, Basel, 1989.

20. M. Demazure, H. Pinkham, and B. Teissier, *Séminaire sur les Singularités des Surfaces* (Palaiseau 1976-77), Lecture Notes in Mathematics, vol. 777, Springer-Verlag, Berlin, 1980.

21. U.V. Desale and S. Ramanan, *Classification of vector bundles of rank 2 on hyperelliptic curves*, Invent. Math. **38** (1976), 161–185.

22. I. Dolgachev, *Algebraic surfaces with $q = p_g = 0$*, in Algebraic Surfaces, C.I.M.E. Cortona 1977, Liguori, Napoli, 1981, pp. 97–215.

23. S.K. Donaldson, *Anti-self-dual Yang Mills connections over complex algebraic surfaces and stable vector bundles*, Proc. London Math. Soc. **50** (1985), 1–26.

24. _____, *Irrationality and the h-cobordism conjecture*, J. Differential Geom. **26** (1987), 141–168.

25. _____, *Infinite determinants, stable bundles and curvature*, Duke Math. J. **54** (1987), 231–247.

26. _____, *Polynomial invariants for smooth four-manifolds*, Topology **29** (1990), 257–315.

27. S.K. Donaldson and P.B. Kronheimer, *The Geometry of Four-Manifolds*, Clarendon, Oxford, 1990.

28. A. Durfee, *Fifteen characterizations of rational double points*, Enseign. Math. **25** (1979), 131–163.

29. G. Ellingsrud and L. Göttsche, *Variation of moduli spaces and Donaldson invariants under change of polarization*, J. Reine Angew. Math. **467** (1995), 1–49.

30. _____, *Wall-crossing formulas, Bott residue formula and the Donaldson invariants of rational surfaces* (to appear).

31. F. Enriques, *Le Superficie Algebriche*, Zanichelli, Bologna, 1949.

32. G. Fernandez del Busto, *Bogomolov instability and Kawamata-Viehweg vanishing*, J. Algebraic Geom. **4** (1995), 693–700.

33. R. Fintushel and R. Stern, *Rational blowdown of smooth 4-manifolds*, Algebraic Geometry e-prints 9505018.

34. M. Freedman, *The topology of four-dimensional manifolds*, J. Differential Geom. **17** (1982), 357–454.

35. R. Friedman, *Rank two vector bundles over regular elliptic surfaces*, Invent. Math. **96** (1989), 283–332.

36. _____, *Vector bundles and SO(3)-invariants for elliptic surfaces*, J. Amer. Math. Soc. **8** (1995), 29–139.

37. _____, *Vector Bundles on Algebraic Varieties*, In preparation.

38. R. Friedman and J.W. Morgan, *On the diffeomorphism types of certain algebraic surfaces* I, J. Differential Geom. **27** (1988), 297–369; II, J. Differential Geom. **27** (1988), 371–398.

39. _____, *Algebraic surfaces and 4-manifolds: some conjectures and speculations*, Bull. Amer. Math. Soc. (New Series) **18** (1988), 1–19.

40. _____, *Smooth Four-Manifolds and Complex Surfaces*, Ergebnisse der Mathematik und ihrer Grenzgebiete 3. Folge, vol. 27, Springer-Verlag, Berlin, 1994.

41. _____, *Algebraic surfaces and Seiberg-Witten invariants*, J. Algebraic Geom. **6** (1997), 445–479.

42. R. Friedman, J.W. Morgan, and E. Witten, *Vector bundles on elliptic schemes* (to appear).

43. R. Friedman and Z.B. Qin, *On complex surfaces diffeomorphic to rational surfaces*, Invent. Math. **120** (1995), 81–117.

44. _____, *Flips of moduli spaces and transition formulas for Donaldson polynomial invariants of rational surfaces*, Comm. Anal. Geom. **3** (1995), 11–83.

45. W. Fulton, *Intersection Theory*, Ergebnisse der Mathematik und ihrer Grenzgebiete 3. Folge, vol. 2, Springer-Verlag, Berlin, 1984.

46. D. Gieseker, *On a theorem of Bogomolov on Chern classes of stable bundles*, Amer. J. Math. **101** (1979), 77–85.

47. _____, *On the moduli of vector bundles on an algebraic surface*, Ann. of Math. **106** (1977), 45–60.

48. _____, *Geometric invariant theory and applications to moduli problems*, Invariant Theory: Montecatini 1982, Lecture Notes in Mathematics, vol. 996, Springer-Verlag, New York Berlin, 1983, pp. 45–73.

49. _____, *A construction of stable bundles on an algebraic surface*, J. Differential Geom. **27** (1988), 137–154.

50. D. Gieseker and J. Li, *Irreducibility of moduli of rank 2 bundles over surfaces*, J. Differential Geom. **40** (1994), 23–104.

51. R. Gompf and T. Mrowka, *Irreducible four manifolds need not be complex*, Ann. of Math. **138** (1993), 61–111.

52. D. Gomprecht, *Rank two bundles on genus two fibrations*, Berkeley thesis, 1993.

53. L. Göttsche, *Modular forms and Donaldson invariants for 4-manifolds with $b_+ = 1$*, J. Amer. Math. Soc. **9** (1996), 827–843.

54. H. Grauert, *Über Modifikationen und exzeptionelle analytische Mengen*, Math. Ann. **146** (1962), 331–368.

55. P. Griffiths and J. Harris, *Principles of Algebraic Geometry*, Wiley, New York, 1978.

56. A. Grothendieck, *Sur la classification des fibrés holomorphes sur la sphère de Riemann*, Amer. J. Math. **79** (1956), 121–138.

57. ———, *Sur une note de Mattuck-Tate*, J. Reine Angew. Math. **200** (1958), 208–215.

58. ———, *Fondements de la Géométrie Algébrique; Extraits du Séminaire Bourbaki 1957–1962*, Sécrétairiat Mathématique, Paris, 1962.

59. A. Grothendieck and J. Dieudonné, *Éléments de géométrie algèbrique*, Publ. Math. Inst. Hautes Études Sci. **4** (1960); **8** (1961); **11** (1961); **17** (1963); **20** (1964); **24** (1965); **28** (1966); **32** (1967).

60. R. Hartshorne, *Ample Subvarieties of Algebraic Varieties*, Lecture Notes in Mathematics, vol. 156, Springer-Verlag, Berlin, 1970.

61. ———, *Algebraic Geometry*, Graduate Texts in Mathematics, vol. 52, Springer-Verlag, Berlin, 1977.

62. A. Hirschowitz and Y. Laszlo, *A propos de l'existence de fibrés stables sur les surfaces*, J. Reine Angew. Math. **460** (1995), 55–68.

63. H.-J. Hoppe and H. Spindler, *Modulräume stabiler Vektorraumbündel vom Rank 2 auf Regelflächen*, Math. Ann. **249** (1980), 127–140.

64. E. Horikawa, *Algebraic surfaces of general type with small c_1^2 I*, Annals of Math. **104** (1976), 357–387.

65. D. Huybrechts and M. Lehn, *The Geometry of Moduli Spaces of Sheaves*, Aspects of Mathematics, Vieweg, Braunschweig, 1997.

66. Y. Kametani and Y. Sato, *0-dimensional moduli spaces of stable rank 2 bundles and differentiable structures on regular elliptic surfaces*, Tokyo J. Math. **17** (1994), 253–267.

67. A. Kas, *Weierstrass normal forms and invariants of elliptic surfaces*, Trans. Amer. Math. Soc. **225** (1977), 259–266.

68. ———, *On the deformation types of regular elliptic surfaces*, in Complex Analysis and Algebraic Geometry (W. L. Baily Jr. and T. Shioda, eds.), Cambridge University Press, Cambridge, 1977, pp. 107–112.

69. S. Kleiman, *Toward a numerical theory of ampleness*, Ann. of Math. **84** (1966), 293–344.

70. S. Kobayashi, *Curvature and stability of vector bundles*, Proc. Japan Acad. Series A **58** (1982), 158–162.

71. K. Kodaira, *On compact analytic surfaces* I, Ann. of Math. **71** (1960), 111–152; II–III, Ann. of Math. **77** (1963), 563–626; IV, Ann. of Math. **78** (1963), 1–40.

72. ———, *On the structure of compact complex analytic surfaces ,I*, Amer. J. Math. **86** (1964), 751–798; *II*, Amer. J. Math. **88** (1966), 682–721; *III*, Amer. J. Math. **90** (1968), 55–83; *IV*, Amer. J. Math. **90** (1968), 1048–1066.

73. ———, *Pluricanonical systems on algebraic surfaces of general type*, J. Math. Soc. Japan **20** (1968), 170–192.

74. ———, *On homotopy K3 surfaces*, in Essays on Topology and Related Topics, Mémoires dédiés à Georges de Rham, Springer-Verlag, Berlin, 1970, pp. 58–69.

75. J. Kollár, *Flips and Abundance for Algebraic Threefolds*, Astérisque, vol. 211, Société Math. de France, Paris, 1992.

76. D. Kotschick, *On manifolds homeomorphic to* $\mathbb{C}P^2\#8\overline{\mathbb{C}P}^2$, Invent. Math. **95** (1989), 591–600.

77. P. Kronheimer and T. Mrowka, *Recurrence relations and asymptotics for four-manifold invariants*, Bull. Amer. Math. Soc. (NS) **30** (1994), 215–221.

78. S. Lang, *Algebra*, 3rd edn., Addison-Wesley, Reading, MA, 1993.

79. N. Leung, *Differential geometric and symplectic interpretations in the sense of Gieseker*, MIT thesis (1993); *Einstein type metrics and stability on vector bundles*, J. Differential Geom. (1997) (to appear).

80. J. Li, *Algebraic geometric interpretation of Donaldson's polynomial invariants of algebraic surfaces*, J. Differential Geom. **37** (1993), 417–466.

81. _____, *Kodaira dimension of moduli space of vector bundles on surfaces*, Invent. Math. **115** (1994), 1–40.

82. W.-P. Li and Z. Qin, *Stable vector bundles on algebraic surfaces*, Trans. Amer. Math. Soc. **345** (1994), 833–852.

83. P. Lisca, *Computations of instanton invariants using Donaldson-Floer theory*, Invent. Math. **119** (1995), 347–359.

84. M. Lübke, *Stability of Einstein-Hermitian vector bundles*, Manuscripta Math. **42** (1983), 245–257.

85. M. Maruyama, *Moduli of stable sheaves II*, J. Math. Kyoto Univ. **18** (1978), 557–614.

86. K. Matsuki and R. Wentworth, *Mumford-Thaddeus principle on the moduli space of vector bundles on a surface*, Internat. J. Math. **8** (1997), 97–148.

87. H. Matsumura, *Commutative Algebra*, 2nd edn., Benjamin/Cummings, Reading, MA, 1980.

88. A. Mayer, *Families of K3 surfaces*, Nagoya J. Math. **48** (1972), 1–17.

89. V.B. Mehta and A. Ramanathan, *Semistable sheaves on projective varieties and the restriction to curves*, Math. Ann. **258** (1982), 213–226.

90. _____, *Restriction of stable sheaves and representations of the fundamental group*, Invent. Math. **77** (1984), 163–172.

91. J.S. Milne, *Jacobian varieties*, in Arithmetic Algebraic Geometry (G. Cornell and J. Silverman, eds.), Springer-Verlag, Berlin, 1986.

92. J. Milnor and D. Husemoller, *Symmetric Quadratic forms* Ergebnisse der Mathematik und ihrer Grenzgebiete 2. Folge, vol. 73, Springer-Verlag, Berlin, 1973.

93. R. Miranda, *The moduli of Weierstrass fibrations over* $\mathbb{P}^1$, Math. Ann. **255** (1981), 379–394.

94. _____, *The Basic Theory of Elliptic Surfaces*, ETS Editrice, Pisa, 1989.

95. Y. Miyaoka, *On the Chern numbers of surfaces of general type*, Invent. Math. **42** (1977), 225–237.

96. _____, *The maximal number of quotient singularities on surfaces with given numerical invariants*, Math. Ann. **268** (1984), 159–171.

97. _____, *The Chern classes and Kodaira dimension of a minimal variety*, Algebraic Geometry (Sendai 1985), Advanced Studies in Pure Mathematics, vol. 10, Kinokuniya, Tokyo; North Holland, Amsterdam, 1987, pp. 449–476.

98. B. Moishezon, *A criterion for the projectivity of complete abstract algebraic varieties*, Amer. Math. Soc. Translations **63** (1967), 1–50.

99. _____, *Complex Surfaces and Connected Sums of Complex Projective Planes*, Lecture Notes in Mathematics, vol. 603, Springer-Verlag, Berlin, 1977.

100. J.W. Morgan, *Comparison of the Donaldson polynomials with their algebro-geometric analogues*, Topology **32** (1993), 449–488.

101. J.W. Morgan and T. Mrowka, *On the diffeomorphism classification of regular elliptic surfaces*, Internat. Math. Research Notices **6** (1993), 183–184.

102. J.W. Morgan and K. O'Grady, *Differential Topology of Complex Surfaces Elliptic Surfaces with $p_g = 1$: Smooth Classification*, Lecture Notes in Mathematics, vol. 1545, Springer-Verlag, Berlin, 1993.

103. S. Mukai, *Symplectic structure of the moduli space of sheaves on an abelian or K3 surface*, Invent. Math. **77** (1984), 101–116.

104. _____, *On the moduli spaces of bundles on K3 surfaces I*, in Vector Bundles on Algebraic Varieties (1987), Oxford Univeristy Press, New York.

105. _____, *Moduli of vector bundles on K3 surfaces, and symplectic manifolds*, Sugaku Exp. **1** (1988), 139-174.

106. D. Mumford, *The topology of normal singularities of an algebraic surface and a criterion for simplicity*, Publ. Math. Inst. Hautes Études Sci. **9** (1961), 5–22.

107. _____, *Lectures on Curves on an Algebraic Surface*, Annals of Mathematics Studies, vol. 59, Princeton University Press, Princeton, NJ, 1966.

108. _____, *Rational equivalence of 0-cycles on surfaces*, J. Kyoto Math. Soc. **9** (1968), 195–204.

109. _____, *Enriques' classification of surfaces in char p: I*, Global Analysis: Papers in Honor of K. Kodaira (D.C. Spencer and S. Iyanaga, eds.), University of Tokyo/Princeton University Press, Princeton, NJ, 1969, pp. 325–339.

110. M. Nagata, *On self-intersection number of a section on a ruled surface*, Nagoya J. Math. **37** (1970), 191–196.

111. M.S. Narasimhan and S. Ramanan, *Moduli of vector bundles on a compact Riemann surface*, Ann. of Math. **89** (1969), 19–51.

112. M.S. Narasimhan and C.S. Seshadri, *Stable and unitary bundles on a compact Riemann surface*, Ann. of Math. **82** (1965), 540–567.

113. P.E. Newstead, *Introduction to Moduli Problems and Orbit Spaces*, Tata Institute for Fundamental Research, vol. 51, Springer-Verlag, Berlin, 1978.

114. A. Ogg, *Cohomology of abelian varieties over function fields*, Ann. of Math. **76** (1962), 185–212.

115. K. O'Grady, *Algebro-geometric analogues of Donaldson's polynomials*, Invent. Math. **107** (1992), 351–395.

116. _____, *Moduli of vector bundles on projective surfaces: Some basic results*, Invent. Math. **123** (1996), 141–207.

117. C. Okonek, M. Schneider, and H. Spindler, *Vector Bundles on Complex Projective Spaces*, Progress in Mathematics, vol. 3, Birkhäuser, Boston, MA, 1980.

118. C. Okonek and A. Teleman, *Les invariants de Seiberg-Witten et la conjecture de van de Ven*, C. R. Acad. Sci. Paris Sér. I Math. **321** (1995), 457–461.

119. C. Okonek and A. Van de Ven, *Stable bundles and differentiable structures on certain elliptic surfaces*, Invent. Math. **86** (1986), 357–370.

120. _____, *Γ-type invariants associated to PU(2)-bundles and the differentiable structure of Barlow's surface*, Invent. Math. **95** (1989), 601–614.

121. U. Persson, *An introduction to the geography of surfaces of general type*, in Algebraic Geometry Bowdoin 1985, Proceedings of Symposia in Pure Mathematics, vol. 46, American Mathematical Society, Providence, RI, 1987.

122. V.Y. Pidstrigach, *Deformation of instanton surfaces*, Math. USSR Izv. **38** (1992), 313–331.

123. _____, *Patching formulas for spin polynomials, and a proof of the Van de Ven conjecture*, Russian Academy of Science Izvestiya Mathematics, Transl. Amer. Math. Soc. **45** (1995), 529–544.

124. V.Y. Pidstrigach and A.N. Tyurin, *Invariants of the smooth structure of an algebraic surface arising from the Dirac operator*, Russian Academy of Science Izvestiya Mathematics, Transl. Amer. Math. Soc. **40** (1993), 267–351.

125. Z.B. Qin, *Moduli spaces of stable rank-2 bundles on ruled surfaces*, Invent. Math. **110** (1992), 615-625.

126. _____, *Equivalence classes of polarizations and moduli spaces of sheaves*, J. Differential Geom. **37** (1993), 397-415.

127. _____, *Complex structures on certain differentiable 4-manifolds*, Topology **32** (1993), 551–566.

128. _____, *On smooth structures of potential surfaces of general type homeomorphic to rational surfaces*, Invent. Math. **113** (1993), 163–175.

129. M. Reid, *Bogomolov's theorem $c_1^2 \leq 4c_2$*, International Symposium on Algebraic Geometry (Kyoto 1977), Kinokuniya, Tokyo, 1978.

130. I. Reider, *Vector bundles of rank 2 and linear systems on algebraic surfaces*, Ann. of Math. **127** (1988), 309–316.

131. W. Rudin, *Functional Analysis*, McGraw-Hill, New York, 1973.

132. I.R. Šafarevič et al., *Algebraic Surfaces*, Proceedings of the Steklov Institute of Mathematics, vol. 75 (1965), American Mathematical Society, Providence, RI, 1967.

133. R.L.E. Schwarzenberger, *Vector bundles on algebraic surfaces*, Proc. London Math. Soc. **11** (1961), 601–622.

134. _____, *Vector bundles on the projective plane*, Proc. London Math. Soc. **11** (1961), 623–640.

135. C. Segre, *Recherches générales sur les courbes et les surfaces réglées algébriques* II, Math. Ann. **34** (1889), 1–25.

136. J.-P. Serre, *Groupes Algébriques et Corps de Classes*, Hermann, Paris, 1959.

137. _____, *Corps Locaux*, Hermann, Paris, 1968.

138. _____, *Cours d'Arithmétique*, Presses Universitaires de France, Paris, 1970.

139. _____, *Algèbre Locale. Multiplicités* (Troisième édition), Lecture Notes in Mathematics, vol. 11, Springer-Verlag, Berlin, 1975.

140. C. S. Seshadri, *Fibrés Vectoriels sur les Courbes Algébriques*, Astérisque, vol. 96, Société Mathématique de France, Paris, 1982.

141. C. Simpson, *Moduli of representations of the fundamental group of a smooth projective variety* I, Publ. Math. Inst. Hautes Études Sci. **79** (1994), 47–129.

142. A. Stipsicz and Z. Szabo, *The smooth classification of elliptic surfaces with* $b^+ > 1$, Duke Math. J. **75** (1994), 1–50.

143. F. Takemoto, *Stable vector bundles on algebraic surfaces*, Nagoya J. Math. **47** (1972), 29–48.

144. _____, *Stable vector bundles on algebraic surfaces* II, Nagoya J. Math. **52** (1973), 173–195.

145. C.H. Taubes, *Self-dual connections on non-self-dual 4-manifolds*, J. Differential Geom. **17** (1982), 139–170.

146. K. Uhlenbeck and S.-T. Yau, *On the existence of Hermitian Yang-Mills connections on stable vector bundles*, Comm. Pure Appl. Math. **39** (1986), 257–293; *A note on our previous paper: On the existence of Hermitian Yang-Mills connections on stable vector bundles*, Comm. Pure Appl. Math. **42** (1989), 703–707.

147. S.-T. Yau, *On the Ricci curvature of compact Kähler manifolds and the complex Monge-Ampère equation*, Comm. Pure Appl. Math. **31** (1978), 339–411.

Index

Abelian surface, *see* surface, abelian
abundance theorem, 279, 293, 306
ad V, 99
adjoint bundle, *see* ad V
adjunction formula, 13
Albanese map, 118, 289
Albanese variety, 288
algebraic function field, 271, 275
allowable elementary modification,
    see elementary modification,
    allowable
almost decomposable, 2
ample cone, 19
Andreotti, A., 278
anti-self-dual connection, *see*
    connection, anti-self-dual
arithmetic genus, *see* genus,
    arithmetic
Artin's criterion, 75, 80
Atiyah's theorem on vector bundles,
    1, 33
Atiyah, M., 1
Atiyah-Singer index theorem, 207
Auslander-Buchsbaum dimension
    formula, 51, 199

Barlow surface, *see* surface, Barlow
Barth, W., 3, 162, 310
base point
  assigned, 69
  infinitely near, 69
  simple, 69

unassigned, 69
basic elliptic surface, *see* surface,
    elliptic, basic
Bauer, S, 240
big divisor, *see* divisor, big
Bloch, S., 312
blowing down, 66
blowing up, 59
Bogomolov number, 274
Bogomolov's inequality, 3, 22, 107,
    149, 163, 245–247, 305
Bogomolov, F., 3, 245
Bogomolov-Miyaoka-Yau inequality,
    1, 282
Bombieri's theorem, 79, 248
Bombieri, E., 1, 248
botany, 282
Brosius, J.E., 3, 150, 310
Brussee, R., 312

Canonical bundle formula, 176
canonical curve, 134, 139
Castelnuovo's criterion, 1, 64–67
Castelnuovo's theorem, 278
Castelnuovo-deFranchis theorem,
    283, 302
Catanese, F., 14
Cayley-Bacharach property, 40, 126
chambers of type (w, p), 100
Chern class, 27–30
  total, 28
Chevalley, C., 275

Universitext   *(continued)*